Compound Semiconductors 2002

Other titles in the series

The Institute of Physics Conference Series regularly features papers presented at important conferences and symposia highlighting new developments in physics and related fields. Previous publications include:

177 **Optical and Laser Diagnostics**
Papers presented at the First International Conference, London, UK
Edited by C Arcoumanis and K T V Grattan

172 **Electron and Photon Impact Ionization and Related Topics 2002**
Papers presented at the International Conference, Metz, France
Edited by L U Ancarani

171 **Physics of Semiconductors 2002**
Papers presented at the 26th International Conference, Edinburgh, UK
Edited by A R Long and J H Davies

170 **Compound Semiconductors 2001**
Papers presented at the 28th International Symposium on Compound Semiconductors, Tokyo, Japan
Edited by Y Arakawa, Y Hirayama, K Kishino and H Yamaguchi

169 **Microscopy of Semiconducting Materials 2001**
Papers presented at the 12th International Conference on Microscopy of Semiconducting Materials, Oxford, UK
Edited by A G Cullis and J L Hutchison

168 **Electron Microscopy and Analysis 2001**
Papers presented at the Institute of Physics Electron Microscopy and Analysis Group Conference, Dundee, UK
Edited by C J Kiely and M Aindow

167 **Applied Superconductivity 1999**
Papers presented at the Fourth European Conference on Applied Superconductivity, Sitges, Spain
Edited by X Obradors and J Fontcuberta

Compound Semiconductors 2002

Proceedings of the Twenty-Ninth International Symposium on
Compound Semiconductors held in Lausanne, Switzerland,
7–10 October 2002

Edited by
Marc Ilegems, Günter Weimann and Joachim Wagner

Institute of Physics Conference Series Number 174

Institute of Physics Publishing
Bristol and Philadelphia

Copyright ©2003 by IOP Publishing Ltd and individual contributors. All rights reserved. No part of this publication may be reproduced, stored in a retrieval system or transmitted in any form or by any means, electronic, mechanical, photocopying, recording or otherwise, without the written permission of the publisher, except as stated below. Single photocopies of single articles may be made for private study or research. Illustrations and short extracts from the text of individual contributions may be copied provided that the source is acknowledged, the permission of the authors is obtained and IOP Publishing Ltd is notified. Multiple copying is permitted in accordance with the terms of licences issued by the Copyright Licensing Agency under the terms of its agreement with the Committee of Vice-Chancellors and Principals. Authorization to photocopy items for internal or personal use, or the internal or personal use of specific clients in the USA, is granted by IOP Publishing Ltd to libraries and other users registered with the Copyright Clearance Center (CCC) Transactional Reporting Service, provided that the base fee of $30.00 per copy is paid directly to CCC, 222 Rosewood Drive, Danvers, MA 01923, USA.

British Library Cataloguing in Publication Data

A catalogue record for this book is available from the British Library.

ISBN 0 7503 0942 3

Library of Congress Cataloging-in-Publication Data are available

Published by Institute of Physics Publishing, wholly owned by the Institute of Physics, London
Institute of Physics Publishing, Dirac House, Temple Back, Bristol BS1 6BE, UK
US Office: Institute of Physics Publishing, The Public Ledger Building, Suite 929, 150 South Independence Mall West, Philadelphia, PA 19106, USA

Printed in the UK by MPG Books Ltd, Bodmin, Cornwall

International Symposium on Compound Semiconductors (ISCS) Heinrich Welker Award

The International Symposium on Compound Semiconductors (ISCS) Heinrich Welker Award was initiated in 1976 to recognize an individual who has carried out outstanding research in the are of III–V compound semiconductors.

The Award consists of US$5000 and a citation honouring the recipient's contribution to the field. The Award was established and has been supported by Siemens AG, Munich, in honour of Professor Heinrich Welker, the foremost pioneer in III–V compound semiconductor development.

The winners of the ISCS Heinrich Welker Award are:

1976	Nick Holonyak	for developing the first practical light-emitting diodes
1978	Cyril Hilsum	for contributions in the fields of transferred electron logic devices (TELDs) and GaAs MESFETs
1980	Hisayoshi Yanai	for his work on TELDs, GaAs MESFETs and ICs, and laser diode modulation with TELDs
1981	Gerald L Pearson	for research and teaching in compound semiconductors physics and device technology
1982	Herbert Kroemer	for his work on hot-electron effects, Gunn oscillators and III–V heterostructure devices
1984	Izuo Hayashi	for development and understanding of room temperature operation of DH lasers
1985	Heinz Beneking	for his contributions to III–V semiconductor technology and novel devices
1986	Alfred Y Cho	for pioneering work on molecular beam epitaxy and his contribution to III–V semiconductor research
1987	Zhores I Alferov	for outstanding contributions in theory, technology and devices, especially epitaxy and laser diodes
1988	Jerry Woodall	for introducing the III–V alloy AlGaAs and fundamental contributions to III–V physics
1989	Don Shaw	for pioneering work on epitaxial crystal growth by chemical vapour deposition

1990	George S Stillman	for the characterization of high-purity GaAs and developing avalanche photodetectors
1991	Lester F Eastman	in recognition of his dedicated work in the field, especially on ballistic electron transport, delta-doping, buffer layer technique, and AlInAs/GaInAs heterostructures
1992	Harry C Gatos	for his contribution to science and technology of GaAs and related compounds, particularly in relating growth parameters, composition and structure to electronic properties
1993	James A Turner	for pioneering the development of GaAs MESFETs, MMICs, circuit fabrication and analytical techniques
1994	Federico Capasso	for leading work on bandgap engineering of semiconductor devices and discovery of many new phenomena in artificially structured semiconductors
1995	Isamu Akasaki	for his pioneering and outstanding contributions in the field of III–V nitride compound semiconductors
1996	Ben G Streetman	for his seminal contribution to the development of ion implantation techniques for both device fabrication and the study of the properties of nitrogen in ternary alloys, and for pioneering work on microcavity emitters and detectors
1997	M George Craford	for his contributions to the development of nitrogen-doped GaAsP technology for yellow and red-orange LEDs
1998	Takashi Mimura	for his pioneering contribution to heterostructure high electron mobility transistors
1999	Claude Weisbuch	for his fundamental contributions to the understanding of the physics of quantum semiconductor structures and the development of novel optoelectronic device concepts
2000	James S Harris	for his pioneering research in GaAs materials and devices over the past 35 years which have been key to realizing an economically viable GaAs technology
2001	Karl Hess	for his outstanding contributions to the theory of electronic transport in semiconductor heterostructures (including real space transfer, the full band Monte Carlo method and quantum well diode laser simulations)

The Award Committee of the 29th International Symposium on Compound Semiconductors has elected Professor Hiroyuki Sakaki of the University of Tokyo as the recipient of the ISCS Heinrich Welker Award for his fundamental contributions to the study of semiconductor quantum heterostructures.

Hiroyuki Sakaki

Hiroyuki Sakaki received the BS in 1968, and MS and PhD degrees in electronic engineering in 1970 and 1973 all from the University of Tokyo. He was appointed associate professor in 1973 and full professor in 1987 at the Institute of Industrial Science of the University of Tokyo. In 1976–77, he was a visiting scientist in Dr Leo Esaki's group of IBM Watson Research Center and in 1999 visiting professor at the Ecole Normale Supérieure in Paris.

In 1968–73, he worked on electrons in MOS-FET channels and clarified the role of quantization at 300 K. Since 1974, he has done pioneering studies on the physics and applications of semiconductor nanostructures; they include seminal works to control electrons with quantum dot (QD) and quantum wire (QWR) structures for new types of transistors (1975–76, 1980) and lasers (1982), the first in-plane transport studies of electrons in type-I and II quantum wells (QWs) and superlattices (SLs) (1976–77), the co-invention of intersubband QW infrared photodetectors (1977), and the MBE synthesis of QWs, QWRs, and QDs and subsequent studies to disclose physics and rich device potentials of confined electrons in such systems (1976–2002).

He received numerous awards such as the Medal of Honor (Purple Ribbon), the IEEE D. Sarnoff Award, the Fujiwara Prize, the Hattori-Hoko Award, the IBM Japan Science Award, the Shimadzu Science Award, and awards from several academic societies. He is a fellow member of the Institute of Electrical and Electronics Engineers, the American Physical Society, and the Institute of Electronics, Information and Communication Engineers.

ISCS Quantum Devices Award

The Quantum Devices Award was initiated in 2000 under the sponsorship of Fujitsu Quantum Devices Ltd. The Award is given to one or more individuals who have made pioneering contributions to the fields of Compound Semiconductor Devices and Quantum Nanostructure Devices, which have made a major scientific or technological impact in the past 20 years.

The award consists of US$5000 (or US$8000 in case the award is shared) and a citation honouring the recipient's contribution to the field.

The Quantum Devices Award past recipients are :

2000 Gerald Bastard and Emilio Mendez — for pioneering work on electric-field induced optic effects in quantum wells and superlattices (Quantum confined Stark effect and Wannier-Stark localization)

2001 Leo P Kouwenhoven, Mark A Reed and Seigo Tarucha — for pioneering contributions to electronic transport and the spectroscopy of quantum dots

The Award Committee of the 29th International Symposium on Compound Semiconductor has elected Professor Yasuhiko Arakawa of the University of Tokyo as the recipient of the ISCS Quantum Devices Award for his pioneering contributions to the development of quantum dot and quantum wire lasers.

Yasuhiko Arakawa

Yasuhiko Arakawa received BS, MS, and PhD degrees in electrical engineering from the University of Tokyo, in 1975, 1977, 1980, respectively. During his graduate course, he worked on communication theory including the proposal of a new transmission code named the extended duo-binary code.

In 1980, he started his academic carrier by joining the University of Tokyo as an assistant professor and was promoted to full professor in 1993. He is now Professor of Research in the Center for Advanced Science and Technology at the University of Tokyo. He is the director of Nanoelectronics Collaborative Research Center at the Institute of Industrial Science, University of Tokyo. He is also working partly as a NTT Research Professor.

Since 1980, his research interests have been focused on semiconductor nanostructures towards quantum photonic devices of the next generation. His current research activities cover growth/fabrication of quantum dots, single dot or femto-second spectroscopy, and device application of quantum dot lasers, microcavities, and photonic crystal with GaAs-based and GaN-based semiconductors. He is the recipient of many awards including the Niwa Memorial Award, the Excellent Paper Award from IECE, the Young Scientist Award, International Symposium on GaAs and Related Compound Semiconductors, the IBM Award, the Distinguished Achievement Award from IEICE, the Hattori Hoko Award, the Sakura–Kenjiro Award from OITDA, the Electronics Award from IEICE, the Nikkei IP Award, and the Nissan Science Award. Professor Arakawa served as an associate editor of the IEEE Journal of Quantum Electronics and is now Editor in Chief of Solid State Electronics, Regional Editor in Chief of the New Journal of Physics of the Institute of Physics, and a board member of the Institute of Pure and Applied Physics.

ISCS Young Scientist Award

The International Advisory Committee of the International Symposium on Compound Semiconductors has established a Young Scientist Award to recognise technical achievements in the field of compound semiconductors by a scientist under the age of forty.

The award consists of a citation honouring the recipients contribution to the field and a financial reward which amount is presently set at US$2000.

The Young Scientist Award past recipients are :

1986	Russell D Dupuis	for work in the development of organometallic vapour phase epitaxy of compound semiconductors
1987	Naoki Yokoyama	for contributions to self-aligned gate technology for GaAs MESFETs and ICs and the resonant tunnelling hot-electron transistor
1989	Russell Fischer	for demonstration of state of the art performance, at DC and microwave frequencies, of MESFETs, HEMTs and HBTs using (AlGa)As on Si
1990	Yasuhiko Arakawa	for pioneering work on low-dimensional semiconductor lasers, showing the superior performance of quantum wire and quantum box devices
1992	Umesh K Mishra	for pioneering and outstanding work on AlInAs–GaInAs HEMTs and HBTs
1993	Young-Kai Chen	for significant advancements in the fields of high-speed III–V electronic and optoelectronic devices
1994	Michael A Haase	for contributions on II–VI based blue LEDs and ZnSe-based electro-optic modulators
1995	John D Ralston	for pioneering and outstanding contributions in the field of high-speed high-power semiconductor lasers
1996	Nikolai Ledentsov	for his pioneering and outstanding contributions to the development of physics and MBE growth of InGaAs–GaAs quantum dots and quantum dot lasers
1997	Fred A Kish Jr	for research on native oxides formed from Al-bearing III–V compound semiconductors with applications to semiconductor laser diodes and the direct wafer bonding on compound semiconductors

1998 Steven P Den Baars   for his contributions to the MOCVD technology of GaN and other compound semiconductors

1999 Jerome Faist   for his fundamental contributions to the realization of high power quantum cascade lasers

2000 Kohki Mukai   for his contributions to pioneering development of quantum dot lasers

2001 Masahiko Kondow   for his pioneering contributions to the epitaxial growth and laser application of GaInNAs

The Award Committee of the 29th International Symposium on Compound Semiconductors has selected Professor Diana Huffaker of The University of New Mexico for her seminal contributions to the development of oxide confined vertical cavity lasers and 1300 nm GaInAs quantum dot lasers.

Diana Huffaker

Diana Huffaker received the BS degree in Engineering Physics from the University of Arizona in 1985, MS in Materials Science and PhD in Electrical Engineering from the University of Texas at Austin in 1990 and 1995, respectively. From 1995 through 2000, she held postdoctoral and research scientist appointments at the University of Texas at Austin. While at the University of Texas, she conducted seminal research in oxide-confined VCSELs and quantum dot lasers. In 2000, she joined Picolight, Inc. as a Senior Research Scientist. She joined the University of New Mexico as an Associate Professor of Electrical and Computer Engineering in 2001. Her research interests include epitaxy and characterization of III/V and III–N quantum dots (QDs) for lasers using MOCVD, physics of nanostructures and coherent processes in QDs. Professor Huffaker has co-authored over 90 refereed journal publications and 2 book chapters, and has reported her work through many invited presentations. She is an Editor of IEEE Circuits and Devices Magazine, IEEE/LEOS Chapter Chair and participates in several other conference committees. She is a senior member of IEEE.

Preface

The 29th International Symposium on Compound Semiconductors (ISCS 2002) was held from October 7 through October 10, 2002 at the hotel Alpha Conference Center in Lausanne, Switzerland. A total of 157 papers were presented, including 2 plenary presentations, 16 invited presentations, and 139 contributed papers of which 32 were selected for oral presentation and 107 for poster presentations.

This symposium, whose history goes back to 1966 under the name 'International Symposium on GaAs and Related Compounds' is the major conference devoted to the field of compound semiconductors science, technology and devices, covering materials growth and processing, physics of electronic and optical processes, and advanced electronic and optoelectronic devices. The materials covered include III–V, II–VI and IV–IV compounds and heterostructures.

The ISCS 2002 symposium highlighted certain subject areas within the general theme of Quantum and Nanostructured Devices, which exploit the interplay between quantum electronic and optical processes and the device characteristics. Several sessions were devoted to presentations of recent results on the physics of quantum wire and quantum dot devices, and on quantum confined emitters and detectors, including quantum cascade lasers and quantum well infrared detectors, in addition to sessions on materials growth and epitaxy, III–V and III–nitride light emitting diodes and lasers, high-frequency high-power electron devices, and industrial aspects of III–V electronics.

The 30th ISCS symposium will be held in San Diego, CA, from 25 to 27 August 2003, while Korea will be the venue for the 2004 meeting.

Finally we would like to thank the members of the Programme Committee for their efforts in selecting and reviewing the papers, and the members of the ISCS Awards Committee and its chairman, Professor Karl Joachim Ebeling, for selecting the nominees for the ISCS Welker Award, the ISCS Quantum Devices Award, and the ISCS Young Scientist Award.

Marc Ilegems, Günter Weimann, Joachim Wagner

ISCS 2002 Symposium Committees

Symposium Co-Chairs
M Ilegems, G Weimann

Programme Committee
J Wagner (Program Chair), M C Amann, G Borghs, B Deveaud, L Eastman, K J Ebeling, J Faist, D Fekete, A Forchel, T Foxon, E Gornik, J Harris, Y Hirayama, N N Ledentsov, P Lugli, J Merz, E Munoz, A Scavennec, M Schlechtweg, F Scholz, K Streubel, G Tränkle, M Walther, J H Wolter, J C Woo, N Yokoyama,

International Advisory Committee
G Weimann (chair), Z Alferov, R. Dawson, H Goronkin, M Ilegems, T Ikegami, T. Nakahara, Y S Park, K Ploog, M Reed, H Sakaki

Symposium sponsoring organizations
Swiss National Science Foundation
Swiss Federal Institute of Technology, Lausanne
Fraunhofer Institute for Applied Solid State Physics, Freiburg
IEEE Lasers and Electro-Optics Society
AIXTRON AG, Aachen
Centre Suisse d'Electronique et de Microtechnique SA, Neuchatel

The symposium organizers gratefully acknowledge the support given by the sponsoring organizations and industries.

Contents

International Symposium on Compound Semiconductors (ISCS)
Heinrich Welker Award   v

ISCS Quantum Devices Award   ix

ISCS Young Scientist Award   xi

Preface   xv

Symposium Committees   xvi

Section 1: Growth

GaAs on silicon using an oxide buffer layer
R Droopad, J Curless, Z Yu, D Jordan, Y Liang, C Overgaard, H Li, T Eschrich, J Ramdani, L Hilt, B Craigo, K Eisenbeiser, J Kulik, P Fejes, J Finder, X Hu, Y Wei, J Edwards, K Moore, M O'Steen and O Baklenov   1

Ultra shallow GaAs sidewall tunnel junctions implemented with low-temperature area selective regrowth
Y Oyama, T Ohno, K Tezuka, K Suto and J-i Nishizawa   9

Growth and properties of polycrystalline GaN on ZnO/Si substrates by ECR-MBE
K Kitamura, H Mamiya, T Araki, T Maruyama and Y Nanishi   13

Single crystalline InN films grown on Si (111) substrates
T Yamaguchi, K Mizuo, Y Saito, T Araki and Y Nanishi   17

Growth and evaluation of CdTe/Si (111) by hot wall epitaxy
G M Lalev, J Wang, S Abe, K Masumoto and M Isshiki   21

A surface reconstruction functioning as a micro mask during in-situ layer-by-layer etching of GaAs(111)B using $AsBr_3$
Y Asaoka, I Ihara, K Yokoyama, S Daicho, N Sano and T Kaneko   25

Comparative study of p-type dopants, Mg and Be in GaN grown by RF-MBE
S Sugita, Y Watari, G Yoshizawa, J Sodesawa, H Yamamizu, K T Liu, Y K Su and Y Horikoshi   29

Area selective epitaxy of anti-dot structure by solid source MBE using MEE deposition sequence
D Okada, H Hasegawa, T Hasegawa, Y Horikoshi and T Saitoh   33

Section 2: Processing and Characterization

Contribution of interface states and bulk traps to GaAs MIS admittance
S Kochowski, B Paszkiewicz and R Paszkiewicz   37

Electrical isolation of p-type InP and InGaAs layers by iron implantation: Effects of substrate temperature
P Too, S Ahmed, B J Sealy and R Gwilliam — 41

Inductively coupled argon plasma enhanced quantum well intermixing in InGaAs/InGaAsP laser structure
H S Djie, J Arokiaraj and T Mei — 45

Annealing studies of Si-implanted GaN by Hall-effect and photoluminescence measurements
J A Fellows, Y K Yeo, M-Y Ryu, R L Hengehold and T D Steiner — 49

Optical evaluation of spatial carrier concentration fluctuations in doped InP substrates
M Baeumler, E Diwo, W Jantz, U Sahr, G Müller and I Grant — 53

An "anomalous" drift of defects under electric field in CdSe and CdS single crystals
L V Borkovska, B M Bulakh, L Khomenkova, N O Korsunska and I V Markevich — 57

Injection energy dependence of electron thermalization length in AlGaAs/GaAs quantum well structures
T Tsuruoka, H Hashimoto, Y Ohizumi, S Ushioda — 61

Optical and structural investigations of $GeSiO_2$ systems
T V Torchynska, J Aguilar-Hernandez, G Polupan and A Kolobov — 65

Thermal quenching of emission of self-assembled InAs quantum dots embedded into InGaAs/GaAs MQW
T V Torchynska, J L Casas Espinola, H M Alfaro Lopez, P G Eliseev, A Stintz, K J Malloy and R Pena Sierra — 69

Guided surface-acoustic-wave modes in AlN layers grown on SiC substrates
Y Takagaki, P V Santos, E Wiebicke, O Brandt, H-P Schönherr and K H Ploog — 73

Photoluminescence from deep levels in Fe-doped InP substrates
M Yamada and M Fukuzawa — 77

Many-body effects as probe of defects presence in heavily doped AlGaAs/InGaAs/GaAs heterostructures
V P Kunets, Z Ya Zhuchenko, H Kissel, U Müller, G G Tarasov and W T Masselink — 81

Two-dimensional mapping of resistivity in semi-insulating GaAs wafers with large diameter using a nondestructive technique
M Fukuzawa and M Yamada — 85

Sonic-stimulated temperature rise around dislocation
R K Savkina and A B Smirnov — 89

The cluster variation method for semiconductor alloys
O V Elyukhina — 93

Exciton formation inhibition in GaInAs/InP Fe doped quantum wells
M Guézo, S Loualiche, J Even, A Le Corre, H Folliot, C Labbé and O Dehaese — 97

Compositional dependence of electron traps in Ga(As,N) grown
by molecular-beam epitaxy
P Krispin, V Gambin, J S Harris and K H Ploog   101

The influence of the quantum lifetime on the width of the quantum Hall plateaus
L Gottwaldt, K Pierz, F J Ahlers, L Schweitzer, E O Göbel and W Stolz   105

Intersubband transitions in strain compensated InGaAs/AlAs quantum well
structures grown on InP
N Georgiev, M Semtsiv, T Dekorsy, F Eichhorn, A Bauer, M Helm, W T Masselink   109

Interfacial and piezoelectric properties of highly strained InGaAs/GaAs quantum
well structures grown on (111)A GaAs substrates by MOVPE
J Kim, S Cho, A Sanz-Hervás, A Majerfeld, G Patriarche and B W Kim   113

Section 3: Quantum-Wires and -Dots

Carrier dynamics in self-organized In(Ga)As/Ga(Al)As quantum dots and their
application to long-wavelength sources and detectors
*P Bhattacharya, A D Stiff-Roberts, S Chakrabarti, S Krishna, C Fischer, T Norris
and J Urayama*   117

Single-electron transistors
P Hadley, G Lientschnig and M-J Lai   125

Selective formation of high-density and high-uniformity InAs/GaAs quantum dots
for ultra-small and ultra-fast all-optical switches
Y Nakamura, H Nakamura, S Ohkouchi, N Ikeda, Y Sugimoto and K Asakawa   133

Mechanical interaction in near-field spectroscopy of single semiconductor
quantum dots
*A M Mintairov, P A Blagnov, O V Kovalenkov, C Li, J L Merz, S Oktyabrsky,
V Tokranov, A S Vlasov and D A Vinokurov*   137

Magnetic properties of (Ga,Mn)N grown directly on 4H-SiC(0001)
by molecular-beam epitaxy
S Dhar, O Brandt, A Trampert, K J Friedland and K H Ploog   141

Selective MBE growth of GaAs ridge quantum wire arrays on patterned (001)
substrates and its growth mechanism
T Sato, I Tamai and H Hasegawa   145

Large transition energy separation at 1.31μm emission from InAs/GaAs quantum
dots
Y Q Wei, S M Wang, F Ferdos, Q X Zhao, J Vukusic, M Sadeghi and A Larsson   149

Optical spectra of quantum dot aggregates in sub-wetting layer region
K Král and P Zdeněk   153

Self-organized growth of InAs quantum dots and reduction of dot density by
in-situ annealing
Y Matsuzaki, T Kobuse, R Ohashi, M Konagai and A Yamada   157

Narrow size-dispersion CdSe quantum dots grown on ZnSe by modified MEE technique
I V Sedova, S V Sorokin, A A Sitnikova, O V Nekrutkina, A A Reznitsky and S V Ivanov — 161

Spectroscopy of high-density assemblage of InAs/GaAs quantum dots
Z Ya Zhuchenko, J W Tomm, H Kissel, Y I Mazur, G G Tarasov, W T Masselink — 165

Properties of InGaAs coupled quantum wire structures grown on vicinal (111)B GaAs with quasi-periodic corrugation
T Noda, N Kondo, Y Akiyama, T Kawazu and H Sakaki — 169

One-dimensional free exciton in CdTe/Cd$_{0.74}$Mg$_{0.26}$Te quantum wires
S Nagahara, T Kita, O Wada, L Marsal and H Mariette — 173

Analysis of self-assembled GaN nanorods on Si(111) substrate
L W Tu, C L Hsiao, T W Chi, J F Wu, I Lo, K Y Hsieh, T T Sheng and C F Hu — 179

Electroluminescence of asymmetric coupled GaAs/AlGaAs V-groove quantum wires
F Karlsson, H Weman, M-A Dupertuis, K Leifer, A Rudra and E Kapon — 183

Effects of nitrogen incorporation in In(Ga)As/GaAs quantum dots
K Park, W G Jeong, J Jang, Y D Jang, N J Kim and D Lee — 187

Photoconductivity of GaAs/AlGaAs quantum wires measured along the wires direction
V Donchev, M Saraydarov, K Germanova, X-L Wang, S-J Kim and M Ogura — 191

Section 4: Electronic Devices

GaAs HBTs with reduced collector capacitance for high-speed ICs and microwave power applications
K Mochizuki — 195

InP/GaAsSb/InP heterojunction bipolar transistors
C Bolognesi, M W Dvorak and S P Watkins — 203

Industrial aspects of III/V electronics: GaAs IC manufacturing in Taiwan for wireless communications
P C Chao — 211

Silicon germanium technologies for high-speed digital and analog applications
H Knapp, J Böck, M Wurzer, K Aufinger and T F Meister — 217

AlGaN/GaN HEMTs grown by molecular beam epitaxy on sapphire, 6H-SiC, and HVPE-GaN templates
N G Weimann, M J Manfra, J W P Hsu, K Baldwin, L N Pfeiffer, K W West, S N G Chu, D V Lang and R J Molnar — 223

High power AlGaN/GaN HEMT's
L F Eastman, V Tilak, R Thompson, B Green, V Kaper, T Prunty, R Shealy, J Smart and H Kim — 227

High breakdown electric field for Npn-type AlGaN/InGaN/GaN heterojunction bipolar transistors
T Makimoto, K Kumakura and N Kobayashi — 231

Multiwafer epitaxy of GaN/AlGaN heterostructures for power applications
K Köhler, S Müller, N Rollbühler, R Kiefer, R Quay and G Weimann — 235

Fabrication and electrical performance of oscillators in GaInP/GaAs-HBT MMIC technology up to 40 GHz
J Hilsenbeck, F Brunner, F Lenk and J Würfl — 239

Experimental demonstration of a resonant tunneling delta-sigma modulator for high-speed, high-resolution analog-to-digital converter
Y Yokoyama, Y Ohno, S Kishimoto, K Maezawa and T Mizutani — 243

InAs/AlGaSb heterostructure displacement sensors
H Yamaguchi, S Miyashita and Y Hirayama — 247

Charge balanced Ga_2O-GaAs interface and application to self-aligned GaAs p-channel enhancement mode MOS heterostructure field-effect transistor
M Passlack, J K Abrokwah, R Droopad, Z Yu, C Overgaard, S I Yi, M Hale, J Sexton and A C Kummel — 251

Investigation of quantum transport phenomena in resonant tunneling structures by simulations with a novel quantum hydrodynamic transport model
J Höntschel1, W Klix and R Stenzel — 255

Correlation between channel temperature and negative resistance in AlGaN/GaN HEMTs
N Shigekawa and K Shiojima — 259

Effect of temperature on the avalanche properties of sub-micron structures
C N Harrison, J P R David, C Groves, M Hopkinson and G J Rees — 263

$In_{0.53}Ga_{0.47}As$ ionization coefficients deduced from photomultiplication measurements
J S Ng, M C Yee, J P R David, P A Houston, G J Rees and G Hill — 267

InGaP/InGaAs/GaAs double channel pseudomorphic high electron mobility transistor
H M Chuang, K H Yu, K W Lin, C C Cheng, J Y Chen and W C Liu — 271

70-nm-gate PHEMT fabricated by a trilayer process of ZEP/P(MMA-MAA)/PMMA resist
S C Kim, B O Lim, H S Lee, S K Kim, H C Park, D-H Shin and J K Rhee — 275

Studies on the low-k BCB passivation of 0.1μm gamma gate PHEMTs
W-S Sul, H-J Han, S-D Lee and J-K Rhee — 279

Design of low loss transmission lines on GaAs substrates using the surface micromachining methods
Y-H Chun, S-C Kim, B-O Lim, H-S Lee, M-K Lee, H-S Kim, D-H Shin, S-K Kim, H-C Park, J-K Rhee and S-W Yun — 283

Correlation of pulsed IV measurements and high power performance of AlGaN/GaN HEMTs
V Tilak, V Kaper, R Thompson, T Prunty, H Kim, J Smart, J R Shealy and L Eastman   287

120 nm gate length e-beam and nanoimprint T-gate GaAs pHEMTs utilising non-annealed ohmic contacts
E Boyd, D Moran, H McLelland, K Elgaid, Y Chen, D Macintyre, S Thoms, C Stanley and I Thayne   291

New composite-emitter HBTs with reduced turn-on voltage and small offset voltage
M K Tsai, Y W Wu, S W Tan, Y J Yang, W S Lour   295

Properties of electronic states at free surfaces and Schottky barrier interfaces of AlGaN/GaN heterostructure
H Hasegawa, T Inagaki, S Ootomo and T Hashizume   299

Direct S-parameter extraction by physical two-dimensional device AC-simulation
V Palankovski, S Wagner, T Grasser, R Schultheis and S Selberherr   303

Section 5: Optoelectronic Devices

High-power blue-violet lasers grown on 3-inch sapphire and GaN substrate
S Uchida, S Ikeda, T Mizuno, S Goto, T Sasaki, Y Ohfuji, T Fujimoto, O Matsumoto, K Oikawa, M Takeya, Y Yabuki and M Ikeda   307

III/V nitride LEDs and lasers
B Hahn, D Eisert, J Baur, M Fehrer, S Kaiser, H-J Lugauer, U Strauss, A Lell and V Härle   315

Recent advances in continuous wave quantum cascade lasers
D Hofstetter, M Beck, S Blaser, T Aellen, J Faist, U Oesterle, M Ilegems, E Gini and H Melchior   323

Edge- and surface-emitting photonic-crystal distributed-feedback lasers
I Vurgaftman, W W Bewley, C L Canedy, J R Lindle, C S Kim and J R Meyer   331

Quantum well infrared photodetectors and thermal imaging cameras
H Schneider, R Rehm, J Fleissner, M Walther, P Koidl, G Weimann, J Ziegler, R Breiter and W Cabanski   339

Temperature sensitivity of high power GaSb based 2μm diode lasers
M Rattunde, C Mermelstein, J Schmitz, R Kiefer, W Pletschen, M Walther and J Wagner   347

Long-wavelength, two-dimensional, WDM vertical-cavity surface-emitting laser arrays fabricated by nonplanar wafer bonding
J Geske, Y L Okuno, J E Bowers and D Leonard   351

High gain, low noise 4H-SiC UV avalanche photodiodes
B K Ng, J P R David, R C Tozer, G J Rees, F Yan, C Qin and J H Zhao   355

Novel microcavity light emitting diodes
R P Stanley, P Royo, U Oesterle, R Joray and M Ilegems   359

High extraction efficiency AlGaInP microcavity light emitting diodes at 650 nm with AlGaAs-AlO$_x$ DBR
R Joray, J Dorsaz, R P Stanley, M Ilegems, M Rattier, C Karnutsch and K Streubel   363

Orientation-mismatched wafer bonding for polarization control of 1.3 µm wavelength vertical cavity surface emitting lasers (VCSEL)
Y L Okuno, J Geske, Y-J Chiu, S P DenBaars and J E Bowers   367

Population inversion enhancement by resonant magnetic confinement in THz quantum cascade lasers
G Scalari, S Blaser, L Ajili, M Rochat, H Willenberg, D Hofstetter, J Faist, H Beere, G Davies, E Linfield and D Ritchie   371

Interferometric temperature mapping of GaAs-based quantum cascade laser
C Pflügl, M Litzenberger, W Schrenk, S Anders, D Pogany, E Gornik and G Strasser   375

Sensitivity of intersubband absorption linewidth and transport mobility to interface roughness scattering in GaAs quantum wells
T Unuma, M Yoshita, T Noda, H Sakaki, M Baba and H Akiyama   379

Improved temperature performance of GaAs/AlGaAs quantum cascade lasers
W Schrenk, S Anders, C Pflügl, E Gornik, G Strasser, C Becker and C Sirtori   385

Electron-phonon strong coupling in intersubband resonators
G Biasiol, L Sorba, D Dini, R Köhler, A Tredicucci and F Beltram   389

Demonstration of 640x512 pixel four-band quantum well infrared photodetector (QWIP) focal plane array
S D Gunapala, S V Bandara, J K Liu, S B Rafol, M Jhabvala and K K Choi   393

Giant polarized photoluminescence and photoconductivity in type-II GaAs/GaAsSb multiple quantum wells induced by interface chemical bonds
Y F Chen, Y S Chiu, M H Ya and T T Chen   397

Efficient nitride-based short-wavelength emitters with enhanced hole injection
J M Zavada, S M Komirenko, K W Kim and V A Kochelap   401

Study of polarization switch in a three-contacts vertical-cavity surface-emitting laser
V Badilita, J-F Carlin, M Ilegems, M Brunner, G Verschaffelt and K Panajotov   405

Analysis of dynamics and intensity noise of semiconductor lasers under strong optical feedback
S Abdulrhmann, M Ahmed, T Okamoto, W Ishimori and M Yamada   409

High performance optically pumped 1.55 µm VCSELs for novel telecom applications
A Syrbu, G Suruceanu, V Iakovlev, A Rudra, A Mereuta, C-A Berseth, A Mircea, C Bungarzeanu and E Kapon   415

GaN-based single mirror light emitting diodes
Ch Zellweger, J Dorsaz, J F Carlin, H J Bühlmann, M Ilegems and R P Stanley   419

Nonlinear semiconductor materials for a fully passive low-loss optical combiner
G Zhao and J J G M van der Tol   423

Optically pumped vertical external cavity semiconductor thin-disk laser with CW operation at 660 nm
M I Müller, C Karnutsch, J Luft, W Schmid, K Streubel, N Linder, S-S Beyertt, U Brauch, A Giesen and G H Döhler   427

Polarization-sensitive photo-detectors based on strained *M*-plane GaN
S Ghosh, P Misra, O Brandt and H T Grahn   431

Investigation of the modulation efficiency of depleted InGaAsP/InP ridge waveguide phase modulators at 1.55 µm
H S Park, J C Yi and Y T Byun   435

Continuous wave operation of far-infrared quantum cascade lasers
L Ajili, G Scalari, H Willenberg, D Hofstetter, M Beck, J Faist, H Beere, G Davies, E Linfield and D Ritchie   439

Resonant phonon-assisted depopulation in type-I and type-II intersubband laser heterostructures
M V Kisin, M A Stroscio, G Belenky and S Luryi   443

Mid-infrared quantum cascade lasers operation above room temperature
Q Yang, C Mann, F Fuchs, R Kiefer, K Köhler and H Schneider   447

Scattering transport and electron temperature evaluation in terahertz GaAs/AlGaAs quantum cascade laser
D Indjin, P Harrison, R W Kelsall and Z Ikonić   451

Graded interface 9.3 µm quantum cascade lasers
T Aellen, M Beck, D Hofstetter, J Faist, U Oesterle, M Ilegems, E Gini and H Melchior   455

Lasing properties of GaAs/(Al,Ga)As quantum cascade lasers as a function of injector doping density
M Giehler, R Hey, H Kostial, T Ohtsuka, L Schrottke and H T Grahn   459

Author Index   463

GaAs on Silicon Using an Oxide Buffer Layer

R. Droopad, J. Curless, Z. Yu, D. Jordan, Y. Liang, C. Overgaard, H. Li, T. Eschrich, J. Ramdani, L. Hilt, B. Craigo, K. Eisenbeiser, J. Kulik*, P. Fejes*, J. Finder, X. Hu, Y. Wei, J. Edwards, K. Moore, M. O'Steen[+], O. Baklenov[+]

Physical Sciences Research Labs - Motorola Labs, 7700 S. River Parkway, Mail Drop ML26, Tempe, AZ 85284
* Motorola Process and Materials Characterization Laboratory, Tempe, AZ 85284
[+]IQE Inc, Bethlehem, PA 18015

Abstract.
Compound semiconductors have been epitaxially grown on Si substrates using an oxide buffer layer and an amorphous interfacial layer formed at the oxide/Si interface during the oxide deposition. The combination of the epitaxial oxide and the amorphous layers serves to help accomodate the lattice and thermal mismatch between the III-V and Si substrate. Majority carrier devices with performance comparable to GaAs/GaAs control have been demonstrated.

1. Introduction

The integration of III-V semiconductors with silicon has been the goal of a tremendous amount of research over the last 30 years. This opens up the possibility of combining the superior electronic and optical properties of compound semiconductors with the mature Si CMOS device technology. This would also allow compound semiconductors, such as InP to be developed on large diameter Si wafers which are readily available, less expensive and mechanically robust. The thermal conductivity of Si is more than 3 times that of GaAs which would result in improved high power microwave devices on Si because of the superior heat dissipation properties of the substrate. Additionally, as microprocessor speeds continue to increase, optical interconnects are being proposed to overcome the limitations of electrical connections. Because of the optical inefficiency of Si, III-V semiconductors can provide the optical elements necessary to implement any optical interconnection scheme.

The successful integration of III-V compound semiconductors with silicon has to overcome the differences in lattice mismatch and also any thermal mismatch that can influence the quality of the layers during thermal cycling either during deposition or processing. For the case of GaAs on Si, the lattice mismatch is 4% while the coefficient of thermal expansion for GaAs is approximately 2.5 times that of Si. For the growth of GaAs directly on Si, these mismatches induce a large number of dislocations in the GaAs layer which can be reduced by increasing the epilayer thickness. Various other methods have been used to reduce the density of threading dislocations which included low temperature buffers, superlattice buffers [1] and thermal cycling [2]. Such approaches require thick buffer layers before the active devices can be grown.

One approach that has been demonstrated to reduce the dislocation density of III-V semiconductors involves the engineering of the lattice constant of the substrate prior to the deposition of the III-V materials [3]. A thick layer of graded Si_xGe_{1-x} layer is grown on Si on which a layer of relaxed Ge is deposited. GaAs layers deposited on such a surface exhibit threading dislocation density around 2×10^6 cm^{-2}. Another method for depositing III-V layers on Si has been the use of a SOI compliant substrate [4]. In this approach, a thin Si membrane layer on top of the oxide layer in a silicon-on-insulator (SOI) structure is used to reduce dislocations in the GaAs layer grown on Si. Strain and dislocation generated during the growth of GaAs layer is relieved in the amorphous oxide layer. GaAs deposited on a SOI structure having a thin Si(511) layer have dislocation density of $<3 \times 10^7$ cm^{-2} for a 3 μm thick GaAs layer.

In this paper we present our approach of utilising a combination of (a) engineering the lattice constant of the substrate and (b) the compliant substrate concept in developing a deposition process for GaAs on Si. Our approach utilises an epitaxial oxide whose lattice constant is approximately midway between that of Si and GaAs and an amorphous layer at the interface between the Si substrate and the epitaxial oxide layer. It is expected that the amorphous layer will help accommodate the lattice mismatch by elastic/plastic deformation. The epitaxial oxide used is a layered perovskite strontium titanate (STO) which is being investigated as a candidate for gate dielectric replacement for SiO_2 in Si CMOS device applications [5].

2. Experimental

Epitaxial layer deposition was carried out using either of two molecular beam epitaxy (MBE) systems configured for both oxide and III-V layers. The first system is a dual chamber production-type multiwafer system capable of deposition on as large as 8" diameter substrates. One chamber is dedicated to the growth of oxides on silicon and is equipped with a 2000 l/s turbo-molecular pump with base pressures $< 5 \times 10^{-10}$ mbar obtained through the use of titanium sublimation pumping and liquid nitrogen cryopanelling. Sr and Ti metals were evaporated from effusion cells and oxygen introduced into the chamber through an RF plasma source. The other chamber is a conventional III-V deposition system equipped with a diffusion pump. Elemental sources including Ga, In, Al, As, and Si and Be dopants are available. Growth rates and temperatures are determined using the reflection high energy electron diffraction (RHEED) intensity oscillations technique and an optical pyrometer, respectively. The second system is comprised of multiple chambers configuration that encompasses a number of in-situ analytical capabilities. These include x-ray photoelectron spectroscopy (XPS), ultraviolet photoelectron spectroscopy (UPS) and scanning probe microscopy (SPM).

The properties of the epilayers were also studied using a number of ex-situ characterisation techniques. X-ray diffraction and atomic force microscopy were used to determine layer crystallinity and surface roughness, respectively. Defect density was determined using an A/B etching method and transmission electron microscopy (TEM).

3. Results and Discussion

The successful deposition of an epitaxial layer depends on the crystalline quality and cleanliness of the substrate. The surface of the substrate should be well ordered and free from contaminants. The first step in our process involves the deposition of a crystalline oxide layer on silicon. The as-received silicon substrate, (100) orientated with a 1-2° miscut towards the <110> direction, is subjected to exposure of ozone generated by a commercial ultraviolet ozone generator prior to loading into the MBE deposition system. This step was critical in the elimination of carbon-containing species on the Si surface. The process of removal the native oxide involved depositing 1-2 monolayers of Sr metal on the surface at temperatures between 400-600°C followed by an annealing at a temperature ≥ 750°C. The resulting surface is a well ordered clean (2x1) reconstructed Si surface [6]. Figure 1 shows an STM image of such a surface. In-situ XPS and RHEED measurements confirm the absence of any carbon on the surface. One advantage of this process is that the silicon oxide can be removed at a lower temperature than that needed for a thermal removal thereby causing less thermal stress on the silicon wafer and lowering the overall thermal budget of the process. This is an important requirement for any integration of compound semiconductor devices with Si CMOS devices since any GaAs deposition has to be carried out after the Si devices have been processed. High temperature exposure to the substrate can affect the performance of the Si CMOS devices.

Figure 1. STM image of a Si (2x1) reconstructed surface after cleaning using a Sr de-oxidation process prior to the oxide growth.

3.1 Oxide growth

The deposition of any oxide on silicon surfaces needs to take into account the kinetics of silicon oxide formation when a clean silicon surface is exposed to oxygen. Once an amorphous oxide forms on the silicon surface there will be a loss of epitaxy for subsequent deposition. One way to transition from silicon to the complex oxide such as STO is through the use of an alkaline earth oxide [7]. Our process of passivating the silicon surface for the STO is by depositing 0.5 monolayers of Sr metal on a de-oxidised Si surface. This acts as a template for epitaxial growth of STO. Oxide growth is initiated by co-deposition of Sr and Ti metal in the presence of oxygen at a substrate temperature of approximately 500°C. Since the sticking coefficients of both Sr and Ti are unity under the growth conditions used, it is important to accurately calibrate the incoming fluxes to maintain stoichiometry. This is done by using the RHEED technique described elsewhere [8]. Also, the oxide surface is monitored throughout the growth and any deviation in stoichiometry can be corrected by interrupting the respective fluxes. Figure 2 shows the RHEED image of an STO surface after the growth of approximately 40Å. A faint (2x) reconstruction along the [110] azimuth is indicative of a slightly Sr-rich surface. The amorphous layer between the epitaxial oxide and the silicon substrate can be formed either during the oxide deposition, or after the deposition via a post growth annealing in the presence of oxygen. In both cases, the amorphous layer is formed through a reaction of silicon and oxygen that have diffused through the crystalline oxide layer. Increasing the growth temperature and/or the oxygen partial pressure can increase the thickness of this layer. The STO layer grown on silicon with the formation of the amorphous layer at the oxide/Si interface is fully relaxed as determined by TEM investigations. The surface of the oxide layer is terminated neutral to slightly Sr rich as determined by RHEED.

STO[100]   STO[110]

Figure 2. RHEED images of a 40Å STO layer grown on Si

3.2 GaAs growth

Nucleation of GaAs layers on the oxide surface was carried out in a conventional III-V deposition chamber. The oxide wafers can be transferred either in vacuum or exposed to atmosphere to be loaded into a stand-alone system. Provided the oxide surface is terminated with a neutral to slightly Sr-rich composition, exposure to the atmosphere does not appear to affect the quality of the GaAs layer. Prior to deposition of GaAs, the substrate was annealed until the RHEED patterns displays a (1x1) reconstruction. GaAs is grown using a 2 step procedure in which initial growth conditions are low temperature and low growth rate for the nucleation layer, followed by higher temperature and higher growth rate for the bulk of the film. With optimised oxide growth and appropriate GaAs conditions, the GaAs layer is single domain by RHEED as shown in figure 3 for a 0.5 µm GaAs on oxide/Si which is the typical (2x4) reconstruction for GaAs at temperatures >525 °C. This is also confirmed by AFM measurements. For double domain GaAs growth, a four-fold reconstruction would be evident along both the orthogonal [110] directions.

Figure 4 shows a high resolution cross sectional transmission electron micrograph of a GaAs/oxide/Si structure. Note the amorphous layer at the oxide/Si interface. This layer forms during the oxide growth and causes the STO layer to be fully relaxed. There is a crystalline transition across the oxide/GaAs interface.

One goal of developing a process for deposition of GaAs on Si is the integration of optoelectronics with silicon circuitry. Large number of defects in the GaAs layer however, can

[110]　　　　　　　　　　　　　[-110]

Figure 3. RHEED images of a single domain GaAs/STO/Si. The thickness is 0.5 μm.

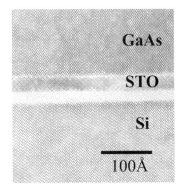

Figure 4. High resolution cross sectional transmission electron micrograph of a GaAs/STO/Si structure

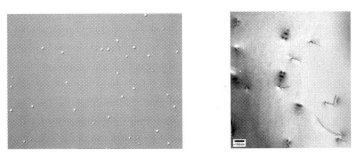

Figure 5. (a) A/B etch GaAs sample (200X magnification), defect density is ~10^5 cm^{-2} and (b) plan-view TEM of GaAs sample, defect density is ~10^8 cm^{-2}

increase the number of non-radiative recombination centers and hence, reduce the optical efficiency. Defects can also reduce minority carrier lifetimes due to recombination at threading dislocations and scattering from point defects.

Film defectivity was characterized by A/B etch and both plan-view and cross-sectional TEM. Chemical etching to reveal dislocations have been used to determine the density of threading dislocations of 2μm GaAs grown on STO/Si. Comparison with commercial GaAs substrates suggested that the defect densities were comparable for both wafers. Figure 5a shows an optical image of a surface of a typical GaAs/STO/Si layer that has been subjected to an A/B etch. The defect density for this wafer was determined to be around 10^5 cm^{-2}. However, careful study by plan wiew TEM, figure 5b, suggest that the density of threading dislocation is about 3 orders of magnitude higher and $> 10^8$ cm^{-2}. One possible explanation for the discrepancy could be due to the difficulty of distinguishing defects that are close to each other and the difficulty of counting the defects.

3.3 MESFET devices

GaAs MESFET structures were grown on a 0.5 μm GaAs layer that had been deposited on STO/Si. A commercial semi-insulating GaAs substrate was also loaded in the same platen as a control for comparison of device properties. A GaAs buffer layer of total thickness of 2 μm followed by a 2000Å low temperature GaAs layer was deposited for isolation from the Si substrate. The device structure consists of a 1500A channel doped at 8×10^{17} cm^{-3} followed by a 500A n+ doped GaAs contact layer. The rms roughness measured for the MESFET wafer using atomic force microscopy was around 10Å.

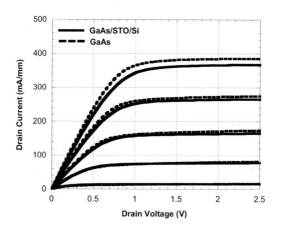

Figure 6. Drain current–drain voltage curves for 0.7μm X 100μm MESFETs. Gate voltage maximum is 0.5V and gate voltage step size is 0.5V.

MESFET devices were fabricated using standard III-V processing techniques. Device isolation was achieved through wet chemical etching of a mesa. Ni/Ge/Au ohmic contacts were then deposited via a liftoff optical resist process and annealed by rapid thermal annealing to form ohmic contacts to the contact layer. A wet gate recess etch was used to remove the n+ contact layer from the gate region. The Ti/Pt/Au gate was then deposited by e-beam evaporation and lifted off using optically defined photoresist patterns. The process used on the GaAs/GaAs control was identical to the process used for the GaAs/STO/Si sample. Devices with gate lengths of 20 μm were used to extract electron mobility from measurements of capacitances and transconductances at low drain bias. The GaAs/GaAs control sample has a peak mobility of 2682 cm^2/Vs while the GaAs/STO/Si sample has an electron mobility of 2524 cm^2/Vs or 94% of the control sample.

The output characteristics for a 0.7 μm x 100 μm MESFETs are shown in figure 6 for the GaAs control and the GaAs/STO/Si samples. The saturation currents at 0.5V forward gate bias are 367mA/mm and 385mA/mm for the GaAs/STO/Si and the GaAs samples respectively with maximum transconductance of 223mS/mm and 240mS/mm respectively. The gate diode characteristics for devices also show no significant difference between the samples with both having forward turn-on voltages of 0.6V and reverse breakdown voltages of 14V and 14.6V for the GaAs/STO/Si and the GaAs samples respectively.

RF characteristics were measured on-wafer using ground-signal-ground probes. Small signal performance was evaluated first. A number of 0.7μm X 3mm FETs were evaluated on the GaAs/STO/Si substrate. Data from six of the devices that are representative of small signal performance is shown in figure 7, along with performance of the same geometry devices on a standard GaAs substrate. The devices are all biased at V_{ds}=3.5V. The maximum available gain of the GaAs/STO/Si devices compares very favorably with the GaAs devices.

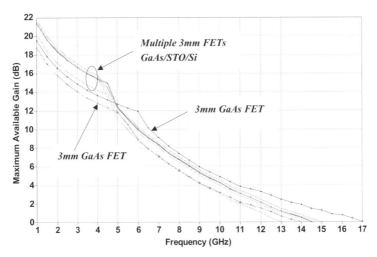

Figure 7. Maximum available gain of six 3mm FETs on GaAs/STO/Si and two on GaAs/GaAs for reference

Two representative devices from different areas on the GaAs control wafer are included as a reference to show the variation of device performances across the wafer. Maximum stable gain of about 18dB was demonstrated for the GaAs/STO/Si 0.7µm x 3mm devices. F_{max} of these devices is 14.5 GHz, which is within the window demonstrated for these devices on a standard GaAs substrate. Differences between the GaAs devices and the GaAs/STO/Si devices are within process variation.

4 Conclusions

We have demonstrated a process for depositing GaAs on silicon using an oxide/amorphous buffer layer. The STO oxide layer is deposited using molecular beam epitaxy on a silicon substrate that has been prepared by a Sr-deoxidation procedure. An amorphous oxide layer is formed at the oxide/Si interface during the growth of the single crystal oxide layer as a result of a reaction between interfacial silicon and oxygen diffusing to the interface. Single domain GaAs growth was achieved with surface roughness around 10Å for a 2 micron thick layer. A large discrepancy has been observed in the determination of defects by A/B etch and plan view transmission electron microscopy. Nevertheless, MESFET devices have been fabricated with performance comparable to GaAs/GaAs control. However, for minority carrier and optical device applications, the defect density needs to be significantly improved.

References

[1] Soga T, Hattori S, Takeyasu M, Umeno M, 1985 J. Appl. Phys. 57 4578
[2] Itoh Y, Nishioka T, Yamamoto A, Yamaguchi M, 1988 Appl. Phys. Lett. 52 1617
[3] Currie M T, Samavedam S B, Langdo T A, Leitz C W, Fitzgerald E A, 1998 Appl. Phys. Lett. 72 1718
[4] Pei C W, Heroux J B, Sweet J, Wang W I, Chen J, Chang M F, 2001 J. Vac. Sci. Tech. B20 1196
[5] Eisenbeiser K, Finder J, Yu Z, Ramdani J, Curless J, Hallmark J, Droopad R, Ooms W, Salem L, Bradshaw S, Overgaard C, 2000 Appl. Phys. Lett. 76 1324
[6] Wei Y, Hu X, Liang Y, Jordan D, Craigo B, Droopad R, Yu J, Demkov A, Edwards J, Moore K, Ooms W, 2002 J. Vac. Sci. Technol. B20 1402
[7] Mckee R A, Walker F J, Chisholm M, 1998 Phys. Rev. Lett. 81 3014
[8] Droopad R, Yu Z, Ramdani J, Hilt L, Curless J, Overgaard C, Edwards J, Finder J, Eisenbeiser K, Wang J, Kaushik V, Ngyuen B-Y, Ooms B, 2001 J. Cryst. Growth 227-228 936

Ultra Shallow GaAs Sidewall Tunnel Junctions Implemented with Low-temperature Area Selective Regrowth

Yutaka Oyama[1,2], Takeo Ohno[1], Kenji Tezuka[1], Ken Suto[1,2] and Jun-ichi Nishizawa[2]

1 Dep. Materials Science and Engineering, Graduate School of Engineering, Tohoku University, Aramaki Aza Aoba02, Sendai 980-8579, Japan

2 Semiconductor Research Institute of Semiconductor Research Foundation, Kawauchi Aoba, Sendai 980-0862, Japan

Abstract. Low temperature (290°C) area selective regrowth by molecular layer epitaxy (MLE) was applied for the fabrication of ultra shallow sidewall GaAs tunnel junctions with the junction depth of 49nm. The tunnel junctions have shown the record peak current density (J_p) up to 31000A/cm^2 and negative differential conductance of -1.4×10^{-5} S at 100µm long strip structure. The mechanisms of strong sidewall orientation depend-ences of J_p are discussed.

1. Introduction

High quality tunnel junction is one of the important semiconductor segments for the application of mm and sub-mm wave source and detectors etc. Among the semiconductor sub-mm wave sources, the TUNNETT (tunnel injection transit time effect diode)[1], which is the transit time effect diode operated by the tunnel injection under the reverse bias, is one of the most promising sub-mm wave source up to terahertz region. In the TUNNETT diode, the most important factor is the quality of p^+n^+ tunnel junction. Reduced junction area is also required in view of the impedance matching[2].

In this work, the low temperature area selective regrowth (ASRe) was applied to form the ultra shallow sidewall GaAs tunnel junctions with the junction depth of 49nm. The ASRe was achieved by MLE at 290°C. The sidewall GaAs tunnel junctions have shown the record peak current density up to 31000A/cm^2, and the strong sidewall orientation dependences are discussed.

2. Experimental

The tunnel diodes were fabricated by MLE[3]. First, Te and S co-doped n^+-GaAs was grown on {100} oriented semi-insulating (S.I.) GaAs at 360°C. Precursors used for Te and S doping were diethyltelluride (DETe) and diethylsulfur (DES). DETe was exposed on the gallium stabilized surface (mode AG)[4] and DES was introduced on the arsenic stabilized surface (mode AA) during MLE. Epitaxial layer thickness of n^+-GaAs is about 49nm, which determines the junction depth of sidewall tunnel junctions. Te and S co-doping[5] is effective to reduce the lattice strain in a heavily donor doped layer. Then, SiN is deposited at 275°C. SiN windows were opened by RIE for ASRe, then, the sidewall mesa was

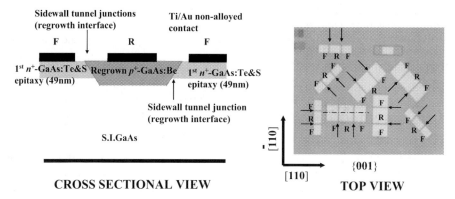

Figure 1. Schematic drawings of the cross sectional and top view of the fabricated sidewall tunnel diodes. "F" and "R" mean the first epitaxial layers and the regrowth layers. The arrows indicate the positions of sidewall regrown tunnel junctions. Large pad has 100mm in length, and small one has 50mm.

formed by H_2SO_4-based solution with the depth of 60nm. Then the surface treatment was carried out at 350°C. The ASRe of p^+-GaAs:Be was carried out at 290°C using Be(MeCp)$_2$. Be(MeCp)$_2$ was introduced by mode AG. Figure 1 shows the schematic drawings of the cross sectional and top view of the fabricated sidewall tunnel junction diodes.

3. Results and discussions

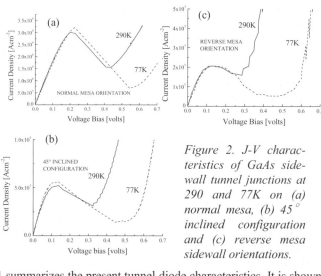

Figure 2. J-V characteristics of GaAs sidewall tunnel junctions at 290 and 77K on (a) normal mesa, (b) 45° inclined configuration and (c) reverse mesa sidewall orientations.

Figure 2 shows the J-V characteristics of GaAs sidewall tunnel junctions. It is shown that the peak current density (J_p) on normal mesa orientation shows the highest value of 31000A/cm^2 with the peak to valley ratio (J_p/J_v) of 2.0. J_p of 45°-inclined configuration and that of reverse mesa orientation are 5200A/cm^2 and 2100A/cm^2. Table 1 summarizes the present tunnel diode characteristics. It is shown that the tunnel junction characteristics have shown strong mesa orientation dependences. In heavily donor doped n^+-GaAs, the donor-defect complex (DX) center[6] is formed, and the excess current is dominated via states. In this case, the valley current density (J_v) is strongly affected by the effective density of states $N^*_{eff} = pn/(p+n)$ [7].

TABLE 1. Sidewall tunnel junction characteristics at RT

	{111}A	{110}	{111}B
J_{peak} (Acm^{-2})	31000	5200	2100
J_{valley} (Acm^{-2})	14900	3110	1530
Peak-to-valley current ratio	2.08	1.67	1.37
V_{peak} (mV)	214	130	136
V_{valley} (mV)	414	282	252
V_{peak}/J_{peak} (Ωcm^2)	6.9×10^{-6}	2.5×10^{-5}	6.5×10^{-5}
Zero-bias specific resistivity (Ωcm^2)	5.3×10^{-6}	1.0×10^{-5}	2.9×10^{-5}
Negative differential conductance (S)	1.4×10^{-5}	7.1×10^{-5}	2.1×10^{-4}

In view of the mechanism for the strong sidewall orientation dependences on J_p, the Be-doping characteristics are investigated by using various surface orientations at the same epitaxial run. The sidewall orientations for normal mesa, 45°-inclined configuration and reverse mesa were determined to be nearly {111}A, {110} and {111}B. Be concentrations on each oriented GaAs is in the order {111}A:{111}B:{110}=1:0.8:0.2. After Kane[8] and Beji et. al. [9], J_p for the direct tunneling is obtained as

$$J_{peak} = 1/36\pi\hbar^2 \sqrt{q^5 m^*_{eff} N^*_{eff} V_0/\varepsilon E_{gap}} \times \exp\left(-\pi/2\hbar \sqrt{\varepsilon E^3_{gap} m^*_{eff}/q^3 N^*_{eff} V_0}\right) D(qV_{peak})$$

where D is an integral in energy units, qV_0 is the built-in potential energy. In addition, in case of trap assisted or phonon assisted inelastic tunneling process, J_p is affected seriously by many parameters[10,11]. The orientation dependence on effective tunneling mass[12] is also one of the important parameters. Actually, $J_v(290K)/J_v(77K)$ is strongly dependent on the sidewall orientations. This indicates that J_v is dominated by the tunneling through the midgap levels in the tunnel junction, especially on reverse mesa orientation.

The reason for the highest J_p on normal mesa orientation can be understood in view of the Be doping characteristics. However, the reason for J_p{110}>J_p{111}B is still not clear. From V_p, the carrier concentration at the tunnel junction is estimated. The electron concentration of first epitaxial n^+-GaAs was about 2x10^{19}cm^{-3}. V_p is obtained by $(\xi_p + \xi_n)/3$. The degeneracy of $n(\xi_n)$ and $p(\xi_p)$ region are calculated from

$$n = N^*_c F_{1/2}(\xi_n/k_b T), \quad p = N^*_v F_{1/2}(\xi_p/k_b T).$$

Hole concentration at interface is estimated to be 2.4x10^{20}cm^{-3} for normal mesa, 5.9x10^{19}cm^{-3} for 45° inclined configuration and 6.8x10^{19}cm^{-3} for reverse mesa orientation, . Whereas the absolute value is different, the magnitude order of estimated hole concentration corresponds well with the Be doping characteristics on each surface. SIMS measurements were performed for the plane structure on {001} surface. Figure 3 shows the SIMS depth profile for the regrown p^+n^+ tunnel junction made on {001} surface. Te and S profiles look very flat, and Be profile shows pile-up at the regrowth interface up to 10^{20}cm^{-3}. It is noticed that this value corresponds well with the hole concentration estimated from V_p. From the SIMS profile, it is considered that the actual doping profile has δ-doping like structure of Be at the tunnel junction interface under the constant heavily Te&S doped n^+-GaAs. This can be one of the reasons for very high J_p of the present tunnel diodes.

.

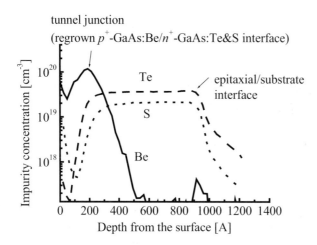

Figure 3. SIMS depth profile for the regrown p^+n^+ tunnel junction made on {001} surface formed by the identical regrowth process and conditions as the sidewall junction.

4. Conclusions

Ultra shallow sidewall GaAs tunnel junctions were fabricated by the low temperature (290°C) area selective regrowth by MLE. The tunnel junctions on the normal mesa orientation have shown the record J_p up to 31000A/cm^2 and differential negative conductance of -1.4x10^{-5} S at 100µm long strip structure. Tunnel junction characteristics have shown strong sidewall orientation dependences due to the orientation-dependent Be doping characteristics and the midgap levels in the tunnel junctions.

References

[1] Nishizawa J and Watanabe Y 1958 Sci. Rep. Res. Inst. Tohoku Univ. **10** 91; Motoya K and Nishizawa J 1988 *Topics in Millimeter Wave Technology* Vol.2 Chapter 1 pp.1-46 (Academic Press, 1988)
[2] Nishizawa J, Makabe H, Matsumoto F, Plotka P and Kurabayashi T 2002 Electronics Letters **38** 660
[3] Nishizawa J and Kokubun Y 1984 Extended Abstracts of 16th Conf. on Solid State Devices and Materials (Kobe Japan) pp.1-5; Plotka P, Kurabayashi T, Oyama Y and Nishizawa J 1994 Appl. Surf. Sci. **82/83** 91-96
[4] Oyama Y, Nishizawa J, Seo K and Suto K 2000 J. Crystal Growth **212** 402
[5] Oyama Y, Tezuka K, Suto K and Nishizawa J 2002 in press J. Crystal Growth
[6] Munoz E and Calleja E 1994 in *DX Centers-donors in AlGaAs and Related Compounds*, ed. by E.Munoz (Scitec, Switzerland, 1994) **Vol.1089** p.135.
[7] Sze S.M 1981 *Physics of Semiconductor Devices*, 2nd ed. (Wiley Inter-science, New York, 1981).
[8] Kane E.O 1961 J. Appl. Phys., **32**, 83
[9] Beji L, Jani B. el, Gilbart P, Portal J.C and Basmaji P 1998 J. Appl. Phys. **83** 5573
[10] Jimenez-Molinos P, Palma A, Gamiz F, Banqueri J and Lopez-Villanueva J.A 2001 J. Appl. Phys. **90** 3396
[11] Rivas C, Lake R, Klimech G, Frensley W.R, Fischetti M.V, Thompson P.E, Rommel S.L and Berger P 2001 Appl. Phys. Lett. **78** 814
[12] Luo L.F, Beresford R, Wang W.I and Mendez E.E 1989 Appl. Phys. Lett. **54** 2133

Growth and properties of polycrystalline GaN on ZnO/Si substrates by ECR-MBE

K Kitamura, H Mamiya, T Araki, T Maruyama Y Nanishi

Dept. of Photonics, Ritsumeikan Univ., 1-1-1 Noji-Higashi, Kusatsu, Shiga 525-8577 JAPAN

Abstract. Growth of polycrystalline GaN on ZnO/Si substrates by ECR-MBE was studied. Desorption of Zn atoms was confirmed over 500°C from ZnO/Si substrates by quadrupole mass spectroscopy. Thus, using a ZnO/Si substrate thermally cleaned at 500°C, polycrystalline GaN grown at 700°C showed smooth surface morphology with grain structure. Three-step growth of polycrystalline GaN was performed, in which GaN was grown at 750°C followed by the growth of a GaN intermediate layer at 700°C on a low-temperature GaN buffer layer. The c-axis orientation, surface morphology and cathodoluminescence property of the polycrystalline GaN were improved by this method.

1. Introduction

Polycrystalline (poly-) GaN including many grain boundaries grown on a variety of substrates has been of particular interest [1-3] in terms of applications, particularly in developments of low cost devices. We have demonstrated that poly-GaN grown on c-oriented polycrystalline ZnO/Si substrates by Electron Cyclotron Resonance plasma-excited molecular beam epitaxy (ECR-MBE) showed strong luminescence [4, 5]. The poly-GaN layer grown on the ZnO/Si had a c-oriented columnar structure. Recently, Bour *et al.* reported fabrication of the first light emitting diode (LED) using poly-GaN grown on glass substrates [3]. However, they indicated that the efficiency of luminescence from poly-GaN based LED was less than that of conventional GaN based LED due to the roughness at the heterointerface. Therefore, improvements in grain size and surface smoothness should be required for device application of poly-GaN. For this purpose, growth at high temperatures is generally performed as a way to enhance surface migration of atoms. But high temperature growth of the poly-GaN on the ZnO/Si is very difficult since the ZnO/Si is thermally unstable compared with sapphire. In this paper, we study on the high temperature growth of the poly-GaN on the ZnO/Si for improving surface morphology of the poly-GaN.

2. Experiments

ZnO was prepared in a separate chamber by magnetron sputtering deposition on a Si (001) substrate prior to growth. The thickness of ZnO was around 2.8 μm. The ZnO was c-axis oriented polycrystalline with random a-axis orientations in c-plane. The ZnO consists of many single crystal domains with random a-axis orientations as shown in Fig. 1.

Poly-GaN films were grown on the ZnO/Si substrate by ECR-MBE. Details of the ECR-MBE system used in this study were reported elsewhere [6]. In this method, metallic Ga was evaporated from the Knudsen effusion cell and the ECR source with mass flow controlled N_2 gas was used to obtain excited nitrogen species. The ECR-MBE system is suitable for the successful growth of GaN on the ZnO/Si substrate due to a low growth temperature (around 700°C) and growth atmosphere without hydrogen both of which suppress the damage of ZnO. A positive bias of 20V was applied to a Mo substrate holder to reduce induced ion damage. Prior to the growth of GaN, ZnO/Si substrates were thermally cleaned in the growth chamber at 720°C (sample A) and 500°C (other samples) for 30 min to obtain a clean surface. A low temperature buffer layer of GaN was grown on the ZnO/Si substrate at 450°C for 20 min. Subsequently poly-GaN was grown at 700-750°C for 2h. The microwave power and nitrogen flow rate were kept constant at 120 W and 30 sccm, respectively. The Ga cell temperatures were 980-990°C. Quadra mass spectroscopy (QMS) was used to investigate surface desorption from ZnO/Si substrates. Growth orientation of the poly-GaN was evaluated by reflection high energy electron diffraction (RHEED). Surface morphology and cross-sectional structure of poly-GaN grown on the ZnO/Si substrate were observed using a scanning electron microscope (SEM : HITACHI S4300SE) and a transmission electron microscope (TEM : JEOL 2010) operated at a 200 keV. Optical property was investigated by cathodoluminescence (CL) using a OXFORD MonoCL2 with an accelerating voltage of 5 keV.

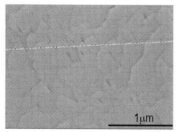

Fig 1. Plane-view SEM image of the ZnO/Si substrate

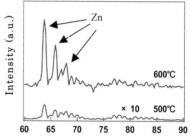

Fig.2 QMS measurement from ZnO/Si

3. Results and Discussion

Prior to the growth of GaN, it is necessary to investigate the influence of a thermal cleaning temperature of the ZnO/Si substrate on the thermal stability. Figure 2 shows results of QMS measurement from the ZnO/Si substrate at 500°C and 600°C. We confirmed desorption of Zn even at 500°C, and the remarkable desorption of Zn at 600°C from the surface. These results suggested that the thermal cleaning of the ZnO/Si should be performed around 500°C for suppressing surface decomposition.

Based on this result, we examined the effect of the thermal cleaning temperature on GaN grown thereon. Figure 3 shows RHEED patterns and SEM images of the poly-GaN grown at 700°C after thermal cleaning at (a) 720°C (Sample A) and (b) 500°C (Sample B), respectively. These poly-GaN films were both c-oriented films with a hexagonal crystal structure. However the c-axis orientation of sample A was slightly distributed as clearly shown in the RHEED pattern. Sample A had the columnar structure with small grain size as we observed in the previous study [5]. On the other hand, the sample B showed smooth surface morphology with large grain size. Streak lines in the RHEED pattern also demonstrate the improvement of surface morphology in the sample B. Figure 4 shows cross-sectional TEM images of sample A and B. It is found that sample A consists of many columnar domains with a size of 50-100 nm. On the other hand, sample B has bigger domains than sample A and the surface of sample B is almost flat. But sample B has a large dislocation density of $4\times10^{10}/\text{cm}^2$. From these results we confirmed that both surface morphology and *c*-axis orientation of the poly-GaN grown on the ZnO/Si were successfully improved by optimizing the thermal cleaning temperature prior to growth.

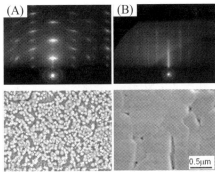

Fig.3 RHEED and SEM images of poly-GaN ; (a) sample A and (b) sample B

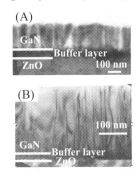

Fig.4 TEM images of poly-GaN ; (a) sample A and (b) sample B

In order to realize further improvement of crystallinity and surface smoothness, poly-GaN was grown at higher temperature (750°C) after thermal cleaning at 500°C (sample C). Figure 5 shows RHEED and SEM images of sample C. Compared with sample B, the *c*-axis orientation and surface morphology of sample C become worse. It might be due to the degradation of the buffer layer while the substrate temperature was raised to the growth temperature. In order to solve this problem, we proposed three-step growth method. The poly-GaN (sample D) was grown at 750°C followed by the growth of an intermediate layer at 700°C on the low-temperature GaN buffer layer. Figure 6 shows a RHEED pattern of the intermediate layer grown at 700°C and RHEED and SEM images of the poly-GaN grown at 750°C. From the RHEED pattern of the intermediate layer, surface of the intermediate layer was considered to be flat as well as sample B. The RHEED pattern of poly-GaN grown at 750°C using three-step growth showed a spot-streak pattern. This result indicated that the *c*-axis orientation of sample D grown at 750°C was also improved compared with sample C. As for the surface morphology shown in the SEM image, however, flat surface was not obtained although the morphology of sample D was improved compared with sample C directly grown at 750°C. This might be because optimum V/III ratio at a growth temperature of 750°C was not obtained due to Ga desorption at this temperature.

Fig.5 RHEED and SEM images of sample C

intermediate layer   grown at 750°C

Fig.6 RHEED and SEM images of sample D

These results confirmed that both surface morphology and *c*-axis orientation of the poly-GaN grown on the ZnO/Si were successfully improved by a three-step growth method at even 750°C.

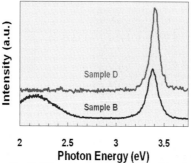

Fig.7 CL spectra measured at RT from sample B and sample D

Figure 7 shows CL spectra from sample B and sample D observed at RT with an acceleration energy of 5 keV. In sample B, although both surface morphology and *c*-axis orientation were improved, yellow luminescence was observed in addition to near band edge luminescence. On the other hand, in sample D grown by the three-step growth method, no deep level luminescence was observed. Intensity and FWHM of the near band edge luminescence were also improved.

4. Conclusions

Growth of poly-GaN on ZnO/Si by ECR-MBE was studied. It is found that desorption of Zn starts over 500°C because of the thermal decomposition of ZnO. In order to suppress this problem, thermal cleaning of the ZnO/Si was carried out at 500°C. Further more, we proposed a three-step growth method, in which poly-GaN was grown at 750°C followed by the growth of the intermediate layer at 700°C on the low-temperature grown GaN buffer layer. Smooth surface morphology together with improved *c*-axis orientation was obtained and CL property was also improved simultaneously.

Acknowledgements

The authors are grateful to Dr. Kadota, Murata Manufacturing Company. Ltd. for the preparation of the ZnO/Si substrates. This work was partly supported by "Academic Frontier Promotion Project" from Ministry of Education, Culture, Sports, Science and Technology.

References

[1] Iwata K, Asahi H, Asami K, Kuroiwa R and Gonda S 1997 Jpn. J. Appl. Phys. 36 L661-L664.
[2] Yagi S, Suzuki S, Iwanaga T 2001 Jpn. J. Appl. Phys. 40 L1349-L1351.
[3] Bour D P, Nickel N M, Van de Walle C G, Kneissl M S, Krusor B S, Ping Mei, and Johnson N M 2000 Appl. Phys. Lett. 76 2182-2184.
[4] Tochishita H, Murata N, Yabe A, Shimizu Y, Nanishi Y and Kadota M 1998 Proceeding of 2nd International Symposium on Blue Laser and Light Emitting Diodes. 174-177.
[5] Araki T, Kagatsume H, Aono H and Nanishi Y 2000 MRS Symposium Proceedings. 639 G6.20 1-6.
[6] Kondo N and Nanishi Y 1989 Appl.Phys. Lett 54 2419-2421.

Single crystalline InN films grown on Si (111) substrates

T Yamaguchi, K Mizuo, Y Saito, T Araki and Y Nanishi
Dept. of Photonics, Ritsumeikan Univ., 1-1-1 Noji-Higashi, Kusatsu, Shiga 525-8577 JAPAN

T Miyajima
Core Technology Development Center, Core Technology & Network Company, Sony Corp., 4-14-1 Asahi-cho, Atsugi, Kanagawa 243-0014, Japan

Abstract. InN films were grown on Si (111) substrates by radio-frequency plasma-excited molecular beam epitaxy (RF-MBE). Use of substrate nitridation and low-temperature buffer layer deposition had the effect of greatly reducing the occurrence of metastable domains in InN films, improving the crystallinity and making the surface flatter. Single crystalline InN films could be successfully grown on Si substrates by using appropriately initial growth processes that consisted of substrate nitridation and low-temperature buffer layer deposition. The band gap of single crystalline hexagonal InN films grown on Si substrates, which are almost the same in local structures to the films on sapphire, is below 1.0 eV, as it is on sapphire.

1. Introduction

InN is an attractive material for photonic and electronic applications. Very recently, high quality, single crystalline InN films have been grown on sapphire substrates [1-6], and several groups have reported that the fundamental bandgap of InN is less than 1 eV [2-6], which is much smaller than the previously reported value of ~1.9 eV [7,8]. If the bandgap of InN is really less than 1 eV, a laser diode (LD) with an $In_xGa_{1-x}N$ active layer can emit over a wider wavelength range. So far, InN films are generally grown on sapphire substrates for these applications even though Si substrates have several advantages over sapphire. For example, Si substrates are easily obtained at high quality and low cost, and Si substrates can be used in opto-electronic integrated circuits (OEICs) with well-developed Si-based devices. For InN, in addition, the lattice mismatch for Si substrate (7.6%) is much smaller than that for sapphire substrate (25.4%). However, growing single crystalline InN films on Si remains difficult [9-14]. Even very recent studies of InN growth on Si (111) show that the InN films are highly c-axis-oriented polycrystalline films [13,14]. In this paper, we report how to obtain single crystalline InN films on Si (111) substrates. We also describe the optical properties and local structures of single crystalline InN films on Si.

2. Experimental

InN films were grown on Si (111) substrates by radio-frequency plasma-excited molecular beam epitaxy (RF-MBE). Si substrates were thermally cleaned at 800 °C for 20 min in vacuum. InN films were then grown at 390 °C for 1 h with the initial growth processes of substrate nitridation and low-temperature buffer layer deposition, hereafter initial growth processes. Substrate nitridation was carried out at 800 °C for 3 min. Low-temperature buffer layer was deposited at 280 °C for 10 min. For comparison, InN films were also directly grown without these initial growth processes. Film thicknesses were ~250 nm. For analysis of surface quality and crystallinity, we carried out *in situ* reflection high-energy electron diffraction (RHEED) measurements during growth, X-ray diffraction (XRD) (Philips X'Pert MRD), transmission electron microscope (TEM) imaging, extended X-ray absorption fine structure (EXAFS), and photoluminescence (PL) measurements using an Ar^+ laser (514.5nm) at 77 K in the spectral range from 600 to 1700 nm.

3. Results and discussions

Figures 1a and 1b show typical RHEED patterns of InN films grown directly on Si substrates. By optimizing the growth temperature and V/III ratio, an InN film highly oriented to the *c*-axis was obtained (Fig. 1b). However, the XRD ω-2θ spectrum (not shown here) showed some metastable tilted hexagonal phases corresponding to ($10\bar{1}1$), ($10\bar{1}2$), and ($10\bar{1}3$), which are consistent with previous reports.[11,13,14] In regard to the *a*-axis orientation, patterns appeared simultaneously along the [$10\bar{1}0$] and [$11\bar{2}0$] azimuths (Fig. 1b). Therefore, the InN films grown directly on Si substrates contained metastable misoriented phases.

Figure 1c shows the RHEED pattern of an InN film grown with the appropriate initial growth processes. A clear streak pattern appears, and simultaneous appearance of patterns with [$10\bar{1}0$] and [$11\bar{2}0$] azimuths do not appear. We found that these initial growth processes improved the crystallinity and produce a very flat surface. Figure 2 shows the XRD ω-2θ spectrum of this film. Existence of metastable tilted hexagonal phases corresponding to ($10\bar{1}1$) was not observed. Existence of other metastable tilted phases was not also observed. Figure 3 shows the result of XRD pole-figure measurement of this film. Metastable rotation domains do not appear. Therefore, the results obtained by RHEED and XRD measurements show that the initial growth processes effectively suppressed the metastable domains and improved both the crystallinity and the surface flatness. The high-resolution TEM image in Fig. 4a and the selected area diffraction (SAD) pattern in Fig. 4b support the results of the XRD measurements, thus confirming microscopically that this InN film had a well-oriented hexagonal structure. These results show that the appropriate initial growth processes produce single crystalline InN films on Si substrates. Further details about initial growth processes will be reported elsewhere.

It has been claimed that the fundamental band gap of InN is ~1.9 eV.[7,8] However, recent studies indicate that the band gap of single crystalline hexagonal InN films grown on sapphire substrates is below 1.0 eV.[2-6] Figure 5 shows the PL spectrum at

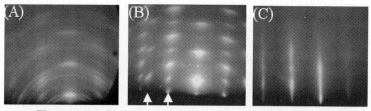

Figure 1. (a),(b) Typical RHEED patterns for InN films grown directly on Si substrates. (c) RHEED pattern of an InN film grown with the appropriate initial growth processes on a Si substrate.

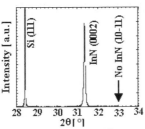

Figure 2. XRD ω-2θ spectrum of an InN film grown with the appropriate initial growth processes. *(left)*

Figure 3. {11$\bar{2}$2} XRD pole-figure of an InN film grown with the appropriate initial growth processes. *(right)*

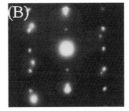
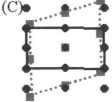

Figure 4. (a) Cross-sectional TEM image and (b) SAD pattern of an InN film grown with the appropriate initial growth processes on a Si substrate. A schematic of the predicted diffraction pattern for InN(●) and Si(■) is shown in (c).

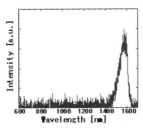

Figure 5. PL spectrum at 77K for a single crystalline hexagonal InN film.

77 K from the single crystalline hexagonal InN film grown on a Si substrate. No peaks near ~1.9 eV (λ~650nm) appear, but a clear peak appears at ~0.8 eV (λ~1550nm). This result agrees with recent reports for single crystalline InN films grown on sapphire substrates [2-6]. According to our EXAFS measurement, the interatomic distances of In-N and In-In were d_{In-N}= 0.213 nm and d_{In-In}= 0.353 nm for a single crystalline InN film grown on Si. These values are close to the values for InN film grown on sapphire [15]. We suggest that local structures in InN films grown on sapphire and Si were almost the same, although the FWHM of the XRD rocking curve for InN films grown on Si substrates was still larger than that for InN films on sapphire substrates. These results support that the band gap of single crystalline hexagonal InN films grown on Si substrates is below 1.0 eV, as it is on sapphire substrates [2-6]. On the other hand, Yodo *et al.* recently observed a strong PL peak at ~1.8 eV for InN films grown on Si substrates [14]. The cause of these different PL peaks is unclear at present.

4. Conclusion

Purely hexagonal, single crystalline InN films could be grown on Si substrates by using the appropriate initial growth processes. In these single crystalline InN films, the peak in the PL spectrum at 77 K appeared at ~0.8 eV, which agrees with recent reports for high quality, single crystalline InN films grown on sapphire substrates. Our EXAFS measurements showed that the interatomic distances of In-N and In-In were d_{In-N}= 0.213 nm and d_{In-In}= 0.353 nm for single crystalline InN film grown on Si substrates, which are almost the same as those for single crystalline InN films grown on sapphire substrates. Therefore, the local structures in single crystalline InN films grown on sapphire and Si are almost the same. These results suggest that the band gap of single crystalline hexagonal InN films grown on Si is below 1.0 eV, as it is on sapphire substrates.

Acknowledgement

The authors would like to express their thanks to Dr. N. Teraguchi and Dr. A. Suzuki for useful discussion. This work was supported in part by the Ministry of Education, Culture, Sports, Science and Technology, Grant-in-Aid for Scientific Research (B), 13450131, 2002.

References

[1] Saito Y, Teraguchi N, Suzuki A, Araki T and Nanishi Y 2001 Jpn. J. Appl. Phys. 40 L91-L93.
[2] Inushima T, Mamutin V V, Vekshin V A, Ivanov S V, Sakon T, Motokawa M and Ohoya S 2001 J. Cryst. Growth 227-228 481-485.
[3] Davydov V Y, Klochikhin A A, Seisyan R P, Emtsev V V, Ivanov S V, Bechstedt F, Furthmuller J, Harima H, Mudryi A V, Aderhold J, Semchinova O and Graul J 2002 phys. stat. sol. (b) 3 R1-R3.
[4] Wu J, Walukiewicz W, Yu K M, Ager III J W, Haller E E, Lu H, Schaff J, Saito Y and Nanishi Y 2001 Appl. Phys. Lett. 80 3967-3969.
[5] Matsuoka T, Okamoto H, Nakao M, Harima H, Kurimoto E and Izutsu 2002 IEICE Technical Report 102, 117 85-88.
[6] Saito Y, Hori M, Yamaguchi T, Teraguchi N, Suzuki A, Araki T and Nanishi Y 2002 IEICE Technical Report 102, 117 89-92.
[7] Osamura K, Nakajima K, Murakami Y, Shingu P H and Ohtsuki A 1972 Sol. Stat. Commun. 11 617-621.
[8] Tansley T L and Foley C P 1986 J. Appl. Phys. 59 3241-3244.
[9] Bello I, Lau W M, Lawson R P W and Foo K K 1992 J. Vac. Sci. Technol A 10 1642-1646.
[10] Bu Y, Ma L and Lin M C 1993 J. Vac. Sci. Technol A 11 2931-2937.
[11] Yamamoto A, Tsujino M, Ohkubo M and Hashimoto A 1994 J. Cryst. Growth 137 415-420.
[12] Morjan R E, Perrone A, Zocco A and Dinescu M 2001 Proceedings of SPIE 4430 310-316.
[13] Yang F H, Hwang J S, Chen K H, Yang Y J, Lee T H, Hwa L G and Chen L C 2002 Thin Solid Films 405 194-197.
[14] Yodo T, Yona H, Ando H, Nosei D and Harada Y 2002 Appl. Phys. Lett. 80 968-970.
[15] Miyajima T, Kudo Y, Liu K –L, Uruga T, Honma T, Saito Y, Hori M, Nanishi Y, Kobayashi T and Hirata S *To be published in phys. stat. sol.*

Growth and evaluation of CdTe/Si (111) by hot wall epitaxy

Georgi M. Lalev[a],*, Jifeng Wang[a], Seishi Abe[b], Katashi Masumoto[b] and Minoru Isshiki[a]

[a]Institute of Multidisciplinary Research for Advanced Materials, Tohoku University, 1–1, Katahira 2–chome, Aoba–ku, Sendai 980–8577, Japan
[b]The Research Institute for Electric and Magnetic Materials, 2-2-1 Yagiama-minami, Sendai 982–0807, Japan

Abstract

In this study, series of experiments were made for producing a high quality CdTe/Si (111) crystal film by hot wall epitaxy (HWE). CdTe epitaxial layers were grown directly on the hydrogen terminated Si (111) substrate without preheating treatment. Two–step growth regime was applied for obtaining high quality CdTe crystal film. X–ray diffraction analysis (powder and four–crystal rocking curves) and photoluminescence (PL) spectra at 4.2 K were employed for assessing the crystal quality. For the selected growth conditions the best FWHM value of 117 arcsec was obtained.

1. Introduction

Since CdTe is one of the most attractive materials for solar cells, photo– and infrared detectors, many attempts were made to produce a high quality epitaxial film on various substrates. GaAs was widely used as a substrate material and excellent crystal films were obtained by different techniques. As one of them, HWE [1,2] is characterized by simple design and low cost of the equipment as well as relatively fast crystal growth. Good results were obtained using As passivated Si(111) by MBE [3,4]. Other attempts for producing high quality CdTe directly grown on hydrogen terminated Si (111) by HWE were made but formation of double domain structure was reported [5,6]. Regardless of the poor thermal properties and the high price of GaAs, for further application as HgCdTe/GaAs the inter–diffusion of Hg and Ga results in voids at the interface [7], and it makes GaAs not applicable for the above–mentioned system. Silicon substrate is proved to be an effective barrier for Hg and due to its good physical properties and low price is expected to be very efficient substrate material.

In this study, our efforts are focused on producing and evaluating a high quality single crystal film on Si(111) with a suitable thickness for further application as a starting material for HgCdTe infrared detectors.

---

*corresponding author: georgi@mail.iamp.tohoku.ac.jp

2. Experimental procedure

Series of experiments were made under different growth conditions using HWE apparatus with four temperature zones: substrate, wall, source and reservoir. Base pressure was kept under 5×10^{-6} Torr.

To ensure high purity epitaxial layer, 6N CdTe and 6N Cd were used as source and reservoir materials, respectively. CdTe single crystal film was grown on the hydrogen terminated Si (111) substrate without any preheating treatment.

Hydrogen termination of the substrate allows not only a single crystal CdTe film to be formed on inactivated Si(111) surface, but it appears to be an essential condition for the desirable low-temperature device fabrication based on CdTe/Si materials. Substrate plates (10x10 mm) were cleaned following modified RCA technology [8]. Two-step etching procedure with fluorine based solutions was applied for ensuring perfect hydrogen termination (1x1) and improved microroughness of the Si surface.

Series of CdTe crystal layers were produced according to two-step growth pattern. Experiments were conducted at $T_{substrate}$ = 350 °C, and two different source temperatures were basically selected (T^1_{source} = 380 °C and T^2_{source} = 440 °C) to provide slow and fast growth rate in the first and second growth step, respectively. Between these two steps the source temperature was gradually increased without crystal growth interruption. For simplicity $T_{wall} = T_{source}$ is adopted hereafter. Reservoir temperature was 180 °C. These four temperatures as well as the growth rate were established as optimum ones in our previous study (not presented here).

3. Results and discussion

Since the lattice mismatch between Si and CdTe is very large, the interfacial epilayer to the Si surface is suffering from extremely high internal stress, resulting from formation of a high-density of dislocations. It was believed that even at very slow growth rates the nucleation process of CdTe on Si follows Volmer-Weber mode. Atomic Force Microscopy (AFM) showed that surface of the CdTe epilayer becomes rough with increasing the thickness. To diminish the effect of critical thickness layer, two-step crystal growing regime was applied. A very slow growth rate of 0.04 µm/h was applied in the first step to form mirror-like surface free of hillocks. Epilayer thickness was considered to be close to the critical one. After this short step, the growth rate was gradually increased until it reached the value of 0.5 ~ 0.7 µm/h. These conditions were controlled by the source temperature.

Powder XRD analysis showed that the epitaxial layers are single crystal with strong preferential orientation of CdTe (111). No other peaks except those related to the CdTe (111) and Si (111) planes were observed.

Four-crystal rocking curves XRD data, plotted in Fig.1, illustrate clear dependence of crystal quality on the epilayer thickness. According to the ω and $\omega-2\theta$ scan, increasing the thickness leads to significant improvement of the crystal quality. FWHM becomes almost constant after 5 µm thickness, and reaches the lowest value of 117 arcsec. This value is close to the best reported [3,4] for MBE CdTe(111)B epilayer grown on As flux treated substrate.

Decreasing the film thickness from 5 to 0.8 µm, the crystal quality becomes strongly limited by mosaic structure and unevenness of the lattice parameters. Almost negligible deviation between ω and $\omega-2\theta$ scan shown in the insert of Fig.1 is a result of very small tilt angle toward the [111] direction of the Si substrate.

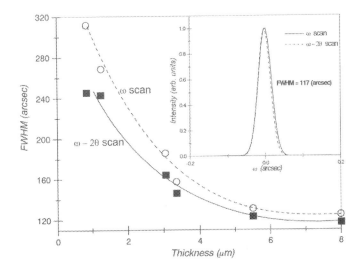

Figure 1. Dependence of FWHM from four–crystal XRD on the epilayer thickness. The insert represents XRD four–crystal rocking curves from 8 μm epilayer.

Photoluminescence (PL) spectra at 4.2 K of the as–grown CdTe crystal films were also measured to investigate the defect–induced emission bands in CdTe shown in Fig. 2.

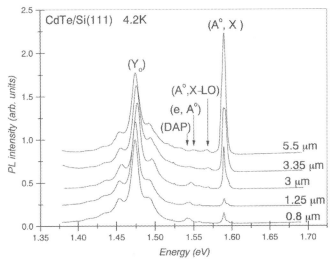

Figure 2. PL spectra of CdTe epilayers with different thickness.

PL spectra show a strong peak at 1.47 eV. This emission band (Y_o) [9], is frequently observed for CdTe crystals, especially for heteroepitaxial films, and its origin is attributed to the electron–hole recombination at extended defects. For the purpose of our analysis this peak was taken as a standard and the intensity of the other peaks were compared. Quality of the epitaxial layer was estimated by the exitonic peak intensity

(A^o, X). A clear correlation was observed between the intensity of exitonic emission band (A^o, X) at 1.5896 eV and the film thickness. The (A^o, X) intensity increases with thickness. The sharpest exitonic peak for 5.5 μm thick epilayer corresponds to the highest crystal quality. This result is in excellent agreement with above discussed XRD data.

Another weak peak at 1.569 eV is LO photon replica of the dominant (A^o, X) transition. (A^o,X–LO) slightly increases with (A^o, X). The bands at 1.541 and 1.549 eV are attributed to the donor–acceptor pair (DAP) and free–to–bound transitions (e, A^o), respectively. The peak intensity of (DAP) decreases and finally almost vanished with increasing the thickness and adding to the (e, A^o) peak. Shin [10] and Hallidey [11] also reported this free to bound (e, A^o) transition at 1.549 eV between a free electron and an acceptor complex consisting of a Cd vacancy and two shallow donors (VCd–2D).

4. Conclusions

High – quality CdTe/Si(111) crystal films were produced by HWE following two–step growth pattern under the selected conditions. This method proved that low cost and relatively fast grown epilayers could be fabricated with suitable thickness for further application in solar cell and IR detectors.

Both XRD and PL results show a significantly increasing crystal quality with increasing epilayer thickness. This conclusion is confirmed by the FWHM value of 117 arcsec obtained from four–crystals rocking curves XRD as well as the dominant exitonic peak from PL spectra. Further adjustment of the growth conditions is considered to optimize these results.

References:

[1] J.F. Wang, K. Kikuchi, B.N. Koo, Y. Ishikawa, W. Uchida, M. Isshiki, J. Crystal Growth 187 (1998) 373-379.
[2] Bounheun Koo, Jifeng Wang, Yukio Isshikawa and Minoru Isshili, Jpn. J. Appl. Phys. 37 (1998) 5674-5679.
[3] Y. Xin, S. Rujirawat, N.D. Browning, R. Sporken, and S. Sivananthan, Appl. Phys. Lett. 75 (1999) 349-351.
[4] H. Schick, F. Bensing, U. Hilpert, U. Richter, L. Hansen, J. Wagner, V. Wagner, J. Geurts, A. Waag, G. Landwehr, J. Crystal Growth 214/215 (2000) 1-4.
[5] S. Seto, S. Yamada, K. Suzuki, J Crystal Growth 214/215 (2000) 5-8.
[6] S. Seto, S. Yamada, T. Miyakawa, K. Suzuki, J Crystal Growth 237-239 (2002) 1585.
[7] J. Giess, J.S. Gough, S.J.C. Irvine, J.B. Mullin and G.W. Blackmore, Mater. Res. Soc. Symp. Proc. 90 (1986) 398.
[8] W. Kern and D.A. Puotinen, RCA Rev. 31 (1970) 187.
[9] P.J. Dean, G.M. Williams and G. Blackmore, J. Phys. D: Appl. Phys. 17 (1984) 2291.
[10] H.Y. Shin, Y. Sun, Mater. Sci. Eng. B 52 (1998) 78.
[11] D.P. Halliday, M.D.G. Potter, J.T. Mullins, A.W. Brinkman, J. Crystal Growth 220 (2000) 30-38.

A surface reconstruction functioning as a micro mask during in-situ layer-by-layer etching of GaAs(111)B using AsBr$_3$

Y. Asaoka, I. Ihara, K. Yokoyama, S. Daicho, N. Sano, T. Kaneko

School of Science and Technology, Kwansei Gakuin University
2-1 Gakuen, Sanda, Hyogo 669-1337, Japan

Abstract. The effect of the surface reconstructions on GaAs(111)B during the layer-by-layer etching using AsBr$_3$ as an etchant gas was investigated. The surface reconstructions strongly affect the etching kinetics, the etching rate observed from reflection high-energy electron diffraction (RHHED) intensity oscillations [1,2]. It is found that the sudden drop in etching rate in the $(\sqrt{19}\times\sqrt{19})$-like temperature region is clearly observed. This means the surface covered with $(\sqrt{19}\times\sqrt{19})$-like reconstruction has the resistance to AsBr$_3$. Moreover, the kinetic-roughening was observed in the temperature region, where both $(\sqrt{19}\times\sqrt{19})$ domain and the (2×2) domain coexist. This means the $(\sqrt{19}\times\sqrt{19})$ domain locally covering the surface acts a micro etching-mask and then the etching mode become not layer-by-layer fashion. These results indicate that this etching process is highly sensitive to the surface reconstruction, and can use the surface reconstruction as a functioning etching mask.

1. Introduction

The *in-situ* layer-by-layer etching using AsBr$_3$ has advantages over the other conventional etching techniques for III-V material systems in controlling nano-scales due to its highly sensitive etching kinetics reflecting chemical and physical properties on the semiconductor surface [1]. The fabrication of 3 dimensional nano-structures utilizing this etching requires the total knowledge on the etching properties against the temperature and the surface orientation. Although the etching properties on (100) surface of III-V material have been studied well, the properties on the other surfaces have been studied little. Therefore, in this study, we investigate the AsBr$_3$ etching on GaAs(111)B surface, which includes a peculiar surface reconstruction that directly associates with the determination of the final surface morphology and crystal quality [3].

The reconstruction on GaAs(111)B surface changes from (2×2) to $(1\times1)LT$, $(\sqrt{19}\times\sqrt{19})R23.4°$, and $(1\times1)HT$ with increasing temperature. The (2×2) is constructed by As-trimmers, while the $(\sqrt{19}\times\sqrt{19})$ by hexagonal-rings with a monolayer height of Ga and As atoms [4]. The $(1\times1)LT$ is the structurally transitional phase between (2×2) and $(\sqrt{19}\times\sqrt{19})$ [5], where both As-trimmers and hexagonal-rings randomly coexist. In this transitional phase, the concentration of As-trimmer and hexagonal-ring changes as a function of substrate temperature and arsenic flux [6]. This surface transition is accompanied with the clear change of the RHEED pattern and specular intensity. In the lower temperature region of $(1\times1)LT$, the RHEED pattern indicates only (1×1) line structure, and the specular intensity decreases with increasing temperature. On the other hand, in the higher temperature region of $(1\times1)LT$, the RHEED pattern indicates (1×1) structure with obscure spots, and the specular intensity increases with increasing temperature. Therefore we have distinguished the $(1\times1)LT$, furthermore, into two different phases; one is hexagonal-ring and As-trimmer coexisting $(1\times1)LT_I$ located in lower temperature region, and another is hexagonal-ring domain dominated $(1\times1)LT_{II}$ located in higher temperature region.

2. Experiment

The experiments were performed in a modified conventional ultra-high vacuum solid source molecular beam epitaxy (MBE) chamber equipped with a gas manifold for AsBr$_3$ to be introduced without carrier gas. The substrates used were epi-ready nominally singular $(0\pm0.5°)$ GaAs(111)B. The substrates cleaned by conventional method were mounted on a molybdenum blocks with Indium and loaded in the MBE. After the out-gassing in the preparation chamber, the substrate was transferred into the growth chamber. The initial surface native oxide was removed by annealing under As$_4$ pressure at 610°C, which was measured with an infrared pyrometer. Before each scan was recorded, to ensure the identical smooth initial-surface, a GaAs buffer layer was deposited on the surface at the constant condition in the same chamber by MBE. The layer-by-layer etching processing was *in-situ* monitored by RHEED with an electron beam energy of 20 keV at an incident angle of $\approx 1°$ along the $[11\bar{2}]$ azimuth. To observe clear RHEED oscillations on arsenic-stabilizing GaAs(111)B surface, the limitation of As$_4$ supply during the etching is essential because the excess arsenic on the surface hinders the layer-by-layer etching [7]. Therefore before the etching the As$_4$ shutter was closed at a targeting temperature to reduce the effective arsenic on the surface. The closing of the As$_4$ shutter induces a RHEED pattern and intensity change that means the transition of the reconstruction into the steady surface at a temperature without As$_4$ flux. Then the etching experiments were conducted by simply supplying an AsBr$_3$ flux to the surface without As$_4$ flux.

3. Results and discussion

A series of RHEED intensity oscillations during the etching of AsBr$_3$ with 1.8×10^{-6} Torr flux is shown in Figure 1 for sample temperatures ranging from 590 to 486°C. The surface during etching exhibited the variety of the reconstructions peculiar to (111)B as a function of temperatures. The etching kinetics, which is directly reflected by the change in RHEED specular beam intensity, revealed a strong dependence on surface reconstructions. In the figure, the symbols A to E correspond to (2×2), $(1\times1)LT_I$, $(1\times1)LT_{II}$, $(\sqrt{19}\times\sqrt{19})-like$, and $(1\times1)HT$, respectively. The layer-by-layer fashion during etching is observed for all reconstructed surfaces except the $(1\times1)LT_{II}$ region (C). The fact that the intensity decreases monotonically as the etching proceeds in the $(1\times1)LT_{II}$ indicates that a reconstruction associated kinetic roughening takes place. The detailed mechanism of this anomalous roughening can be understood by considering the surface reconstruction dependence of etching rates.

The inset of the Figure 1 plots the temperature dependence of the etching rates deduced from the period of the RHEED intensity oscillations at a constant AsBr$_3$ flux. A non-linear effect in etching rate is clearly observed with a sudden drop in etching rate in the $(\sqrt{19}\times\sqrt{19})-like$ region (D). Such a non-linear effect has never been observed for the other crystallographic orientations of GaAs surfaces [8]. Taking into account of the supply rate limited temperature regime for those surfaces using AsBr$_3$ [8], the etching rates in the

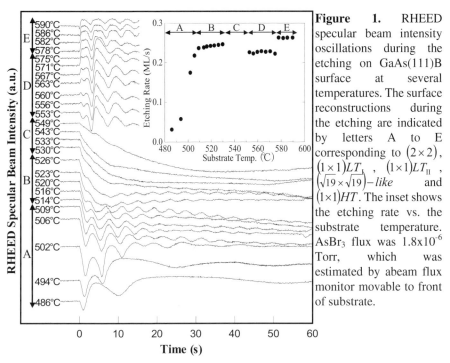

Figure 1. RHEED specular beam intensity oscillations during the etching on GaAs(111)B surface at several temperatures. The surface reconstructions during the etching are indicated by letters A to E corresponding to (2×2), $(1\times 1)LT_I$, $(1\times 1)LT_{II}$, $(\sqrt{19}\times\sqrt{19})-like$ and $(1\times 1)HT$. The inset shows the etching rate vs. the substrate temperature. AsBr$_3$ flux was 1.8×10^{-6} Torr, which was estimated by abeam flux monitor movable to front of substrate.

regions C to D should have reached the value in the E region (the A region corresponds to a reaction rate limited situation). However, the fact that the etching is reduced in the region D implies that the hexagonal-ring domain plays a dominant role being a stable micro structure to hinder the removal of surface Ga from the AsBr$_3$ etching.

The difference in the etching characteristics of the regions C and D can be explained by the difference in the effective coverage of the hexagonal-ring domains. In the region C due to the coexistence of hexagonal-ring domains and As-trimmers, spatially inhomogeneous etching occurs with the hexagonal-ring domain functioning as micro-mask. In the $(1\times 1)LT_I$ region (B), that is the other $(1\times 1)LT$, because the hexagona rings don't form "domains" that is an etching mask due to its small concentration, th etching rate shows supply rate limited mode.

These surface reconstructions that were a function of the substrate temperature were observed in the different AsBr$_3$ flux too. The surface reconstructions are a function of the effective arsenic on the surface. Therefore the surface reconstructions as function of the substrate temperature are also the function of AsBr$_3$ flux. The surface phase map during the etching on GaAs(111)B, generated from the RHEED data in this study, is displayed in Figure 2. While the surface during the etching indicates (2×2) reconstruction at a large AsBr$_3$ flux and a low substrate temperature, it indicates $(1\times 1)HT$ reconstruction at a little AsBr$_3$ flux and a high substrate temperature. The activation energies for the surface phase transitions during the AsBr$_3$ etching were obtained as 1.9 eV (region (A) $\Leftrightarrow$ region B)), 2.0 eV (region (B) $\Leftrightarrow$ region (C)), 3.5 eV (region (C) $\Leftrightarrow$ region (D)) and 6.0 eV (region (D) $\Leftrightarrow$ region (E)). These values are nearly equal to the values obtained on th GaAs(111)B under As$_4$ flux [5]. Therefore the reconstruction during the etching strongly depends on the arsenic kinetics on the surface. Also by utilizing this map, the etching mode is perfectly controllable.

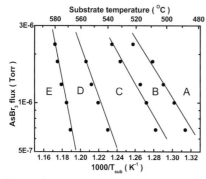
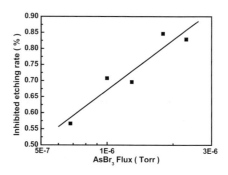

Figure 2. Phase diagram of GaAs(111)B surface reconstruction during AsBr$_3$ etching. The letters A to E indicate the reconstructions same as Fig.2.

Figure 3. The inhibited etching rate, which is a ration in etching rate on the region D to the region E, is plotted against AsBr$_3$ flux.

The inhibited etching rate in the region D, which is the ratio of etching rate in the region D to the region E, is plotted as a function of the AsBr$_3$ flux in Figure 3. The inhibited etching rate on the region D increases with the decreasing AsBr$_3$ flux. In other words the hexagonal-ring domain acts as strong mask where using a little AsBr$_3$ flux. Therefore if the position and the size of the hexagonal-ring domains are controlled, the reconstruction could become an efficient micro mask to fabricate nano-structures.

4. Conclusion

The etching kinetics on GaAs(111)B using AsBr$_3$ was investigated by RHEED observation. The surface reconstructions during the etching strongly affect the etching kinetics. The surface reconstructions were a function of temperature and AsBr$_3$ flux. The layer-by-layer fashion during etching is observed on all reconstructed surfaces except the $(1 \times 1)LT_{II}$. The sudden drop in etching rate is clearly observed in the $(\sqrt{19} \times \sqrt{19})-like$ region. From these results, the inhibition of the etching on the surface covered with a hexagonal-ring domain was revealed. And it was shown that the hexagonal-ring domain acts as strong mask where using a little AsBr$_3$ flux rather than the large flux.

References

[1] H. Schuler, T. Kaneko, M. Lipinski, K. Eberl, Semicond. Sci. Technol. 15 (2000) 169
[2] Y. Asaoka, T. Arai. N. Sano, T. Kaneko, Inst. Phys. Conf. Ser. 170 (2002) 331
[3] Y. Park, M. J. Cich, R. Zhano, P.Specht, E. R. Weber, E. Stach, S. Nozaki, J. Vac. Sci. Technol. B 18(2000)
[4] D. K. Biegelsen, R. D. Bringans, J. E. Northrup, L. E.Swartz, Phys. Rev. Lett. 65 (1990) 452
[5] D. A. Woolf, D. I. Westwood, R. H. Williams, Appl. Phys. Lett. 62 (1993) 1370
[6] A. R. Avery, E. S. Tok, T. S. Jones, Surf. Sci. Lett. 376(1997) L397
[7] M. Ritz, T. Kaneko, K. Eberl, Appl. Phys. Lett. 71 (1997) 695
[8] T. Kaneko, P. Smilauer, B.A. Joyce, T. Kawamura, D. D. Vvedensky, Phys. Rev. Lett. 74 (1995) 3289

Comparative study of p-type dopants, Mg and Be in GaN grown by RF-MBE

S. Sugita[1], Y. Watari[1], G. Yoshizawa[1], J. Sodesawa[1], H. Yamamizu[1], K. T. Liu[2], Y. K. Su[2] and Y. Horikoshi[1]

[1]School of Science and Engineering, Waseda University, 3-4-1 Okubo, Shinjuku-ku, Tokyo 169-8555, Japan
[2]Department of Electrical Engineering, National Chen Kung University, Taiwan, R.O.C

Abstract. We have compared the characteristics of Be-doped and Mg-doped GaN layers grown by RF-MBE. When Be-doping is performed in GaN grown on sapphire (0001), the surface polarity changes from N to Ga during the growth. When this polarity change takes place, the resistivity of grown layer is high. If the AlN buffer layer is used, GaN layer can be grown under the Ga surface polarity condition from the beginning. Thus, we have successfully grown Be-doped p-type GaN without inducing polarity change. We have confirmed that the Be acceptor level is shallower than that of the Mg acceptor by comparing the activation energies.

1. Introduction

Group III-nitride semiconductors have proved useful for light emitting devices in the blue and ultraviolet wavelength region. The improvement of p-type GaN conductivity is the key issue in developing these devices. Mg is the most commonly used dopant to obtain p-type GaN. However, large energetic depth of the Mg acceptor level (200meV) results in low efficiency of the electrical activation of the Mg acceptor. Photoluminescence (PL) data[1] and theoretical calculations[2] of Be-doped GaN have suggested that Be has a shallower acceptor level than Mg. Nevertheless, experimental evidences indicate that deep levels are formed in GaN by Be-doping. Several authors have reported photoluminescence due to Be-related deep centers[3,4]. In this work, we have successfully grown Be-doped p-type GaN and have compared the characteristics of Be-doped and Mg-doped GaN layers.

2. Experimental details

The samples used in this study are grown by MBE on sapphire (0001) substrate. Nitrogen is supplied by RF plasma source. After the growth of 100nm-thick undoped buffer layer at 820°C, the 0.5-2.0μm-thick GaN layer is grown. The nitrogen plasma source is operated at an RF power of 470W with N_2 flow rate of 8sccm. Ga Flux density is fixed at 1.0×10^6 [Torr] while the Be-cell temperature is changed between 760 and 900.

Samples are investigated by PL using 325nm He-Cd laser and Hall measurements. We use NaOH solution (3N) for the polarity tests[5].

3. Results and discussions

3.1. Polarity change

Figure.1 compares the photoluminescence spectrum of Be-doped GaN with that of a Mg doped sample. Be acceptor related emission peaks appear at higher energies than those of Mg acceptor, suggesting that the Be acceptor level is much shallower than the Mg acceptor level. However, the Be-doped GaN shows high resistivity. Such highly resistive Be-doped GaN exhibits prominent red emission centered at approximately 2.0eV which keeps its intensity even at room temperature. The existence of the defect level responsible for the red emission probably causes high resistivity, because these two phenomena appear simultaneously.

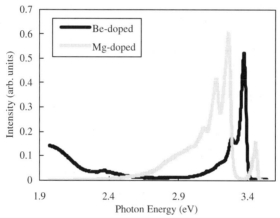

Fig.1 Low temperature (10K) PL spectra of Be-doped and Mg-doped samples.

Figure.2 compares the RHEED patterns observed during the growth of Mg-doped and Be-doped GaN. Mg-doped GaN surfaces always show streaky patterns, and the resulting surfaces are mirror-like. However, during the growth of Be-doped GaN, spotty patterns appear and the resulting surfaces are slightly rough.

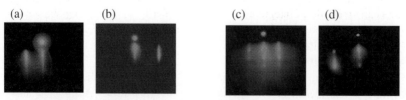

Fig.2 RHEED patterns after the growth of GaN. (a), (b) indicate the patterns measured in the (10-10) and (11-20) azimuth of Mg sample, respectively. (c), (d) indicate those of Be doped samples.

In order to investigate the difference in the crystal quality between the Mg-doped and Be-doped GaN, we examined the surface polarity. It is known that when the top surface is N-terminated, the GaN layer is soluble in alkaline solution[5], while when the top surface is Ga-terminated, the GaN layer is insoluble. Therefore, the surface polarity can be easily tested by immersing the sample to the solution. The results of the test show both undoped and Mg-doped GaN layers grown on sapphire (0001) substrates have the N-surface. However, we have found that the Be-doped GaN layer grown on sapphire (0001) substrate shows a Ga-surface polarity, regardless of the polarity of buffer layer surface. This result implies that during the growth of Be-doped GaN, polarity change takes place from N to Ga surface. If this polarity change is the main cause of the difficulty of p-type layer growth by Be-doping, it can be circumvented by using the buffer layer with the Ga-surface polarity. Thus, the Ga-surface polarity can be kept throughout the growth of Be-doped GaN layer. Usually, GaN buffer layer surface shows N-surface polarity, while when AlN buffer layer with appropriate thickness is used, the over grown GaN layer shows Ga surface polarity[6]. In order to find the optimum AlN buffer layer thickness, undoped GaN layers are grown on AlN buffer layers with various thicknesses. The grown layers are etched in alkaline solution for 30min. After the etching, the X-ray diffraction measurement is performed, and the integrated diffraction intensity of GaN layer is plotted as a function of the AlN thickness (Fig.3). No discernible etching is observed when the AlN layer thickness is approximately 500 Å. Therefore, this value is used in the Be-doped GaN growth. Indeed, when 500Å-thick AlN buffer layer is used, Be-doped GaN layers show p-type conductivity and no red emission (Fig.4).

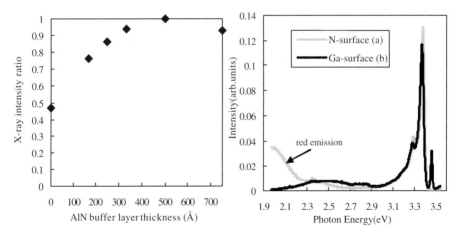

Fig.3 The relationship of AlN thickness and the ratio of integrated diffraction intensity of GaN layers before and after alkaline etching for 30min.

Fig.4 PL spectra of Be-doped samples. The polarity of the starting surface is (a) N-surface, (b) Ga-surface, respectively.

3.2. Hall effect measurement

We have performed Hall effect measurement for several samples and confirmed that the formation of a p-type GaN layer by Be-doping. The maximum hole concentration obtained so far is approximately 2×10^{16} [cm^{-3}] with the hole mobility of approximately 25 [cm^2/Vs].

The Be concentration in grown GaN layer is estimated by chemical method (ICP-MS) to be approximately 3×10^{19} [cm^{-3}]. The activation energy of Be acceptor is estimated to be 120meV which is favourably compared with the activation energy of Mg acceptor (198meV) as shown in Fig.5. These values almost coincide with the acceptor levels estimated by PL data.

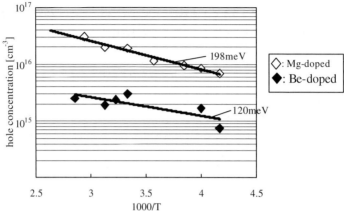

Fig.5 Temperature dependence of hall concentration for the samples of Mg-doped and Be-doped.

4. Conclusion

We have successfully grown Be-doped p-type GaN on sapphire (0001) substrate by conserving the surface polarity during growth. When Be-doping is performed in GaN grown on sapphire (0001), the surface polarity changes from N to Ga during the growth. When this polarity change takes place, red emission (2.0eV) appears and the grown layer is highly resistive. If the AlN buffer layer is used, GaN layer can be grown under the Ga surface polarity condition from the beginning. Thus, we have successfully grown Be-doped p-type GaN without using the co-doping technique[7]. The activation energy of this sample is 120meV which almost coincides with the PL data. We have confirmed that Be acceptor is shallower than Mg acceptor by comparing with activation energy of Mg and Be acceptor.

References

[1] F J Sanchez, F Calle, M A Sanchez-Garcia, E Calleja, E Munoz, C H Molloy, D J Somerford, J J Serrano and J M Blanco 1998 *Semicond. Sci. Technol.* **13** 1130
[2] Fabio Bernardini and Vincenzo Fiorentini 1997 *Appl. Phys. Lett.* **70** 2990
[3] D J Dewsnip, A V Andrianv, I Harrison, J W Orton, D E Lacklison, G B Ren, S E Hooper, T S Cheng and C T Foxon 1998 *Semicond. Sci. Technol.* **13** 1998
[4] F J Sanchez, F Calle, M A Sanchez-Garcia, E Calleja, E Munoz, C H Molloy, D J Somerford, J J Serrano and J M Blanco 1998 MRS Internet J. Nitride Semicond. Res. 3, 19
[5] J L Weyher, S Muller, I Grzegory and S Porowski 1997 *J. Cryst. Growth* **182** 17
[6] S Mikroulis, A Georgakilas, A Kostopoulos, V Cimalla and E Dimarkis 2002 *Appl. Phys. Lett.* **80** 2886
[7] Brandt O, Yang H, Kostial H and Ploog K H 1996 *Appl. Phys. Lett.* **69** 2707

Area selective epitaxy of anti-dot structure by solid source MBE using MEE deposition sequence

D. Okada[1,2], H. Hasegawa[1,2], T. Hasegawa[1,2], Y. Horikoshi[1,2] and T. Saitoh[3]

[1]Scholl of Science and Engineering, Waseda University, 3-4-1 Okubo, Shinjuku-ku Tokyo 169-8555, Japan

[2]Kagami Memorial Laboratory for Materials Science and Technology, Waseda University, 2-8-26 Nishiwaseda, Shinjuku-ku, Tokyo 169-0051, Japan

[3]NTT Basic Research Laboratories, 3-1 Morinosato Wakamiya, Atsugi-shi, Kanagawa 243-0198, Japan

Abstract. We have carried out area selective epitaxial growth of GaAs anti-dot structures using solid source molecular beam epitaxy (MBE) which makes it possible to achieve "damage-free" structures. However, area selective epitaxy by MBE is very difficult unless the substrate temperature is very high. This problem has been solved by using migration-enhanced epitaxy (MEE) deposition sequence. To achieve well-defined anti-dot network structures, control of sidewall formation is very important. In addition, lateral growth beyond the SiO_2 mask boundaries has to be strictly prohibited. By MEE method, anti-dots with vertical sidewalls can be fabricated without shrinking holes, even though the mask diameter is as small as 30nm.

1. Introduction

Quantum and nano-structured devices have received considerable attention recently. Quantum dot/wire network structures have been reported to show new functions for further applications. We use area selective epitaxy to fabricate these structures because this method can produce less damaged structures than the other methods such as etching techniques and focused ion beam implantation. Another "damage-free" method, self-assembled technique suffers from the size fluctuation of dots and their uneven distribution. The growth conditions of well-defined area selective epitaxy of GaAs have already been established by the growth method using gas sources such as metalorganic vapor phase epitaxy (MOVPE) [1,2] and metalorganic molecular beam epitaxy (MOMBE) [3,4]. However, area selective epitaxy by solid source MBE is very difficult unless the substrate temperature is very high. This problem has been solved by using a migration-enhanced epitaxy (MEE) deposition sequence [5-7]. The MEE deposition sequence has proved useful to achieve complete selectivity using 2s annealing after Ga deposition [8]. In this study, we performed selective epitaxial growth of GaAs anti-dot networks with hole diameter as small as 30 nm by using solid source MBE.

2. Experimental Procedure

A 30-nm-thick SiO_2 film is deposited on GaAs(111)B substrate by a conventional sputtering technique. Prior to SiO_2 deposition, the substrate is lightly etched using an alkaline etchant. Using photolithography or electron beam (EB) lithography followed by wet chemical etching, SiO_2 masks with hole array structures are fabricated. The diameters of holes are varied between 30nm-300nm.

The masked GaAs substrate is mounted on Mo holder alongside unmasked GaAs(001) substrate. The latter is used to monitor the substrate temperature using a pyrometer and to observe reflection high energy electron diffraction (RHEED) patterns. The pyrometer reading is calibrated at the temperature of native oxide evaporation from (001) GaAs (580°C). The beam equivalent pressure of the As_4 flux is 2.0×10^{-5} Torr and the Ga flux intensity is 1.58×10^{-14} cm^{-2} s^{-1}. The substrate temperature is fixed at 590°C. The grown structures are evaluated by scanning electron microscopy (SEM).

3. Experimental Result and discussion

The area selective epitaxial growth is performed by MBE at 590°C on GaAs(111)B substrates. The MEE deposition sequence with 2s annealing after Ga deposition is applied. The structure is composed of GaAs buffer layer (50nm) and Si-doped GaAs layer (150nm) with an electron density of 5.0×10^{17} cm^{-3}. Figure 1 shows a SEM micrograph of the grown anti-dots with a hole diameter of 300nm.

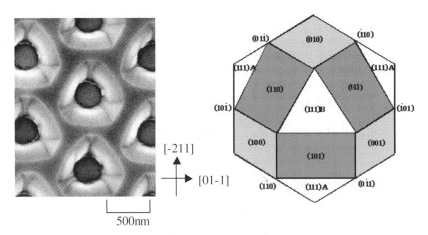

Figure 1. SEM photograph of GaAs anti-dots with a diameter of 300 nm

Figure 2. Expected facets of anti-dot structure on (111)B substrate

The shape of the anti-dot walls is determined so as to minimize the surface energy. Surface energy can be lowered by forming the facets which have low surface energies, and also by reducing the surface area [9]. When the diameters of mask holes are above 150nm, their sidewalls are composed of three major {110} facets and three minor {100} facets, and show apparent coincidence with stable facets on (111)B shown in Fig. 2. In III-V compound semiconductors with zinc blend structure, the {110} surface is one of the lowest surface energy planes. Therefore, six vertical {1-10}-type facets and three inclined {110}-type facets tend to appear. The six vertical {1-10} facets are most dominant when the As pressure during the growth is high. However, when the growth is performed by the MEE deposition sequence, as in the present experiments, the As pressure is relatively low because As is supplied alternately. Therefore, in the present case, inclined {110}-type facets appear rather than vertical ones.

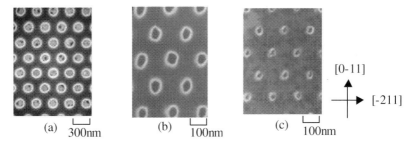

Figure 3. SEM photographs of GaAs anti-dots on (111)B substrates (a)d=90nm (b)d=60nm (c)d=30nm

When the diameters of mask holes are reduced below 150nm, the facets disappear and vertical sidewalls appear instead. Figure 3 shows SEM photographs of grown ant-dots (d=30-90nm). The density of dots is approximately 5.0×10^9 cm^{-2}. For smaller diameter holes, the curvature is large, and as a result, the surface tension becomes stronger, because the surface tension is inversely proportional to the radius of holes [9]. Since the surface tension in this case enhances the growth toward the hole center, it proceeds until vertical sidewalls appear. However, when vertical sidewalls are formed, the growth almost stops without further lateral growth beyond SiO_2 mask. This is an important characteristic when the MEE deposition sequence is employed. Indeed, holes with round shapes which coincide with patterned mask are successfully grown. This phenomenon is based on the facet that no substantial growth takes place on the {110} surface when the MEE deposition sequence is used. {110} surfaces consist of equal number of Ga and As atoms as shown in Fig.4. When As atom is supplied, it is captured by only one bond, and therefore, easily re-evaporated. So the lateral growth beyond SiO_2 mask is difficult to proceed because Ga and As are supplied alternately by the MEE deposition sequence.

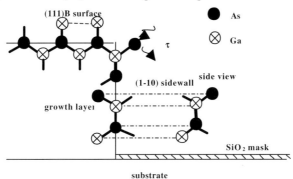

Figure 4. Schematic illustration of the area selective growth boundary on (111)B substrate

4. Conclusions

We have carried out area selective epitaxy of GaAs anti-dot structure by solid source MBE. The diameters of mask holes are 30-300nm. When the mask diameter is above 150 nm, sidewalls consist of inclined {110} facets together with small {100} facets. However, anti-dots with vertical sidewalls can be fabricated when the mask diameter is smaller than 150nm. This phenomenon is tentatively explained by considering the surface tension of smaller dot structures. The absence of lateral growth beyond SiO_2 mask is explained by considering the characteristics of {110} surfaces and the MEE deposition sequence.

5. Acknowledgements

This work is partly supported by Grand-in Aid for Scientific Research on Priority Areas program *Photonics Based on Wavelength Integration and Manipulation* from MEXT, and COE Program *Molecular NanoEngineering* from MEXT.

References

[1] S. Ando and T. Fukui: J. Cryst. Growth **98** (1989) 646.
[2] T. Fukui, S. Ando, Y. Tokura and T. Toriyama: Appl. Phys Lett. **58** (1991) 2018
[3] E. Tokumitsu, Y. Kudou, M. Konagi and K. Takahashi: J. Appl. Phys. **55** (1984) 3163.
[4] H. Heinecke, A. Brauers, F. Grafahred, C. Plass, N.Pütz, K. Werner, M. Weyers, H. Lüth and P. Balk: J. Cryst. Growth. **77** (1986) 309.
[5] K. Suzuki, M. Ito and Y. Horikoshi: Jpn. J. Appl. Phys. **38** (1999) 6197.
[6] M. Ito, K. Suzuki and Y. Horikoshi: Inst. Phys. Conf. Ser. **166** (2000) 39.
[7] H. Kuriyama, M. Ito, K. Suzuki and Y. Horikoshi: Jpn. J. Appl. Phys. **39** (2000) 2457
[8] H. Hasegawa, H. Kuriyama, M. Ito and Y. Horikoshi: J. Cryst. Growth **227-228** (2001) 1078.
[9] H. Hasegawa, D. Okada, Y. Horikoshi and T. Saitoh: Jpn. J. Appl. Phys. **41** (2002) 2505

Contribution of interface states and bulk traps to GaAs MIS admittance

S Kochowski

Institute of Physics, Silesian University of Technology, Gliwice, Poland

B Paszkiewicz and R Paszkiewicz

Faculty of Microsystem Electronics and Photonics, Wroclaw University of Technology, Wroclaw, Poland

Abstract. Measurements of Au/Ti/Pd-SiO$_2$–n-GaAs structure characteristics have revealed two maxima of normalized MIS conductance vs. frequency and large frequency dispersion of MIS capacitance. The position of one maximum is dependent on gate voltage and the second is not dependent on a bias. A model of insulator-semiconductor interface as a disordered system with interface states distributed in energy and in space including the presence of semiconductor bulk traps allows us to describe these characteristics. The possibility of separating the interface state contribution and bulk trap contribution have been presented and their parameters estimated.

1. Introduction

The localised electron states (the insulator-semiconductor interface states and the semiconductor bulk traps) play a crucial role in electron processes at GaAs-dielectric structures and are an essential factor limiting the performance of GaAs based devices. It causes investigations of GaAs-dielectric systems to be performed continuously. Various phenomena associated with the localised states in these systems may be analysed with the measurements of the electrical characteristics of MIS structures.

In the previous paper [1] we presented measurements of the frequency dependence of MIS capacitance and conductance performed for the Au/Pd/Ti -SiO$_2$-GaAs structures at different gate voltages. We observed a large dispersion of MIS capacitance at positive biases and the broad maxima of normalized MIS conductance G_m/ω versus frequency with position dependent on gate voltage as well as additional peaks on G_m/ω curves with position not dependent on a bias. We connected these peculiarities with the contribution of insulator-semiconductor interface states and the contribution of deep traps in the semiconductor, respectively. Unfortunately, no quantitative analysis was performed in paper [1].

The purpose of this work is the detailed analysis of those characteristics and an estimation of SiO_2-GaAs interface state parameters as well as GaAs bulk trap parameters.

2. Results and discussion

The Au/Pd/Ti-SiO_2-(n) GaAs structures with PECVD obtained SiO_2 dielectric layers have been investigated. Prior to the SiO_2 deposition, the (100) GaAs surface was exposed to the $(NH_4)_2S_x$ solution for 16 h. The details of MIS structure preparation and admittance measurements are described in paper [1].

Figure 1 shows the MIS capacitance C_m and normalized MIS conductance G_m/ω vs. frequency characteristics recorded at different gate voltages. The conductance peaks with a position dependent on gate voltage occur. Additionally seen are peaks with position not dependent on a bias. The broadening of conductance maximum caused by overlapping two peaks is also emphasized. A large capacitance dispersion exists in the frequency range where G_m/ω peaks are observed. In order to describe the essential features of the frequency behaviour presented in Figure 1, we employed a small signal equivalent circuit of MIS structure which contains: insulator capacitance C_{ox} in series with a parallel combination of surface space charge capacitance C_{sc}, interface state capacitance C_{it}^p and conductance G_{it}^p as well as the bulk trap capacitance C_t^p and conductance G_t^p. The series resistance R_s has also been included. The frequency response of the interface state capacitance and conductance have been described in the frame of the surface disorder model [2-4]. It has been considered that continuum of acceptor and donor type electron states occurs at insulator-semiconductor (I-S) interface. The functional form of their density distribution has been assumed as:

$$D_{it}^{acc}(E,x) = A\exp\left[a\frac{(E-E_o)}{kT}\right]\exp\left(-\frac{x}{x_o}\right), \quad (1)$$

$$D_{it}^{don}(E,x) = B\exp\left[-b\frac{(E-E_o)}{kT}\right]\exp\left(-\frac{x}{x_o}\right), \quad (2)$$

for acceptor and donor states, respectively. E is the state energy (zero at the Fermi level of intrinsic semiconductor E_i), k and T have their usual meanings. x is a distance in the insulator (x=0 at the semiconductor surface). The parameters A, a, B, b and E_O describe energetic distribution of states, x_O their spatial distribution. In this model the electrons are exchanged by tunneling between the semiconductor and spatially distributed I-S interface states. The time constant of this process τ_o can be written [3,4] in the form

$$\tau_o = f_o[\sigma_n\exp(-2\kappa_n x)v_n n_b\exp(v_s)]^{-1}, \quad (3)$$

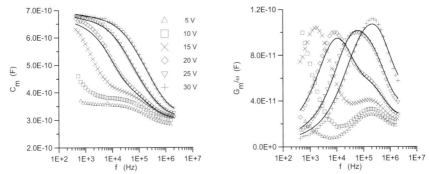

Figure 1. Capacitance C_m and normalized conductance G_m/ω as function of the frequency f at different gate voltages for the Au/Pd/Ti-SiO$_2$-GaAs structures. Symbols – the experimental data, full lines represent theoretical curves calculated with use of parameters discussed in text.

where f_o is the Fermi-Dirac function, σ_n is electron capture cross-section at x=0, κ_n is the quantum mechanical decay constant of electron wave function. v_s is the surface potential (dimensionless), v_n and n_b represent the thermal velocity and bulk density of electrons, respectively. The broad spectrum of time constants due to an exponential decay of the effective capture cross section with the tunneling distance will be observed, resulting in frequency dispersion of electrical characteristics.

For the calculation of the trap capacitance and conductance contribution we used the Dharival and Deoraj theory [5]. The existence of monoenergetic deep traps with density N_t, energy E_t and capture cross-sections σ_t have been assumed. The time constant of deep trap τ_t may be presented as

$$\tau_t = \left[\sigma_t v_n N_c \exp\left(\frac{E_c - E_t}{kT}\right)\right]^{-1}, \qquad (4)$$

where N_c and E_c are effective density of states and edge of the conduction band, respectively. From (3) and (4) it results that the peak of normalized conductance evoked by electrons exchange between the semiconductor and interface states occurs at the frequency dependent on surface potential (so, on the gate voltage), and the peak evoked by interaction with deep traps is not dependent on a bias, as we can see in Figure 1.

The analysis of the parameters of the model used have been performed by the least squares fitting the theoretical dependencies (calculated with formulae from [3,4]) and experimental data. Both MIS capacitance and conductance versus frequency characteristics have been simultaneously fitted at different gate voltages, using the procedure described in the work [3] and the results are shown in Figure 1 and in Table 1. The full lines in this figure represent the best fit of experimental data for parameters given in Table 1. We can see that the observed frequency behaviour of MIS capacitance and conductance is rather well reproduced with the model used in the gate voltage region where depletion of semiconductor surface space charge occurs. The obtained interface state parameters (A, B, a, b, E_o, x_o) are consistent with our estimates in work[4] and support validity of the surface disorder model. The rather large values of

Table 1
Values of interface state and bulk trap parameters obtained by fitting of experimental results presented in Fig.1

U_G (V)	20	25	30
v_s	-17.2	-15.7	-14.5
$A=B$ (cm^{-3}eV^{-1})	$(5.44\pm0.09)\times10^{18}$	$(5.39\pm0.08)\times10^{18}$	$(5.39\pm0.06)\times10^{18}$
$a=b$	0.190 ± 0.002	0.190 ± 0.003	0.190 ± 0.002
x_o (Å)	113*	113*	113*
E_o-E_i	-0.20*	-0.20*	-0.20*
σ_n (cm^2)	$(1.0\pm0.1)\times10^{-11}$	$(6.5\pm0.1)\times10^{-12}$	$(3.7\pm0.2)\times10^{-12}$
κ_n (Å^{-1})	0.61 ± 0.02	0.63 ± 0.02	0.63 ± 0.02
N_t (cm^{-3})	$(1.1\pm0.2)\times10^{15}$	$(0.9\pm0.2)\times10^{15}$	$(1.1\pm0.2)\times10^{15}$
E_t-E_c (eV)	-0.27 ± 0.02	-0.27 ± 0.02	-0.27 ± 0.02
σ_t (cm^2)	$5.1\times10^{-15*}$	$5.1\times10^{-15*}$	$5.1\times10^{-15*}$
C_{ox} (F)	$(712\pm9)\times10^{-12}$	$(712\pm9)\times10^{-12}$	$(710\pm6)\times10^{-12}$
R_s (Ω)	8.1 ± 0.4	8.2 ± 0.3	8.1 ± 0.3

* The parameter value was changed in the calculations, but not calculated in last iteration i.e not very significant parameter

interface state capture cross-sections have also been found in other work [3,4,6] and their explanation is possible by reduction of thermal activation energy in agreement with the surface disorder model [2,4]. The existence of deep trap in GaAs is a known fact [7]. The values presented in Table 1 of the trap energy as well as their capture cross-sections are close to values for ET2 traps (0.3 eV, 2.5×10^{-15} cm^2) [7].

3. Conclusions

The contribution from SiO$_2$-GaAs interface states and the semiconductor bulk traps to the measured frequency dependence of MIS capacitance and conductance has been confirmed. The model used and the method of analysis of experimental data allows us to separate these contributions as well as to estimate the interface state parameters and the bulk traps parameters.

This work is supported financially by the research program from Silesian University of Technology and Wroclaw University of Technology Advanced Materials and Nonotechnology Centre program.

References

[1] Kochowski S, Paszkiewicz B and Paszkiewicz R 2000 Vacuum 57 157-162
[2] Hasegawa H and Sawada T 1983 Thin Solid Films 103 119-140
[3] Kochowski S and Nowak M 1999 Vacuum 54 183-188
[4] Kochowski S, Nitsch K and Paszkiewicz R 1999 Thin Solid Films 348 180-187
[5] Dharival S R and Deoraj B M 1993 Solid State Electron. 36 1165-1174
[6] Bayraktaroglu B and Johnson R L 1981 J. Appl. Phys. 52 3515-3519
[7] Martin G M, Mitonneau A and Mircea A 1977 Electron. Lett. 13 191-193

Electrical isolation of p-type InP and InGaAs layers by iron implantation: Effects of substrate temperature

P Too[*], S Ahmed, B J Sealy, and R Gwilliam

Surrey Centre for Research in Ion Beam Applications, University of Surrey, Guildford, GU2 7XH, United Kingdom

Abstract. 1MeV Fe^+ implantations at a fluence of 5×10^{14} /cm^2 were performed into p-type doped InP and InGaAs epilayers at different substrate temperatures: 77K, room temperature (RT), 100°C and 200°C to obtain high-resistivity regions. The evolution of the sheet resistance in layers implanted at different temperatures and the stability of the formed isolation during post implantation annealing was studied. In both types of material, the conductivity converts to n-type after implantation for all substrate temperatures. In the case of p-type InP, samples implanted at 77K and RT a thermally stable maximum resistivity of $\sim 1 \times 10^7$ Ω/□ up to a post-implant annealing temperature of 500°C was found. A lower thermally stable resistivity of $\sim 3 \times 10^6$ Ω/□ is obtained for 100°C and 200°C implants up to an annealing temperature of 550°C. Similar post-implant annealing characteristics are observed after iron bombardment into p$^+$ InGaAs. For this material, a maximum sheet resistivity of $\sim 2.2 \times 10^7$ Ω/□ was obtained for a 77K implant after annealing at 500°C. For the first time, such a high sheet resistivity value is reported for isolating both p-type InP and InGaAs using iron implantation. These results are expected to find applications in the isolation of In-based devices.

1. Introduction

Ion implantation has two major applications in compound semiconductor technology, doping and interdevice electrical isolation [1]. The latter application is becoming more popular and has significant advantages over standard mesa etching for maintaining surface planarity, better reproducibility and less intrusion under the mask edges. Implant isolation can be achieved by either damage-induced compensation or chemical related deep-levels. Implantation with the right ion species, dose, energy and substrate temperature is important to convert a conductive layer into a highly resistive one or to improve the isolation between devices in integrated circuits.

Extensive research has been conducted on implant isolation of n-type InP and InGaAs layers. High thermally stable isolation ($\sim 10^7$ Ω/□) in n-type InP was reported by implantation of He^+, B^+ and Fe^+ [2,3]. In the case of n$^+$ InGaAs, O^+ or Fe^+ irradiation was used to create high resistivity layers of $\sim 5 \times 10^6$ Ω/□ [4,5]. However fewer investigations have been published on the electrical isolation of p-type InP and InGaAs using the ion implantation technique [6,7]. Many InP-based Heterojunction Bipolar Transistors (HBTs) and lasers have their base and top region doped p-type. Thus information is needed on the production of high resistivity in these materials. Here, we have studied the formation of high resistivity regions produced by Fe^+ bombardment in both p-type InP and InGaAs layers. The effects of substrate temperature and annealing temperature have been quantified by sheet resistance measurements.

---
[*]Fax : +44 1483534139, E-mail: p.too@eim.surrey.ac.uk

2. Experimental

Semi-insulating Fe-doped InP wafers of (100) orientation of 1μm in thickness were used as substrates throughout this work. Both p-type InGaAs and InP epilayers, with the (100) axis 2^0 off normal orientation were grown using a Solid Source Molecular Beam Epitaxy (SSMBE) reactor. An undoped InP buffer layer of thickness 1μm was first grown below the p-type layers. Zinc was used as the dopant with a concentration $1 \times 10^{18}/cm^3$. The wafers were cleaved to obtain several samples of approximately 1 cm² for the preparation of the resistors. All samples were cleaned in organic solvents and the clover-leaf pattern was printed using optical lithography. Both InP and InGaAs samples were etched using standard etching solutions. The photoresist was then removed in acetone leaving the cloverleaf Hall pattern on the samples.

The samples were divided into four different groups with implant isolation at temperatures of 77K, 25^0C, 100^0C, and 200^0C using a 2MV High Voltage Engineering Europa (HVEE) implanter. The accuracy in the temperature control was $\pm 3^0C$. In order to minimize channelling effects, the sample was tilted by 7^0 to the surface normal with respect to beam incident direction. The centre of the Hall pattern for all the samples was irradiated with Fe^+ using a dose and energy of 5×10^{14} cm⁻² and 1MeV respectively, with a beam current density < 0.33 μA/cm². Post-implant annealing was performed in a rapid thermal annealing furnace in the temperature range from 100^0C to 800^0C ($\pm 5^oC$) for a duration of 60s in a nitrogen atmosphere, following a ramp up to temperature of 60s. Ohmic contacts to the InGaAs and InP samples were fabricated by using indium and Au/Zn/Au respectively. The sheet resistivity, sheet carrier concentration and Hall mobility were measured using a Bio-Rad HL5500 Hall effect system employing Van der Pauw geometry at 300K under a magnetic field strength of 0.32 T. All measurements were performed at RT.

3. Results and discussions

The damage distribution resulting from iron implants into InP and InGaAs has been simulated using Transport of Ions in Matter (TRIM) [8]. The projected range of 1MeV iron in InP and InGaAs is about 0.6μm and 0.4μm respectively. An average maximum damage of 5×10^{22} vacancies/cm³ is created in both epilayers by using a dose of 5×10^{14} /cm². An energy of 1MeV for the iron beam is chosen to place most of the iron atoms well inside the doped layer. In this way, the chemical compensation will be more effective for the electrical isolation of the InP and InGaAs epilayers.

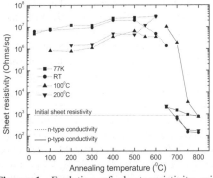
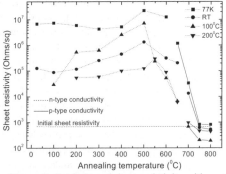

Figure 1. Evolution of sheet resistivity with annealing temperature for iron implanted p-type InP layers irradiated with 5×10^{14} cm⁻² at 1MeV, as a function of implant temperature.

Figure 2. Evolution of sheet resistivity with post-implant annealing temperature for iron implanted p-type InGaAs layers irradiated with 5×10^{14} cm⁻² at 1MeV at 77K, RT, 100°C and 200°C.

The evolution of sheet resistivity as a function of post-implant annealing temperature for p-type InP is shown in figure 1. It is observed that the initially p-type layers convert to n-type conductivity for all as-implanted samples. The initial sheet resistivity of the Zn-doped InP layer before implant isolation is 872 $\Omega/\square$. After iron implantation, the sheet resistivity for 77K and RT implants is $\sim 6 \times 10^6$ $\Omega/\square$ and that for 100^0C and 200^0C implants is $\sim 10^6$ $\Omega/\square$. Thus there is an increase in R_s by at least three orders of magnitude for all samples. Maximum sheet resistivities of 2.7×10^7 $\Omega/\square$ and 1.5×10^7 $\Omega/\square$ are obtained after a post-implant annealing temperature of 500^0C for 77K and RT implants respectively.

This type conversion is also observed by other authors [9,10] after proton bombardment in p-type InP. They have used the bombardment created deep levels model to explain this behaviour. They have suggested that in p-type InP, the Fermi level moves close towards the intrinsic Fermi level with increasing proton dose as both donor- and acceptor-like defects are created. When it reaches a point near and above the intrinsic Fermi level where the defect related free electron concentration just exceeds the hole concentration, the resistivity becomes maximum and p-type bombarded layers convert to n-type. This agrees with the interpretation of the sheet resistivity data presented in this work. We believe that the iron dose is close to the threshold value which occurs when p-type material converts to n-type. This is the reason why as implanted samples produce high sheet resistivity ($\sim 10^6$ $\Omega/\square$).

Hot implants show one order of magnitude lower sheet resistivity as compared to 77K and RT implants until a post-implant annealing temperature of 500^0C. Less damage is produced for 100^0C and 200^0C implants due to enhanced dynamic annealing and this is most likely the reason why lower temperature implants produce higher resistivities. Bahir et al [11] have used Rutherford Backscattering spectroscopy (RBS) to demonstrate that less damage was formed for hot implants of Si into SI InP using a fluence of 3.3×10^{14} /cm^2 at 180keV. We believe that most of the defects responsible for the high sheet resistivity during 77K and RT implants are annealed out during 100^0C and 200^0C implants. From figure 2, 77K and RT implants show similar post-implant annealing behaviour and the changing resistivity may therefore be due to the same isolation mechanism in both cases. Another possible explanation is that the relaxation mechanism during 77K implants and the subsequent measurement at RT may affect the isolation behaviour of the samples. Most of the defects formed at 77K may have annealed out as the samples were warmed up to room temperature.

A broad thermally stable region up to a post-implant annealing temperature of 500^0C is obtained for all four substrate temperatures. Such a wide annealing window is quite useful from the technological point of view. With continued annealing above 600^0C, the carriers are converted back to p-type as defects responsible for trapping of mobile carriers are gradually annealed out and the sheet resistivity recovers towards its initial unimplanted value. Similar behaviour is observed for all implant temperatures. From the other data not shown here, we observed a gradual recovery of the sheet hole concentration above an annealing temperature of 600^0C. The out-diffusion of Fe atoms from their substitutional sites towards the surface or deep into the bulk at temperatures between 650^0C and 800^0C is another possible explanation for the decrease in the sheet resistivity.

The sheet resistance versus post-implant annealing temperature for Fe$^+$ implanted p-type InGaAs samples is shown in figure 2. Conductivity type conversion from p to n is also observed for all as-implanted samples. The post-implant annealing behaviour is quite similar to that of p-type InP. However only samples implanted at 77K show better and thermally stable isolation as compared to those implanted at RT, 100^0C and 200^0C.

One possible explanation of this result is that the type of defects formed during 77K implants is different from RT implants in the case of InGaAs and this may be the explanation for the two orders of magnitude difference in the sheet resistivity between 77K and RT as-implanted samples. For 77K implants, a thermally stable isolation is maintained up to 600^0C. A maximum sheet resistivity of $\sim 2.2 \times 10^7$ $\Omega/\square$ is achieved for 77K implanted samples which have been annealed at 500^0C for 60s. For higher annealing cycles, the sheet resistivity decreases sharply and the initial p-type conductivity is restored by 700^0C even for the other three implantation temperatures.

The implant and anneal behaviour of p^+ InGaAs samples resemble that of p^+ InP samples, so a generally similar explanation of the results would be expected. P- to n-type conductivity conversion was also reported by Sargunas et al [12] after nitrogen implantation into both p-type InGaAs and InGaAsP. Depending on the ion species and initial doping concentration, there is a threshold dose at which the p-type InGaAs converts to n-type conductivity and the Fermi level is pinned in the upper half of the bandgap. We infer that the iron dose used is close to the threshold value and hence the high sheet resistivity observed for as implanted p-type InGaAs samples. The lower isolation values obtained for RT, 100^0C and 200^0C implants are due to the effect of enhanced dynamic annealing as seen in the case of p^+ InP.

4. Conclusion

In summary, high energy implantation of Fe at 77K following low temperature annealing (400^0C –500^0C) has been shown to be an effective means to render both types of epilayers highly resistive. Using a single MeV implant, we have obtained high resistivities and a wide thermally stable isolation region up to 500^0C in both materials. These results offer a better route to the isolation of InP-based devices than mesa etching. This single implant isolation scheme is inherently simple and is compatible with conventional processing steps. For both InP and InGaAs epilayers, p to n-type conductivity conversion occurs after iron implantation for all substrate temperatures.

References

[1] Li G, Han W, Luo Y, Han D and Ji C 1995 4th Intl. Conf. on Sol. Stat. Int. Cir. Tech. 399-401
[2] Too P, Ahmed S, Gwilliam R and Sealy B J 2002 Nucl. Instrum. Meth. B 188 205-209
[3] Pearton S J, Abernathy C R, Panish M B, Hamm R A and Lunardi L M 1989 J. Appl. Phys. 66 656-662
[4] Akano U G, Mitchell I V, Shepherd F R, Miner C J, Margittai A and Svilans M 1996 Can. J. Phys. 74 S59-S63
[5] Almonte M, Yu K M, Haller E E, Ridgway M C, Hou H and Mirecki-Millunchick J 1998 Proc. of the 10th Conf. on Semi. and Insul. Mat. 29-32
[6] Ridgway M C, Jagadish C, Elliman R G and Hauser N 1992 4th Intl. Conf. on Ind. Phos. and Rel. Mat. 294-297
[7] Focht M W and Schwartz B 1983 Appl. Phys. Lett. 42 970-972
[8] Ziegler J F, Biersack J P and Littmark U 1985 The Stopping and Range of Ions in Solids (Oxford: Pergamon)
[9] Donnelly J P and Hurwitz C E 1977 Sol. Stat. Elec. 20 727-730
[10] Boudinov H, Tan H H and Jagadish C 2001 J. Appl. Phys. 89 5343-5347
[11] Bahir G, Merz J L, Abelson J R and Sigmon T W 1989 J. Appl. Phys. 65 1009-17
[12] Sargunas V, Thompson D A and J G Simmons 1995 J. Appl. Phys. 77 5580-5583

Inductively Coupled Argon Plasma Enhanced Quantum Well Intermixing in InGaAs/InGaAsP Laser structure

H.S. Djie*, J. Arokiaraj, and T. Mei

Photonics Research Group, School of Electrical and Electronic Engineering
Nanyang Technological University, Singapore 639798
* corresponding email: hery@pmail.ntu.edu.sg

Abstract. The bandgap shift and mechanism of argon (Ar) plasma enhanced quantum well intermixing in Inductively Coupled Plasma (ICP) machine were studied by using photoluminescence (PL) and Raman spectroscopy. The linewidth broadening of 1.9 nm was observed in the PL spectrum of the intermixed sample exposed under the plasma with ICP power in contrast to the linewidth broadening of 7.8 nm for the sample exposed without ICP power. This is the first time to use the high-density plasma enhanced intermixing technique to demonstrate the result of bandgap blue-shift with the preservation of crystalline quality as compared to plasma generated from pure RF power. This technique provides a promising approach of bandgap tuning for photonic integrated circuits, which demand high crystalline quality.

1. Introduction

Quantum Well Intermixing (QWI) has been adopted as an effective technique the photonic integration due to its ability to fine tune the energy of the epitaxial layer structure in locally selected area of a single photonic chip. QWI is a post-growth technique of bandgap tailoring, thus considerable interest due to the simplicity, compatibility, and effectiveness of the technique for many photonics applications [1]. To date, the ICP machine has been exploited for high rate, low damage etching process and the deposition of a wide variety of semiconductor materials including dielectrics and III-V semiconductors. An ICP reactor allows a high-density of free radicals to be produced without generating a large number of high energetic ionic species. The collision of the Ar ions with the sample appears to create vacancies, interstitial and other point defects on the quantum well (QW) structure [2]. Under certain conditions, Ar plasma in ICP can be used to generate point defects to promote QWI in QW samples [3].

In this paper, we highlight a QWI technique using ICP generated Ar plasma on InGaAs/InGaAsP QW structures. PL and Raman spectroscopy measurements were adopted to investigate the effect of the Ar plasma exposure and subsequent annealing on the QW samples. Being exposed to the Ar plasma in ICP machine, a QW material experiences both high-density plasma damage as well as physical ion bombardment damage, which enhances intermixing in the subsequent annealing process. We demonstrate that the high-density plasma generated in ICP machine enhances the intermixing and results in the preservation of crystalline quality of samples.

2. Experiment

The samples used in the present investigations [3] were grown by metal-organic vapor phase epitaxy on a (100) oriented n^+-type S-doped InP substrates. The lattice-matched InGaAs/InGaAsP laser structure consists of five periods of 55 Å $In_{0.53}Ga_{0.47}As$ quantum wells with 120 Å InGaAsP barriers. The active region was bounded by a stepped graded index waveguide core consisting of InGaAsP confining layers. The thickness and compositions of these layers were 500 Å and 800 Å respectively. The structure was completed by an upper cladding of 1.4 μm with Zn doping of 5×10^{17} cm^{-3}. The contact layers consist of 500 Å InGaAsP (Zn-doped of 2×10^{18} cm^{-3}) and 1000 Å InGaAs (Zn-doped of 2×10^{19} cm^{-3}) respectively. The samples resulted in low temperature PL peak at 4 K and 77 K at 1.42± 0.02 μm.

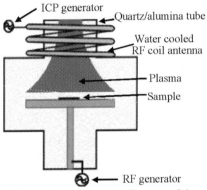

Figure 1. A schematic diagram of the ICP reactor with remote inductive coil.

The plasma source generator ICP180 used in this experiment (Figure 1) was built by Plasmalab System100. The system uses inductive coil to generate high-density "remote" plasma with no direct contact between the plasma and the substrate. The 13.56 MHz RF and ICP power supply can provide the independent control of ion bombardment energy and ion current density with power up to 500 W and 3000 W respectively. The ICP parameter settings were: 100 sccm Ar flow rate, 80 mTorr chamber pressure, and 480 W RF power. After Ar plasma exposure, the samples were annealed at at 600 °C for 120 s in a flowing nitrogen ambient. The annealing conditions were obtained from a thermal stability test performed on the as-grown samples. GaAs proximity caps were used to provide As over-pressure during the annealing step. PL measurements were then performed to assess the degree of bandgap shift and linewidth broadening. The PL measurements were carried out at 300 K and 4 K using an Nd:YAG laser (1.064 μm) and an Ar-ion (514.5 nm) laser respectively for excitation, a monochromator and a TE-cooled InGaAs photodetector associated with a SR-830 lock-in amplifier. The micro-Raman measurements were carried out at 300 K using the Argon laser of 514.5 nm and power of 25 mW. The predicted penetration depth of the light is about 100-150 nm below the sample surface.

3. Results and discussions

The Ar plasma exposure and subsequent annealing was found considerably modifying the optical properties of the InGaAs/InGaAsP QW structures. A control sample, which was annealed without Ar exposure, resulted in very small bandgap energy blue-shift 6 nm mainly due to the thermal interdiffusion. With the absence of ICP power (pure RF mode), the ionic species in plasma sheath region is accelerated at normal incidence by electric field and is expected to dominantly induce highly energetic ion bombardment damage. The plasma exposure results in the ion bombardment dominated QWI. The annealing after exposure induces the propagation of created point defects towards QWs and promotes intermixing. From Figure 2 (a), it can be seen that the PL peak at 300 K from QW active region shifts towards higher energy with an increase in the exposure time up to 15 min. The maximum bandgap shift at this condition is 79.8 nm as compared to the as-grown sample for 10 min exposure. This bandgap shift is comparable with QWI using RIE machine [4] adopting the same structure. The PL linewidth does not increase significantly as the exposure time increases, thus indicating that no significant degradation of the optical properties occurred in the intermixed QW samples. This implies that the high temperature annealing is also helping to minimize the ion bombardment induced damage in addition to the point defect propagation.

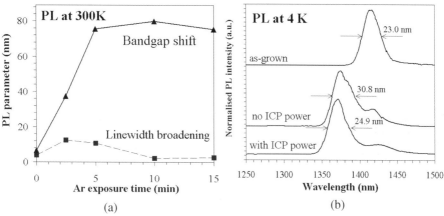

Figure 2. (a) Plot of normalised PL at 300 K spectra for different Ar ion exposure times. The ICP power was set to 0 W (pure RF mode). (b) The comparison of PL spectra at 4 K on the intermixed samples with the application of ICP power.

A comparative study was made to identify the intermixing effect on higher density plasma exposure with ICP power application. Increasing the ICP power (with the RF power constant) results in the increase in ion flux and a decrease in the ion bombardment energy. Figure 2 (b) shows the PL spectra of Ar exposed samples to the Ar plasma with the same process conditions with the exception of different ICP powers of 0 W (pure RF power) and 250 W. Reproducible results were obtained from other pieces of the same sample. It shows the comparable bandgap shift of 61.5 nm (28 meV) after intermixing for 10 min exposure time. With the increase of ICP power at 250 W, the ICP machine produces higher ion current density and lower RF bias operation. The RF bias operation was -840V at pure RF power and was reduced to –780 V when the ICP power was 250 W. The RF bias reduction leads to less ion bombardment induced damage. The comparable bandgap shift achieved using 250 W ICP power indicates that the high-density plasma application is also playing a strong role in intermixing the QW structures besides the pure ion bombardments. The application of ICP power induces more damage due to the plasma radiation together with high ion current density on the samples such as the relevant observation reported elsewhere [5]. The linewidth broadening as compared to the as-grown samples was observed as low as 7.8 nm and 1.9 nm for ICP power at 0 W and 250 W respectively. These values are the lowest reported linewidth broadening compared to techniques such as high energy [6] and low energy [7] ion implantation induced intermixing with similar annealing temperature. The improvement in preserving QW quality is shown by smaller linewidth broadening as compared to pure RF power exposure.

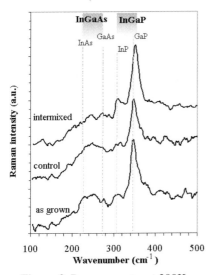

Figure 3. Raman spectra at 300K obtained from the QWs samples.

Raman measurements have been carried out to study the intermixing mechanism on the near sample surface through low energy ion exposure using ICP machine. From Figure 3, the 210-270 cm^{-1} band corresponds to the InGaAs related LO and the 300-370 cm^{-1} band corresponds to the InGaP related LO were observed. The presence of these bands in the as-grown sample indicates the presence of InGaAs/InGaAsP cap layer. The GaAs-like LO phonon peak displays a small shift from 270.2 cm^{-1} (the as-grown sample) to 265 cm^{-1} (the control and the intermixed sample) and a decrease in peak intensity. Simultaneously the InAs-like LO phonon peak shift to higher frequency. The decrease in GaAs-like LO phonon in control sample as compared to the as-grown sample is attributed due to GaAs outdiffusion during annealing. The Raman spectra of control sample show the similar Raman spectra to that of the as-grown sample except a slight frequency shift at the GaP-like LO phonon (~343.5 cm^{-1}), indicating the thermal diffusion effect, consistent with the result of a small bandgap shift from PL measurements. The intermixed sample shows a clear Raman shift towards GaP-like LO phonon at higher frequency (345.2 cm^{-1}) and the increase of InP-like LO phonon peak (304.3 cm^{-1}), indicating that the interdiffusion on the cap layers has occurred and Phosporus from the InP upper cladding diffused into the InGaAs/InGaAsP cap layers. The similar Raman spectra and intensity level of InGaP related LO peak as compared to as-grown sample implied that the surface properties of intermixed samples were preserved.

4. Conclusions

We have demonstrated a way to achieve bandgap using the Ar plasma exposure in ICP machine. The maximum bandgap shift was 79.8 nm under 10 min Ar exposure with pure RF power. A comparable bandgap shift between ICP power of 0 W and 250 W indicated the high-density plasma enhanced QWI mechanism with a small linewidth broadening of 1.9 nm. The result of Raman measurements are consistent with PL measurements indicated the surface quality preservation after intermixing process. This technique provides a promising approach of bandgap tuning for the photonic integrated circuits which demand high crystalline quality.

References

[1] J. H. Marsh 1993 Semicon. Sci. Technol. 8 1136-1155.
[2] O. P. Kowalski et.al 1998 Appl. Phys. Lett. 72 581-583.
[3] H.S. Djie, T. Mei, J. Arokiaraj and P. Thilakan 2002 Jpn. J. Appl. Phys. 41 L867-869
[4] D.Leong, H.S. Djie and L.K. Ang 2002 *Proc.IEEE/LEOS Workshop Fibre and Optical Passive Components* 148-152.
[5] M. Rahman et. al. 2001 J. Phys. D: Appl. Phys. 34 2792-2797.
[6] S. Charbonneau et. al. 1995 Appl. Phys. Lett. 67 2954-2956.
[7] J. Oshinowo et. al. 1993 J. Appl. Phys. 74 1983-1986.

Annealing studies of Si-implanted GaN by Hall-effect and photoluminescence measurements

James A Fellows[1], Yung Kee Yeo[2], Mee-Yi Ryu[2], Robert L Hengehold[2], and Todd D Steiner[3]

[1]Air Force Research Laboratory, Wright-Patterson AFB, Ohio, USA
[2]Air Force Institute of Technology, Wright-Patterson AFB, Ohio, USA
[3]Air Force Office of Scientific Research, Arlington, Virginia, USA

Abstract. Electrical and optical activation studies of Si-implanted GaN have been made as a function of ion dose and anneal temperature. Si implantation was done at 25 and 800 °C with doses ranging from 5×10^{14} to 5×10^{15} cm^{-2} at 200 keV. The samples were proximity cap annealed from 1050 to 1350 °C with a 500-Å-thick AlN cap in a nitrogen environment. A 100% electrical activation was obtained for the samples implanted with doses of 1×10^{15} and 5×10^{15} cm^{-2} and annealed at 1350 °C. The mobility generally increased with anneal temperature and with decreasing dose, and the highest mobility obtained at 25 °C was 117 cm^2/V·s for this dose range. Photoluminescence measurements show a reasonably sharp bound exciton peak after annealing at 1350 °C for 17 s, indicating an excellent implantation damage recovery.

1. Introduction

Group III-nitride semiconductors have received significant attention over the past decade for application to blue-UV light-emitting diodes, laser diodes, and UV detectors as well as high-power, high-frequency, and high-temperature electronic devices. Despite the recent progress in crystal growth and device fabrication, one serious hurdle that must still be overcome is efficient doping of III-nitride materials. Doping during epilayer growth is often very expensive and suffers from limitations. Doping via diffusion is impractical due to the extremely chemically robust nature of the III-nitrides. Ion implantation is an alternative doping technique, which offers many advantages including selective area doping and precise lateral and depth control of impurity species; however, annealing must be used to remove the implantation-induced crystalline defects. Recent studies of ion-implanted GaN have shown that the activation of implanted impurities is much more difficult than for conventional compound semiconductors such as GaAs and InP. Zolper *et al.* [1] implanted GaN with 100 keV Si ions at doses ranging from 5×10^{13} to 1×10^{16} cm^{-2}, and annealed at 1100 °C for 15 s. They reported negligible activation for doses below 5×10^{15} cm^{-2}, but reported 50% electrical activation for a dose of 1×10^{16} cm^{-2}. Dupuis *et al.* [2] reported only 19% activation for GaN implanted with 100 keV Si ions at a dose of 5×10^{14} cm^{-2} and annealed at 1150 °C for 5 min. Molnar *et al.* [3] implanted a total dose of 4.4×10^{14} cm^{-2} Si ions with multiple energies, annealed at 1150 °C for 2 min, and obtained only 1% donor activation. Edwards *et al.* [4] implanted GaN at 300 °C with Si at a total dose of 4.4×10^{14} cm^{-2} and annealed at 1150 °C for 2 min, obtaining a 27% electrical activation and a mobility of 55 cm^2/V·s. Parikh *et al.* [5] implanted 160 keV Si with doses ranging from 1×10^{14} to 1×10^{15} cm^{-2} at room temperature and 550 °C into GaN, and annealed at 1000 °C for 60 s. They reported that hot-implantation produced less damage than the corresponding room-temperature implantation. In this study, we have performed electrical and optical activation studies of Si-implanted GaN as a function of ion dose and anneal temperature.

2. Experiments

The GaN layers were grown 2 μm thick by molecular beam epitaxy on sapphire substrates with a 500 Å thick AlN cap layer. The as-grown unintentionally-doped *n*-type GaN has a background carrier concentration less than 1×10^{15} cm^{-3}. This material was implanted at 25 and 800 °C with Si through the AlN cap layer with an ion energy of 200 keV at doses ranging from 5×10^{14} to 5×10^{15} cm^{-2}. The implanted wafers were cut into 5mm x 5mm samples and annealed from 1050 to 1350 °C from 5 min to 17 s in a flowing nitrogen environment to activate the Si. An open-tube furnace was used to anneal the samples at temperatures from 1050 to 1200 °C for 5 min, while an Oxy-Gon furnace was used to anneal the samples from 1250 to 1350 °C for 17 s. During annealing, the samples were tightly wrapped face-to-face using 5 mil thick Ta wire. Because of the large thermal mass of the graphite heating elements and the SiC-coated graphite pedestal in the Oxy-Gon, the sample spent almost 4 min above 1200 °C during the 1350 °C anneal for 17 s. After selectively removing the AlN cap layer in a hot KOH solution, 400 Å Ti/1200 Å Al contacts were deposited on the samples and annealed at 900 °C for 30 s to produce ohmic behavior. The Hall-effect measurements were made using the standard van der Pauw technique, and the photoluminescence (PL) measurements were made using the 275 nm line from an argon-ion laser at a typical power of 180 mW. The PL signals were dispersed with a ¾-m spectrometer using a 5000 Å blazed grating and a liquid nitrogen cooled GaAs PMT detector.

3. Results and Discussion

The sheet electron concentrations for GaN implanted at 25 and 800 °C with 200 keV Si ions at doses ranging from 5×10^{14} to 5×10^{15} cm^{-2} and annealed from 1050 to 1350 °C for 5 min to 17 s are shown in Fig. 1(a). The results obtained from the GaN implanted at 25 °C with doses ranging from 1×10^{13} to 5×10^{15} cm^{-2} have previously been published [6]. Here, we must note that the actual implanted doses after taking into account implantation through the 500 Å-thick AlN cap layer are 4.75×10^{14}, 9.51×10^{14}, and 4.75×10^{15} cm^2 for the nominal doses of 5×10^{14}, 1×10^{15}, and 5×10^{15} cm^2, respectively. The electron concentration is highly dependent upon annealing temperature, and increases steadily up through 1350 °C for each of the three doses. For a dose of 5×10^{15} cm^{-2}, good electrical activation occurs even after annealing at 1050 °C for 5 min, and carrier concentration continues to increase monotonically as anneal temperature increases to 1350 °C. For a dose of 5×10^{14} cm^{-2}, carrier concentration increases more than two orders of magnitude as anneal temperature increases from 1100 to 1150 °C. The concentrations of the samples implanted at 800 °C with a dose of 5×10^{15} cm^{-2} are an average of 44% greater than those of the sample implanted at 25 °C over all anneal temperatures. On the other hand, for doses of 5×10^{14} and 1×10^{15} cm^{-2}, the sheet concentrations of the samples implanted at both 25 and 800 °C are about the same. The electrical activation efficiencies obtained after annealing at 1350 °C are 92, 100, and 91% for the sample implanted at 25 °C and 78, 102, and 112 % for the samples implanted at 800 °C with a dose of 5×10^{14}, 1×10^{15} and 1×10^{15} cm^{-2}, respectively.

In order to see the effect of the hot temperature implantation, the Hall effect measurement was made on GaN implanted at 800 °C with Argon at a dose of 1×10^{15} cm^{-2} and annealed at 1350 °C for 17 s. The result shows an electron concentration of only 9×10^{11} cm^{-2}. The Hall measurement was also made on as-grown sample annealed at 1350 °C in order to see the effects of high-temperature annealing on the GaN background carrier

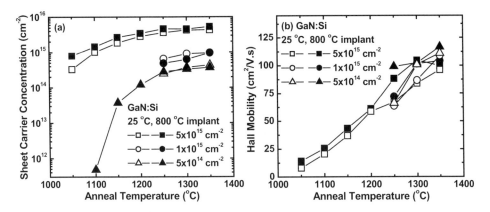

Figure 1. Room-temperature sheet electron concentrations (a) and Hall mobilities (b) for GaN implanted at 25 and 800 °C with 200 keV Si ions at doses ranging from 5×10^{14} to 5×10^{15} cm^{-2} and annealed at 1050 to 1350 °C from 5 min to 17 s.

concentration. The result shows a carrier concentration of only 3×10^{11} cm^{-2}.

Figure 1(b) shows the electron Hall mobilities for all three doses as a function of anneal temperature. The mobilities increase considerably with anneal temperature up to 1350 °C for all doses, indicating successively greater recovery of implantation damage after each successive anneal. After annealing at 1350 °C, the mobility values are higher for the lower doses. The highest mobility obtained at 25 °C is about 117 cm^2/V·s for the sample implanted at 800 °C with a dose of 5×10^{14} cm^{-2}.

Figure 2(a) shows the PL spectra taken at 3 K for unimplanted GaN annealed from 1250 to 1350 °C. The primary features of these spectra are a neutral-donor-bound exciton (D°,X) peak at 3.487 eV and a donor-acceptor-pair (DAP) peak at 3.28 eV with phonon replicas. The intensities of bandedge peaks of the as-grown sample increase by a factor of 5 after annealing at 1250 °C, and then they remain mostly unaffected when anneal temperature is increased. This behavior of PL signals indicates that the annealing damage on GaN even after annealing at 1350 °C is very minimal. The PL spectra taken at 3 K for GaN implanted at 800 °C with 200 keV Si ions with doses ranging from 5×10^{14} to 5×10^{15} cm^{-2} and annealed at 1350 °C for 17 s in a flowing nitrogen environment are shown in Fig. 2(b). For comparison, the spectrum for the unimplanted sample annealed at 1350 °C for 17 s is also shown in this figure. The sample implanted at 5×10^{14} cm^{-2} shows a strong (D°,X) peak and a very weak DAP peak. The presence of this reasonably sharp (D°,X) peak indicates an excellent implantation damage recovery after annealing at 1350 °C for 17 s. Also, the near band-edge broadening begins to show on the low energy side of the (D°,X) peak for the sample implanted with a dose of 5×10^{14} cm^{-2} due to band tailing. At a dose of 5×10^{14} cm^{-2}, most of the implanted Si concentrations exceed the Mott concentration of about 1×10^{18} cm^{-3} causing random band-edge fluctuations and band tailing. At this dose, the free carrier concentration is also high enough to initiate noticeable band filling. This is seen as a broadening on the high energy side of the (D°,X) peak above the bandgap. Luminescence from the band tailing and filling is much more evident for a dose of 1×10^{15} cm^{-2} and the most intense on the sample implanted with a dose of 5×10^{15} cm^{-2} extending the very broad band from 3.1 to 3.7 eV. All evidence of DAP transitions have disappeared on the spectra from the two highest doses because the donor band has merged with the conduction band.

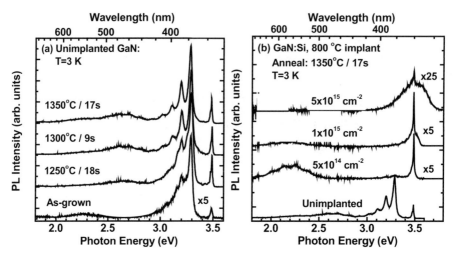

Figure 2. PL spectra taken at 3 K (a) for as-grown GaN and annealed at various temperatures from 9 to 17 s and (b) for GaN implanted at 800 °C with 200 keV Si ions at doses ranging from 5×10^{14} to 5×10^{15} cm^{-2} along with unimplanted GaN and annealed at 1350 °C for 17 s.

A broad yellow band centered near 2.2 eV appeared for the samples implanted with doses of 5×10^{14} and 1×10^{15} cm^{-2}.

4. Conclusion

Electrical and optical activation studies were performed for the GaN implanted with Si at 25 and 800 °C with 200 keV at doses ranging from 5×10^{14} to 5×10^{15} cm^{-2} and annealed from 1050 to 1350 °C for 5 min to 17 s. The electrical activation efficiencies increase with the anneal temperature. For the two highest dose samples of 1×10^{15} and 5×10^{15} cm^{-2}, an electrical activation efficiency of 100% was obtained after annealing at 1350 °C for 17 s. In general, the electrical activation efficiencies for samples implanted at 800 °C are higher than those for samples implanted at 25 °C, but the improvement is not decisive at present, and further studies are required. The mobility generally increases with anneal temperature and with decreasing dose. The presence of a reasonably sharp (D^o,X) peak on the Si-implanted samples indicates an excellent implantation damage recovery after annealing at 1350 °C for 17 s.

References

[1] J. C. Zolper, H. H. Tan, J. S. Williams, J. Zou, D. J. H. Cockayne, S. J. Pearton, M. Hagerott Crawford, and R. F. Karlicek, Jr., Appl Phys Lett, **70**, 2729 (1997).
[2] R. D. Dupuis, C. J. Eiting, P. A. Grudowski, H. Hsia, Z. Tang, D. Becher, H. Kuo, G. E. Stillman, and M. Feng, J. Electron. Mater. **28**, 319 (1999).
[3] B. Molnar, A. E. Wickenden, M. V. Rao, Mat. Res. Soc. Symp. Proc., **423**, 183 (1996).
[4] A. Edwards, M. V. Rao, B. Molnar, A. E. Wickenden, O. W. Holland, and P. H. Chi, J Elect Mat, **26**, 334 (1997).
[5] N. Parikh, A. Suvkhanov, M. Lioubtchenko, E. Carlson, M. Bremser, D. Bray, R. Davis, and J. Hunn, Nuc Instr & Meth in Phys Res B, **127/128**, 463 (1997).
[6] J. A. Fellows, Y. K. Yeo, R. L. Hengehold, and D. K. Johnstone, Appl. Phys. Lett. **80**, 1930 (2002).

Optical evaluation of spatial carrier concentration fluctuations in doped InP substrates

M. Baeumler[1], E. Diwo[1], W. Jantz[1], U. Sahr[2], G. Müller[2] and I. Grant[3]

[1]Fraunhofer Institut Angewandte Festkörperphysik, Freiburg, Germany
[2]Crystal Growth Laboratory, Universität Erlangen-Nürnberg, Germany
[3]Wafer Technology Ltd., Tongwell, Milton Keynes, UK

Abstract. Nondestructive electrical characterization of LEC and VGF grown InP:S substrates is achieved with innovative photoluminescence line shift topography. The optical data, satisfactorily in agreement with band-gap renormalization and band filling theory, are absolutely calibrated against Hall effect measurements in the range $2 \cdot 10^{17}$ cm^{-3} up to $2 \cdot 10^{19}$ cm^{-3}. The 70 μm lateral resolution and 1% sensitivity of the full wafer carrier concentration imaging allow to discern characteristic variations, such as doping striations.

1. Introduction

Doped InP substrates are needed to fabricate optoelectronic devices in the near infrared spectral range. Manufacturers of lasers and photodiodes specify carrier concentrations n in the range 10^{18} cm^{-3} with high lateral and wafer-to-wafer homogeneity. The progress of crystal growth technology towards meeting these demands is supported by developing application-oriented characterization techniques to determine absolute values and spatial variations of n. High resolution imaging also helps to improve the design of the crystal growth system and the control of the solidification process.

For absolute evaluation of n one would of course prefer an electric measurement technique. However, available eddy current mapping systems have limited lateral resolution of about 10 mm. Conversely, photoluminescence topography (PLT) is a convenient, high-resolution, nondestructive material assessment technique, but the conventional PL *intensity* mapping yields qualitative information only.

We have recently shown that mapping the PL line *shift* is possible with a sensitivity in the percent range [1]. The spectral position of the band-to-band (BB) recombination is correlated with n, predominantly due to band-gap renormalization and conduction band filling, allowing to generate full wafer n topograms with 70 μm lateral resolution. Experimentally, one exploits the advantage that a small spectral shift results in a strong variation of the PL intensity recorded at the steep *slope* of the emission line. The results are intended to demonstrate the practical value of the technique. We do not pursue a detailed comparison of growth techniques or material of different suppliers.

2. Experimental

Commercial and proprietary VGF and LEC 2" InP:S wafer with n ranging from $3 \cdot 10^{17}$ cm^{-3} to $6 \cdot 10^{18}$ cm^{-3} were investigated, including three pairs of wafer cut from one ingot [1,2] and wafer from other sources. For PL spectroscopy (PLS) and PLT the wafer was mounted in a 2 K He bath cryostat with a 3" window. 1 mW Ar$^+$ laser emission at 514 nm was focused onto a 70 µm sample spot. The PL emission was dispersed by a 1m monochromator and detected by a liquid Nitrogen cooled Ge-detector. For PLT, the excitation beam was scanned with a pair of moving mirrors. Comparative sheet conductance maps were recorded with a Lehighton 1500 system.

3. Results and discussion

Two emission lines due to band-to-band (BB) and band-to-acceptor (BA) transitions are observed [1] for n above the "Mott" concentration ($\approx 1 \cdot 10^{17}$ cm^{-3} for InP [3]). Fig. 1 illustrates how the line shift data are obtained from the measured intensity variation at ½ of the peak intensity (see insert of Fig. 1, showing the BB line for $n = 4.7 \cdot 10^{17}$ cm^{-3}).

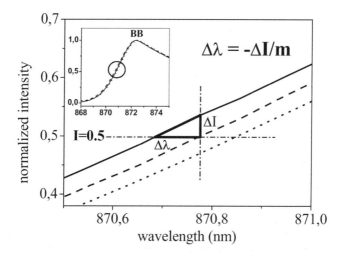

Fig. 1 High-energy edge (slope m) of normalized PL spectra recorded at three different wafer positions along one radius of the wafer.

In Fig. 2 these data are plotted against n, together with published data obtained with PL and PL excitation spectroscopy [3]. The solid line is calculated according to

$$E_{1/2} = 1.4355 - 5.06 \cdot 10^{-8} \cdot n^{\frac{1}{3}} + 4.61 \cdot 10^{-14} \cdot n^{\frac{2}{3}} - 1.236 \cdot 10^{-27} \cdot n^{\frac{4}{3}}$$

where $E_{1/2}$ is the maximum recombination energy of a conduction band electron with a valence band hole. The first two terms describe the band-gap shrinkage with increasing n, including the exchange and correlation energy as well as electron-donor and electron-hole interaction [4,5]. These effects are expressed in a simplified form $a+b \cdot n^{1/3}$, with a and b adjusted to fit the data points[1]. The last two terms of the formula account for the n dependence of the Fermi energy for a non-parabolic conduction band[2]. These terms have

---

[1] For the exchange energy $2.5 \cdot 10^8 \cdot n^{1/3}$ eV has been reported by Ref. [5]
[2] For n< 7×10^{17} cm^{-3} the data can be fitted with an empirical formula omitting the non-parabolicity of the conduction band [1]

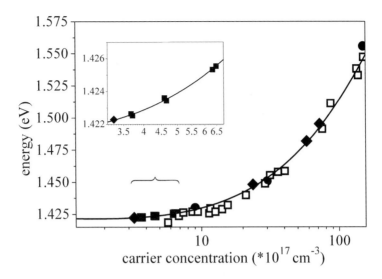

Fig. 2 High-energy edge value at half intensity of the BB peak plotted against the carrier concentration: (■,◆) VGF and LEC InP:S, (●) from Ref. 3. □ onset from PL excitation spectra in Ref. 3.

been calculated according to the theory of by Raymond et al. [6][3]. Fig. 2 shows that the sensitivity of the technique increases with n.

Topographic results for a LEC InP:S wafer with average $n = 5.9 \cdot 10^{18}$ cm^{-3} are shown in Fig. 3. First, for normalization purposes a topogram was recorded at the peak wavelength $\lambda = 842.9$ nm of the BB emission (left). Very satisfactory homogeneity with a standard deviation of 5.5% was obtained. The center part of Fig. 3 shows the n topogram, resulting from the topogram taken at $\lambda = 836.5$ nm, normalized using the BB peak intensity topogram and converted to n with the two-step procedure outlined above. The variation of n across the wafer is below 10%. However, the high spatial resolution of the technique allows to resolve striations of 3% doping variation, shown magnified in the right part of Fig. 3. In line with the present findings, Raman microprobe investigations have revealed n variations of about 10% across growth striations in Bridgman-grown GaAs:Si [9].

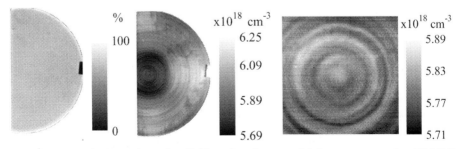

Fig. 3 PL intensity (left) and n (center, right) topograms of a 2" LEC InP:S wafer. The right image shows a 5x5 mm² image from the center.

---

[3] Formulas (2)-(5) in Ref. 6: $m_e^* = 0.0791 \cdot m_o$ [7], E_o (2K)=1.4236eV [8], Δ=0.108eV [8]

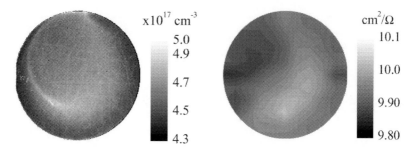

Fig. 4 Topograms of carrier concentration (left) and conductance (right) of a 2" VGF InP:S wafer.

Fig. 4 demonstrates the characterization of a VGF wafer with low $n = 4.7 \cdot 10^{17}$ cm^{-3}. The n topogram (left) is very homogeneous; however, characteristic features of the residual lateral variation of n are clearly identified. The pattern of the n variation is roughly reproduced and thus basically confirmed by the sheet conductance topogram shown on the right side of Fig. 4. But, due to the strong averaging effect caused by the limited lateral resolution (14 mm) of the eddy current measurement, the topographic information is substantially inferior. For instance, the narrow curved line, which is the most prominent feature in the left image, is not discerned. Moreover, the total variation amplitude of the conductance mapping is only 3%, as opposed to 15% of the optically generated topogram. The underestimation of the carrier fluctuations resulting from inadequate lateral resolution underscores the necessity and value of the optical n analysis. A contribution of local strain fluctuations to the line shift pattern cannot be excluded; however, the agreement with the Raman microprobe results as well as the consistency with the Hall effect and sheet conductance data over a wide n range supports the conclusion that PL line shift topography as a suitable tool to detect local fluctuations of the carrier concentration.

Acknowledgement

The authors would like to thank Dr. G. Guadalupi from Venezia Tecnologie S.p.a. (Italy) for supplying the LEC InP:S wafer. They further like to thank J. Wagner and R. Fornari for stimulating discussions and the Hall effect measurements.

References

[1] U. Sahr, G. Müller, I. Grant, M. Baeumler and W. Jantz, Conf. Proc., 14th Int. Conf. on Indium Phosphide and Related Materials, Stockholm, Sweden, (2002), 405-408
[2] U. Sahr, I. Grant, G. Müller, Conf. Proc., 13th International Conference on Indium Phosphide and Related Materials, Nara, Japan, (2001), 533-536
[3] R. Schwabe, A. Haufe, V. Gottschalch, K. Unger, Solid State Commun. 58, 485 (1986)
[4] J. Wagner, Proc. 31st Scottish Universities Summer School in Physics, St. Andrews, (1986), 343
[5] M. Bugajski and W. Lewandowski, J. Appl. Phys. 57, 521 (1985)
[6] A. Raymond, J.L. Robert, and C. Bernard, J. Phys. C 12, 2289 (1979)
[7] E.P.O'Reilly, in Properties of Indium Phosphide, (INSPEC, IEE, 1991) ch. 6.1, p. 93.
[8] Landolt and Börnstein, Numerical Data and Functional Relationships in Science and Technology, Vol 22a (Springer, Berlin, 1987) p. 107.
[9] M. Herms, G. Irmer, J. Monecke, and O. Oettel, J. Appl. Phys. 71, 432 (1992).

An "anomalous" drift of defects under electric field in CdSe and CdS single crystals

L V Borkovska, B M Bulakh, L Yu Khomenkova, N O Korsunska, I V Markevich

Institute of Semiconductor Physics at NAS of Ukraine, Kyiv 03028, Ukraine

Abstract. An "anomalous" defect drift in external electric field, namely, transport of acceptor-like centres from the anode to the cathode, has been observed in CdS:Cu, CdS:Ag and nominally undoped CdSe crystals at 400-700 K. The effect is accounted for by transformation of acceptors into donors under heating. Drifting centres are concluded to be Cu- or Ag-related defects in CdS and defects related to some residual impurity in CdSe crystals. Two mechanisms of centre transformation are discussed: i) the change of impurity atom position in the crystal lattice; ii) the change of centre charge from negative to positive due to double ionization. The first mechanism is stated to take place in CdS:Cu and CdS:Ag crystals. For CdSe crystals both first and second mechanisms are supposed to be probable.

1. Introduction

Drift of lattice defects under electric field is a well-known phenomenon. When an external electric field is applied to a semiconductor at a suitable temperature, redistribution of charged mobile defects along the sample occurs, donors being collected near the cathode and acceptors being accumulated near the anode [1]. In n-type semiconductor this redistribution must result in the increase of sample conductivity near the cathode and its decrease near the anode. Such "normal" effect was observed, in particular, in undoped and doped with Li CdS crystals, where drift of mobile shallow donors Cd_i and Li_i in external electric field caused the increase of conductivity and photosensitivity near the cathode region [2]. It has been found, however, that sometimes an "anomalous" effect, namely, the accumulation of acceptors near the cathode, can take place. This effect that has been observed in nominally undoped CdSe, as well as in CdS crystals doped with copper and silver, is described and investigated in the present work.

2. Experimental procedure and results

Bulk and platelet CdS crystals doped with Cu or Ag and nominally undoped CdSe platelets were investigated. The crystals were high-resistivity: $\rho > 10^5$ Ohm·cm for CdSe and $\rho > 10^9$ Ohm·cm for CdS crystals. All crystals were n-type. Ohmic In or Cd electrodes were applied to the crystals as shown in fig.1. In the initial state, i.e. before action of electric field, dark current, DC, photocurrent, PC, and photoluminescence, PL, spectra were measured at 300 or 77K between electrodes A, A' and B, B' (fig.1,a). Then electrodes A, A' and B, B' were closed (fig.1,b), the sample was heated to $T_d = 350\text{-}700K$

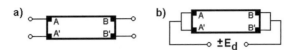

Figure 1. Appearance of the sample mounted for measurement of DC and PC spectra (a) and for drift carrying out (b).

and direct electric field E_d=50-100 V/cm was applied to it. After a time interval Δt the sample was quickly cooled to room temperature, the electric field was switched out, electrodes A, A' and B, B' were opened, and above-mentioned characteristics were again measured. To control DC and PC in any sample region additional indium electrodes were applied between A, A' and B, B' ones.

The influence of electric field on CdSe crystal characteristics was found to take place at T_d=370 K already. After application of E_d=50-100 V/cm during Δt=30-40s at this temperature DC and PC values increase at the anode and decrease at the cathode (fig.2,a). Simultaneously the only band at λ_m=0.93 μm present in PL spectrum strengthened near the anode and quenched near the cathode (fig.2,b). With Δt increase the high-conductivity region spread to the cathode, and at last only thin low-conductivity strip created near this electrode, while the rest crystal became highly conductive. If then electric field of opposite direction was applied to the sample at T_d=370 K, the low-conductivity region created near the new cathode. This process could be repeated many times and the results were reproduced.

In PL spectra of CdS:Cu crystals in the initial state an infrared band at λ_m=1.0μm was the most intensive (fig.3,a) and in PC spectra a strong extrinsic maximum peaked at nearly 0.75 μm was present. Application of an electric field at T_d=650-700 K during Δt=3-5 min resulted in sharp drop of IR-band intensity near the anode and its rise near the cathode, while the intensity of the other PL band did not change noticeably. Simultaneously PC extrinsic maximum value decreased near the anode and increased near the cathode. The measurements of PL and PC in various sample regions between electrodes A, A' and B, B' showed that with Δt increase the region with quenched IR band and PC extrinsic maximum spread along the sample to the cathode. The effect was reversible: when the sample was heated again and electric field of opposite direction was applied to it, the strengthening of IR band and PC extrinsic maximum near the new cathode and their quenching near the new anode occurred.

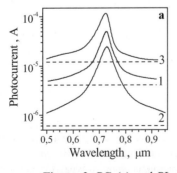
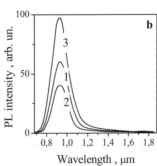

Figure 2. PC (a) and PL (b) spectra of CdSe crystal at 300 K (a) and 77 K (b) measured between the electrodes A, A' before (1) and after (2,3) the action of electric field, when the electrodes A, A' were the cathode (2) and the anode (3) (E_d=100 V/cm, T_d=400K, Δt_d=1 min). The values of dark current are shown as dashed lines.

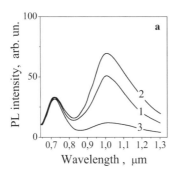 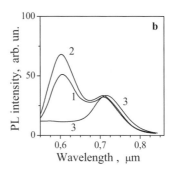

Figure 3. PL spectra of CdS:Cu (a) and CdS:Ag (b) at 77K measured between electrodes A, A' before (1) and after (2,3) the action of electric field, when the electrodes A, A' were the cathode (2) and the anode (3).
a) E_d=70V/cm, T_d=650K, Δt_d=5min; b) E_d=90V/cm, T_d=700K, Δt_d=10min.

Similar changes of PL and PC spectra took place also in CdS:Ag crystals, where in the initial state PL band λ_m=0,61 µm dominated in PL spectra (fig.3,b) and PC extrinsic maximum peaked at about 0.6 µm was observed. The dark currents in CdS crystals both before and after action of electric field were too small to be measured exactly.

In the initial state CdS:Cu and CdS:Ag crystals had reddish-brown and bright brown colour correspondingly. Described above changes in PL and PC spectra were accompanied by change of crystal colour: after switching on of electric field at first a thin bright yellow (like undoped CdS) strip appeared near the anode. Then this strip broadened with Δt increase, and at last only a thin intensively coloured region was observed near the cathode. Under electric field of opposite direction the crystal acquired reddish-brown or bright brown colour again and then bright yellow strip appeared near the new anode.

Changes induced by electric field remained stable for many months at 300 K both for CdS and CdSe crystals.

3. Discussion

Analysis of above results shows that in investigated crystals transport of acceptors from the anode to the cathode occurs under electric field. Really, redistribution of dark conductivity along the sample in CdSe crystals indicates directly that the change of acceptor density with respect to donor one takes place. The increase of DC in the most part of the sample and its sharp decrease near the cathode testifies that this change is due to extraction of acceptors from the anode-side region and their accumulation at the cathode. The latter process is not accompanied by the rise of any PL band intensity, so, one can think that acceptors under consideration are nonradiative recombination centres. The increase of such centres density must lead to the drop of PC and PL intensity [3], which is observed indeed (fig.2, a,b).

In CdS:Cu and CdS:Ag crystals electric field induces considerable changes in IR (for CdS:Cu) and orange (for CdS:Ag) PL band intensities, while the intensities of other bands remain almost unchanged. This is the evidence that observed PL spectrum alteration is due to changes of densities of radiative centres responsible for λ_m=1,0 µm and λ_m=0.61 µm bands [3]. These bands are known to result from recombination of free electrons on Cu_{Cd} and Ag_{Cd} acceptors respectively, and the observed extrinsic maxima in PC spectra were shown to be due to photoionization of electrons from these acceptors to c-band [3,4]. Since acceptors under consideration are "sensitizing" centres [3,4], the value of the intrinsic PC maximum correlates with this acceptor density too. Thus, the densities of Cu_{Cd} and Ag_{Cd} centres decrease in the anode-side region and increase near the cathode.

To explain described above "anomalous" drift effects, some defect species that are acceptors in the stable state at 300 K but can transform into donors at elevated temperatures should be supposed to exist; these donors must be ionized at drift temperatures and must have higher diffusivity than the acceptors. One can imagine two ways of acceptor/donor transformations: i) the change of the atom position in the crystal lattice; ii) the change of the centre charge from negative to positive due to double ionization. Since in semiconductors, including CdS and CdSe, diffusivity of interstitials is usually much higher than that of sites and vacancies [1,2], in the first case only site-to-interstitial transitions are actual.

It had been shown that Cu and Ag incorporate in CdS and diffuse in the crystal lattice interstitially as donors Cu_i^+ and Ag_i^+ with following reactions $Cu_i+V_{Cd} \rightarrow Cu_{Cd}$ and $Ag_i+V_{Cd} \rightarrow Ag_{Cd}$ superimposed on the diffusion processes [5,6]. Obtained results testify that at 650-700 K in previously doped crystals reverse reactions $Cu_{Cd} \rightarrow Cu_i+V_{Cd}$ and $Ag_{Cd} \rightarrow Ag_i+V_{Cd}$ are intensive enough. Because of these processes, the impurity atoms acting at 300 K as acceptors, at elevated temperatures drift in electric field as donors and become acceptors again under cooling. So, the impurities are extracted from the anode-side sample region and accumulated near the cathode. The change of crystal colour after the action of electric field confirms this conclusion.

Experimental results relative to CdSe crystals make no hints about the chemical nature of defects responsible for the "anomalous" drift. In general, these defects can be both native and impurity-relative. A distinctive feature of drifting defects, however, is their high diffusivity. The only native defect that is mobile enough at 370-400K is shallow donor Cd_i [2,7]. Thus, one must think that defects under consideration are connected with some residual impurity, each of described above transformation mechanisms being able to realize. The nature of drifting defect in CdSe crystals is under investigation.

In conclusion, an "anomalous" defect drift in electric field, namely, transport of acceptors from the anode to the cathode was observed in CdS:Cu, CdS:Ag and nominally undoped CdSe crystals at 400-700K. The effect was accounted for by transformation of acceptors into donors under heating. Transformation process in CdS crystals was stated to consist in replacement of Cu or Ag atoms from Cd sites to interstitial positions. In CdSe crystals drifting defects are supposed to be residual impurity atoms that change the sign of their charge due to either the change of position in the lattice or to double ionization.

References

[1] Boltaks B J 1961 Diffuziya v poluprovodnikah (Moskva: Fizmatgiz)
[2] Korsunskaya N E Markevich I V Shably I Yu and Sheinkman M K 1981 Fiz.Tech.Poluprovod. 15 279-282
[3] Lashkarev V Ye Lyubchenko A V and Sheinkman M K 1981 Nonequillibrium processes in photoconductors (Kiev: Naukova Dumka)
[4] Bube R H 1960 Photoconductivity of solids (New York- London: Eds.John Willey and Sons)
[5] Sullivan J A 1969 Phys. Rev. 184 796-805
[6] Timan B L and Zagoruiko Yu A 1979 Fiz. Tverd. Tela 21 2949-2851
[7] Aven M and Prener J S (Eds) 1967 (Amsterdam: North-Holland publishing company)

Injection energy dependence of electron thermalization length in AlGaAs/GaAs quantum well structures

T Tsuruoka[1,2], H Hashimoto[1], Y Ohizumi[1], and S Ushioda[1,2]

[1]Research Institute of Electrical Communication, Tohoku University, Sendai 980-8577, Japan
[2]RIKEN Photodynamics Research Center, Sendai 980-0845, Japan

Abstract. We have investigated the transport properties of hot electrons in AlGaAs/GaAs quantum well structures in real space by using a scanning tunneling microscope (STM). The electrons were injected from the STM tip at different distances from a target well. Then by measuring the light emission intensity from the target well, the thermalization and diffusion lengths of the injected electrons were determined. We found that the thermalization length increases with the increase of the initial energy of injection, while the diffusion length is independent of the injection energy.

1. Introduction

The transport properties of charge carriers in quantum well (QW) structures are important in the design and characterization of nanostructured electronic devices. Several techniques based on cathodoluminescence (CL) and microphotoluminescence have been used to measure the carrier diffusion length of various semiconductors [1]. However, the spatial resolution of these methods is typically not better than a few hundred nm, due to the spot size and the penetration depth of the incident probes. Recently we have demonstrated that scanning-tunneling-microscope light-emission spectroscopy (STM-LES) has a potential for measuring the electron transport parameters in QW structures with a spatial resolution of a nm scale [2]. By using this method one can determine the thermalization length as well as the diffusion length of the electrons injected from the STM tip. In this paper we present some of the results and discuss the dependence of the transport parameters on the injection energy that is controlled by the sample bias voltage of the STM.

2. Experiments

The samples were grown on p-type GaAs(100) substrates by molecular-beam epitaxy. They had GaAs wells of widths 3.1, 5.1, and 10.2 nm sandwiched between $Al_{0.3}Ga_{0.7}As$ layers of 50 nm width. All the layers were Be doped uniformly, with the hole concentration of 1.4×10^{19} cm^{-3} for the AlGaAs layers. An atomically flat (110) surface was prepared by cleaving the sample in ultrahigh vacuum. The STM-LE measurements were performed on this surface at room temperature. The light emitted from the sample was collected by an optical fiber mounted in vacuum. It was then analyzed with a grating spectrograph equipped with a liquid-nitrogen-cooled charge coupled device (CCD) camera.

3. Results and Discussion

The STM image of the cleaved (110) surface showed contrasts reflecting the QW structures. The light emission spectra were measured by locating the tip at different points along the [001] direction with an increment of ~6 nm over the QW structures. The sample bias voltage V_S was set at +2.1 and +2.5 V with the fixed tunneling current of 0.5 nA. Figure 1 shows typical examples of the STM-LE spectra that were measured for V_S = +2.5 V. The spectra consist of four peaks located at 1.44, 1.53, 1.63, and 1.84 eV. The peaks at 1.44, 1.53, and 1.63 eV were assigned to the transitions between the electron and heavy-hole ground states in the wells of widths 10.2, 5.1, and 3.1 nm, respectively [3]. The peak at 1.84 eV corresponds to the transition between the bottom of the conduction band to the acceptor state in the AlGaAs barrier. We see that the intensity ratio between the emission peaks change continuously with the position of the tip.

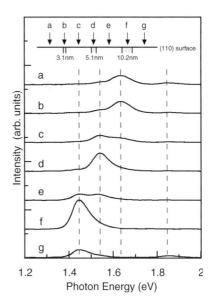

Figure 1. STM-LE spectra measured at different points over the QW structures, as illustrated in the inset.

The observed spectra can be decomposed into separate peaks associated with the three wells. Figure 2 shows the emission intensity from the well of 10.2 nm width plotted as a function of the tip position for V_S = +2.5 V. The intensity was corrected for the energy-dependent sensitivity of our detection system. The emission intensity was found to decay with two distinct decay constants L_1 and L_2. The constant L_1 is identified with the thermalization length through which the injected hot electrons rapidly relax to the bottom of the conduction band [2]. L_2 is the diffusion length by which the minority electrons drift toward the well after thermalization [4]. By fitting exponential decay functions to the data as shown in Fig. 2, the two decay constants were estimated.

Figure 3 shows the emission intensities from individual wells plotted as a function of the tip position for two sample bias voltages. The solid curves are the superpositions of two exponential decay functions fitted to the data as shown in Fig. 2. From this comparison between the experimental results and the theoretical curves, the thermalization length L_1 was estimated to be 6±1 nm for +2.1 V and 15.5±1.5 nm for +2.5 V. The diffusion length L_2 was ~200 nm for both sample bias voltages.

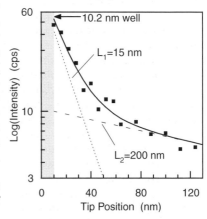

Figure 2. Determination of the decay constants from the emission intensity plotted as a function of the tip position. The dotted and dashed lines represent the exponential decay lines with decay constants L_1 and L_2, respectively. The solid curve is the superposition of the dotted and dashed lines.

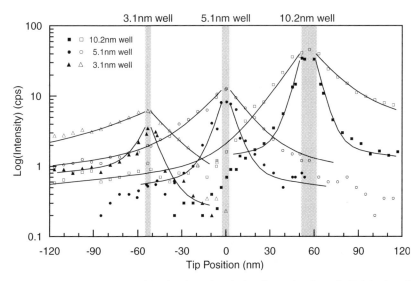

Figure 3. Logarithmic plot of emission intensity from individual wells as a function of the tip position. The solid and open symbols correspond to the data for V_S = +2.1 and +2.5 V, respectively. The solid curves are the superpositions of two exponential decay lines with different decay constants as shown in Fig.2.

Our values of L_2 is about one third of that reported by Zarem et al. who measured it for $Al_{0.25}Ga_{0.75}As$ layers using the CL technique [5]. We believe that the short diffusion length results from the high hole concentration of our samples. In Fig. 3 the emission intensity profile becomes asymmetric with increasing sample bias voltage. This result suggests that the presence of the adjacent QW influence the transport process of the injected electrons. A detailed discussion will be given in a separate paper [6].

To understand the bias voltage dependence of the thermalization length, we have carried out Monte Carlo simulations for hot electron relaxation in p-type $Al_{0.3}Ga_{0.7}As$ based on the model of conduction band with nonparabolic Γ, L, and X valleys. From the peak energy of the emission from the AlGaAs barrier, we know that the band gap of $Al_{0.3}Ga_{0.7}As$ is 1.84 eV. Figure 4(a) illustrates the energy diagram of a STM junction with the AlGaAs(110) surface. For the hole concentration of 1.4×10^{19} cm^{-3}, the Fermi level lies ~8 meV below the top of the valence band [7]. The STM tip induces upward band-bending at the sample surface for positive sample bias voltage [8]. By solving the Poisson's equation with relevant boundary conditions, the values of upward band-bending were estimated to be 0.15 and 0.19 eV for V_S = +2.1 and +2.5 V, respectively. Thus the electrons are injected with the initial energy of 0.1 eV for +2.1 V and 0.46 eV for +2.5 V. Using these injection energies, we calculated the spatial distribution of the injected electrons that relaxed below the thermal energy at the bottom of the Γ valley. The simulation included intervalley phonon scattering, scattering by coupled plasmon-phonon modes, ionized impurity scattering, and alloy scattering [6].

Figure 4(b) plots the electron density as a function of the distance from the injection point, which was obtained from the calculated spatial distribution. The electron density decreases with decay constants 4 and 20 nm for the injection energies 0.1 and 0.46 eV, respectively. Since the emission intensity is proportional to the electron density, these decay constants are equivalent to the thermalization length. Thus we see that the

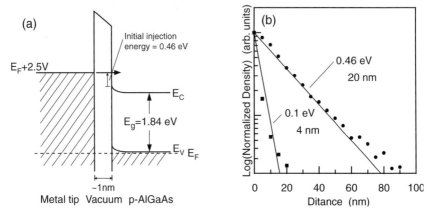

Figure 4. (a) Energy diagram of the STM junction with an p-AlGaAs(110) surface for V_S = +2.5 V. (b) Electron densities after thermalization in p-Al$_{0.3}$Ga$_{0.7}$As plotted as a function of the distance from the injection point for the two injection energies.

calculated thermalization lengths 4 and 20 nm agree reasonably well with the corresponding experimental values of 6±1 and 15.5±1.5 nm. By examining the contribution of each scattering mechanism, it was found the scattering by coupled plasmon-phonon modes has the most significant effect on the thermalization length, because of the strong coupling between holes and optical phonons for the used hole concentration. In addition, the intervalley phonon scattering also contributes to the thermalization length for the injection energy greater than the separation energy of the Γ - L valleys (0.106 eV).

4. Conclusion

We have measured the thermalization and diffusion lengths of the hot electrons injected from the STM tip into AlGaAs/GaAs QW structures by light emission measurements. From the emission intensity for the individual wells as a function of the tip position, it was found that the thermalization length increases with increasing initial energy of injection, while the diffusion length is independent of the injection energy. The injection energy dependence of the thermalization length was explained by a Monte Carlo simulation for hot electron relaxation in AlGaAs.

References

[1] Gustafsson A, Pistol M –E, Montelius L and Samuelson L 1998 J. Appl. Phys. **84** 1715-1775, and references cited therein
[2] Tsuruoka T, Tanimoto R, Ohizumi Y, Arafune R and Ushioda S 2002 Appl. Phys. Lett. **80** 3748-3750; Appl. Surf. Sci. **190** 275-278
[3] Tsuruoka T, Ohizumi Y, Tanimoto R and Ushioda S 1999 Appl. Phys. Lett. **75** 2289-2291
[4] Alvarado S F, Renaud Ph and Meier H P 1991 J. Phys. IV **1** 271-275
[5] Zarem H A, Lebens J A, Nordstrom K B, Sercal P C, Sanders S, Eng L E, Yariv A and Vahala K J 1989 Appl. Phys. Lett. **55** 2622-2624
[6] Tsuruoka T, Hashimoto H, and Ushioda S (to be published)
[7] Seeger K 1973 Semiconductor Physics (Wien: Springer-Verlag)
[8] Feenstra R M and Stroscio J A 1987 J. Vac. Sci. Technol. B **5** 923-929

Optical and Structural Investigations of GeSiO$_2$ Systems

T.V. Torchynska, J. Aguilar-Hernandez, G. Polupan and A. Kolobov[1]

SEPI-National Polytechnic Institute, Mexico D.F. 07738, México,
ttorch@esfm.ipn.mx
[1]LAOTECH-National Institute for Advanced Industrial Science and Technology, Tsukuba, Ibaraki 305-8562, Japan

Abstract. This paper presents the results of a photoluminescence investigation of silicon oxide films enriched by Ge. Photoluminescence peculiarities are analyzed both for "as prepared" silicon oxide films enriched by Ge (without Ge-nanocrystalls) and for films with Ge-nc, created during film annealing in inert atmosphere at 800°C. Raman scattering spectra and high-resolution transmission electronic microscopy are used for the confirmation the Ge-nc plane existing in the structures, and for an estimation of their sizes. The mechanism of photoluminescence is discussed as well.

1. Introduction

Semiconductor quantum dots based on Si and Ge have recently been the subject of numerous investigations and are of interest both for the basic physics of quantum confinement and for their application in optoelectronic devices of new generation. This is due to the possibility that the radiation efficiency of indirect optical transitions may be significantly increased if the size of the semiconductor is in the nanometer-scale range [1, 2]. A number of groups have reported the formation of Ge and Si nanocrystallites embedded in SiO$_2$ and intense photoluminescence (PL) in the spectral range of 1.7-2.4 eV has indeed been observed [2,3]. The photoluminescence of these structures in the spectral range of 1.7-2.4 eV is attributed both to radiative recombination via Si (Ge) quantum-confined states [3], and to defects in silicon oxide [4] or at the Si/SiO$_x$ interface [5]. Thus, the origin of visible PL of Si (Ge) – nano-crystallites (nc) embedded in silicon oxide still remains to be clarified.

2. Experimental details

Silicon oxide films enriched by Ge were prepared by co-deposition of Ge and silicon oxide onto Si (100) or quartz glass substrates by radio frequency magnetron sputtering [3]. The thickness of the samples was in the 500nm-1µm range and the Ge concentration varied from 25 to 60-mol % (Table 1). After deposition, some of the samples were isothermally annealed for 1h at 800 °C in an argon atmosphere, which produced nano-crystals with a characteristic size of 6-30 nm.

Photoluminescence (PL) and Raman scattering spectra have been measured by exciting the samples with an Ar laser tuned at a wavelength of 514.5 nm. The laser power was 80 mW. The PL and Raman signals were measured with a model 1403 spectrometer in backscattering geometry. The formation of Ge nano-crystallites was checked by high-resolution transmission electron microscopy (HR-TEM) to examine their crystallinity. The images were taken using a Hitachi –9000 microscope NAR with an accelerating voltage of 300 kV.

Table 1. Characteristic of the investigated samples

N, samples	%Ge	T-anealing (time)	Wafer	Size of Ge- nc
1	25	800 °C (1h)	Si (100)	no detected
2	42	800 °C (1h)	Si (100)	5-10nm
3	60	800 °C (1h)	Si (100)	15-30nm
4	25	as-deposit	Si (100)	no detected
5	42	as-deposit	Si (100)	no detected
6	60	as-deposit	Si (100)	no detected
7	25	800 °C (1h)	Quartz glass	no detected
8	42	800 °C (1h)	Quartz glass	no detected
9	60	800 °C (1h)	Quartz glass	4-5nm
10	25	as-deposit	Quartz glass	no detected
11	42	as-deposit	Quartz glass	no detected
12	60	as-deposit	Quartz glass	no detected

3. Experimental results

A PL band peaked at 2.1 eV with the full width at half maximum (FWHM) equal to 600 meV is revealed in the samples with 25 and 42 mol % Ge concentrations (Fig.1b). The intensity of this band decreases with increasing Ge concentration and in films with 60 mol % Ge this band does not appear. The annealing at 800 °C during 1 hr initiates a shift of the maximum of the PL band to a somewhat higher energy of 2.2 eV, and a decrease of the FWHM to 500 meV (Fig.1a). The PL intensities of the bands in films, with and without annealing, are practically the same.

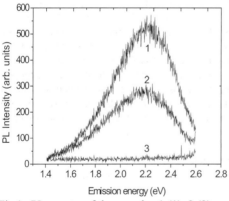

Fig.1a PL spectra of the samples: 1 (1), 2 (2), 3(3).

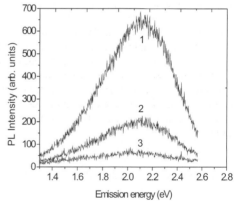

Fig.1b PL spectra of the samples: 4(1), 5(2), 6(3).

Raman scattering studies have shown that Ge-SiO$_2$ films in as-prepared state are amorphous with the broad feature characteristic of Ge-Ge vibrations, found in the Ge-SiO$_2$ samples with 60-mol % Ge. Upon annealing, a crystalline peak centered at ≈ 290 cm^{-1} appears in samples created on Si substrate with 60 and 42 mol. %. Ge. Raman scattering have not detected the Ge-Ge bond vibrations in samples with lower Ge concentration (25 mol%) for both types of substrates.

The existence of Ge-nc in the silicon oxide matrix after thermal annealing additionally has been confirmed by TEM investigation. TEM images of the annealed samples reveal the presence of nanocrystals with a typical size of ~ 5-10 nm in the 42 mol % Ge sample and 15-30 nm in the 60 mol % Ge sample for the case of Si (100) substrate. For the case of quartz-glass substrate, the formed nano-crystals are considerably smaller (~ 5 nm for the 60 mol. % Ge).

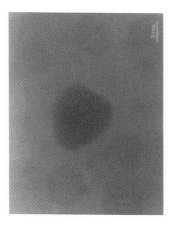

Fig.2 TEM image of the annealed Ge <60%> Si(100) sample. TEM image magnification is 10^6. .

4. Discussion

It is shown, that thermal treatment of Ge-SiO$_2$ films in inert atmosphere at 800 °C during 1hr creates the Ge-nc in the films with high concentrations of Ge equal to 40-60 mol %. Thus it is possible to investigate the influence of Ge-nc on PL of Ge-SiO$_2$ films. The presented results reveal that the visible PL band peaked at 2.1-2.2 eV in Ge-SiO$_2$ films does not correlate with the presence of Ge-nc in the sample. Actually, the intensity of this PL band decreases with Ge concentration increasing from 20 up to 60 mol %, and does not depend on thermal treatment. The maximum intensity of this PL band is in the samples with 20mol % Ge. In these samples Raman scattering and HR-TEM investigations do not confirm the presence of Ge-nc. So, we conclude that the PL band peaked at 2.1-2.2 eV in Ge-SiO$_2$ films is connected with defects in silicon oxide films.

Acknowledgment
This work was supported by CONACYT (Project 33427-U) and CGPI – IPN Mexico.

References
[1] T. Takagahara and K.Takeda, 1992 Phys. Rev. B **46**, 15578-15585.
[2] S.Okada and Y.Kanemitsu, 1996 Phys. Rev. B **54**, 16421-16424.
[3] Y.Maeda, 1995 Phys.Rev.B, **51,** 16581663.
[4] 6. S.Charvet, R.Madelon, R.Rizk, B.Garrido, O.Gonzalez-Varona, M.Lopez, A.Perez-Rodriguez, J.R.Morante, 1999 J. Lumines. **80,** 241-247.
[5] J.Linnros, N.Lalic, A.Galeckas, V.Grivickas, 1999 J. Appl. Phys. **86,** 6128-6133.

Thermal Quenching of Emission of Self-Assembled InAs Quantum Dots Embedded into InGaAs/GaAs MQW

T. V. Torchynska, J. L. Casas Espinola, H.M.Alfaro Lopez, P. G. Eliseev[1], A. Stintz[1], K. J. Malloy[1] and R. Pena Sierra[2]

ESFM - National Polytechnic Institute, Ed. 9,Mexico D.F., 07738, MEXICO,
ttorch@esfm.ipn.mx
[1]Center of High Technology Material, Univ. New - Mexico, Albuquerque, USA
[2]CINVISTAV-IPN, México D.F., 07738, MÉXICO

Abstract. This paper presents the investigation of PL bands connected with ground (GS) and multi excited states (ES) in highly uniform self-assembled InAs QDs using variable temperatures. Investigated QDs are embedded in $In_{0.15}Ga_{0.85}As/GaAs$ multi-quantum-well (MQW) structure. The types of GS and ES optical transitions are discussed as well.

1. Introduction

The quantum-dot (QD) heterostructures are found to be important version of active medium for semiconductor lasers. The realization of GaAs-based optoelectronic devices for 1.3-µm wavelength range is a subject of strong interest due to matching the transparency window of optical fibers. GaAs-based structures operating at 1.1-1.3 µm are currently developed using self-organized InAs quantum dots in GaAs, AlGaAs or InGaAs matrix [1-3]. The temperature dependence of optical properties of self-assembled QDs is of importance for their use in devices operating at room and higher temperatures. A lot of articles have discussed the photoluminescence (PL) temperature dependencies of InAs QDs in different types of matrixes. These investigations, in general, have discussed the temperature dependence of the PL band connected with ground-state optical transition mainly. This paper presents the investigation of PL bands, connected with ground (GS) and multi excited states (1-4ES), using variable temperatures.

2. Experiment details

The solid-source molecular beam epitaxy (MBE) in V80H reactor is used for growth of InAs self-organized QDs inserted into $In_{0.15}Ga_{0.85}As/GaAs$ MQW. The individual dots are of 15 nm in the base diameter and ~7 nm in height. The dot density is determined by atomic force microscope observation of parallel wafer that have not been overgrown by QW and by cladding layers. The in-plane dot density is 2.5×10^{10} cm^{-2}. The luminescence signal is dispersed by a Hilger-Watts 0.75-m Monospeck 600 monochromator and detected by a S1 type photomultiplier with Ag-O-Cs photocathode and lock-in amplifier (Princeton Applied Research) model 124. Ar ion laser with light wavelength 514.5 nm excites PL; power is up to 100 mW (excitation power density is up to 600 W/cm^2). PL is measured in the 12-220 K temperature ranges in a cryostat (Cryomech) model 510.

3. Experimental results and discussion

Fig.1 shows the PL spectra at different temperatures, recorded at the excitation level of 600 W/cm^2. A small full width at half maximum (FWHM) equal to 30 meV is measured for the GS transition in investigated structure at 12K. For the GS optical transition the FWHM did not change at low temperature 12-150 K and then increases with temperature rising above 150 K. This stability of the FWHM at low temperature indicates a uniform distribution of QD sizes and of the level of strain in the investigated MQW system.

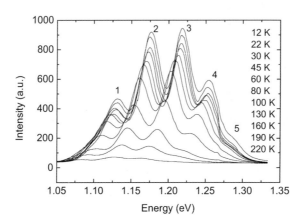

Figure 1 PL spectra measured at different temperatures, indicated on the figure.

The GS in the QD's is characterized by the low energy PL band centered at 1.131 eV. The four higher energy PL bands appear at an excitation level > 150 W/cm^2. The red shift of the PL peaks, connected with the GS and all ES, in temperature range 12-200 K is the same and equal to ~50 meV. The energy difference $\Delta h\nu_{max}$ between optical transitions decreases with increase excited state numbers and is equal to 45.3, 41.8, 37.0 and 29.7 meV. So the localized levels in QD's are not equidistant. The value of $\Delta h\nu_{max}$ did not change in the temperature range of 12-150K.

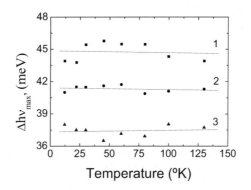

Figure 2 Temperature dependences of the $\Delta h\nu_{max}$ for optical transitions: 1ES-GS (1), 2ES-1ES (2), 3ES-2ES (3).

The information concerning electron (hole) binding energies can be received from the analysis of the thermal quenching process for GS and ES emission bands (Fig.3).

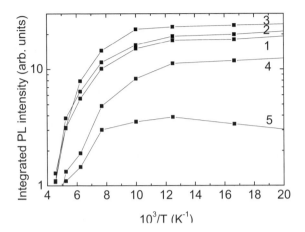

Figure 3. The dependences of spectral integrated intensities on temperature for PL bands connected with transitions: GS (1), 1ES (2), 2ES (3), 3ES (4) and 4ES (5).

The temperature dependence of the PL intensity could be associated with a thermal ionisation of carriers from localized levels into the wetting layer and a substantial contribution of non-radiative recombination. It is found that the activation energies of thermal quenching for GS, 1ES and 2ES bands are the same and equal to 52-54 meV. The corresponding values for 4ES and 5ES optical transitions are smaller and equal to 43 and 30 meV, respectively.

A similarity of the quenching of the GS, 1ES and 2ES emission bands suggests that one common localized level is involved in these spectral lines. Most probably this common level is a single GS electron level. In this case, the ground state line is associated with transitions from that level to the ground state level of holes. An excited state emission is generated by the transitions from the same electron level to excited state levels of holes. For final conclusions concerning the QD energy diagram it is necessary to continue this investigation.

Acknowledgments

This work was supported by CONACYT (Project 33427-U and International cooperation project Mexico-USA), as well as CGPI – IPN Mexico.

References

[1] V. A. Egorov, V. N. Petrov, N. K. Polyakov, G. É. Tsyrlin, B. V. Volovik, A. E. Zhukov and V. M. Ustinov, 2000, *Tech. Phys. Lett.* **26**, 631-635.
[2] A. Polimeni, A. Patane, M. Henini, L. Eaves, P. C. Main, 1999 *Phys. Rev.* B **59**, 5064-68.
[3] G. T. Liu, A. Stintz, H. Li, K. J. Malloy, L. F. Lester, 1999 *Electron. Lett.* **35**, 1163-68.

Guided surface-acoustic-wave modes in AlN layers grown on SiC substrates

Y Takagaki, P V Santos, E Wiebicke, O Brandt, H-P Schönherr and K H Ploog

Paul Drude Institute for Solid State Electronics, 10117 Berlin, Germany

Abstract. The operation of surface-acoustic-wave (SAW) delay lines is demonstrated at frequencies around 10 GHz using AlN layers on a SiC substrate. The SAW propagation is mediated by fundamental and confined Rayleigh modes, the latter of which originates from the mismatch of sound velocities between AlN and SiC. From the dispersion curves of the guided modes, we estimate the elastic constant c_{44} of SiC to be 1.7×10^{11} N/m^2.

1. Introduction

High-frequency filters based on a surface acoustic wave (SAW) are a vital component in communication systems. The everlasting demand to increase the transmission frequency requires the SAW devices to operate at higher frequencies. To raise the operation frequency, one has to either reduce the wavelength λ_{SAW} of the SAW or use a material having a large SAW velocity v_{SAW}. For the latter approach, among those materials having the highest sound velocities after diamond, AlN [1-3] and SiC are at present available with good quality. In addition, AlN is suited for high-frequency SAW devices because of its relatively large electromechanical coupling coefficient k_{eff}^2. In table 1, we compare v_{SAW} and k_{eff}^2 in AlN, SiC, LiNbO$_3$, and GaAs. Notwithstanding the large v_{SAW}, the disappointingly weak electromechanical coupling in SiC cannot be overlooked.

In this paper, we investigate the SAW propagation in AlN layers grown on SiC substrates. The fact that SAWs travel faster in SiC than in AlN results in a confinement of acoustic waves in the slow-velocity overlayer. The guiding of the SAWs gives rise to the appearance of higher-order Rayleigh modes as well as the normal SAW mode in piezoelectric materials. We demonstrate that the present heterostructure allows us to exploit both the large v_{SAW} of SiC and k_{eff}^2 of AlN.

2. Experimental results

Plasma-assisted molecular-beam epitaxy was employed to grow AlN layers on semi-insulating 4H-SiC(0001) substrates. The growth conditions are described in Ref. [4]. All

Table 1. SAW velocity v_{SAW} and electromechanical coupling coefficient k_{eff}^2 in c-oriented AlN and 4H-SiC, Y-cut LiNbO$_3$, and (001)-oriented GaAs.

		AlN	SiC	LiNbO$_3$	GaAs
v_{SAW}	(m/s)	5790	6832	3488	2867
k_{eff}^2	(10^{-3})	2.5	0.112	23.2	0.593

the layers exhibited no free carriers and were electrically highly resistive. We fabricated single-finger interdigital transducers using electron-beam lithography and lift-off techniques. The electrodes were made of a 25-nm-thick Al film and a 6-nm-thick Ti adhesion layer. Two identical transducers were arranged to form a delay line. The SAW transmission is isotropic in the C-plane of the hexagonal crystals. In our devices, the SAW propagation was set to be along the [1$\bar{1}$00] crystallographic direction as it is one of the cleavage directions. The centre-to-centre distance between the transducers was 0.5 mm. The transmission and reflection characteristics of the delay lines were evaluated using an HP8720D network analyser. All the data presented in this paper were obtained from delay lines fabricated on a 1-μm-thick AlN film.

In figure 1, we show the transmission and reflection characteristics in delay lines having λ_{SAW} = 1.0 and 0.5 μm. Two clear resonances are found in both of the devices. The surface of a bulk material normally supports one surface mode (first-order Rayleigh mode). In contrast, multiple resonances are found in the AlN/SiC heterostructure as the interference of the reflection of acoustic waves from the fast-velocity substrate produces guided surface modes, see the inset of figure 1(b). These guided modes are an acoustic counterpart of the confined electromagnetic modes in a waveguide. The most noticeable similarity is that the ith-order Rayleigh mode emerges when λ_{SAW} is less than $d/(i-1)$.

When the mass-loading effect [5] is negligible, v_{SAW} can be estimated using λ_{SAW} and the resonance frequency f_{SAW} as $v_{SAW} = \lambda_{SAW} f_{SAW}$. In figure 2, we plot v_{SAW} (open circles) as a function of λ_{SAW}. Here, we have normalized λ_{SAW} by the AlN layer thickness d (= 1.0 μm). It becomes clear that three branches of the Rayleigh modes have been observed in our devices. As expected, the second- (labelled R_2) and third- (labelled R_3) order Rayleigh modes are present for λ_{SAW}/d below approximately 1 and 0.5, respectively.

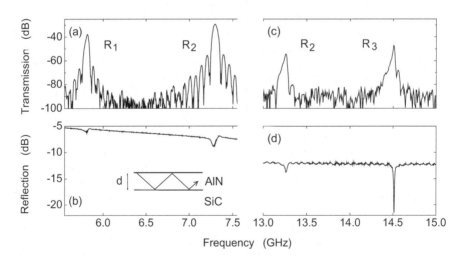

Figure 1. Transmission and reflection characteristics of delay lines fabricated on a 1-μm-thick AlN layer grown on SiC. The SAW wavelength λ_{SAW} is 1.0 μm for (a) and (b) and 0.5 μm for (c) and (d). The number of finger pairs is 100 and 160 for λ_{SAW} = 1.0 and 0.5 μm, respectively. The ith-order Rayleigh modes is labelled R_i. The inset in (b) illustrates the confinement of acoustic waves in the AlN overlayer.

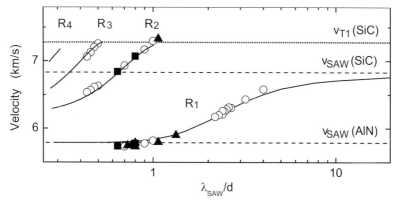

Figure 2. Velocities of the SAW modes in a 1-µm-thick AlN layer grown on SiC. The SAW wavelength λ_{SAW} is normalized by the thickness d of the AlN layer. The circles, triangles, and squares correspond to the fundamental mode and the third and fifth harmonics, respectively. The results of a numerical calculation are shown by the solid lines. The dashed lines indicate the SAW velocities v_{SAW} in SiC and AlN. The slow bulk transverse velocity v_{T1} in SiC is indicated by the dotted line.

3. Discussion

The solid lines in figure 2 show the numerically calculated dispersion curves. The SAWs decay exponentially with distance into the substrate. The mode energy is hence localized at the surface within a depth of about one wavelength. As a consequence, v_{SAW} decreases when the slow-velocity overlayer becomes more influential for smaller λ_{SAW}. The dashed lines indicate the SAW velocities in AlN and SiC. The SAW velocity of the first-order Rayleigh mode (R_1) changes from that of SiC to that of AlN when λ_{SAW}/d is varied from about 10 to about 1. Contrary to this dispersion of the first-order Rayleigh mode, the higher-order Rayleigh modes can propagate faster than the SAW mode in SiC. Thus, the higher-order Rayleigh modes are advantageous to achieve high operation frequencies, given the fact that the SAW excitation efficiency appears to be almost identical for all the orders of modes or even better for the higher-order modes, see figure 1.

The SAW velocity whenever the ith-order Rayleigh mode emerges at $\lambda_{SAW}/d = (i-1)^{-1}$ is identical to the bulk transverse velocity v_{T1} with polarization parallel to the c-axis in SiC, as indicated by the dotted line in figure 2. This slowest bulk sound velocity, in principle, sets an upper limit to the velocity of nonleaky SAWs since an acoustic wave can escape into the bulk when its velocity exceeds v_{T1}. As v_{T1} is determined solely by the elastic constant c_{44}, $v_{T1} = (c_{44}/\rho)^{1/2}$ with ρ being the mass density, the value of this parameter in SiC is deduced to be 1.7×10^{11} N/m^2.

The experimental data in figure 2 contain the velocities associated with the third and fifth harmonic modes (filled symbols). In spite of the fact that the transducers were equipped with single-finger gates, the transmission associated with the harmonic modes was surprisingly strong in our devices. An example of such a transmission spectrum is shown in figure 3. Here, the period of the transducers, i.e., λ_{SAW} for the fundamental mode, is 3.2 µm, so that the wavelength for the third harmonic mode (1.07 µm) approximately satisfies the condition for the appearance of the second-order Rayleigh mode. However, λ_{SAW} for the fifth harmonic mode (0.64 µm) is still too large to have the third-order Rayleigh mode generated. For comparison, we also show the transmission

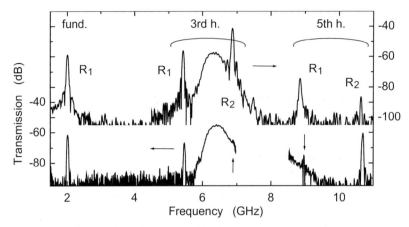

Figure 3. Transmission spectra of delay lines having $\lambda_{SAW} = 3.2$ μm and $d = 1$ μm. The lower (upper) curve was obtained using single- (split-) finger transducers. The SAW wavelength of the 3rd and 5th harmonics is a third and a fifth in comparison to that for the fundamental mode, respectively.

spectrum obtained using split-finger transducers (the upper curve). The roughly comparable amplitude of the transmission peaks at 5.45 GHz is remarkable considering that the single-finger transducers are not supposed to generate the third harmonic mode. In any case, the feasibility to excite the harmonic modes evidences the strong electromechanical coupling in AlN.

4. Conclusions

We have demonstrated SAW transmission in the SHF frequency range using transducers having submicron wavelengths fabricated on AlN/SiC structures. The guided Rayleigh modes in the composite system can propagate even faster than the Rayleigh mode in the fast-velocity substrate. The higher-order Rayleigh modes are hence advantageous for high-frequency SAW devices. The present materials system allows us to exploit the large SAW velocity of SiC while utilizing the large electromechanical coupling coefficient of AlN.

Acknowledgements

Part of this work was supported by the Deutsche Forschungsgemeinschaft and by the NEDO collaboration program.

References

[1] Okano H, Tanaka N, Takahashi Y, Tanaka T, Shibata K and Nakano S 1994 Appl. Phys. Lett. **64** 166
[2] Deger C, Born E, Angerer H, Ambacher O, Stutzmann M, Hornsteiner J, Riha E and Fischerauer G 1998 Appl. Phys. Lett. **72** 2400
[3] Khan A, Rimeika R, Ciplys D, Gaska R and Shur M S 1999 phys. stat. sol. (b) **216** 477
[4] Brandt O, Muralidharan R, Waltereit P, Thamm A, Trampert A, von Kiedrowski H and Ploog K H 1999 Appl. Phys. Lett. **75** 4019
[5] Takagaki Y, Wiebicke E, Santos P V, Hey R and Ploog K H 2002 Semicond. Sci. Technol. **17** 1008

Photoluminescence from deep levels in Fe-doped InP substrates

M. Yamada and M. Fukuzawa

Department of Electronics and Information Science,
Kyoto Institute of Technology, Kyoto 606-8585, Japan

Abstract. Photoluminescence (PL) measurements have been carried out at room temperature in commercially-available 4-inch-diameter Fe-doped semi-insulating InP substrates with a high-sensitivity PL mapping system using photon-counting technique. Not only a sharp PL line due to band-to-band transition but a broad PL line due to recombination via deep levels are successfully detected at room temperature. It is found from a comparison between band-to-band and deep-level PL maps measured near striation pattern that there is a correlation between their PL intensities.

1. Introduction

Semi-insulating (SI) InP substrates are technologically becoming of great interest because they are key materials for high-speed electronics and optoelectronic applications. In response to the strong demands from market, the size of commercially available wafer becomes up to 4 inch [1]. A key point that such large-diameter wafer is successfully provided is to remain uniform throughout the individual substrate and from substrate to substrate. The uniformity of SI InP is, however, currently limited by the need of doping with Fe at a level close to the solubility limit ($\sim 10^{17} cm^{-3}$) in order to obtain sufficient high resistivity material. In Fe-doped SI InP crystals grown by the conventional LEC techniques, a pronounced doping gradient between the top and tail is found due to the low Fe distribution coefficient (~ 0.001). Therefore, the Fe content is a function of the axial position of crystal and hence electrical properties are different from wafer to wafer sliced from the same crystal. In addition to that, other types of nonuniformities are usually detected in Fe-doped SI InP crystals: growth striations, dislocations decorated by microdefects, microprecipitates, embedded in the crystal matrix and sometimes inclusions. These defects have a direct effect on the quality of the devices grown or implanted on the substrates that contain them. Therefore, the characterization of these Fe related defects as well as the Fe content are of great importance in order to fabricate good quality devices.

Both photoluminescence (PL) spectroscopy and scanning photoluminescence (sPL) mapping have proved themselves to be very useful for characterizing undoped and doped InP crystals [2-11]. From the low-temperature PL spectroscopy measurements [2-4,8,9,11], there were found several sharp lines due to a group of nonresolved bound excitons (BE) together with the free exciton, due to conduction band-to-acceptor transition (eA), and due to donor-to-acceptor transition (DA) near the band edge, and a broad line due to recombination via deep levels (DLs). The sPL measurements connecting with BE, eA, and DA near the band edge were widely made both at low temperature [4,11] and at room temperature [4-8,10] to assess uniformity. On the other hand, the sPL measurement connecting with DLs was made only at low temperature [11]. The PL connecting with DLs has not been observed yet at room temperature, because the PL intensity is generally decreased at room temperature and furthermore the recombination via DLs is extremely weak compared with that near the band edge. Therefore, it is very useful to develop a sPL system with high sensitivity in the near infrared region, with which we can measure the PL at room temperature not only near the band edge but at DLs in Fe-doped SI InP substrates.

In this paper, we will introduce the high-sensitivity sPL system using the photon counting technique recently developed by us and then demonstrate the DL spectra and maps measured at room temperature in commercially-available 4-inch-diameter Fe-doped SI InP substrate.

2. Experimental

Figure 1 shows the block diagram of the high-sensitivity sPL system using the photon counting technique recently developed by us. The exciting light source used here is a He-Ne laser operating at $\lambda=633$ nm with the power of 17 mW. The laser beam is introduced through a bending mirror, an interference filter, a beam splitter, and an objective lens of microscope, onto a sample mounted on a X-Y-Z stage. The interference filter is used to eliminate unwanted discharge light from the He-Ne laser. The PL light is collected with the same objective lens and then introduced through the beam splitter, a low-pass filter, and a focusing lens, into a monochromator (f =100mm, F =3) with a 600 grooves/mm grating blazed at $\lambda=800$ nm. The low-pass filter is used to stop the exciting light reflected from the sample. The output light is detected with a cooled (-80 °C) photomultiplier whose spectral response is fairly flat from $\lambda=400$ nm to 1600 nm. The output pulse signal from the photomultiplier is counted with a photon counter. It should be noted here that the photon counting technique is the best one for detecting very weak light. A personal computer is used to control the photon counter, the monochromator, and the X-Y stage.

The PL spectroscopy and sPL measurements were carried out in air at room temperature. The 20x objective lens was mainly used to focus the laser beam down to about 5 µm diameter on the sample. The accumulation time of photon for a fixed wavelength and a fixed X-Y position was varied depending on the PL intensity. The typical accumulation times for I_{BB} and I_{DL} later shown in Fig.4 were 5 msec and 1 sec, respectively.

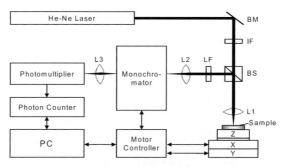

Figure 1. Brock diagram of high-sensitivity scanning photoluminescence mapping system. BM: bending mirror, IF: interference filter, BS: beam splitter, LF: low-pass filter, L1-L3: lens.

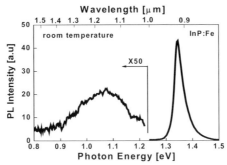

Figure 2. Typical PL spectrum measured at room temperature in a commercially- available 4-inch-diameter Fe-doped SI InP substrate.

3. Results and Discussion

Typical PL spectrum measured at room temperature is shown in Fig. 2. It is found that a strong sharp line and a weak broad line are observed at 1.34 eV and at 1.08 eV, respectively, which may be identified to be due to band-to-band transition and due to recombination via DLs. It was found from the low-temperature PL experiment [11] that as increasing the Fe content, the DL intensity (I_{DL}) was increased whereas the band-to-band intensity (I_{BB}) was decreased. In order to check this result, we have made the room-temperature PL experiment in the front substrate with low Fe content and the tail substrate with high Fe content sliced from the same ingot. It is found contrarily that both I_{DL} and I_{BB} are decreased as increasing the Fe content. This leads us to suggest that the recombination centers are more complex and the recombination processes at room temperature are different from those at low temperature. The origin of DLs is not clear at the present stage, although it seems to be associated with Fe.

The sPL whole-wafer map was measured while fixing the monochromator energy to 1.34 eV (I_{BB}). The result is shown in Fig. 3. It is found that a concentric striation pattern is observed besides bright spots. The concentric striation pattern may be considered to be due to doping fluctuation during crystal growth [6, 10]. The bright spots can be either attributed to a reduced density of recombination centers, e.g. Fe or due to an enhanced concentration of shallow defects involved in radiative transitions [5-7,10]. The sPL maps in the narrow striation region marked by A were also measured both at 1.34 eV (I_{BB}) and at 1.08 eV (I_{DL}). The results are shown in Fig. 4. It is found that there is a macroscopic correlation between I_{BB} and I_{DL}; that is, both I_{BB} and I_{DL} are enhanced or reduced while

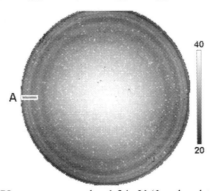

Figure 3. sPL map measured at 1.34 eV (I_{BB}: band-to-band transition).

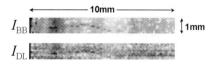

Figure 4. sPL maps measured at 1.34 eV (I_{BB}) and at 1.08 eV (I_{DL}) in the narrow region marked by A in Fig. 3.

corresponding to the striation pattern. This result is opposite to the feature of bright spots that was observed in a microscopic low-temperature sPL map of I_{BB} [11]. The origin of the bright spot is considered to be a Fe-depleted region where Fe is gettered by defects, most probably dislocations, through their strain field in the microscopic region. However, we cannot extend this idea applied in the microscopic region to the macroscopic region. The interactions among Fe dopants, residual impurities, and crystal defects could be more complex, and therefore the macroscopic feature observed here might be explained as their average effect. Besides striation pattern and bright spots, there are found dark spots and lines, which are probably caused by surface scratches, inclusions, and/or drift of dark count from photomultiplier.

4. Concluding Remark

We have successfully performed room-temperature PL spectroscopy and mapping of recombination via DLs as well as of band-to-band transitions in Fe-doped SI InP substrates. The room-temperature PL spectroscopy and mapping are more convenient compared with the low-temperature one. The deep level PL will give us more information on Fe-dopants, residual impurities, and crystal defects. Along this direction, further study is in progress.

Acknowledgements

The authors would like to express their sincere thanks to Mr. M. Suzuki and Mr. H. Kakui for their assistance in performing the present experiment and in preparing the present manuscript.

References

[1] Hosokawa Y, Yabuhara Y, Nakai R, and Fujita K 1998 *Proc. of 10th Int. Conf. on InP and Related Materials,* IEEE Catalog #98H36129, 34.
[2] Temkin H and Bonner W A 1981 J. Appl. Phys. **52** 397-401
[3] Temkin H, Dutt B V, Bonner W A, and Keramidas V G, 1981 J. Appl. Phys. **53** 7526-7533
[4] Erman M, Gillardin G, Bris J LE, and Renaud M 1989, J. Crystal Growth **96** 469-482
[5] Longère J Y, Schohe K, Krawczyk K 1990 J. Appl. Phys. **68** (2) 755-759
[6] Miner C J, Knight D J, Zorzi J M, and Ikisawa M1995 *Inst. Phys. Conf. Ser.* **135** 181-186
[7] Hirt G Hoffman B, Kretzer U, Woitech A, Zemke D, and Müller 1995 *Proc. of 7th Int. Conf. on InP and Related Materials* IEEE Catlog #95CH35720 33-36
[8] Kang J, Matsumoto F, and Fukuda T 1997 J. Appl. Phys. **81** 905-909
[9] Doğan S, Tűzemen, Gűrbulak B, Ateş A, and Yildrim M 1999 J. Appl. Phys. **85** 6777-6781
[10] Fornari R, Gőrőg T, Jimenez J, De la Puente E, Avella M, Grant I, Brozel M, and Nicholls M 2000 J. Appl. Phys. **88** 5225-5229
[11] Wakahara M, Uchida M, Warashina M, Oda O, and Tajima M 2000 J. Crystal Growth **210** 226-229

Many-body effects as probe of defects presence in heavily doped AlGaAs/InGaAs/GaAs heterostructures

Vas P Kunets[1], Z Ya Zhuchenko[1], H Kissel[2], U Müller[1], G G Tarasov[1], and W T Masselink[1]

[1]Department of Physics, Humboldt-Universität zu Berlin, 10115 Berlin, Germany
[2]Ferdinand-Braun-Institut für Höchstfrequenztechnik, D-12489 Berlin, Germany

Abstract. Strikingly strong many-body enhancement of the oscillator strength for interband transitions is observed in the photoluminescence (PL) of heavily doped pseudomorphic $Al_xGa_{1-x}As/In_yGa_{1-y}As/GaAs$ heterostructures under condition of the $n = 2$ subband filling. The many-body excitations reveal a remarkable stability with respect to thermal and excitation density decay. Such behaviour is addressed the intersub-band coupling in high-density one-component plasma of InGaAs quantum well in the presence of defects localising the heavy holes.

1. Introduction

Modulation-doped pseudomorphic $Al_xGa_{1-x}As/In_yGa_{1-y}As/GaAs$ heterostructures, besides the excellent transistor characteristics with respect to power gain, current, transconductance, and noise performance at high frequencies, occur to be ideally suited for the study of a number of effects fundamental for the semiconductor physics such as the Fermi edge singularity (FES). The defect presence is of extreme importance for the realization of the preferable conditions for the FES observation [1,2]. Thus, on the contrary, the FES origination can serve as a sensitive probe for the detection of the defects presence. Here we present the results of spectroscopic study of the interband optical transitions in a close vicinity of Fermi energy in the conduction band of $Al_xGa_{1-x}As/In_yGa_{1-y}As/GaAs$ modulation-doped structures and their behavior under temperature and excitation density variation. The results give evidence of the many-body nature of a strong complementary PL stabilized by the defects.

2. Experimental details

Pseudomorphic modulation-doped $Al_xGa_{1-x}As/In_yGa_{1-y}As/GaAs$ samples have been grown on semi-insulating (100)-oriented GaAs substrates in a Riber 32-P gas-source molecular-beam epitaxy (GSMBE) system. The $Al_xGa_{1-x}As$ and $In_yGa_{1-y}As$ layer compositions were intentionally held invariable at the level of $x \approx 0.20$ and $y \approx 0.1$, respectively. The Si-doping and the InGaAs quantum well (QW) width d_w were adjusted to meet requirements of second electronic subband ($n_e = 2$) filling. The set of samples with invariable Si doping and the d_w value changing within the range 18 nm < d_w < 27 nm has been grown. The PL was excited by the 514.5-nm line of a cw Ar laser. The excitation density was varied in the range 0.05÷20 W×cm^{-2}. The PL signal was dispersed through a 3/4-m Czerny-Turner scanning spectrometer, with a spectral resolution better than 0.1 meV. The samples were mounted in an Oxford Spectromag 4000 system, which allows measurements at temperatures from 1.7 to 300 K.

3. Results and Discussion

A principally new behaviour in PL of heavily doped pseudomorphic modulation–doped Al$_x$Ga$_{1-x}$As/In$_y$Ga$_{1-y}$As/GaAs ($x = 0.2$, $y \le 0.1$) heterostructures is detected in the case of occupied $n = 2$ electronic subband. Fig.1 depicts the low temperature spectra measured at low excitation density for the set of samples with varying quantum well width d_w (from 8 nm up to 27 nm) at invariable Si doping of supplier Al$_x$Ga$_{1-x}$As layer. The energy origin is placed at the maximum of the high-energy peak in each particular spectrum. The d_w variation allows to change the population of $n = 2$ electronic subband. The ratio ξ of magnitudes for the E_{21} and E_{11} peaks varying within the range 0.1 to 50 mirrors the

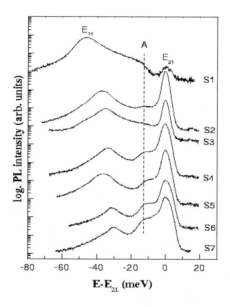

Figure 1. Low temperature PL spectra for samples with varying d_w value: S1 = 8 nm, S2 = 18.2 nm, S3 = 19.7 nm, S4 = 21.4 nm, S5 = 21.5 nm, S6 = 25.4 nm, S7 = 26.7 nm.

change of $n = 2$ subband population. It is clearly seen that if the ξ value reaches nearly 20 a complementary feature A arises below E_{21} transition. This feature grows by intensity, exceeding even E_{11} transition, if ξ takes the value ≈ 30. In order to ascertain the nature of this complementary peak the temperature and excitation density elevation has been performed. Fig.2 depicts the modification in the PL spectrum of S7 sample with temperature increase. It occurs that the strength of A feature grows initially with temperature increase and this peak becomes even more distinct. If temperature reaches 80 K the feature A decays completely revealing the binding energy the order of 7 meV. The similar behaviour is detected also at the excitation density elevation.

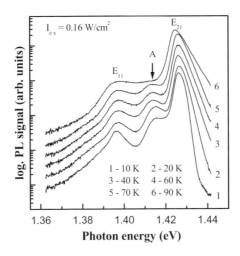

Figure 2. Temperature dependent spectra for sample S7.

Such behaviour of complementary feature A is observed for the first time. Indeed for the case of empty $n = 2$ electronic sub-band the PL study has shown convincingly that the carrier concentration and the excitation density strongly influence the additional feature appearance, immediately below the $n = 2$ bare exciton, in the presence of hole localization. This particular feature has been addressed to the FES manifestation. At least the FES evolves in PL spectra under appropriate light illumination being absent at the lowest excitation density. The FES excitation density dependence cannot be addressed to the presence of inner electric field and has to be bounced back to the properties of 2DEG unexplored. The model of Fano resonance between the many-body excitations of Fermi sea and the discrete excitonic state of $n = 2$ conduction sub-band is also not very plausible for the explanation of both the excitation density and temperature dependences of the feature observed experimentally.

An additional peak has been also detected earlier in the PL spectra of pseudomorphic GaAs/$In_{0.2}Ga_{0.8}As$/$Al_{0.25}Ga_{0.75}As$ modulation-doped single QW's [3]. It was ascribed to the strong coupling between the InGaAs QW and the potential well formed in the AlGaAs barrier due to planar doping. This coupling could produce the hybrid states in the conduction band giving rise to additional transitions in the emission spectrum. One of these transitions between hybrid electronic state and heavy-hole state could be responsible for the extra PL peak observed below the E_{21} transitions. The strongly nonmonotonic temperature dependence of the peak strength, however, makes such an assignment somewhat questionable. Indeed, the calculated electron-hole wave function overlaps exhibit no substantial variation within the investigated temperature range (2 K- 51 K), thus no reasonable explanations remain for the strong quenching of this peak at temperatures as high as 40 K.

Our novel findings concerning the spectroscopy of the Fermi edge states in pseudomorphic modulation-doped AlGaAs/InGaAs/GaAs heterostructures are as follows: i) The temperature behaviour of the E_{21} transition provides unequivocal evidence of the FES presence for the $n = 2$ electron subband. The E_{21} peak amplitude falls abruptly down under the temperature increase within the range of 20 K. Then it becomes weakly temperature dependent up to the temperature $T = 80$ K; ii) The large strength of the additional feature is observed even at the lowest temperatures. It can not be of impurity nature while the doping level in the AlGaAs supplier layer is held constant under variation of the InGaAs QW width; iii) The strength of additional PL feature is strongly dependent on the $n = 2$ subband filling. The 2DEG concentration was derived from the low temperature Hall measurements; iv) The high stability of the feature under the temperature and excitation density elevation allows to evaluate the binding energy. It is about of 7 meV; v) The A feature reveals the non-monotonous temperature and excitation density behaviour, non-typical to the well-known FES manifestations.

Nevertheles this feature can be related to the many-body effects. Energetically this feature can be addressed the transition from the Fermi edge states in the $n = 1$ electronic band to the localized state of heavy holes in the valence band. In this case the binding energy of localized hole can be directly estimated from the temperature dependent PL (see Fig.2). Indeed in the PL spectrum the corresponding FES features will arise at the E_{21} transition energy and at the energy, shifted toward lower energies by value of the localization energy for heavy holes, as follows from Fig.2. The energy by which the A feature is shifted with respect to the energy of the E_{21} transition is found to be 14 meV. We assume, that the temperature and excitation density dependences of the feature A observed from the samples under investigation, can be caused by a strong enhancement of the oscillator strength for the transitions at the Fermi edge in the E_1 band due to the inter-subband Coulomb scattering of electrons [4]. Besides the non-Coulomb scattering can also efficiently control the formation of FES in multi-subband heterostructures [5]. The alloy dependent inter-subband scattering may easily prevail over multiple diffusions from charged valence holes expected by many-body processes. The hole localization is of importance for several reasons. The localized hole states may be viewed as being constructed from states with a range of momenta and therefore allow the PL from the vicinity of the Fermi level to be observed. The heavy holes can be localized by a form of disorder, typically alloy fluctuations in InGaAs QW. The PL enhancement is due to the mixing of the subband states by the photoexcited hole and due to virtual excitations of electron-hole pairs involving the higher subbands.

So, the main points of novelty are: i) the FES enhancement from E_F of $n = 1$ electronic subband is observed under condition of the $n = 2$ subband population; ii) heavy hole localization energy is directly observed in the FES development; iii) the many-body enhancement is nonmonotonicly dependent on temperature and excitation intensity. The behavior observed is assigned to the inter-subband scattering in one-component plasma in the presence of strong electron-hole correlations. The defects localizing the heavy hole become reachable for investigation through the many-body effects.

References
[1]   Skolnick M S, Whittaker D M, Simmonds P E et al 1991 Phys. Rev. B 43 7354-7
[2]   Kissel H, Zeimer U, Maaßdorf A et al 2002 Phys. Rev. B 65 235320-6
[3]   Abbade M L F, Iikawa F, Brum J A et al 1996 J. Appl. Phys. 80 1925-30
[4]   Mueller J F, Ruckenstein A E, and Schmitt-Rink S (unpublished)
[5]   Mélin T and Laruelle F 2000 Phys. Rev. Lett. 85 852-4

Two-dimensional mapping of resistivity in semi-insulating GaAs wafers with large diameter using a nondestructive technique

M. Fukuzawa and M. Yamada

Dept. of Electronics and Information Science, Kyoto Institute of Technology
Matsugasaki, Sakyo-ku, Kyoto 606-8585, Japan

Abstract. By using a nondestructive resistivity measurement (NDRM) technique, which is based on frequency analysis of charge response in metal-insulator-semiconductor-metal (MISM) structure, the two-dimensional map of resistivity has been measured in 6-inch-diameter semi-insulating GaAs wafer. The NDRM map exhibits not only macroscopic variation over whole wafer but also local fluctuations originating from lineage and cell structures of dislocations, which are found by infrared transmittance measurement.

1. Introduction

According to the strong demand from the telecommunication market, the production volume of semi-insulating (SI) GaAs wafer, which is widely used to fabricate a variety of high-speed electronic devices, is required to increase. Consequently, the diameter of the wafer is increasing up to 6 inch for increasing the production number of device without the increase in process cost. It is important for such large-diameter wafer to achieve good uniformity of resistivity over whole wafer and suppress its variation from wafer to wafer. In this case, nondestructive method is very useful to characterize the profile and two-dimensional (2D) distribution of resistivity in the wafer before the process of device fabrication. However, it was difficult to measure the resistivity in the wafer to be processed, because the resistivity was conventionally measured by destructive methods such as van de Pauw method.

In 1991, Stibal et al. [1] developed a sophisticated nondestructive method to quantitatively measure the resistivity; that is, so-called as time-dependent charge measurement (TDCM) method. At present, the contactless resistivity measurement system based on the TDCM method is commercially available [2] and it is being recognized that nondestructive method is very useful to measure the profile and 2D distribution of resistivity in the large-diameter wafer. Since the charge response of the sample is directly measured at specific timing by a charge amplifier in the TDCM method, unideal transfer function of charge amplifier may affect the performance of TDCM apparatus. By making major modification in signal processing of charge response in the TDCM technique, we have recently developed another nondestructive resistivity measurement (NDRM) method [3,4] based on frequency analysis of charge response in order to eliminate the influence of unideal transfer function of the charge amplifier essentially. In this paper, we present the features of our NDRM technique and demonstrate the performance of the NDRM apparatus with experimental results made in 6-inch-diameter SI GaAs wafer.

2. Comparison between TDCM and NDRM methods

A metal-insulator-semiconductor-metal (MISM) structure, which consists of a metal probing electrode, an insulating air-gap, a thin semi-insulating wafer to be examined, and a metal back electrode on which the wafer is placed, is used in both NDRM and TDCM methods. Figure 1 shows (a) a schematic of MISM structure, (b) its equivalent circuit and (c) a schematic of charge measurement with a charge amplifier. Here, $C_a=\varepsilon_0 A/d_a$, $C_s=\varepsilon_0\varepsilon A/d_s$, and $R_s=d_s\rho/A$, where A is the area of probing electrode, ε_0 is the permittivity of the vacuum, d_a is the air-gap thickness, d_s, ε and ρ are the thickness, the dielectric constant and the resistivity of the sample, respectively.

In the TDCM, a unipolar step wave is applied to the MISM structure and the charge response $Q(t)$ is measured by a charge amplifier just after the step wave is applied ($t=0$) and both when the C_a is fully charged and when C_s is discharged ($t=\infty$). The output voltage of the charge amplifier $V(t)$ is assumed to be equivalent to $Q(t)/C_F$ although it is modulated by the transfer function $G(\omega)$ of the charge amplifier. The resistivity is derived from the $V(0)$, $V(\infty)$, and the charge relaxation time τ_e. This method has an advantage that it is not necessary to estimate d_a and d_s. It is noted that the $V(0)$ and $V(\infty)$ is influenced by $G(\omega)$ because they are the fastest and slowest responses of the step wave, respectively. Furthermore, the unipolar step wave accumulates a certain amount of charge in the capacitances of C_a and C_s in the MISM structure at the end of each measurement and therefore it should be forced to discharge before starting the next measurement. Therefore, the TDCM apparatus should be carefully designed to overcome these issues discussed above.

On the other hand, we employ a bipolar square wave in the NDRM method instead of the unipolar step wave used in the TDCM, because it does not accumulate any charge at the end of each measurement. Instead of measuring the $V(0)$, $V(\infty)$ and τ_e, we continuously digitize the $V(t)$ as a function of a discrete time with a fast A-D converter and then analyze it in frequency domain by Fourier transform. The contribution of $Q(t)$ will appear only in the odd terms in the discrete frequency transform $V(\omega_i)$ because the bipolar square wave is odd function and there is no initial charge in the MISM structure. Therefore, we can calculate the resistivity by the best fitting only of the odd terms to the charge response function in the frequency domain not affected by the $G(\omega)$ which appears in the even terms of frequency component. The measurable range of resistivity of the NDRM apparatus is from 10^5 to 10^{10} Ωcm, the spatial resolution is down to 500 μm, the repeatability is less than 1%, which meets the performance of commercial TDCM ones.

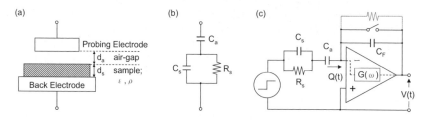

Figure 1. (a) a schematic of MISM structure, (b) its equivalent circuit and (c) a schematic of charge measurement by a charge amplifier.

3. Experimental Results

Figure 2 shows (a) 2D map of resistivity measured in a 6-inch-diameter wafer of SI GaAs crystal and (b) an enlarged map of the 30×30mm white rectangle region shown in the

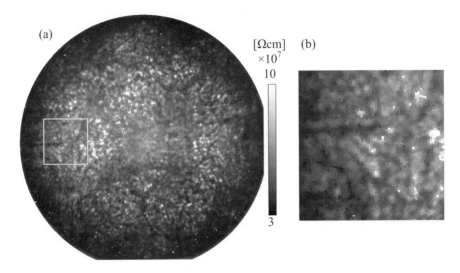

Figure 2. (a) 2D map of resistivity measured in a 6-inch-diameter wafer of SI GaAs crystal and (b) an enlarged map of the 30×30mm white rectangle region shown in the figure (a).

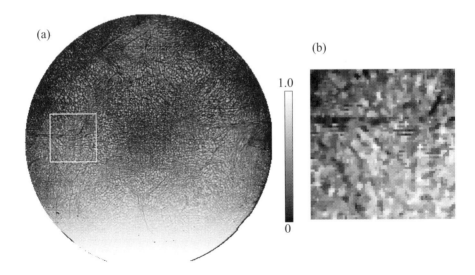

Figure 3. (a) infrared (1.3µm) transmittance topograph and (b) its enlargement of the 30×30mm region in the same wafer shown in Fig.2.

figure (a). The resistivity mapping was done at 500 μm scanning step. The measurement and positioning time per data point was about 400 ms. It is found in Fig.2 (a) that the whole-wafer distribution of resistivity revealed weak four-fold symmetry. The amount of resistivity averaged over the wafer was 7.5×10^7 Ωcm. The resistivity was decreased at the peripheral regions compared with that at the center. Figure 2 (b) exhibits local fluctuations with characteristic forms. For examples, we can clearly find that the lineage-like patterns with low resistivity were originated from the $\langle 011 \rangle$ peripheral regions of the wafer and extended to the center branched off in several directions. Furthermore, there were small regions with high resistivity surrounded by low resistivity regions, which formed cell-like structure.

Figure 3 shows infrared (1.3μm) transmittance topograph measured in the same wafer shown in Fig.2. It is well known that the infrared transmittance of GaAs wafer reflects the lineage and cell structures of dislocations decorated by deep levels such as EL2, which may cause the fluctuation of resistivity. From the Fig. 3 (a), the infrared transmittance also reveals weak four-fold symmetry, which is in good agreement with the whole-wafer distribution of resistivity shown in Fig. 2 (a). The lineage and cell structures of dislocations are clearly found in Fig.3 (b), which are clearly correlated with the local fluctuation of resistivity shown in Fig. 2 (b). It should be noted that not all the defect structure causes the resistivity fluctuations because the processes of crystal growth and wafer fabrication may strongly vary from wafer to wafer. However, it was confirmed from the comparison between Fig. 2 (b) and Fig. 3 (b) that the 2D resistivity map measured by NDRM method could detect not only the macroscopic variation over whole wafer but also local fluctuations associated with defect structures.

4. Concluding Remark

By using a nondestructive resistivity measurement (NDRM) technique, which is based on frequency analysis of charge response in metal-insulator-semiconductor-metal (MISM) structure, the 2D map of resistivity has been successfully measured in 6-inch-diameter SI GaAs wafer. It was confirmed that the NDRM technique has enough sensitivity and spatial resolution to detect not only macroscopic variation over whole wafer but also local fluctuations associated with the lineage and cell structures of dislocations. Therefore, the NDRM technique, as well as TDCM technique, can be used as powerful tool to characterize the resistivity uniformity of the large-diameter SI GaAs wafers to be processed.

References

[1] Stibal R, Windscheif J, Jantz W 1991 Semicond. Sci. Technol. 6 995-1000
[2] iaf-COREMA 2000, supplied by Hologenix, 15301 Connector Lane, Huntington Beach CA 92649, USA.
[3] Yamada M, Fukuzawa M, Akita M, Herms M, Uchida M and Oda O, 1999 Proc. of the 10th Conf. on Semiconducting and Insulating Materials (SIMC-X) 45-48
[4] Fukuzawa M, Yoshida M, Yamada M, Hanaue Y, and Kinoshita K 2002 Materials Science and Engineering B91-92 376-378
[5] Fukuzawa M and Yamada M, in preparation.

Sonic-stimulated temperature rise around dislocation

R.K. Savkina, A.B. Smirnov
Institute of Semiconductor Physics of NAS of Ukraine, Kiev

Abstract. The nonuniform temperature distribution in crystal surface during ultrasonic loading was detected. This effect was associated with sonic-stimulated temperature rise around dislocations and heating of the nonperfect regions of investigated samples. The dislocation moved in ultrasonic field was considered as a linear thermal source and the temperature distribution around dislocation was calculated. We determined conditions of the discrete and continuous distribution of thermal sources. The discrete distribution of thermal sources is realised for $Cd_xHg_{1-x}Te$ solid solutions at average dislocation density $\sim 10^{10}$ m^{-2}. We also discuss the possibility to use the investigated effect as the base of the non-destructive technique for structural perfection control of $Cd_xHg_{1-x}Te$ alloys.

1. Introduction

The physical origin of the ultrasonic oscillation effect on the crystal with dislocations is connected with intensive sonic-dislocation interaction which can be explained by vibrating string model of Granato-Luecke [1]. Such interaction has resulted in effective transformation of the absorbed ultrasonic energy to the internal vibration states of the crystal and has stimulated numerous defect reactions [2-8].

Previous studies of the effect of ultrasonic loading on electrical and photoelectric parameters of $Cd_xHg_{1-x}Te$ alloys have shown essential sensitivity this material to sonic vibrations. Processes of ultrasonic-induced transformation of crystal defect structure and the modification of charge-carrier scattering conditions in $Cd_xHg_{1-x}Te$ crystals were studied in [6,7]. The research of correlation between the value of sonic-stimulated effects and the state of the defect system in this material are established in [8]. The result of these investigation has allowed to formulate the general regularities of sonic-stimulated processes in this material and to propose an effective procedure of a direct influence on device performance which consists in ultrasonic treatment.

In this paper we want to present the result of the sonic-stimulated temperature rise investigation for $Cd_xHg_{1-x}Te$ solid solutions and to discuss the possibility of such effect application.

2. Experimental procedure and results

The *n*-type $Cd_xHg_{1-x}Te$ (x=0.2) alloys were subject of the investigation. The samples were cut from a single-crystal ingot grown by the Bridgman method. The linear dimensions of the samples after polishing and chemical etching were 8x2 mm^2 and the thickness was about 1 mm. The dislocation density was controlled by the optical microscope NV2E (Carl Zeiss Jena). Its value varies from $10^9 m^{-2}$ to $10^{11} m^{-2}$ for all investigated samples.

It was used an *in-situ* regime of ultrasonic loading. Longitudinal ultrasonic vibrations with frequency f_{US} = (5÷7) MHz and intensity $W_{US} \leq 0.5$ W/cm² were generated by a LiNbO₃ transducer (35° Y-cut) and fed to the sample. The pre-threshold intensity regime was used. Switching off the ultrasonic loading led to a relaxation of all sample parameters to their original value in 10^3 s.

We placed several thermocouples along investigated samples and detected the nonuniform macroscopic temperature distribution in their surface during *in situ* ultrasonic loading. The value of the sonic-stimulated temperature deviation from the average temperature in crystal was a 10÷20 K at ultrasonic intensity $W_{US} \leq 0.5$ W/cm². We made the chemical selective etching of samples and determined by optical microscope investigation that nonhomogeneous distribution of extended crystal defects such as dislocations is characteristic for investigated samples and the sonic-stimulated heating is observed in the nonperfect regions.

3. Discussion

We suggest that the physical origin of the nonuniform heating of the crystal is a selective absorption of the ultrasonic energy at dislocations. According to Granato-Luecke model the dislocation moves in ultrasonically loaded crystal as a vibrating string and periodical compression and expansion of the crystal around dislocation takes place. During fast ultrasonic deformation, a vast amount of energy is put into the material during a relatively short period time. Through the damping of dislocation motion by electrons and phonons the kinetic energy of a dislocation is dissipated into heat [9]. Since the processing time (period of ultrasonic wave is ~10^{-7}s) is shorter than the relaxation time of sonic-stimulated heating, the heat may not have sufficient time to become dissipated throughout the sample or radiated into the environment. So, the temperature around dislocation can become considerably high.

We consider the dislocation moved in ultrasonic field as a linear thermal source. Temperature distribution around dislocation line L can be written as:

$$T = T_0 + \int_t d\tau \int_L \frac{W_0}{\rho c} G dl \qquad (1)$$

In the above expression, $G(r,t)$ is Green function of the point thermal source, W_0 is the dissipated ultrasonic energy, ρ is the crystal density, C is the crystal heat, T_0 is an average equilibrium temperature in the crystal. If $L \to \infty$ and $t \to \infty$, the stationary temperature field around dislocation moved with velocity v along x-direction is :

$$T(x,z) = T_0 + \frac{W_0}{2\pi x} \exp\left(\frac{x}{R_0}\right) K_0\left(\frac{r}{R_0}\right)$$
$$r = \left(x^2 + z^2\right)^{1/2} \qquad (2)$$

where χ is thermal conductivity, K_0 is zeroth order modified Bessel function, $R_0 = 2\chi/\rho Cv$. Dislocation line is normal to plane xz. The regularity of the temperature distribution around dislocation for cases $r << R_0$ and $r >> R_0$ can be written as:

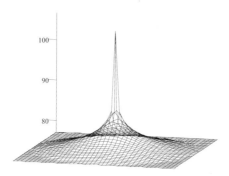

Figure 1. Sonic-stimulated temperature distribution around dislocation line calculated by (3): equilibrium crystal temperature is $T_0=77$ K, sonic-stimulated temperature increase is $\Delta T = T - T_0 = 25$ K, intensity of the ultrasonic loading is $W_{US}=0.4$ W/cm^2.

$$T = T_0 + \frac{W_0}{2\pi\chi} \ln\frac{R_0}{r}, \qquad (3)$$

and

$$T = T_0 + \frac{W_0}{2\pi\chi}\left(\frac{\pi R_0}{2r}\right)^{1/2} \exp\left(\frac{r-x}{R_0}\right), \qquad (4)$$

accordingly. At $r > R_0$ the temperature value decreases exponentially and as consequence R_0 can be considered as a stationary heating radius of the dislocation. Figure 1 shows the temperature distribution around dislocation during ultrasonic loading calculated by (3). The calculated value of the temperature rise (~25 K) is in a good agreement with experimental one (10÷20 K).

If the average distance between dislocations d exceeds the doubled heating radius R_0 the discrete distribution of thermal sources is realised. Otherwise, the thermal source distribution is continuous. Figure 2 shows regions of discrete (1) and continuous (2) distribution of thermal sources separated by $d=2R_0$ regularity and allows to determine the limit value of the average dislocation density. It was determined that the heating radius of the linear thermal source estimated by expression $R_0 = 2\chi/\rho C v$ is ~0.5 μm for $Cd_xHg_{1-x}Te$ crystals if dislocation velocity v is $0.01 v_{aw}$ [10], where v_{aw} is velocity of acoustic wave. So, the discrete character of thermal source distribution will be realised for $Cd_xHg_{1-x}Te$ solid solutions up to dislocation density $N_{DIS} < 10^{11}$ m^{-2}.

$Cd_xHg_{1-x}Te$ alloy is the main material for infrared (8-14 μm) detector technology. The quality and reliability of $Cd_xHg_{1-x}Te$ based devices can be dramatically changed by linear defects such as dislocations and low angle boundaries. We think that investigated effect of ultrasonically stimulated heating can be used for structural perfection control of $Cd_xHg_{1-x}Te$ solid solutions.

Really, we made attempt to investigate several group of $Cd_xHg_{1-x}Te$ crystals which differed in structural defects and determined that the effect of ultrasonic-stimulated heating manifests itself by a next manner. For samples with uniform distribution of dislocations and $N_{DIS} \leq 10^8$ m^{-2} sonic-stimulated temperature distribution has discrete

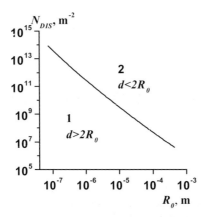

Figure 2. The region of the discrete (1) and continuous (2) character of thermal source distribution.

character. The macroscopic effects were not detected. We observed the effect of sonic-stimulated macro-heating for $Cd_xHg_{1-x}Te$ crystals with nonuniform dislocation distribution. For such samples the average value of the dislocation density varies from $10^9 m^{-2}$ to $10^{10} m^{-2}$ and some "hot" regions with $N_{DIS} \geq 10^{11}$ m^{-2} were found. Finally, we investigated samples with mechanically produced damage layer with $N_{DIS} \sim 10^{12}$ m^{-2}. The total heating of such samples during ultrasonic loading was took place. It is need to underline also that such method allows to detect low angle boundaries presence in crystal.

4. Conclusions

In this paper we report the results of an experimental study of ultrasonically stimulated thermal effect in $Cd_{0.2}Hg_{0.8}Te$ crystals. The nonuniform temperature distribution in crystal surface during ultrasonic loading was detected. This effect was associated with sonic-stimulated temperature rise around dislocation and heating of the nonperfect regions of investigated samples. It is possible to use the investigated effect as the base of the non-destructive technique for structural perfection control of $Cd_xHg_{1-x}Te$ alloys.

5. References

[1] Granato A.V., Luecke K. 1966 The vibrating string model of dislocation damping, in Physical Acoustic, ed. by Mason W.P. (New York: Academic).
[2] Sheinkman M.K., Korsunskaya N.E., Ostapenko S.S. 1999 Romanian Journal of Information Science and Technology 2 173-187.
[3] Ostapenko S.S.,. Bell R.E 1995 Journal Appl. Phys.77 5458-5460.
[4] Ostapenko S.S., Jastrebski L., Lagovski J. 1994 Appl. Phys. Letters. 65 1555-1557.
[5] Tartachnyk V.P., Gontaruk O.M., Vernydub R.M. 1999 Proc. SPIE 3890 559-563.
[6] Olikh Ya.M., Savkina R.K., Vlasenko O.I. 1997 Proc. SPIE 3359 259.
[7] Olikh Ya.M., Savkina R.K., Vlasenko O.I. 1999 Semiconductors 33 398-401; 2000 Semiconductors 34 644-649.
[8] Savkina R.K., Vlasenko O.I. 2002 Phys. Stat. Sol.(b) 1 275-278.
[9] Roos A., De Hosson J. Th. M. 2001 Phys. Rev. B 53
[10] Loktev V.M., Khalak J. 1997 Ukrainian Physical Journal 42 343-352.

The Cluster Variation Method for Semiconductor Alloys

O V Elyukhina
Ioffe Physical-Technical Institute, St.-Petersburg

Abstract. Order-disorder phenomenon of semiconductor alloys is studied by the cluster variation method. The free energy of the binary alloys A_xB_{1-x} and the quaternary alloys $A_xB_{1-x}C_yD_{1-y}$ are estimated. The miscibility gap and the short-range order of binary alloys are shown. Strong correlations in atom arrangement in quaternary alloys below the critical point are shown.

1. Introduction

Atomic structure and phase stability of semiconductor alloys have been studied for many years. According to the experimental and theoretical studies strong three- or two-dimensional clusterization, as well as phase separation occur in many types of semiconductor alloys [1]. These results show that semiconductor alloys are not random and have a tendency to decomposition. The stability problem of the semiconductor alloys is usually described by the regular solution model. The cluster variation method (CVM) was developed by R. Kikuchi [2] for the description of decomposition and order-disorder phenomena in the Ising model. It was showed that the conventional strictly regular and quasichemical methods are more approximate cases of the CVM method [1]. Hijmans and De Boer [3] reformulated CVM in a more systematic way and showed that further advances of quasichemical method can be also considered as special cases of the Kikuchi method.

A number of approximation schemes in CVM were developed to avoid cumbersome calculations and give the theoretical foundation [4, 5, 6, 7]. It made the CVM a powerful approximate method for description of the different lattice systems often used in statistical physics. Rigorous theoretical foundation of CVM was made by Morita [8, 9]. The convergence of the CVM to an exact solution for the infinite lattice systems was proved in [10, 11]. CVM was applied to the study of various lattice systems, including the binary metallic alloys, ferromagnets and antiferromagnets. The study proposes the application of CVM to some of the widely used types of semiconductor alloys.

2. Cluster variation method

CVM describes the lattice system as a partially ordered set of the clusters, i.e. groups of atoms. A set of r lattice points defining arbitrary geometrical figure is termed r-point cluster [5]. The level of approximation is defined by the biggest considered set of atoms or the basic cluster. In cases where sublattices are not required, such as the Ising ferromagnet, two clusters related by the symmetry operations of the lattice (point group plus translations) are considered to be identical. All distinct r-point clusters are classified and labelled by the index t ($t=1,...$). Thus the pair of integers (r,t) refers to a specific r-point cluster in the classification scheme.

In a binary system [5], the points of a given (r,t) cluster can be occupied by any one of the two "atomic species," thus determining 2^r distinguishable configurations of (r,t). In principle, a particular (r,t) cluster configuration can be specified by a set of r numbers that take values 1 and -1 for each of the two components in the system. However, all configurations of (r,t) cluster which are related by the symmetry operations have the same probability (concentration in the crystal). Thus we can label the (r,t) cluster configurations by a single index i where, in general, $i=1,..., s$ with $s \leq 2^r$. The configuration i of (r,t) will have associated with it a degeneracy factor α_i^r defined as the number of distinguishable configurations which can be generated by the symmetry operations of the (r,t) cluster.

The configurational entropy is given by

$$S = R \sum_r \lambda^r \gamma^r \sum_i \alpha_i^r x_i^r \ln x_i^r$$

where R is the universal gaseous constant and x_i^r the concentration of the (r,t) cluster in the i-th configuration. If the largest clusters to be considered, henceforth referred to as the basic clusters, contain n points, the coefficients γ^n are given by

$$\gamma^n = -\frac{N^n}{N}$$

and

$$\gamma^r = -\frac{N^n}{N} - \sum_{q=r+1}^{n} \sum_s M(r,t;q,s)\gamma(q,s)$$

where N^n is the total number of (r,t) clusters in the system, N is the total number of lattice points, and $M(r,t;q,s)$ is the number of (r,t) clusters contained in (q,s) cluster. Different sublattices are taken into account by the coefficient λ^r that is equal to the number of sublattices that are equivalent for this cluster.

3. Binary alloys

In the description of binary alloys A_xB_{1-x} with the diamond lattice (Ge_xSi_{1-x} alloys) the basic cluster consisting of six atoms was chosen. For this system the Helmholtz free energy of mixing may be written as follows

$$F = U - RT\{-3(x\ln(x) + (1-x)\ln(1-x))$$
$$-2\sum_{i=1}^{3} \alpha_i^2 x_i^2 \ln(x_i^2) + 6\sum_{i=1}^{6} \alpha_i^3 x_i^3 \ln x_i^3 - 2\sum_{i=1}^{13} \alpha_i^6 x_i^6 \ln x_i^6\}$$

where U is the internal energy of the system including the strain energy, x_i^n and α_i^n are the concentration and degeneracy factor of the i-kind of n-point cluster, respectively.

The miscibility gap, estimated for the binary alloys with the diamond lattice, is shown on Fig. 1. It is presented in the strictly regular (SR) and quasichemical (QC) approximations and in CVM. The degree of clusterization defined through the short-range order equal to the concentrations of different pairs of atoms is shown on Fig. 2. The estimated values show that there are large clusters enriched with atoms of one type in the alloy.

Ternary alloys of binary compounds $A_xB_{1-x}C$ ($In_xGa_{1-x}As$, $In_xGa_{1-x}N$ alloys) and AB_xC_{1-x} ($GaSb_xAs_{1-x}$ alloys) are considered as binary alloys with the structure of mixed sublattice. Ternary alloy with zinc-blende structure corresponds to the binary alloy with face-centered-cubic (fcc) lattice, as it was described in [12]. Ternary alloy with wurtzite structure is considered as binary alloy with hexagonal-close-packed (hcp) lattice in mixed sublattice [12].

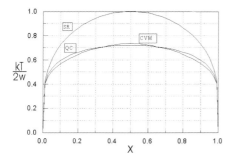

Figure 1. The miscibility gap of A_xB_{1-x} alloys.

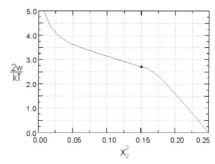

Figure 2. The short-range order of A_xB_{1-x} alloys.

4. Quaternary alloys

4.1. Bulk

Two sublattices are taken into account at the description of the configuration entropy of quaternary alloys $A_xB_{1-x}C_yD_{1-y}$ ($In_xGa_{1-x}As_yP_{1-y}$, $In_xGa_{1-x}N_yAs_{1-y}$ alloys). These quaternary alloys of four binary semiconductor compounds AC, AD, BC and BD have zinc-blende structure consisting of two geometrically equivalent the face-centered cubic sublattices, filled with A,B and C,D atoms, respectively.

The group of six atoms is a basic cluster. The Helmholtz free energy of mixing is given as

$$F = U - RT\left[\begin{array}{l} -8\sum_i \alpha_i^6 x_i^6 \ln x_i^6 + 6\sum_i \alpha_i^3 (x_i^3 \ln x_i^3 + x_i'^3 \ln x_i'^3) - 8\sum_i \alpha_i^2 x_i^2 \ln x_i^2 \\ -3(x\ln x + (1-x)\ln(1-x) + y\ln y + (1-y)\ln(1-y)) \end{array}\right]$$

where U is the internal energy, including strains and all types of interactions, x and y are the concentrations of cations and anions, respectively. Symbols of x_i^3 and $x_i'^3$ correspond to two types of different triangles.

4.2. Surface

The epitaxial semiconductor films are usually grown as layer by layer. Therefore, the alloy epitaxial films are formed by "freezing" of the surface arrangement of the atoms and can be treated as two-dimensional systems. The correlations in the atomic arrangement in the two-dimensional system should be substantially larger than in three-dimensional one. The Helmholtz free energy of mixing of $A_xB_{1-x}C_yD_{1-y}$ alloys with simple square lattice is given by

$$S = R\left[\begin{array}{l} -4\sum_i \alpha_i^4 x_i^4 \ln x_i^4 + 8\sum_i \alpha_i^2 x_i^2 \ln x_i^2 - x\ln x - (1-x)\ln(1-x) \\ -y\ln y - (1-y)\ln(1-y) \end{array}\right]$$

The concentration of AC pairs in $A_{0.5}B_{0.5}C_{0.5}D_{0.5}$ alloys is shown of Fig.3 in the strictly regular (SR) and quasichemical (QC) approximations and in CVM. At the critical temperature the value is $x_{AC} = 0.110$, while it is $x_{AC} = 0.25$ at the random arrangement.

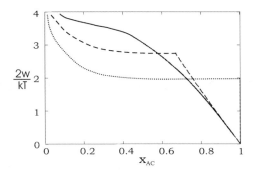

Figure 3. The concentration of AC pairs in $A_{0.5}B_{0.5}C_{0.5}D_{0.5}$ alloys.

The concentrations of other types of pairs are $x_{BD} = 0.110$ and $x_{AD} = x_{BC} = 0.390$. Although this result is common, it describes quaternary alloys with the zinc-blende structure ($In_xGa_{1-x}As_yP_{1-y}$ alloys) that show strong atomic ordering and phase separation in the epitaxial layers [1].

5. Conclusion

The obtained results can be applied to the explanation of strong correlations in the semiconductor compounds grown near thermodynamic equilibrium. The experimental studies showed that $In_xGa_{1-x}As_yP_{1-y}$ alloys consist of the domains enriched with InP and GaAs compounds [1]. The typical size of these domains is 5 nm. The observed strong correlations are usually explained as disintegration of the alloy. However, this phenomenon in the developed model can be explained as one-phase system with the strong correlations in the arrangement of the atoms. Even above the critical temperature the arrangement of pairs of atoms is not random. Thus, the formation of domains enriched with one type of atoms in the binary system and two types of pairs should take place in the one-phase system.

References

[1] McDevitt T L et al 1992 Phys. Rev. B 45 6614-6622
[2] Kikuchi R 1951 Phys. Rev. 81 988-1003
[3] Hijmans J and de Boer J 1955 Physica 21 471-499
[4] Morita T 1957 J. Phys. Soc. Japan 12 1060
[5] Sanchez J M and De Fontaine D 1978 Phys. Rev. B 17 2926-2936
[6] An G 1988 J. Stat. Phys. 52 727-734
[7] Matic V M and Milosevic S 1999 Physica A 262 215-231
[8] Morita T and Tanaka T 1966 Phys. Rev. 145 288
[9] Morita T 1994 Progr. Theoret. Phys. 92 1081-1093
[10] Schlijper A G 1983 Phys. Rev. B 27 6841-6848
[11] Kikuchi R 1995 Progr. Theoret. Phys. Suppl. 115 1
[12] Elyukhin V A 1997 J. Cryst. Growth. 173 69-72

Excitonic formation inhibition in GaInAs/InP Fe doped quantum wells

M. Guézo, S. Loualiche, J. Even, A. Le Corre, H. Folliot, C. Labbé, O. Dehaese

Laboratoire de Physique du Solide - INSA de Rennes
20, avenue des Buttes de Coësmes 35043 RENNES Cedex

Abstract. Fe-doped InGaAs/InP multiple quantum wells for ultrafast saturable absorption applications are measured by pump-probe experiments. Iron-doping allows us to control sample photoresponse time from nanosecond range to a value as short as 0.45 ps for an iron concentration of 6×10^{18} cm^{-3}. The Differential Transmission Ratio (DTR) of samples with increasing Fe-doping at moderate optical excitation levels is studied. We observe a DTR decrease below 1ps which we attribute to an inhibition of exciton formation at high iron doping level when the exciton Bohr radius is comparable to the mean distance between Fe related traps.

1. Introduction

All-optical switching devices are based on materials having fast nonlinear optical response to satisfy high-bit-rate communication systems. The GaInAs/InP multiple quantum wells (MQWs) are extensively used as active layers in optical saturable absorber (SA) based on the saturation of the excitonic peak absorption. They provide several advantages for telecommunication technologies such as a low switching energy and a high contrast when they are inserted in resonant microcavities [1]. However, the recovery time of absorption of these saturable absorbers (SAs) is situated in the nanosecond range which reduces their usefullness in the all optical regeneration devices. Several approaches have been undertaken to minimize this photoresponse time, which corresponds to the photogenerated carrier lifetime.

A first way consists in irradiating MQWs with light or heavy ions that leads to create point or extended defects in the sample acting as trapping centers [2-4]. This introduction of recombination centers reduces the response time of the SAs to a minimum value of 2 ps. Another way consists of doping with beryllium low-temperature-grown MQWs and to anneal them [5]. In the case of InGaAs/InAlAs MQWs, this method leads to a very short response time (1.3 ps at 1.54 µm) [6].

We have proposed and demonstrated the use of iron doped structures to control efficiently and easily the SA recovery time constant from nanosecond range for undoped samples down to picosecond range for high Fe doping levels [7]. Iron doping creates, in InP as well as in InGaAs, localized deep level which acts as carrier traps and fast recombination centers.

In this paper we present pump-probe measurements of the decay time from MQWs structures having increasing iron concentrations. We show that the Differential Transmission Ratio (DTR) presents two regimes when plotted versus the time decay and we discuss this phenomenon.

2. Experiments

The samples are grown by gas source molecular beam epitaxy (MBE) on a (001)InP epi-ready substrate at 450°C. They consists of 40 periods of InGaAs/InP (8.5 nm $In_{0.53}Ga_{0.47}As$ well and 10 nm InP barrier). Iron is uniformly incorporated during the growth of the samples. The iron concentration has been varied from from 2×10^{17} cm^{-3} to 6×10^{18} cm^{-3}.

The wavelength of the excitonic resonance peak is precisely determined by Fourier transform infrared spectroscopy and the nonlinear absorption dynamic is measured at this wavelength using a transmission mode degenerate pump-probe technique (PPT) which has been described elsewhere [7]. The change in the transmission of the probe beam induced by the pump, taking into account the probe signal without pump beam on the sample, leads to a measure of the DTR ($\Delta T/T_0 = (T-T_0)/T_0$, where T ($T_0$) corresponds to the transmission value of the probe beam with (without) the presence of the pump beam) as a function of the time delay between the probe and the pump.

3. Results and discussion

Samples having different iron concentrations ranging from 1×10^{17} cm^{-3} to 6×10^{18} cm^{-3} present decay times from 250 ps down to 0.45 ps [8]. The temporal evolution of the normalized differential transmission ratio for samples with iron doping levels of 9×10^{17} cm^{-3}, 2×10^{18} cm^{-3} and 6×10^{18} cm^{-3} are shown on figure 1.

When extracting the decay times from the pump-probe experiments, two time constants are usually obtained. The fastest component (subpicosecond range) is almost constant for all the samples and corresponds to the exciton ionization. The second is related to the carrier decay time controlled by the Fe concentration. Figure 1 emphasizes that the increase of iron concentration clearly reduces the decay time. Response times in the picosecond range confirms that iron is an efficient trap center for electrons and holes. No saturation of the decay time at high iron doping level is observed.

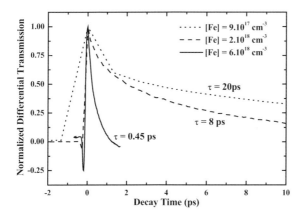

Figure 1. Normalized differential transmission ratio versus pump-probe delay, measured on Fe-doped $In_{0.53}Ga_{0.47}As$ / InP Multi Quantum Wells.

A relaxation mechanism has been proposed previously to describe this effect [9,10]. When electrons and holes are photogenerated, the neutral iron state Fe^{3+} traps an electron, and therefore becomes ionized to Fe^{2+}, which captures a hole and then becomes neutral Fe^{3+} state to be ready to capture a new electron.

For the highest iron doping level, the decay time falls into subpicosecond range with a minimum value of 0.45 ps and only one time constant is obtained. Moreover, in this case an important reduction of the DTR is observed (figure 2). A close look to figure 2 show that, at low excitation (10 µJ.cm^{-2}), the DTR exhibits three different values for undoped, Fe-doped or heavily Fe-doped InGaAs/InP MQWs. The DTR is around 5% for undoped samples where the decay times is greater than 1 ns. Iron doping induces a slight decrease of DTR to 2%. This value stays almost constant for Fe concentrations varying from 5×10^{16} cm^{-3} to 2×10^{18} cm^{-3} and decay times from 400 ps to 3 ps. This DTR decrease could be explained by the reduction of the exciton absorption peak due to the material degradation induced by iron doping. A second decrease of the DTR to 0.4% appears for the highest Fe concentration where the subpicosecond range is reached for the recovery time.

For the highest doping level of 6×10^{18} cm^{-3}, the mean distance between Fe traps is around 7 nm, which is comparable to the exciton Bohr radius [11]. Exciton is known to be sensitive to coulomb interactions (free carriers or fixed charges) [12]. So exciton formation can be perturbed by these iron related defects. The inhibition of exciton formation implies a decrease of the absorption coefficient and thus of the DTR. So, the highly doped sample, where only iron assisted carrier recombination is observed and the DTR reduced, shows that we are reaching the limit of exciton stability.

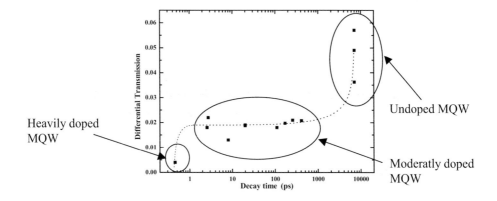

Figure 2. : Differential transmission ratio versus decay time for undoped and doped-samples at 10µJ.cm^{-2}. Dotted line stays for eye guide.

4. Conclusion

The control of the decay time of Fe-doped InP-based MQW saturable absorbers is demonstrated, and no saturation of the phenomenon is observed at high iron concentrations where a record value of 0.45 ps was reached. The DTR decreases slowly with decay time down to subpicosecond range, corresponding to high iron concentrations. At this high doping regime, the DTR decrease is explained by the assumption of having reached the limit where the exciton formation is inhibited.

Aknowledgments: This work is supported by France Telecom Research and Development.

[1] A. Hirano et al. 1998 Electr. Lett. 34, 198
[2] Y. Sildeberg et al. 1985 Appl. Phys. Lett. 46, 701
[3] J. Mangeney et al. 2000 Appl. Phys. Lett. 76, 1371
[4] J. Mangeney et al. 2001 Appl. Phys. Lett. 79, 2722
[5] R. Takahashi et al. 1994 Appl. Phys. Lett. 65,1790
[6] Y. Chen et al. 1998 Appl. Phys. Lett. 72,439
[7] A. Marceaux et al. 2001 Appl. Phys. Lett. 78,4065
[8] M. Guezo et al. submitted for publication in Appl. Phys. Lett.
[9] D. Södertröm et al. 1997 Appl. Phys. Lett. 70,3374
[10] B. Srocka et al. 1994 Phys. Rev. B 49,10259
[11] D. Miller et al 1982 Appl. Phys. Lett. 41,679
[12] P. Holtz et al 1998 Phys. Rev. B 58,4624

Compositional dependence of electron traps in Ga(As,N) grown by molecular-beam epitaxy

P Krispin[1], V Gambin[2], J S Harris[2], K H Ploog[1]

[1]Paul-Drude-Institut für Festkörperelektronik, Hausvogteiplatz 5-7, 10117 Berlin, Germany
[2]Solid State and Photonics Laboratory, Stanford University, Stanford, California 94305, USA

Abstract. Two predominant electron traps at about 0.80 and 1.1 eV above the valence band edge E_V are observed in Ga(As,N), which do not depend on composition. The gap level at $E_V+1.1$ eV is due to defects associated with nitrogen atoms on As sites (N_{As}). For more than 2.5% N, it is resonant with the conduction band. The level at $E_V+0.80$ eV is connected with nitrogen dimers, i.e., two N atoms on a single As site [$(NN)_{As}$]. The dimer defect occurs at the growing Ga(As,N) surface and can be removed by rapid thermal annealing, in contrast to the stable N_{As}-related gap state in the bulk.

1. Introduction

Group-III-arsenide-nitrides grown on GaAs substrates are promising materials for optoelectronic devices. Ga(As,N) layers are of special interest, because the bandgap is drastically reduced by increasing the GaN mole fraction. It is known that crystal quality, luminescence efficiency, and electrical characteristics of Ga(As,N) layers deteriorate for larger GaN mole fractions. The properties can be partly improved by postgrowth heat treatment. However, the degradation and annealing mechanisms are still controversially discussed. The reason for the low Si doping efficiency is also not clear till now. Nitrogen-related deep-level defects in Ga(As,N) have been proposed as compensating and nonradiative recombination centers, but have not been explored yet. In particular, there is no study up to now about the compositional dependence of both the level energy and the concentration of deep-level defects in Ga(As,N). We have previously shown that the electronic states found in the lower half of the Ga(As,N) bandgap are not due to N-related defects, but mainly due to impurities [1].

In this contribution, we concentrate on *strained* Ga(As,N) layers with compositions from the dilute limit (≤0.1% N) to alloys with 3% N in order to identify the dominant electron traps and their spatial distributions. Metal-semiconductor (MS) contacts on Si-doped Ga(As,N)/GaAs heterojunctions grown by molecular-beam epitaxy are investigated by deep-level transient spectroscopy (DLTS) to search for electron traps in the upper half of the Ga(As,N) bandgap. Ga(As,N) was grown at 465°C with a thickness of 500 nm for ≤0.5% N and 50 nm for 3% N in order to avoid lattice relaxation.

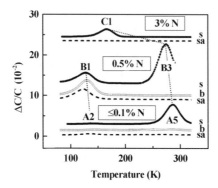

Figure 1. DLTS signal $\Delta C/C$ for three Ga(As,N) compositions (1 s period, 100 ms pulse width). The spectra labeled s and sa were measured close to the surface of as-grown and annealed layers, respectively. The b scans originate from the Ga(As,N) bulk. The main peaks are indicated. Dotted lines connect the peaks of electron traps with similar features.

2. Results and discussion

Typical DLTS signals of the dominant electron traps are plotted in Fig. 1. Surface- and bulk-related deep-level spectra of as-grown Ga(As,N) layers (labeled s and b, respectively) as well as the ones of annealed Ga(As,N) surfaces (labeled sa) are depicted. The bulk of the layer with 3% N could not be inspected due to the limited thickness. For the compositions studied, the dominant peaks A5, B3, and C1 are confined close to the as-grown surface (cf. spectra s and b). They disappear completely after rapid thermal annealing (RTA) at 720°C for 120 s (see spectra sa). Based on these similar features, we believe that the traps A5, B3, and C1 are due to the same lattice defect near the surface of Ga(As,N) layers (see dotted line in Fig. 1).

The electron traps A2 and B1 are present in bulk- as well as in surface-related spectra of Ga(As,N). These levels cannot be annihilated by RTA (see spectra sa). We therefore assume that the deep levels A2 and B1 originate from the same bulk defect in Ga(As,N). The peak height, i.e., the concentration of the underlying defect, increases remarkably from the dilute N limit to the sample with 0.5% N. Although an even higher level concentration is expected for larger N content, a trap with comparable features cannot be found in the 3% N sample.

The compositional dependence of the dominant deep levels in the Ga(As,N) bandgap is compiled in Fig. 2. The GaAs-related band offsets for 3% N content have recently been determined independently for the conduction and valence band to be +400 and −11 meV, respectively [2,3]. Concerning Ga(As,N) in the dilute N limit, we have previously shown [4] that the levels A2 and A5 are linked with the intrinsic defects E3 and E4, respectively, which are known from electron-irradiated GaAs. E3 and E4 are due to displacements in the As sublattice and mainly associated with arsenic vacancies. Since the traps A2 and A5 are detected only in regions with larger than the nominal composition [4], the measured level energies are shifted in Fig. 2 toward the conduction band of Ga(As,N) in the dilute N limit. The common origins of A5, B3, and C1 as well as of A2 and B1 are indicated in Fig. 2 by dotted lines. It is apparent that the E3- and E4-related levels at different compositions are energetically fixed with respect to the top of the valence band at about $E_V+1.1$ and $E_V+0.80$ eV, respectively.

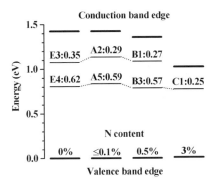

Figure 2. Compositional dependence of dominant deep levels in the Ga(As,N) bandgap. All trap energies are given in eV with respect to the conduction band edge. Dotted lines connect levels of the same origin.

Nitrogen occupying an arsenic site (N_{As}) is the dominant defect in the dilute N limit due to the low formation energy [5,6]. In calculations [5], a double acceptor charge state (2−/−) is predicted particularly in n–type layers at E_V+1.04 eV, which perfectly fits the experimental energy of the E3-related levels. We therefore identify the dominant bulk levels A2 and B1 in Ga(As,N) as isolated N_{As} defects or related N_{As} complexes. In accordance with such an assignment, the concentration strongly increases with larger N content (see Fig. 1), and both levels cannot be removed from the bulk by annealing. In Ga(As,N) with more than about 2.5% N, the N_{As}-related electron trap is resonant with the conduction band (see Fig. 2) and therefore no longer detectable, in agreement with the experiment. It should be noted that the incorporation of nitrogen on As sites generates even in Ga(As,N) with 0.5% N *localized* electronic states.

Nitrogen dimers, i.e., two N atoms on a single As site [$(NN)_{As}$], are also likely to form under As-rich growth conditions [6]. Midgap levels are theoretically predicted for such defects in the dilute limit [6]. We therefore believe that the traps A5, B3, and C1 are associated with $(NN)_{As}$. The annealing mechanism at the surface is probably related to reactions with As vacancies, which lead to the dissociation of the $(NN)_{As}$ defect accompanied by the generation of two substitutional N_{As} sites.

For the dominant electron traps in Ga(As,N) layers, depth profiles of the concentration are displayed in Fig. 3. The depth profiles of the Si dopant determined from capacitance-voltage measurements are also plotted in Fig. 3. For each composition, the $(NN)_{As}$ density in as-grown layers (dots in Fig. 3) rises exponentially towards the surface and reaches values above 10^{18} cm^{-3}, i.e., higher than the doping level n_{Si}. After annealing, the surface density of the $(NN)_{As}$ defect is decreased by more than two orders of magnitude (circles in Fig. 3). With increasing N content, the density of this defect is apparently enhanced at earlier growth stages of the 500 nm thick layers. The strong accumulation of this native defect at the Ga(As,N) surface is linked with increasing strain during growth. The $(NN)_{As}$ formation leads to less tensile strain as compared to the formation of N_{As} [7]. Approaching the critical layer thickness, the increasing stress may thus be partially compensated by the formation of more and more $(NN)_{As}$ defects.

The concentration of the N_{As}-related defect does not vary with depth. It increases drastically from about 5×10^{15} cm^{-3} in the dilute limit (not shown) to about 1×10^{17} cm^{-3} for 0.5% N [squares in Fig. 3(b)]. Because the N_{As}-related trap cannot be annealed (see Fig. 1), this intrinsic level controls the properties of annealed Ga(As,N) up to N contents

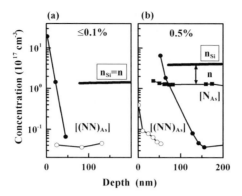

Figure 3. Depth profiles of Si dopant and deep level concentrations in Ga(As,N) measured in (a) for the dilute N limit and (b) for 0.5% N. The distribution of the dimer defect $(NN)_{As}$ is plotted for the as-grown and annealed layers by dots and circles, respectively. In (a), the electron density n in the bulk is given by the concentration n_{Si} of Si dopants. In (b), n is determined by the difference between n_{Si} and the concentration $[N_{As}]$ of the N_{As}-related defect (squares) as indicated by an arrow.

of 2.5%. The electron concentration n is therefore given by the difference between the doping density n_{Si} and the concentration $[N_{As}]$ of the dominant N_{As}-related defect [see arrow in Fig. 3(b)].

3. Conclusions

In conclusion, the two dominant gap states in Ga(As,N) originate from N-related point defects in the As sublattice. When the composition is changed, their electronic levels are fixed with respect to the valence band edge. The level at $E_V+1.1$ eV is due to a N_{As}-related defect in Ga(As,N) and lowers the Si doping efficiency. It cannot be removed by RTA. For more than 2.5% N, this state is resonant with the conduction band. The other dominant level at $E_V+0.80$ eV originates from the dimer defect $(NN)_{As}$, which is increasingly generated during growth approaching the critical layer thickness. The corresponding trap leads to strong carrier depletion at the as-grown surface, which, however, can be strongly reduced by annealing.

Acknowledgments

The authors are very grateful for the technical assistance of E. Wiebicke. They would like to thank H. T. Grahn for comments and a careful reading of the manuscript.

References

[1]   Krispin P, Spruytte S, Harris J S, Ploog K H 2001 J. Appl. Phys. 89 6294-6301
[2]   Krispin P, Spruytte S, Harris J S, Ploog K H 2001 J. Appl. Phys. 90 2405-2410
[3]   Krispin P, Spruytte S, Harris J S, Ploog K H 2000 J. Appl. Phys. 88 4153-4158
[4]   Krispin P, Gambin V, Harris J S, Ploog K H 2002 Appl. Phys. Lett. 81 accepted
[5]   Orellana W, Ferraz A C 2001 Appl. Phys. Lett. 78 1231-1233
[6]   Lowther J E, Estreicher S K, Temkin H 2001 Appl. Phys. Lett. 79 200-202
[7]   Li Wei, Pessa M, Likonen J 2001 Appl. Phys. Lett. 78 2864-2866

The influence of the quantum lifetime on the width of the quantum Hall plateaus

L. Gottwaldt[1,2], K. Pierz[1], F.J. Ahlers[1], L. Schweitzer[1], E.O. Göbel[1], W. Stolz[2]

[1] Physikalisch-Technische Bundesanstalt, Bundesallee 100, 38116 Braunschweig, Germany
[2] Materials Sciences Center and Department of Physics, Philipps-University, 35032 Marburg, Germany

Abstract We report on the quantum and transport lifetime dependence of the width of the quantum Hall plateaus. The width of the spin-split i = 3 plateau rises continuously with increasing quantum lifetime and reveals remarkable changes. The width of the Landau-split plateaus are a consequence of the pronounced changes of the spin-split plateaus. The broadening of the i = 3 plateau can be attributed via the exchange interaction of the electrons in different Landau levels to the quantum lifetime. In contrast, the transport lifetime and thus the electron mobility do not allow precise predictions of the widths of quantum Hall plateau.

1. Introduction

It is generally believed that the integer quantum Hall effect (QHE) is due to a localization-delocalization transition caused by disorder [1]. This model is often used to explain the mobility dependence of the widths of the quantum Hall plateaus. With increasing electron mobility the delocalized range of the density of states (DOS) is broadened and as consequence the localized range of the DOS and thus the widths of the quantum Hall plateaus decrease [2].

We have systematically studied the lifetime dependence of the width of the quantum Hall plateaus. It is necessary to distinguish between the transport lifetime, which determines the electron mobility, and the quantum lifetime. Whereas the quantum lifetime counts every scattering event equally, the transport lifetime favours large-angle scattering over small-angle scattering [3].

2. Sample structure

For our study it was necessary to compare samples with the same carrier density, since free carriers screen the Coulomb potential of the ionized impurities. Thus the measurements were performed on $GaAs/(Al_{0.3}Ga_{0.7})As$ modulation doped structures grown by molecular beam epitaxy with a 10 nm thick quantum well (QW) in which electron density could be varied between zero and $4 \cdot 10^{11}$ cm^{-2} by varying the voltage applied to a semi-transparent top Schottky gate. The samples studied were grown under nominally identical growth conditions, but with systematic changes in the growth interrupts before and after the QW.

3. Experimental results

For the measurements a fixed carrier density of $2 \cdot 10^{11}$ cm^{-2} was adjusted. For this carrier density the quantum lifetime of the samples is between 0.15 ps and 1 ps and the transport lifetime between 0.2 ps and 1.8 ps. As the widths of the Hall plateaus we determine the region of magnetic field for which

$$\rho_{xy} = h/ie^2 \pm 0.1 k\Omega. \tag{1}$$

3.1 The quantum Hall plateau with filling factor four

Figure 1 shows the widths of the plateaus with filling factor $i = 4$ as a function of the quantum lifetime (solid symbols in fig. 1a). The short lifetimes represent the onset of the quantum Hall effect, in which the first plateau are formed and a deviation from the classical straight line occurs. As a consequence the width of the $i = 4$ plateau rises in this range. For lifetimes larger than 0.45 ps the widths of the plateau reveal the expected decrease.

However, one can see an apparent contradiction, if the slope between two adjacent plateaus is also considered. For quantum lifetimes where the width of the $i = 4$ plateau decreases the slope between the $i = 4$ and the $i = 3$ plateau still increases (fig. 1a, open symbols), which means that the magnetic field range between the plateaus also decreases. This contradiction casts doubt on the simple explanation for the width of the plateaus as being due to with an enhanced delocalized energy interval of the DOS due to the longer lifetime. In this model the assumption is made that the slope of the DOS remains unchanged, we believe this assumption not to be valid.

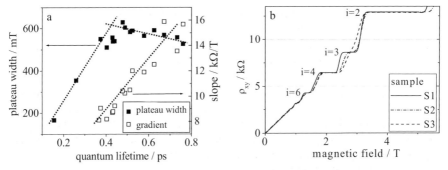

Figure 1. (a) The width of the $i = 4$ plateau and the slope between the $i = 4$ and the $i = 3$ plateaus versus quantum lifetime. The dashed lines are only guides to the eye. (b) Shown are three typical QHE spectra exhibiting pronounced changes for $i = 3$ plateau.

3.2. The quantum Hall plateau with filling factor three

To resolve this apparent contradiction we turn our attention to the plateau with filling factor 3. In comparison with this the changes of the plateau with even filling factors are negligibly small (fig. 1b). In figure 2 the changes of the $i = 4$ plateau, as already shown in figure 1, are compared with the quantum lifetime dependence of the width of the $i = 3$ plateau.

The broadening of the $i = 4$ plateaus only takes place in the range of small lifetimes, where no $i = 3$ plateau exists. For quantum lifetimes larger than 0.4 ps the widths of the two plateaus show a contradictory dependence. The continuous and pronounced increase of the width of the $i = 3$ plateau is opposite to the concurrent decrease of the width of the i = 4 plateau. Due to the broadening of the i = 3 plateaus a smaller magnetic field range is available for the neighboring filling factors. Therefore the increase of the width of the i = 3 plateau is responsible for the slight decrease of the width of the i = 4 plateau.

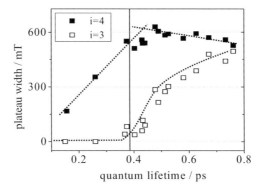

Figure 2. Plateau width versus quantum lifetime for the i = 3 and i = 4 plateau. The increase of the width of the i = 3 plateau is responsible for the decrease in the width of the i = 4 plateau.

4. Physical causes for the width of the plateaus

In the following the physical causes for the behavior of the spin-split i = 3 plateau are discussed. The spin-splitting is determined by the Zeeman energy E_S

$$E_S = s \cdot g^* \cdot \mu_B \cdot B \qquad (2)$$

where g^* is the enhanced effective g-factor, composed of the bare g-factor and the exchange energy E_{ex} between electrons in different levels such that

$$g^* \cdot \mu_B \cdot B = g \cdot \mu_B \cdot B + E_{ex} \qquad (3)$$

The spin splitting due to the bare g-factor for all magnetic fields is so small that an i = 3 plateau is not to be expected (g = 0.5). This is made possible only by the exchange interaction of the electrons in the two levels, whereby the effective g-factor g^* is increased [4]. The magnitude of the exchange energy determines the maximum magnetic field range for a spin-split quantum Hall plateau.

The exchange energy is of the form

$$E_{ex} \propto n_{N\uparrow} - n_{N\downarrow} \qquad (4)$$

where $n_{N\uparrow}$ and $n_{N\downarrow}$ are the relative populations in the two spin states of a given Landau level. The difference of the relative populations depends on the broadening of the Landau levels and increases for narrower Landau levels.

In the ideal case without scattering and without disorder discrete Landau levels would be expected. With increasing scattering the Landau levels effectively broaden, and the relative population difference decreases. With the relative population difference the exchange energy between the electrons and accordingly the effective g-factor are reduced. Therefore the magnetic field range for the spin-split Landau level is reduced and, together with only a slight change of the slope between the plateaus, the width of the spin-split plateau decreased. Thus its width is linked directly with the broadening of the effective Landau levels.

The collision broadening of the Landau levels is related to the quantum lifetime:

$$\Gamma_q = \frac{\hbar}{2\tau_q} \qquad (5)$$

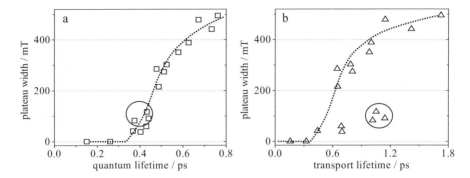

Figure 3. (a) With increasing quantum lifetime the width of the i = 3 plateau increases continuously. (b) Plotting the width of the plateau as a function of transport lifetime does not produce a curve with a simple correlation. Samples grown with lower arsenic pressure for example show more small-angle scattering events and therefore the quantum lifetime is relatively large in comparison to the transport lifetime (these special samples are marked with a circle).

In contrast to the quantum lifetime the transport lifetime, which determines the mobility, represents only a fraction of the actual number of collisions. Thus the transport lifetime is not directly linked with the broadening of the Landau levels and the width of the i = 3 plateaus (figure 3). Instead, as clearly demonstrated by our results, the quantum lifetime determines the width of the i = 3 plateau and hence that of the neighbouring plateaus.

5. Summary

In summary, our results demonstrate that the width of the i = 3 plateau, which increases significantly with increasing quantum lifetime, is a useful quantity to characterise the quantum Hall effect. The changes of the i = 3 plateau are responsible for the dependence of the plateaus of neighbouring filling factors. The width of the i = 3 plateau can be attributed, via the effective broadening of the Landau levels, to the quantum lifetime, which determines the width of the i = 3 plateau. In contrast to the quantum lifetime the transport lifetime is not directly related to the broadening of the Landau levels nor to the width of the i = 3 or any other plateau.

Acknowledgement

We thank R.J. Haug (University of Hanover) for helpful discussions.

References

[1] B. Huckestein, Rev. Mod. Phys. 67, 357 (1995)
[2] J. E. Furneaux, T. L. Reinecke, Phys. Rev. B 29, 4792 (1984)
[3] J. P. Harrang, R. J. Higgins, et. al, Phys. Rev. B 32, 8126 (1985)
[4] T. Englert, D. C. Tsui, A. C. Gossard, C. Uihlein, Surface Science 113, 295 (1982)

Intersubband transitions in strain compensated InGaAs/AlAs quantum well structures grown on InP

N Georgiev[1], M Semtsiv[2], T Dekorsy[1], F Eichhorn[1], A Bauer[1], M Helm[1], W T Masselink[2]

[1]Institute of Ion Beam Physics and Materials Research, Forschungszentrum Rossendorf, P.O.Box 510119, D-01314 Dresden, Germany
[2]Humboldt-University Berlin, Department of Physics/FET, Invalidenstrasse 110, D-10115, Berlin, Germany

Abstract. We report a X-ray diffraction (XRD) and Fourier transform infrared (FTIR) spectroscopy study of highly strain compensated $In_xGa_{1-x}As/AlAs/In_yAl_{1-y}As$ multiple quantum well (MQW) structures grown pseudomorphicaly grown on InP substrate by gas-source molecular beam epitaxy (GSMBE). XRD shows compositional grading of the interfaces that markedly influence the absorption spectra. Intersubband transitions (ISBT) at wavelengths shorter than 2.0 µm for MQWs with different InGaAs well thickness are presented.

1. Introduction

There is considerable interest in developing shorter wavelength (< 3 µm) optical devices based on ISBT such as ultrafast switches [1], modulators, and quantum cascade lasers [2]. Pseudomorphic $In_xGa_{1-x}As/AlAs$ heterostructures on InP or GaAs have emerged as excellent candidates for short-wavelength ISBT because of their large conduction band offset. However, the growth of appropriate structures on both GaAs or InP substrates requires to grow strained well and/or barrier layers, respectively. The influence of the In concentration on the intersubband absorption in the InGaAs/AlAs MQW on GaAs has been investigated, theoretically and experimentally, by several groups [3-5]. It was demonstrated that in ultrathin, strained $In_xGa_{1-x}As/AlAs$ MQW on GaAs substrate, x should be more than 40% in order to confine the carriers in the well region. The carrier loss was mainly attributed to the leakage of carriers from the Γ-minimum of the well to the X-minimum of the barrier. However, island formation due to strain relaxation would be a problem in these MQW with high In content. One way to overcome this problem is by growing InGaAs QWs with high In composition on InP substrate. In this letter, we report structural and optical characterization of the $In_xGa_{1-x}As/AlAs/In_yAl_{1-y}As$ (x>0.6, y~0.55) QW structures grown on InP substrates. In these structures, we utilise the high barrier provided by thin AlAs layers. An increased In content in the well and in the $In_yAl_{1-y}As$ barrier layers helps to compensate the large AlAs tensile strain. Additionally, it provides a smaller InGaAs band gap that results in a shift of the first Γ-like well subband to lower energies relative to the X-minimum in the barrier layers even in very narrow wells [4].

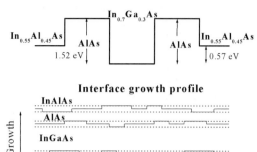

Figure 1. Schematic presentation of the interface roughness and its influence on the $In_xGa_{1-x}As/AlAs/In_yAl_{1-y}As$ QW band-edge profile.

2. Experimental

The samples are grown on semi-insulating InP(001) substrates using a GSMBE. Solid sources of In, Ga and Al are used; the P and As are supplied from thermally cracked PH_3 and AsH_3, respectively. Oxide desorption was done at 480 °C at PH_3 overpressure. To provide a smooth growth front a 200 nm undoped InP buffer was grown first at 465 °C followed by an undoped InAlAs (y~0.55) buffer grown at 480 °C. Then the temperature was ramped down during a 4 min growth interruption for the growth of the MQW structure at 400 °C. The substrate temperature was measured with a pyrometer above 450 °C and a gauge thermocouple below. Streaky reflection high energy electron diffraction (RHEED) patterns are observed during the growth of the strained layers with 2x3, 1x2 and 2x3 reconstructions for InAlAs, AlAs and InGaAs layers, respectively. The MQW structure consists of 30 periods of an InGaAs well (Si-doped to ~$1 \times 10^{+19}$ cm^{-3}) between two AlAs/InAlAs barriers, as shown schematically in Fig. 1.

High-resolution XRD was used to study the crystal quality and to determine the actual thickness and compositions of the layers. The measurements are performed at the ESRF, Grenoble, France, at the Rossendorf beamline (ROBL), using a wavelength of 1.54 Å. Scans around symmetric (004) and asymmetric (115) reflections are recorded. The intersubband absorption spectra are measured at room temperature using a FTIR using a multiple-reflection waveguide geometry fabricated by polishing two end facets at an 45° angle.

3. Results and Discussion

The X-ray diffraction pattern of the MQW sample with 7 monolayer (ML) thick $In_xGa_{1-x}As$ (x~0.7) and AlAs layers, and 20 nm $In_yAl_{1-y}As$ (y~0.55) layer is presented in Fig.2. The main peak at the lower angle side of the InP(004) peak originates from the slight compressively strained $In_yAl_{1-y}As$ buffer layer. The zero-order peak of the MQW is located close to the substrate peak which indicates that compressive strain of the InGaAs and the InAlAs layers nearly compensate the tensile strain of the AlAs layers. In spite of the large misfit of the AlAs layers the average mismatch in the MQW is small, because the AlAs layers are much thinner than the InGaAs and InAlAs layers in the MQW. Note that the choice of an appropriately strained second barrier layer and buffer layer is essential for growing pseudomorphic QW structures. Indeed, if we grow these layers lattice matched we observe both a broadening and a reduced number of the satellite peaks. This is an indication for the deterioration of the crystalline and interfacial quality in MQW due to the increased strain from strongly mismatched AlAs layers. The experimental curves are fitted with theoretical profiles obtained from simulation based on dynamical diffraction theory (Fig.2). This proves that the sample is pseudomorphic and does not show any relaxation. However, a model allowing 1-2 ML compositional grading at the interfaces provides an even better fit to the experimental data than the model assuming abrupt interfaces. This is connected with the difficulties to control precisely the quality of the interfaces in such strained structures. There are several mechanisms that contribute to the deterioration of the interfacial abruptness: (i) interface interdiffusion; (ii) In segregation; (iii) a rough growth front. While the relatively low growth temperature of ~ 400 °C to great extent suppresses (i) and (ii) [5], an increased surface roughening at the growth front may occur due to the growth of highly strained layers [6]. Indeed, throughout the growth, RHEED shows streaky patterns indicating an "island"-like two-dimensional growth mode. These rough layers at the interfaces may be modelled as layers with compositional grading defined by the appropriate weight of the inclusions of the well and barrier constituents. As we show below, the optical measurements strongly support the above structure model of interface modification.

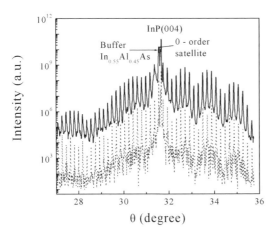

Figure 2. XRD pattern of the $In_{0.7}Ga_{0.3}As/AlAs/In_{0.55}Al_{0.45}As$ MQW around the InP (004) reflex. Experimental data (solid line), and simulated (dotted line) using a model that includes interfacial compositional grading.

The transmission spectra of the sample discussed above and another sample with 6 ML thick wells are shown in Fig.3. The experimental (theoretical) ISB peaks are at 0.64 (0.80) eV and 0.71 (0.88) eV for the samples with 7 ML and 6 ML thick wells, respectively. The calculation was conducted in the framework of the effective mass approximation including band nonparabolicity and strain effect using the band offsets determined by model-solid theory. There is a relatively large deviation between the calculated and measured ISBT energies. The observed red shift of the ISBT indicates that a significant change in the shape of the QW profile has occurred. Due to the large conduction band offset in this system, ~1.52 eV, the composition grading at the interfaces deduced from the X-ray data analysis can explain the shift to lower energies of the ISBT. Thus the inhomogeneity along the interfaces is mainly responsible for the observed relatively broad peaks.

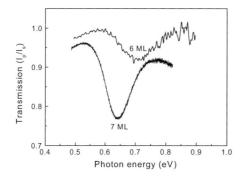

Figure 3. Room temperature transmission spectra of the $In_xGa_{1-x}As/AlAs/In_yAl_{1-y}As$ MQW with 6 ML and 7 ML thick InGaAs wells.

4. Conclusion

We have grown strain compensated $In_xGa_{1-x}As/AlAs/In_yAl_{1-y}As$ MQWs on InP substrate by GSMBE. X-ray diffraction patterns indicated that the structures were grown pseudomorphically to the substrate with 1-2 ML compositional grading at the interfaces. This modification of the QW profile caused a shift of the ISBT to lower energy and substantial broadening of the absorption peak profiles. Intersubband absorption at a wavelength as short as 1.74 µm was observed. These results are very promising for realising optical devices based on ISBT working in the communication wavelength region.

Acknowledgements

This work was supported by the Deutsche Forschungsgemeinschaft within the Forschergruppe FOR 394.

References

[1] Yoshida H, Mozume T, Neogi A and Wada O 1999 Electr Lett 35 1109
[2] Gmachl C, Ng H M and Cho A Y 2000 Appl Phys Lett 77 3722
[3] Jancu J M, Pellegrini V, Colombelli R, Mueller B, Sorba L, Franciosi A, 1998 Appl Phys Lett 73 2621
[4] Asano A, Noda S, Abe T, Sasaki A 1997 J Appl Phys 82 3385
[5] Smet J H, Peng L H, Hirayama Y and Fonstad C 1994 Appl Phys Lett 64 736
[6] Drouot V, Gendry M, Santinelli C, Letart X, Tardy J, Viktorovitch P, Holliger G, Ambri M and Pitaval M 1996 IEEE Trans. Electr. Devices 43 1326

Interfacial and piezoelectric properties of highly strained InGaAs/GaAs quantum well structures grown on (111)A GaAs substrates by MOVPE

Jongseok Kim[1], Soohaeng Cho[1], Alfredo Sanz-Hervás[1], Arnoldo Majerfeld[1,a], Gilles Patriarche[2], and B. W. Kim[3]

[1] Department of Electrical and Computer Engineering, University of Colorado, Boulder, CO 80309-0425, U.S.A.
[2] Laboratoire de Photonique et de Nanostructures, CNRS-UPR 20, Route de Nozay, F-91460 Marcoussis, France
[3] Electronics and Telecommunications Research Institute, P.O. Box 106, Yusong, Taejon 305-600, Korea

[a] Author to whom correspondence should be addressed.
E-mail: arnoldo.majerfeld@colorado.edu

Abstract We achieved InGaAs/GaAs strained Quantum Well (QW) structures on [111]A-oriented substrates with excellent interfacial properties by metal organic vapor phase epitaxial growth. From photoluminescence and Transmission Electron Microscopy (TEM) analyses it is found that the interfacial roughness of QWs with up to 23 % of In is only ±(1-2) monolayers. High resolution X-ray diffractometry analyses indicate that the InGaAs layers are pseudomorphic and that there is no evidence of dislocation formation, which was also confirmed by TEM observations. The abruptness of the QW interfaces and the strain induced piezoelectric field in the well were determined from an analysis of the high-order QW transitions obtained by photoreflectance spectroscopy.

1. Introduction

Strained InGaAs/GaAs Quantum Well (QW) structures on <111>-oriented substrates have recently received considerable attention not only because of the presence of a strong Piezoelectric (PE) field in the wells [1], but also because their optical, electrical, and structural properties, which include possibly a larger critical layer thickness than [100] structures and the suppression of 3-D growth on (111)A substrates, show potential advantages for device applications [2-4]. In order to exploit the unique properties of [111]A QW structures for device applications it is necessary to achieve high crystalline and interfacial quality and to determine their structural and physical properties in detail.

Table 1. Structural parameters of double confinement QW structures

Sample	Structure	Well width	In content
JK152	n-i-p InGaAs/GaAs/AlGaAs	110 Å	21 %
JK247	undoped InGaAs/GaAs/AlGaAs	85 Å	23 %
JK254	undoped InGaAs/GaAs/AlGaAs	90 Å	19 %

The growth and properties of strained InGaAs/GaAs QWs on <111>-oriented GaAs substrates have been studied during the last few years mostly on [111]B-oriented substrates by employing the molecular beam epitaxy process [2,5,6]. Regarding Metalorganic Vapor Phase Epitaxy (MOVPE) grown structures, there are only a few early reports on the optical properties of [111]B InGaAs/(Al,Ga)As QWs with In contents up to 15 % [7] and two recent reports on [111]A-oriented InGaAs/GaAs QW structures [8], besides very recent publications by our group [4,9].

In this paper, we report the optical, structural and PE properties of high quality MOVPE grown highly strained [111]A-oriented InGaAs/GaAs/AlGaAs double confinement QW structures, some of which are centered in n-i-p structures, as needed for laser structures. The properties of these [111]A-oriented QW structures were extensively analyzed by Photoluminescence (PL) and Photoreflectance (PR) spectroscopy, High Resolution X-ray Diffractometry (HRXRD) and Transmission Electron Microscopy (TEM).

2. Experimental

The InGaAs/GaAs/AlGaAs double confinement single QW structures were grown by MOVPE in an atmospheric pressure horizontal quartz reactor. Nominally exactly oriented [111]A GaAs substrates were used. The sources for the chemical elements were trimethylindium, trimethylgallium, trimethylaluminum, and 100 % arsine. The In contents of the QWs are in the range of 19-23 %, which result in up to 1.64 % of in-plane strain on GaAs; the well widths are in the range of 85-110 Å. The principal structural parameters for the samples presented are shown in Table 1. Detailed MOVPE conditions and the QW structures can be found in our previous Publications [4,9]. The optical and structural properties of intentionally undoped InGaAs/GaAs/AlGaAs QW structures were obtained by low temperature PL spectroscopy, TEM and HRXRD analyses. A double confinement structure in an n-i-p configuration was also analyzed by PR spectroscopy in order to obtain the PE properties of [111]A strained layers.

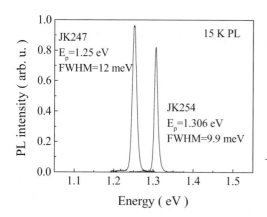

Fig. 1 Photoluminescence spectra taken at 15 K for [111]A InGaAs/GaAs/AlGaAs QW structures with different In contents. E_p denotes the peak energy.

3. Results and discussion

Figure 1 shows the PL spectra taken at 15 K for two different [111]A InGaAs QW structures with 19 % (JK254) and 23 % (JK247) of In. The Full Width at Half Maximum (FWHM) values are 9.9 meV and 12 meV, which correspond to well width fluctuations of only ±1 monolayer (ML) and ±1.5 ML, respectively. It should be noted that for [111] strained QWs and conventional [100] QWs with the same interfacial roughness, relatively larger PL FWHM values should be expected for the [111] QWs due to the presence of the PE field in these wells. The excellent interfacial properties of the [111]A QWs are confirmed by TEM observations. Figure 2 shows the interfaces of the double confinement QW with a PL FWHM of 9.9 meV obtained in the 002 dark field condition. The heterointerfaces between the InGaAs and GaAs layers show low roughness and the interfaces between the AlGaAs and GaAs are excellent. The TEM image shows that there is no evidence of dislocation formation in the strained InGaAs layer. The HRXRD analysis also confirms that the [111]A strained InGaAs QW is pseudomorphic. Figure 3 shows the HRXRD profiles and the theoretical curves for JK247 whose PL spectrum is shown in Fig. 1. From both the asymmetric (422) and symmetric (333) reflections it is proven that the strained QW is pseudomorphic. The HRXRD simulations provided the well width and In content of the structures given in Table 1.

Photoreflectance spectroscopy was performed for JK152, an n-i-p double confinement laser structure. Three optical transitions between the first electron confined state (E1) in the conduction band and the first, third and fifth heavy hole confined states (H1, H3, and H5) in the valence band, were observed at 25 K. From the Franz-Keldysh oscillations observed in the PR spectrum an electric field in the GaAs barrier layers of 21 kV/cm was obtained, which allowed us to determine the PE field in the well [10]. The transition energies obtained from PR spectroscopy were compared to the energies calculated theoretically for the PE QW profile simulated using the measured GaAs barrier field and 123 kV/cm PE field in the well.

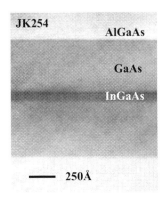
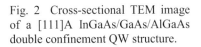

Fig. 2 Cross-sectional TEM image of a [111]A InGaAs/GaAs/AlGaAs double confinement QW structure.

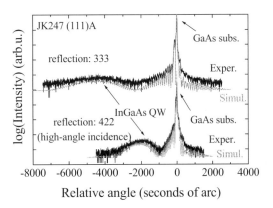

Fig. 3 HRXRD experimental profiles and theoretical curves for a [111]A InGaAs/GaAs/AlGaAs double confinement QW structure.

Table 2. Comparison of transition energies obtained from PR spectroscopy at 25 K and theoretical calculations for the n-i-p InGaAs/GaAs/AlGaAs QW structure (JK152)

Transition	PR	Calculation
E1-H1	1.274 eV	1.276 eV
E1-H3	1.315 eV	1.324 eV
E1-H5	1.343 eV	1.345 eV
Barrier field	21 kV/cm	
PE field		123 kV/cm

The experimental and calculated transition energies are in excellent agreement for all the observed transitions, as shown in Table 2. This agreement indicates that the PE QW band profile is essentially abrupt, as assumed in the theoretical simulations. The difference between the experimental and calculated transition energies corresponds to only ±1.5 ML width change across the actual QW profile.

4. Summary and conclusions

We achieved very high quality [111]A piezoelectric QW structures grown by MOVPE containing highly strained InGaAs layers with up to 23 % In. The excellent interfacial characteristics, abrupt QW profiles and that, even for highly strained InGaAs layers, the structures are pseudomorphic with the absence of dislocations were demonstrated by extensive PL, TEM, HRXRD and PR experimental and theoretical analyses. The PE field in the QW determined from an analysis of the Franz-Keldysh oscillations in the PR spectrum lead to an excellent agreement between the experimental and calculated PR transitions. The successful MOVPE growth of high quality strained QW structures on (111)A GaAs substrates and the unique PE properties of strained QWs grown in this particular orientation show good prospects for the fabrication of novel or enhanced optoelectronic devices exploiting PE properties.

Acknowledgment

The work at the University of Colorado was partly supported by the Electronics and Telecommunications Research Institute, Korea.

References

[1] Smith D L 1986 Solid State Commun. **57** 919
[2] Anan T et al. 1992 Appl. Phys. Lett. **60** 3159
[3] Yamaguchi H et al. 1996 Appl. Phys. Lett. **69** 776
[4] Kim J et al. 2001 J. Cryst. Growth **225** 415
[5] Sánchez J J et al. 1999 Microelectron. J. **30** 363
[6] Vaccaro P O et al. 1994 J. Appl. Phys. **76** 8037
[7] Cartwright A N et al. 1993 J. Appl. Phys. **73** 7767
[8] Sauncy T et al. 1999 Phys. Rev. B **59** 5049; Phys. Rev. B **59** 5056
[9] Kim J et al. 2000 J. Cryst. Growth **221** 525
[10] Cho S et al. 2002 Phys. Stat. Sol. (a) (to appear)

Carrier Dynamics in Self-Organized In(Ga)As/Ga(Al)As Quantum Dots and Their Application to Long-Wavelength Sources and Detectors

P Bhattacharya, A D Stiff-Roberts, S Chakrabarti, S Krishna, C Fischer, T Norris and J Urayama

Department of Electrical Engineering and Computer Science, University of Michigan, Ann Arbor, MI 48109-2122, USA

Abstract. The unique hot-carrier dynamics and intersubband transitions in quantum dots play important roles in defining the device characteristics for mid- and far-infrared sources and detectors. The properties of these devices are described and discussed.

1. Introduction

Long-wavelength (3-20 µm) coherent sources and detectors are required for a variety of applications, such as automobile effluent detection, remote sensing, chemical spectroscopy, and laser radar heterodyne detection. Intersubband transitions in quantum dots can be used for mid- and far-infrared sources and detectors. The unique hot-carrier relaxation rates in quantum dots play an important role in defining the device characteristics. Extensive theoretical and experimental studies of carrier dynamics in In(Ga)As/(Ga(Al)As self-organized quantum dots demonstrate that the intersubband electron relaxation rates, which are strongly temperature dependent, are determined by electron-hole scattering in the dots [1,2]. It has also been demonstrated that a phonon bottleneck exists in the dots for very weak excitations [3-5], with intersubband electron relaxation rates $\geq$ 100 ps. Quantum dot infrared photodetectors (QDIPs) benefit from normal-incidence operation, as well as low dark current and multi-wavelength response [6-9]. The long intersubband relaxation time of electrons in these quantum dots improves the responsivity of the detectors, contributing in part to better high-temperature performance. Quantum dot infrared sources also benefit from the carrier dynamics in quantum dots. Also, surface emission is made possible due to the three-dimensional confinement and the polarization selection rules. We have investigated bipolar electrically injected intersubband sources in which the electrons that relax from the excited states to the ground state are removed by coherent photons, provided in the same cavity by simultaneous interband lasing [10]. We are currently investigating unipolar intersubband quantum dot light emitters designed for surface emission.

In what follows, we will first review the important results related to the dynamics of hot carriers in dots, and then describe the properties of infrared detectors and light emitters based on intersubband transitions in quantum dots.

2. Carrier dynamics in self-organized quantum dots

Theoretical bandstructure calculations [1] using an 8-band $k \cdot p$ description of both conduction and valence band states and a variety of experimental techniques, including photoluminescence and electroluminescence, reveal that there are a number of near-degenerate hole states and a few discrete electron states in $In_{0.4}Ga_{0.6}As/GaAs$ quantum dots. Of these, the dominant ground and first excited states are separated by 50-100 meV, depending on dot material, composition, heterostructure used, and growth parameters. Since the separation between these bands is greater than the LO phonon energy, the LO phonon scattering mechanism, which is mainly responsible for rapid carrier relaxation in bulk semiconductor and quantum wells, is suppressed. Therefore, we expect longer carrier relaxation times in quantum dots. The ground state is two-fold degenerate, whereas the excited state is four-fold degenerate.

Two- and three-pulse femtosecond pump-probe differential transmission spectroscopy (DTS) measurements have been conducted by us on $In_{0.4}Ga_{0.6}As/GaAs$ quantum dot heterostructures, similar to SCH lasers, at different temperatures and for a range of excitation levels [3-5]. A 100 fs, 250 kHz, amplified Ti:Sapphire laser is used to generate a white light source from which 10 nm wide pulses are spectrally selected for the pump beam. The probe pulse consists of a dispersion-compensated near-infrared band between 820 and 1050 nm, selected with a long-pass filter. The DT signal measures the change in the carrier occupation of the levels that are in resonance with the probe spectrum. When the pump and probe pulses are delayed with respect to each other, the transient dot level population is resolved directly. In a three-pulse experiment, a "gain" pulse first creates a non-equilibrium carrier population and this is followed by the pump and probe pulses.

Figure 1(a) shows the excited state (n=2) differential transmission signal for weak excitation (1 e-h pair per 20 dots) in the GaAs barrier at T = 10 K. The population of the excited state shows a fast decay, followed by a very slow one. We have recorded similar data at temperatures as high as 300 K. The hole relaxation times in quantum dots are < 1 ps [4]. The fast excited-state-to-ground-state relaxation is ~ 5-6 ps, and this component decreases as the sample temperature is raised. The slow component is due to non-geminate capture of electrons and holes amongst the dots and reflects the thermal equilibrium that is reached among carriers in higher lying electronic states together with carrier remission and non-radiative recombination from the excited state [3]. Further evidence of injected electrons residing in higher lying states of the dot heterostructure are shown in Fig. 1(b), the three-pulse differential transmission signal corresponding to gain recovery in a laser heterostructure. A gain pulse to establish interband population inversion is followed 14 ps later by a pump pulse to deplete the e-h pairs in the ground state. The fast recovery is due to the excited-to-ground-state transition, and the slow component is due to carrier capture from higher lying states. Figure 1(c) shows the differential transmission signal at GaAs barrier energy ± 20 meV with the pump tuned to the barrier energy. It is evident that as the temperature is increased, most of the carriers remain in the barrier and the dot excited states.

The unique carrier dynamics in the In(Ga)As/Ga(Al)As quantum dot system are beneficial to the operation of both intersubband detectors and sources. For detectors, the long lifetime of the carriers in the higher energy dot states ensures that photoexcited

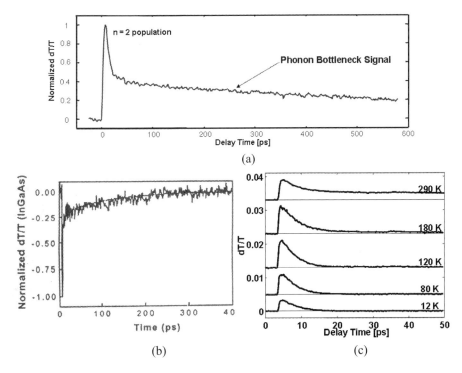

Figure 1. (a) Excited state (n=2) differential transmission signal for weak excitation (1 e-h pair per 20 dots) in the GaAs barrier at T = 10 K; (b) three-pulse differential transmission signal corresponding to gain recovery in a laser heterostructure; and (c) differential transmission signal at GaAs barrier energy ± 20 meV with the pump tuned to the barrier energy.

carriers will contribute to the photocurrent. In intersubband light sources, a long effective lifetime in the upper states implies that the intersubband stimulated emission time can also be large, and therefore the threshold current can be low.

3. Quantum dot infrared photodetectors

3.1. Characteristics of discrete devices

The responsivity of InAs/GaAs vertical QDIPs must be increased in order to be competitive with other infrared (IR) detectors. In order to achieve this, we have increased the absorption region in QDIPs by growing a large number of uncoupled quantum dot layers. Growth of these heterostructures is achieved by molecular beam epitaxy (MBE). The most important parameter that had to be calibrated to increase the number of quantum dot layers was the GaAs barrier thickness required to prevent dislocations. We have grown device heterostructures, shown in Fig. 2, with 30 to 70 dot

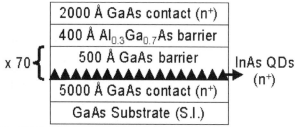

Figure 2. QDIP device heterostructure with 70 dot layers.

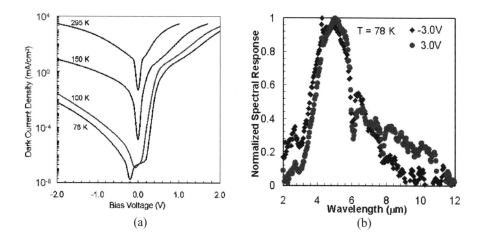

Figure 3. (a) The dark current density at various temperatures and (b) the MWIR spectral response at 78 K for typical QDIPs.

layers. Cross-sectional transmission electron microscopy images indicate that the heterostructures are almost defect-free. The device heterostructure consists of a 5000 Å bottom GaAs contact layer ($n=2\times10^{18}$ cm^{-3}) and a 2000 Å top GaAs contact layer ($n=2\times10^{18}$ cm^{-3}). A 500 Å GaAs intrinsic buffer was grown at 620 °C after the bottom contact layer and before the first QD layer. The InAs QDs were 2.0 monolayers thick and doped with Si to a nominal concentration of $n=1\times10^{18}$ cm^{-3}. A 400 Å Al$_{0.3}$Ga$_{0.7}$As current-blocking barrier was grown after the last GaAs barrier and before the top GaAs contact layer to help reduce dark current. Mesa-shaped devices, 100 to 300 µm in radius, and with top and bottom (ring) contacts are fabricated by standard photolithography and metallization techniques.

The devices are characterized by the following measurements: dark current, Fourier transform infrared (FTIR) spectral response, and 800 K calibrated blackbody response. Typical dark current densities, shown in Fig. 3(a) are very low ($J_{dark} = 6\times10^{-10}$ A/cm^2, T = 100 K, V$_{bias}$ = 0.1 V), as is usually seen in QDIPs. The spectral response, shown in Fig. 3(b), was obtained at 78 K. The spectral response, which we believe originates from electron transitions from the ground state to the continuum in the dots,

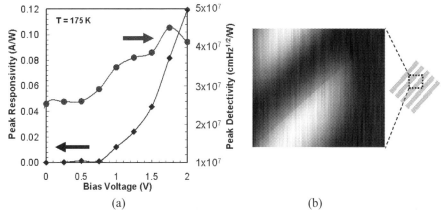

Figure 4. (a) The peak responsivity and detectivity for a 70-layer QDIP measured at 175 K; and (b) the raster-scanned image of a 500 °C heating element from a hot plate (shown schematically in the inset) partially showing two metal strips which was obtained from a (13x13) QDIP array at 80 K.

peaks at approximately 5 μm, which is in the MWIR atmospheric window. Figure 4(a) shows the peak responsivity and detectivity for the 70-layer QDIP measured at 175 K. For V_{bias} = 2.0 V, the responsivity is 0.12 A/W. The maximum detectivity obtained for the 70-layer QDIP was 4.51×10^7 cmHz$^{1/2}$/W for V_{bias} = 1.75 V and T = 175 K, which is very large for such high temperature operation. The variation of responsivity with bias is asymmetric with respect to positive and negative biases, and the responsivity increases greatly with increasing forward bias as band bending across the absorption region and the $Al_{0.3}Ga_{0.7}As$ barrier allows more photocurrent to be collected at the contacts. The peak detectivity increases with responsivity until some maximum is reached, beyond which the detectivity begins to decrease because of excess noise current.

3.2. QDIP arrays

Ultimately, it is of interest to incorporate QDIPs in focal plane arrays (FPAs), which is a hybrid, bump-bonded circuit comprised of an IR detector array to sense the infrared light and a silicon read-out circuit to access each pixel of the detector array and initiate digital signal processing.

As a preliminary investigation into the imaging capabilities with dot arrays, we have performed a raster-scanning experiment with a single detector [11]. In this technique, the field of view is simultaneously scanned in the x- and y- directions by two gold-coated mirrors. A small (13x13) interconnected array is used because it is easier to collect infrared light over a larger area. The image of a portion of a heating element of a hot plate at 500 °C is shown in Fig. 4(b). Note that the field of view is limited by the exit aperture of the bracket used to mount the mirrors.

4. Quantum dot infrared sources

4.1. Bipolar quantum dot intersubband light sources

We have reported spontaneous emission at 12 µm, resulting from intersubband transitions in $In_{0.4}Ga_{0.6}As/GaAs$ quantum dots [10]. The output from the surface-emitting device was measured at 80 K and 300 K. The device was essentially a 4-dot layer interband laser heterostructure. The interband lasing occurring in the device at 1 µm due to ground state electron-hole recombination helped to deplete the electron ground state in the dots and to establish a population inversion between the excited states and the ground state. The far-infrared signal was enhanced after the interband transition reached threshold at 300 K. In order to determine the possibility of stimulated emission and lasing with this scheme, we established the condition for intersubband population inversion from the carrier and photon rate equations, and also calculated the intersubband gain. The intersubband gain was dependent on the inhomogeneous linewidth broadening of the transition. Gains as high as 150 cm^{-1} can easily be achieved.

For observing stimulated emission, plasmon-enhanced waveguides were designed and grown by MBE. Edge-emitting devices were fabricated using standard photolithography, lift-off, and a combination of dry and wet-etching, followed by ohmic contact formation. The width of the waveguide was varied from 20-60 µm, and the length was varied from 800-1200 µm. Therefore, the devices are multimode laterally. The output is broad and multimode, with a more distinct peak at 13 µm which, in all probability, originates from the electron transitions from higher quantum dot excited levels to the ground state, separated by ~ 90-100 meV. Theoretical calculations have confirmed the presence of these excited states [2]. The measured light (power)-current characteristics are shown in Fig. 5(a). The threshold in the FIR output occurs at 1.6 times the interband laser threshold. The additional carriers injected after the interband laser reaches threshold recombine to provide for high coherent photon density required for

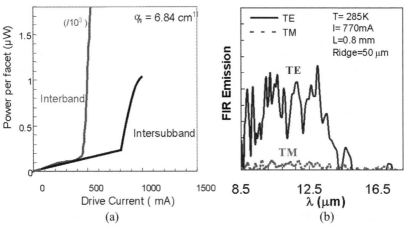

Figure 5. (a) The measured light (power)-current characteristics, and (b) the polarization dependence of the far-infrared emission showing the intersubband transition to be strongly TE polarized.

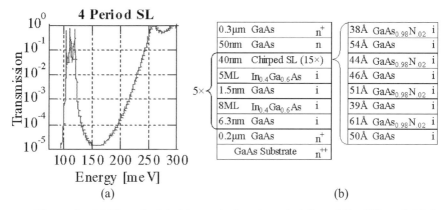

Figure 6. (a) The calculated transmission characteristics of the CSL and (b) the device heterostructure for a light-emitting diode grown by MBE.

intersubband gain. In essence, the device converts the more readily available near-IR photons to the more difficult to obtain mid-IR photons. Unfortunately, device heating prevents us from making measurements at higher injection currents. We believe, however, that these devices demonstrate intersubband gain and stimulated emission. The polarization dependence of the far-infrared emission, measured using a far-infrared polarizer, is found to be strongly TE polarized, as is shown in Fig. 5(b).

4.2. Quantum cascade lasers with quantum dot active regions

It is apparent that the bipolar device described above, while adequate to demonstrate intersubband light emission, is not suitable for achieving significant stimulated emission and lasing. A unipolar design, similar to a quantum cascade laser with quantum wells, was therefore investigated. We have made preliminary studies of surface-emitting quantum cascade devices with $In_{0.4}Ga_{0.6}As/GaAs$ quantum dots. To facilitate tunneling between ground and excited states of adjacent quantum dot layers, a chirped superlattice (CSL), consisting of alternate GaAsN and GaAlAs layers, is inserted, and the complete device is designed with multiple periods of QD/CSL. The calculated transmission characteristics of the superlattice are shown in Fig. 6(a). The heterostructure, for a light-emitting diode, grown by MBE, is shown in Fig. 6(b). Preliminary light-current characteristics are shown in Fig. 7. Refinement in the design of the device and of material characteristics and the design of Bragg reflectors for a surface emitting laser are in progress.

5. Conclusion

The carrier dynamics in quantum dots provide advantages for infrared sources and detectors based on intersubband transitions within the dots. QDIPs with state-of-the-art performance and novel quantum dot infrared sources have been described.

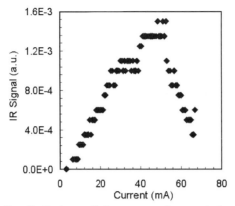

Figure 7. Preliminary light-current characteristics for the $In_{0.4}Ga_{0.6}As/GaAs$ quantum dot cascade light emitter with a chirped superlattice.

6. Acknowledgements

This work is supported by the Army Research Office under Grants DAAD19-01-1-0462 and DAAD19-00-1-0394, the Air Force Office of Scientific Research under Grant F49620-00-1-0328, and the National Science Foundation under Grant ECS-9820129.

References

[1]   Jiang H and Singh J 1997 Phys. Rev. B 56 4696-4701
[2]   Bhattacharya P, Kamath K, Singh J, Klotzkin D, Phillips J, Jiang H, Chervela N, Norris T, Sosnowski T, Laskar J and Murty M R 1999 IEEE Trans. Electron. Devices 46 871-883
[3]   Urayama J, Norris T B, Singh J and Bhattacharya P 2001 Phys. Rev. Lett. 86 4930-4933
[4]   Sosnowski T, Norris T, Jiang H, Singh J, Kamath K and Bhattacharya P 1998 Phys. Rev. B 57 R9423- R9426
[5]   Urayama J, Norris T, Jiang H, Singh J and Bhattacharya P 2002 Appl. Phys. Lett. 80 2162-2164
[6]   Berryman K W, Lyon S A and Segev M 1997 Appl. Phys. Lett. 70 1861-1863
[7]   Pan D, Towe E and Kennerly S 1998 Appl. Phys. Lett. 73 1937-1939
[8]   Phillips J, Bhattacharya P, Kennerly S W, Beekman D W and Dutta M 1999 IEEE J. Quantum Electron. 35 936-943
[9]   Stiff A D, Krishna S, Bhattacharya P and Kennerly S 2001 IEEE J. of Quantum Electron. 37 1412-1419
[10] Krishna S, Bhattacharya P, Singh J, Norris T, Urayama J, McCann P J and Namjou K 2001 IEEE J. Quantum Electron. 37 1066-1074
[11] Stiff-Roberts A D, Chakrabarti S, Pradhan S, Kochman B and Bhattacharya P 2002 Appl. Phys. Lett. 80 3265-3267

Single-Electron Transistors

Peter Hadley,[1] Günther Lientschnig,[1] and Ming-Jiunn Lai[2]

[1]Department of Nanoscience, Delft University of Technology,
Lorentzweg 1, 2628 CJ Delft, The Netherlands.
[2]ERSO/ITRI Chung Hsing Road, Chutung, Hsinchu, Taiwan 310, R.O.C.

Abstract. Single-electron transistors (SET's) are often discussed as elements of nanometer scale electronic circuits because they can be made very small and they can detect the motion of individual electrons. However, SET's have low voltage gain, high output impedances, and are sensitive to random background charges. This makes it unlikely that single-electron transistors would ever replace field-effect transistors (FET's) in applications where large voltage gain or low output impedance is necessary. The most promising applications for SET's are charge-sensing applications such as the readout of few electron memories, the readout of charge-coupled devices, and precision charge measurements in metrology.

1. Introduction

Single-electron transistors [1] have been made with critical dimensions of just a few nanometers using metals, [2] semiconductors, [3] carbon nanotubes, [4] and individual molecules. [5-7] Some of the smallest transistors operate at room temperature. In this paper, first some basics of single-electron transistors are introduced and then a few different kinds of SET's are described. The real problems preventing the use of SET's in most applications are the low gain, the high output impedance, and the background charges. Each of these problems is discussed and the circuits where SET's show the most promise are described.

2. SET Basics

A single-electron transistor consists of a small conducting island coupled to source and drain leads by tunnel junctions and capacitively coupled to one or more gates. The geometry of a SET is shown in Fig. 1(a) and the equivalent electrical circuit is shown in Fig. 1(b). A stray capacitance C_0 from the island to ground and a random background charge on the island Q_0 are also included in the model. There are two gates in the equivalent circuit because two gates are often used in practice. For example, one gate can be used to tune the background charge while the other is used as the input of the SET. A straightforward electrostatic calculation shows that the voltage of the island as a function of the number of electrons on the island is,

$$V(n) = (-ne + Q_0 + C_1V_1 + C_2V_2 + C_{g1}V_{g1} + C_{g2}V_{g2})/C_\Sigma. \qquad [1]$$

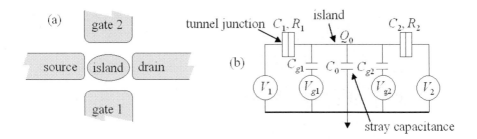

Fig. 1 (a) The geometry of a SET with two gates. (b) The equivalent circuit for a SET.

Here n is the number of electrons on the island, e is the positive elementary charge, and C_Σ is the total capacitance of the island $C_\Sigma = C_1 + C_2 + C_{g1} + C_{g2} + C_0$. The energy it takes to move an infinitesimally small charge dq from ground at a potential $V = 0$ to the island is Vdq. As soon as charge is added to the island, the voltage of the island changes. The energy needed to take a whole electron from ground and put it on the island is,

$$\int_0^{-e} Vdq = -eV(n) + \frac{e^2}{2C_\Sigma}. \qquad [2]$$

Here n is the number of electrons on the island before the final electron is added. The term $E_c = e^2/(2C_\Sigma)$ is called the charging energy and it sets the energy scale for single-electron effects. The charging energy is typically in the range 1 - 100 meV.

There are four single-electron tunneling events that can take place. An electron can tunnel left through junction 1, right through junction 1, left through junction 2 or right through junction 2. The energies associated with these four tunnel events can be calculated using Eq. 2,

$$\Delta E_{1R} = eV_1 - eV(n) + E_c, \quad \Delta E_{1L} = -eV_1 + eV(n) + E_c, \qquad [3]$$
$$\Delta E_{2R} = -eV_2 + eV(n) + E_c, \quad \Delta E_{2L} = eV_2 - eV(n) + E_c.$$

If any of these energies are negative, an electron will tunnel. If all four of these energies are positive, a condition known as the Coulomb blockade is achieved and no electrons will tunnel. Figure 2 shows a calculation of the conductance of a SET as a function of the gate voltage and the bias voltage. The straight lines that form the edges of the diamond shaped regions are given by setting $\Delta E = 0$ in Eqs. 3. The diamond labeled 0 is a region of Coulomb blockade where there are zero excess electrons on the island and the diamond labeled 1 is a region of Coulomb blockade with one excess electron on the island, etc. When a bias voltage is applied that is great enough to overcome the Coulomb blockade, current flows as electrons tunnel from the source onto the island and then from the island to the drain. In the region labeled (0,1), only one electron at a time can pass through the SET at low temperatures. Exact formulas for the current in each diamond can be derived in the limit of low temperature. For all of the diamonds labeled with just a single number n, the current is zero. In the diamonds labeled by two numbers $(n,n+1)$ the charge on the island is alternately n and $n+1$ as current flows through the SET. The formula for the current in this case is,

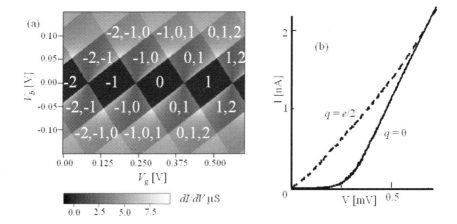

Fig. 2 (a) The calculated conductance (dI/dV) through a SET plotted as a function of the gate voltage and the bias voltage. The diamonds are labeled with the charge states that are occupied at low temperature. (b) The experimental current-voltage characteristics of a SET are shown for two values of the gate. The solid line is for an induced gate charge of zero; this is the condition for maximum Coulomb blockade. The dashed line is for an induced gate charge of $e/2$; this is the minimum Coulomb blockade.

$$I = \frac{ab}{C_\Sigma(R_2 a + R_1 b)}, \qquad [4]$$

where $a = C_\Sigma V_1 + ne - Q_0 - C_1 V_1 - C_2 V_2 - C_{g1} V_{g1} - C_{g2} V_{g2} + e/2$ and $b = -C_\Sigma V_2 - ne + Q_0 + C_1 V_1 + C_2 V_2 + C_{g1} V_{g1} + C_{g2} V_{g2} - e/2$. This is a useful formula because in most applications a SET is biased in the $(n,n+1)$ diamond. From this formula, the voltage gain can be determined. For a current biased SET, the maximum voltage gain is $-C_{g1}/C_1$ if junction 2 is grounded and $-C_{g1}/C_2$ if junction 1 is grounded. The maximum transconductance (dI/dV_{g1}) at low temperatures can also be determined from Eq. 4. Near the boundary between the diamonds labeled n and $n,n+1$; the quantity $b \approx 0$ and the transconductance is $dI/dV_{g1} \approx C_g/(R_2 C_\Sigma)$. Near the boundaries between the diamond labeled $n,n+1$ and $n+1$; the quantity $a \approx 0$ and the transconductance is $dI/dV_{g1} \approx -C_g/(R_1 C_\Sigma)$. The individual resistances of the two junctions in a SET are typically determined by measuring the transconductance.

As more charge states are included, the formula for the current gets more complicated. The formula for the current in the diamonds labeled $(n-1,n,n+1)$ is,

$$I = \frac{R_1(a+(n-1)e)(b-ne)(b-(n-1)e) + R_2(a+(n-1)e)(a+ne)(b-ne)}{C_\Sigma\left(R_2^2(a+(n-1)e)(a+ne) + R_1 R_2(a+(n-1)e)(b-ne) + R_1^2(b-(n-1)e)(b-ne)\right)}. \qquad [5]$$

3. Types of single-electron transistors

Single-electron transistors can be made using metals, semiconductors, carbon nanotubes, or single molecules. Aluminum SET's made with Al/AlO$_x$/Al tunnel junctions are the SET's that have been used most often in applications. This kind of SET is used in metrology to measure currents, capacitance, and charge. [8] They are used in astronomical measurements [9] and they have been used to make primary thermometers. [10] However, many fundamental single-electron measurements have been made using GaAs heterostructures. The island of this kind of SET is often called a quantum dot. Quantum dots have been very important in contributing to our understanding of single-electron effects because it is possible to have just one or a few conduction electrons on a quantum dot. The quantum states that the electrons occupy are similar to electron states in an atom and quantum dots are therefore sometimes called artificial atoms. The energy necessary to add an electron to a quantum dot depends not just on the electrostatic energy of Eq. 2 but also on the quantum confinement energy and the magnetic energy associated with the spin of the electron states. By measuring the current that flows thorough a quantum dot as a function of the gate voltage, magnetic field, and temperature allows one understand the quantum states of the dot in quite some detail. [11]

The SET's described so far are all relatively large and have to be measured at low temperature, typically below 100 mK. For higher temperature operation, the SET's have to be made smaller. Ono et al. [3] used a technique called pattern dependent oxidation (PADOX) to make small silicon SET's. These SET's had junction capacitances of about 1 aF and a charging energy of 20 meV. The silicon SET's have the distinction of being the smallest SET's that have been incorporated into circuits involving more than one transistor. Specifically, Ono et al. constructed an inverter that operated at 27 K. Postma et al. [4] made a SET that operates at room temperature by using an AFM to buckle a metallic carbon nanotube in two places. The tube buckles much the same way as a drinking straw buckles when it is bent too far. Using this technique, a 25 nm section of the nanotube between the buckles was used as the island of the SET and a conducting substrate was used as the gate. The total capacitance achievable in this case is also about 1 aF. Pashkin et al. [2] used e-beam lithography to fabricate a SET with an aluminum island that had a diameter of only 2 nm. This SET had junction capacitances of 0.7 aF, a charging energy of 115 meV, and operated at room temperature. SET's have also been made by placing just a single molecule between closely spaced electrodes. Park et al. [5] built a SET by placing a C_{60} molecule between electrodes spaced 1.4 nm apart. The total capacitance of the C_{60} molecule in this configuration was about 0.3 aF. Individual molecules containing a Co ion bonded to polypyridyl ligands were also placed between electrodes only 1-2 nm apart to fabricate a SET. [6] In similar work, Liang et al. [7] placed a single divanadium molecule between closely spaced electrodes to make a SET. In the last two experiments, the Kondo effect was observed as well as the Coulomb blockade. The charging energy in the molecular devices was above 100 meV.

One of the conclusions that can be drawn from this review of SET devices is that small SET's can be made out a of variety of materials. Single electron transistors with a total capacitance of about 1 aF were made with aluminum, silicon, carbon nanotubes and individual molecules. It seems unlikely that SET's with capacitances smaller than the capacitances of the molecular devices can be made. This sets a lower limit on the smallest capacitances that can be achieved at about 0.1 aF. Achieving small capacitances such as this has been a goal of many groups working on SET's. However, while some of the device characteristics improve as a SET is made smaller, some of the device characteristics get worse as SET's are made smaller. For some applications, the single molecule SET's are too small to be useful. As SET's are made smaller, there is an increase in the operating temperature, the operating frequency, and the device packing density. These are desirable consequences of the shrinking of SET devices. The undesirable consequences of the shrinking of SET's are that the electric fields increase, the current densities increase, the operating voltage increases, the energy dissipated per switching event increases, and the power dissipated per unit area increases, the voltage gain decreases, the charge gain decreases, and the number of Coulomb oscillations that can be observed decrease.

4. The problems: gain, high output impedance, and background charges

Voltage gain is one of the properties of a SET that decreases as SET's are made smaller. This is because voltage gain decreases with decreasing gate capacitance. It is difficult to achieve a large gate capacitance when the island of a SET consists of a single molecule. For the single molecule devices, the gate capacitance can be as small as a few zeptoFarads. In this case, tens of volts have to be applied at the input to modulate the output by tens of millivolts. This results in a voltage gain on the order of 0.001; the transistors attenuate the signals by a factor of about 1000. The voltage gain in a SET is the ratio of the gate capacitance to the junction capacitance. As the gate capacitance is increased for fixed junction capacitance and fixed temperature, the voltage gain first increases and then it decreases. This is illustrated in Fig. 3(a) where the voltage gain is plotted as a function of gate capacitance for different temperatures. In all of the curves the junction capacitance is assumed to be 0.1 aF. The voltage gain increases with increasing gate capacitance until the charging energy is on the order of k_BT and then the voltage gain drops sharply.

Thus for every junction capacitance and temperature, there is a maximum voltage gain. Figure 3(b) is a plot of the maximum gain. To determine the maximum gain, both the gate capacitance and the bias current of the SET were varied. The graph shows that it will be very difficult to make SET's with voltage gain greater than one that operate at room temperature. It will be even harder to get them to operate in a dense integrated circuit, which usually has a temperature of about 400 K. For room temperature voltage gain, the junction capacitance will have to be about 0.1 aF with a gate capacitance of 0.3 aF. This kind of SET has not yet been fabricated. So far, the largest voltage gain that has been observed is 5.2 and that was measured at 100 mK. [12] The highest temperature

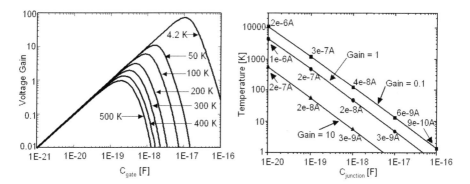

Fig. 3(a) The voltage gain is plotted as a function of the gate capacitance and temperature for a junction capacitance of 0.1 aF. The voltage gain depends on the bias current. The bias current was adjusted to achieve maximum gain. The bias currents that were used were I = 1 nA at 4.2 K, I = 20 nA at 50 K, I = 50 nA at 100 K, I = 100 nA at 200 K, I = 200 nA at 300 K, I = 200 nA at 400 K, I = 300 nA at 500 K. (b) The maximum voltage gain possible for a given junction capacitance is plotted as a function of temperature. The currents which resulted in the maximum gain are given in the plot.

where voltage gain greater than one was observed is 27 K. [3] While voltage gain at room temperature seems difficult, a voltage gain of 10 would be possible at 77 K, a voltage gain of 100 should be possible at 4.2 K, and a voltage gain of 1000 should be possible at 100 mK. The low temperature SET's could be useful for sensitive low temperature measurements.

Some of the circuit architectures that have been proposed for single-electron transistors are basically copies of the semiconducting architectures and require SET's with voltage gain. [13] Because of the limited gain of room temperature SET's, it now seems unlikely that dense integrated circuits based on these principles will ever be made. However, a transistor does not necessarily have to exhibit voltage gain to be useful. For single-electron transistors it is more relevant to consider the charge gain. The charge gain is the modulation of the charge that passes through the SET divided by the change in charge on the gate. This is a frequency dependent quantity. By waiting long enough, it is always possible to transport more charge through the SET than was added to the gate. Charge gain greater than one can easily be achieved at room temperature. The charge gain is maximum at low bias voltages and low temperatures where it is $g_{charge} = (dI/dV_{g1})/(2\pi f C_g) = 1/(2\pi f R C_\Sigma)$. Here f is the frequency at which the charge is modulated and R is the lower of the two junction resistances. This result can be derived from Eq. 4.

Charge gain would be used in situations where charge needs to be measured, for instance to readout a memory cell or to readout a charge coupled device. The speed of the charge readout would be limited by the RC-delay formed by the resistance of the SET and the capacitance at the output of the SET. This capacitance depends on the parasitic capacitance of the wire at the output of the SET and the input capacitance of any devices connected to the SET. The parasitic capacitance of a wire is approximately 100 aF/μm. The large output impedance of a SET of at least 100 kΩ makes the SET an intrinsically slow device. To speed up the charge measurement, conventional field-effect transistors (FET's) should be placed as close as possible to the SET. The field-effect transistor can then buffer the high output impedance of the SET. Using both SET's and FET's in a circuit is a retreat from the idea that small SET's will someday replace FET's entirely. However, it now seems unlikely that SET's could ever replace FET's. Perhaps twenty different logic schemes that utilize single-electron transistors exclusively have been proposed but none is widely accepted as being practical. Until a practical scheme in developed, the best way to proceed is to use SET's only as sensitive charge sensors and perform all other functions with conventional FET's.

The background charge problem is another important issue that is inhibiting the widespread use of SET's. The origin of the background charge problem is the extreme charge sensitivity of SET's. A single charged vacancy or an interstitial ion in the oxide near a SET can be enough to switch the transistor from the being conducting to being nonconducting. The same kinds of charged defects are present and move in field-effect transistors but most field-effect transistors are not as sensitive to charge so the consequences of these background charges are not as great. The only effective way to compensate for this problem is to use field-effect transistors to tune the background charges away. One circuit that accomplishes this is the charge-locked loop shown in Fig. 4(a). A charge-locked loop uses feedback to keep the charge on the island of a SET constant. This improves the speed, the linearity and the dynamic range of the charge measurements.

The three problems of low gain, high output impedance, and sensitivity to background can all be remedied by combining SET's with FET's. In such hybrid

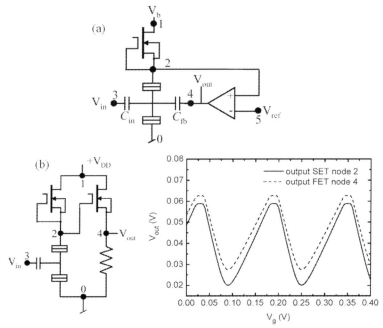

Fig. 4. (a) A charged lock loop that automatically tunes away the background charge. Provided that the amplifier has enough gain, the voltage gain in this circuit is set by the ratio of the input and output capacitors C_{in}/C_{out}. The linearity and the dynamic range of the charge measurement are also improved by the charge-locked loop. The FET is used to current bias the SET. (b) The schematic of a current biased SET with a FET output stage and the corresponding SPICE simulation. The solid line is the voltage at the output voltage of the SET stage (node 2), and the dashed line is the voltage at the output of the FET stage (node 4). A voltage of 0.4 V has been subtracted from the voltage at node 4 to remove a dc offset. The SPICE source code for simulating the circuits shown can be found at http://qt.tn.tudelft.nl/research/set/spice/.

circuits, the SET's provide the charge sensitivity while the FET's provide the gain and the low output impedance. In order to design SET/FET circuits, it is important to have a simulation package that will model both SET's and FET's. The simulation package SPICE is one of the few packages that will do this. [14] Figure 4(b) shows a single-electron transistor that is current biased by a FET and the corresponding SPICE simulation of the circuit. A second FET is used to buffer the output of the SET which increases the speed of the circuit.

5. Conclusions

Single-electron transistors are the most sensitive charge-measuring devices presently available. They have become an important tool in the field of fundamental measurements. The fact that most SET's only perform at low temperature is not seen as a disadvantage for fundamental measurements because these measurements are often performed at low temperature anyway to reduce noise. In fact, very low temperature operation (less than 100 mK) is seen as an advantage because many semiconducting devices don't work in this temperature range. However for mass-market applications, room temperature operation is necessary. The SET's that operate at room temperature have the problems of low gain, high output impedance, and background charges. No room temperature SET logic or memory scheme is now widely accepted as being practical. The most promising room temperature applications for SET's are in charge sensing circuits where the problems of low gain, high output impedance, and background charges can be solved by integrating SET's with field-effect transistors.

References

[1]. For an overview of single-electron devices and their applications, see: K. K. Likharev, Proceedings of the IEEE **87** p. 606 (1999).
[2] Yu. A. Pashkin, Y. Nakamura, and J.S. Tsai, Appl. Phys. Lett. **76** 2256 (2000).
[3] Yukinori Ono, Yasuo Takahashi, Kenji Yamazaki, Masao Nagase, Hideo Namatsu, Kenji Kurihara, and Katsumi Murase, Appl. Phys. Lett. **76** p. 3121 (2000).
[4] H. W. Ch. Postma, T.F. Teepen, Z. Yao, M. Grifoni, C. Dekker, Science, **293** p. 76 (2001).
[5] H. Park, J. Park, A. K. L. Lim, E. H. Anderson, A. P. Alivisatos and P. L. McEuen, Nature **407** p. 57 (2000).
[6] Jiwoong Park, Abhay N. Pasupathy, Jonas I. Goldsmith, Connie Chang, Yuval Yaish, Jason R. Petta, Marie Rinkoski, James P. Sethna, Héctor D. Abruña, Paul L. McEuen, and Daniel C. Ralph, Nature 417 p. 722 (2002).
[7] Wenjie Liang, M. P. Shores, M. Bockrath, J. R. Long, Hongkun Park, Nature 417 p. 725 (2002).
[8] N. M. Zimmerman and M. W. Keller, "Electrical Metrology with Single Electrons", to be published in IOP J. Phys.
[9] T. Stevenson, A. Aassime, P. Delsing, R. Schoelkopf, K. Segall, C. Stahle, IEEE Transactions on Applied Superconductivity, Vol. 11. No. 1, pp. 692-695, March 2001.
[10] K. Gloos, R. S. Poikolainen, and J. P. Pekola, Appl. Phys. Lett. 77, 2915 (2000).
[11] L. P. Kouwenhoven, D. G. Austing, S. Tarucha, Reports on Progress in Physics **64** pp. 701-736 (2001).
[12] C. P. Heij and P. Hadley, Review of Scientific Instruments **73** pp. 491-492 (2002).
[13] J. R. Tucker, J. Appl. Phys. 72 4399 (1992).
[14] G. Lientschnig, I. Weymann, and P. Hadley, submitted to Jap. J. Appl. Phys.

Selective Formation of High-density and High-uniformity InAs/GaAs Quantum Dots for Ultra-small and Ultra-fast All-optical Switches

Y. Nakamura[1,2], H. Nakamura[1], S. Ohkouchi[1], N. Ikeda[1,2], Y. Sugimoto[1], and K. Asakawa[1]

[1]*The Femtosecond Technology Research Association, 5-5 Tokodai, Tsukuba 300-2635, Japan*
[2]*The New Energy and Industrial Technology Development Organization,
1-1-3 Higashi Ikebukuro, Tokyo 170, Japan*

Abstract We report on two types of quantum dots (QDs) to realize ultra-fast monolithic optical switches. One type is selectively formed by controlling the nucleation sites on patterned substrates with nano-hole array, showing two-dimensional ordering with a 100nm-periodicity and high selectivity in the QD density giving dense QDs ($\sim 1 \times 10^{10} \text{cm}^{-2}$) on nano-holes and less QDs ($\sim 1 \times 10^9 \text{cm}^{-2}$) on a flat region. The other type is formed on flat substrates by the two-step-growth method, which has a photoluminescence emission at 1.3µm with satisfying both high density ($3.5 \times 10^{10} \text{cm}^{-2}$) and high uniformity (photoluminescence linewidth of 29meV).

1. Introduction

Ultra-fast demultiplexing has been reported with symmetric-Mach-Zehnder-type (SMZ) all-optical switch by Tajima et al [1, 2], whose switching speed is not restricted to the carrier lifetime of optical nonlinear materials. However, this device is as large as 20mm-long hybrid system, it is important to integrate it into monolithic ultra-small devices within 1mm-square. To approach this target, we have been studying photonic-crystal-based SMZ (PC-SMZ) all-optical switch with quantum dots (QDs) as an optical nonlinear material [3, 4]. For this purpose, uniform high-density QDs should be selectively formed only in a required region of the devices. These requirements for QDs are essential also for other devices such as lasers and semiconductor optical amplifiers.

Figures 1 (a) and (b) show a schematic plan view and a cross-section of our proposed PC-SMZ all-optical switch, respectively [3], where InAs QDs should be selectively formed as the optical nonlinear material only in the phase-shift regions of the waveguide defined by photonic crystal (see two hatched regions in Fig. 1 (a)). As shown in Fig. 1 (b), multi-QD-layer has to be stacked in these regions. In this paper, we report how to approach this structure with our special growth methods of QDs. One method is a selective formation of QDs by controlling the nucleation sites on patterned substrates [5-9]. The other is the two-step-growth method on flat substrates, which enabled high-density and high-uniformity QDs with an emission wavelength of 1.3µm for optical communications [10].

2. Site-controlled quantum dots on patterned substrates

Figure 2 (a) shows a schematic diagram of selectively grown QDs on the mesa stripe where the PC-waveguide is formed. At first, we fabricated a patterned substrate with a nano-hole array having a 100nm-periodicity and a hole-diameter of 50nm by electron beam lithography and reactive ion beam etching as indicated in the hatched regions in Fig. 2 (a).

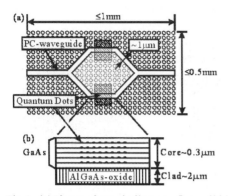
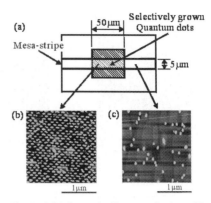

Fig. 1 (a) Planar schematic diagram of monolithic all-optical switch based on photonic crystal and (b) its cross-sectional structure.

Fig. 2 (a) Schematic diagram of waveguide. AFM images of (b) selectively grown InAs QDs on patterned area and (c) less QDs on flat area.

Next, a mesa-stripe with a width of 5μm and a height of 65nm was formed on this substrate by wet chemical etching, which restricts nano-holes only in the area of 50μm x 5μm for selective nucleation of QDs as shown in Fig. 2 (a). The patterned substrates were cleaned by an irradiation of atomic-hydrogen and InAs QDs were grown by molecular beam epitaxy. We deposited a three-monolayer (ML) -thick InAs at 400°C with a growth rate of 0.02ML/s and desorbed it at 525°C to embed a part of the deposited InAs into the nano-holes, which is expected to work as a template for an upper QD layer with a periodic lateral strain modulation. Then, we grew a 10nm-GaAs spacer and a 2ML-InAs at 477°C with a growth rate of 0.01ML/s to form QDs. Figures 2 (b) and (c) show atomic force microscope (AFM) images of the InAs QDs formed on the nano-holes and the flat region on the mesa-stripe, respectively. The InAs QDs on the nano-holes are two-dimensionally ordered with the 100nm-periodicity, corresponding to a QD density of $\sim 1\times 10^{10} cm^{-2}$, though the shape of the QDs are not a circle and should be improved. On the other hand, a QD density on the flat region on the mesa is as low as $\sim 1\times 10^{9} cm^{-2}$, demonstrating high selectivity in the QD density. In this sample, the mesa edges indicated only a slight effect as an absorber for diffused In adatoms to reduce the QD density on a flat region of the mesa, but the effect was enhanced when we repeated the combination of the InAs deposition and the desorption in a separate experiment. In addition, very recently, we have improved the shape of InAs QDs with almost perfect ordering by using a different growth condition, which will be reported later.

3. Uniform high-density quantum dots on flat substrates

It is well known that, in the conventional Stranski–Krastanov growth-mode on flat surfaces, there is a trade-off relation between the uniformity and the density of QDs on the flat surfaces [11]. Namely, inhomogeneous small QDs with a high density are formed at a high growth-rate, while homogeneous large QDs with a low density are formed at a low growth-rate. To combine both merits, we tried the two-step-growth, which includes the first growth, an InAs is grown at the high rate until just above the critical thickness to form high-density nuclei as a template, and the second growth, the InAs is continuously grown at the low rate to enlarge and homogenize the QDs. For example, just after depositing a

2.0ML-InAs with the high rate (0.2ML/s), the rate was quickly reduced to 0.02ML/s by shutter control and the InAs was added by 0.6ML with the low rate.

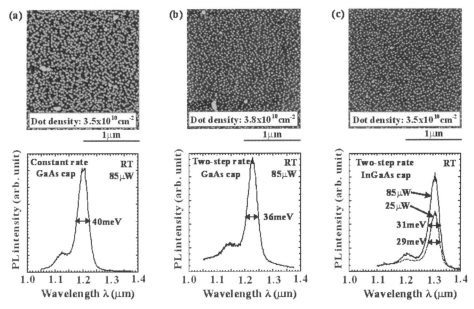

Fig. 3 AFM images and PL spectra of QDs. GaAs-capped QDs were grown with (a) constant rate and (b) two-step rate. InGaAs-capped QDs were grown with two-step rate in (c).

We characterized the QD density by AFM and the uniformity of QDs by photoluminescence (PL) at room temperature with an excitation power of 85μW in micro-PL system. Figure 3 (a) shows an AFM image and a PL spectrum of reference InAs QDs grown at a conventional constant-rate of 0.1ML/s, which shows a QD density of $3.5 \times 10^{10} cm^{-2}$ and a broad PL linewidth of 40meV. In contrast, the two-step-growth method provided a narrower PL linewidth of 36emV with a similar QD density of $3.8 \times 10^{10} cm^{-2}$ as shown in Fig. 3 (b). This tendency was confirmed by systematic investigation for other samples with various QD densities, which indicates that the two-step-growth method is effective to improve the uniformity of high-density QDs. As a next step, we tried extending the PL wavelength of QDs to 1.3μm for the optical communications by covering the QDs with a 4nm-$In_xGa_{1-x}As$ (x=0.18) as reported by other groups [11-13]. In Fig. 3 (c), the PL spectrum and the AFM image show a peak at 1.31μm with a linewidth of 31meV and a high density of $3.5 \times 10^{10} cm^{-2}$, respectively. This linewidth was reduced to 29meV at a lower excitation of 25μW. From our inspection, the uniformity of our QDs is regarded as high level in reported high-density QDs emitting at 1.3μm.

In addition, they also reported such high quality QDs by stacking double QD layers in Ref. 10, where the first layer and the second one played roles to set a density and to make them large and uniform, respectively, but we have used two kinds of growth rates in a single QD layer for the similar purposes.

4. Conclusion

We have demonstrated two important techniques to grow InAs QDs for the application to the optical devices such as the PC-SMZ all-optical switch, that is, highly selective growth of QDs by using patterned substrates and two-step-growth of QDs emitting at 1.3μm with high density and high uniformity.

Acknowledgement

This work was supported by The New Energy and Industrial Technology Development Organization within the framework of the Femtosecond Technology Project. The authors express their thanks to S. Kohmoto, O. G. Schmidt, and K. Eberl for their useful discussions.

References

[1] K. Tajima, Jpn. J. Appl. Phys. **32**, L1746 (1993).
[2] S. Nakamura, Y. Ueno, K. Tajima, J. Sasaki, T. Sugimoto, T. Kato,T, Shimota, M. Ito, H. Hatakeyama, T. Tamanuki, and T. Sasaki, IEEE Photon. Technol. Lett. 12, 425 (2000).
[3] Y. Sugimoto, N. Ikeda, N. Carlsson, K. Asakawa, N. Kawai, K, Inoue, J. Appl. Phys. **91**, 922 (2002).
[4] H. Nakamura, S. Nishikawa, S. Kohmoto, and K. Asakawa, Proceedings of LEOS2001, San Diego, Wu2 (2001).
[5] S. Kohmoto, H. Nakamura, T. Ishikawa, and K. Asakawa, Appl. Phys. Lett. **75**, 3488 (1999).
[6] W. Seifert, N. Carlsson, A. Petersson, L. -E. Wernersson, and L. Samuelson, Appl. Phys. Lett. 68 (1996) 1684.
[7] H. Lee, J. A. Johnson, M. Y. He, J. S. Speck, and P. M. Petroff, Appl. Phys. Lett. 78, 105 (2001).
[8] O. G. Schmidt, N. Y. Jin-Phillipp, C. Lange, U. Denker, K. Eberl, R. Schreiner, H. Gräbeldinger, and H. Schweizer, Appl. Phys. Lett. 77 (2000) 4139.
[9] Y. Nakamura, O. G. Schmidt, N. Y. Jin-Phillipp, S. Kiravittaya, C. Müller, and K. Eberl, H. Gräbeldinger and H. Schweizer, J. Cryst. Growth **242**, 339 (2002).
[10] I. Mukhametzhanov, R. Heitz, J.Zeng, P. Chen, and A. Madhukar, Appl. Phys. Lett. 73 (1998) 1841.
[11] Y. Nakata, K. Mukai, M. Sugawara, K. Ohtsubo, H. Ishikawa, N. Yokoyama, J. Cryst. Growth **208**, 93 (2000).
[12] K. Nishi, H. Saito, amd S. Sugou, Appl. Phys. Lett. **74**, 1111 (1999).
[13] J. X. Chen, U. Oesterle, A. Fiore, R. P. Stanley, and M. Ilegems, Appl. Phys. Lett. **79**, 3681 (2001).

Mechanical Interaction in Near-Field Spectroscopy of single Semiconductor Quantum Dots.

A. M. Mintairov[1], P. A. Blagnov[2], O. V. Kovalenkov[2], C. Li[3], J. L. Merz[1], S. Oktyabrsky[4], V. Tokranov[4], A. S. Vlasov[2], D. A. Vinokurov[2].

[1]Electrical Engineering Department, University of Notre Dame, Notre Dame, IN 46556, USA; [2]Ioffe Physico-Technical Institute, RAS, 194021 St. Petersburg, Russia; [3]Aerospace and Mechanical Engineering Department, University of Notre Dame, IN 46556, USA; [4]UAlbany Inst. for Materials University at Albany-SUNY, 251 Fuller Rd. Albany, NY 12203

Abstract. We have studied high-energy shifts of single quantum dot (QD) emission lines induced by contact pressure exerted by a near-field optical fiber tip. "Pressure" coefficients of 0.65-3 meV/nm have been measured for self-organized InAs/GaAs, InAs/AlAs and InP/GaInP QDs in agreement with numerical calculations of the local strain field. We found an increase of the tip-induced pressure with increasing aperture diameter from 50-300 nm. A correlation between the shift rate and QD stiffness was obtained. We also observed an increase (x10) of QD emission intensity with increased pressure.

1. Introduction

Near-field scanning optical microscopy allows one to extend spatial resolution of optical experiments far beyond the light diffraction limit, which opens new possibilities for nanoscale characterization. In the present paper we have studied the shift of the emission lines of single InAs and InP quantum dots (QDs) induced by the pressure produced by a near-field tip [1, 2]. Using 50-300 nm tips we observed an increase of the energy shift with increasing aperture diameter. Unexpectedly, a strong increase of the emission intensity of single QDs has also been observed.

2. Experimental details

The InAs QDs were formed within a short-period superlattice consisting of two/eight

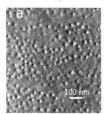
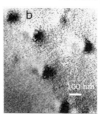

Fig. 1. TEM images (plan-view) of InAs/AlAs (a) and InP/GaInP (b) QDs and SEM picture of 90 nm near-field probe (c).

monolayers (ML) of AlAs/GaAs grown by MBE at 500 °C [3]. The thickness of the cap layer was 40 nm. In the first sample, called InAs/AlAs, InAs QDs were grown between 2ML of AlAs. In the second, called InAs/GaAs, the dots were grown between 8 ML of GaAs. The InAs/AlAs and InAs/GaAs dots have base 22 and 14 nm and density 300 and 100 μm^{-2}, respectively. The height of the dots is ~5 nm. As can be seen from the plan-view transmission electron micrograph of the InAs/AlAs sample presented in Fig.1a, there is a bimodal distribution of dot sizes - in addition to the 20 nm base dots, the QD layer also contains smaller dots having base ~10nm. The density of these small dots is half that of the larger ones. The InAs/GaAs sample has a similar bimodal distribution. Some measurements were done using InP QDs embedded in GaInP, which were studied by us [4]. The dots have a base ~100 nm, height 10 nm and density 20 μm^{-2} (see Fig.1b).

The near-field photoluminescence (NPL) spectra were taken in collection-illumination mode at 10 K under excitation of the 514.5 nm Ar-laser line. The spectra were measured using a CCD detector together with a 280 mm focal length monochromator. The excitation power was below the emission threshold of QD excited states (<20 W/cm^2). An Oxford Instruments CryoSXP cryogenic positioning system was used for 3D scanning. We used 50-300 nm diameter tapered fiber tips with an Al coating of 50-200 nm thickness (see Fig.1c) and transmission 10^{-4}-10^{-2}.

Contact of the tip to a sample surface (zero vertical tip position – z) was controlled by a tuning fork and further increase of the z position produces pressure on the sample. We used 7 nm steps and total vertical displacement up to 70 nm. The damage threshold of

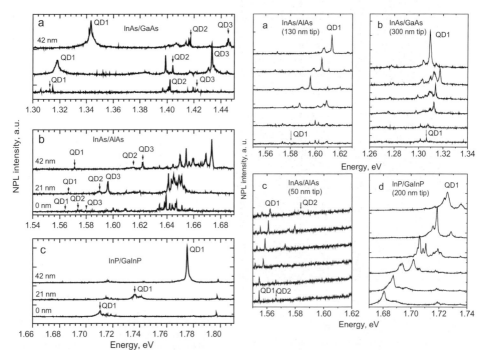

Fig. 2. NPL spectra of InAs/GaAs(a - 130 nm tip), InAs/AlAs (b - 130 nm tip), and InP/GaInP QDs (c - 200 nm tip); z =0, 21 and 42 nm.

Fig.3. NPL spectra of InAs/AlAs (a – 50, c –130 nm tip), InAs/GaAs (b), and InP/GaInP (d) QDs taken for z=0, 7, 14…35 nm (z increased from lower to upper spectra in 7 nm steps).

QDs (no recovery and reproducibility of the spectra after pressure) was found to be ~300 nm. To locate the lateral position of the QDs we measured the intensity of the QD line in NPL spectra using several line x- and y-scans, with a step 50 nm. The uncertainty in the determination of the QD location was 50 nm.

The near-field spectra of InAs QDs for z~0 (no pressure) consist of a number of the sharp lines (up to twenty) corresponding to ground state emission of single QDs. The lines appear in two distinct energy ranges (1.31 and 1.42 eV for InAs/GaAs QDs and 1.55 and 1.64 eV for InAs/AlAs), which reflect the bimodal distribution of the dot sizes. Our magneto-PL [5] measurements show that the dots having higher emission energy correspond to the dots of the larger sizes and must, therefore, have an admixture of Ga. The halfwidth of the sharp lines is equal to the spectral resolution of our setup (0.2-0.4 meV).

3. Results and discussion

In Fig. 2 a-c and 3 a-d we present selected spectra of our InAs/GaAs, InAs/AlAs and InP/GaInP samples showing "pressure" behavior of single QD emission lines. The main effect is a strong high energy shift with z increase. The lines also broaden and their intensity can increase more then one order of magnitude (QD1 in Fig. 2a, c and in Fig 3a, b). For most of the lines the increased halfwidth does not exceed one meV (Fig. 3a) and

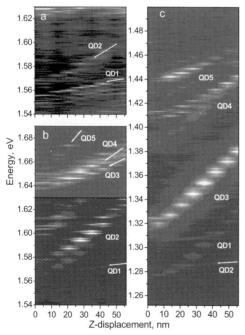

Fig. 4. Dependence of the NPL spectra (3D plot, low contrast features marked by arrows) of InAs/AlAs (a, b) and InAs/GaAs (c) QDs on z. Aperture diameters are 50 (a) and 130 (b and c) nm. Plot (b) has different intensity contrast for small and large dots. Lateral tip-dot separation R is equal: a - 0 for QD1 and 200 nm for QD2 (see also Fig. 3c), b and c - see Fig. 5.

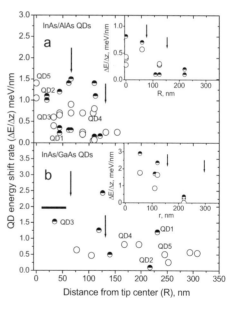

Fig. 5. "Pressure" coefficients of the emission energy of InAs/AlAs (a) and InAs/GaAs (b) QDs (half-filled circles: small dots; empty circles: large) versus lateral tip-QD separations R (aperture 130 nm). Inserts show data for 140 (a) and 300 (b) nm tips. Vertical arrows show position of aperture and Al coating edge, horizontal bar shows the uncertainty of QD position determination.

most probably is connected with change of the radiative life time. In some cases the broadening is much stronger (up to 6 meV, see QD1 in Fig. 2a). This can be related to the coupling between dots.

Fig. 4a-c shows pressure dependence of the energy of individual InAs/AlAs and InAs/GaAs QDs for aperture diameters 130 and 50 nm. The dependence can be traced by curves created by bright dashes corresponding to individual QD lines. The pressure dependence of the QD emission energy has linear character, which is agreed well with numerical calculations.

Some dots (e.g., QD1 in Fig. 4c) have a quadratic region up to z~40 nm. The slope of the z dependence ("pressure" coefficient $S=\Delta E/\Delta z$) is directly related to the pressure experienced by QDs and its value can be used to probe a distribution of the strain field produced by the tip in the QD structure.

Fig. 5a-b summarizes the measurements of the S values for InAs/AlAs (130 and 140 nm tips) and InAs/GaAs QDs (130 and 300 nm tips) versus lateral separation (R) between QD and tip center. One can see strong inhomogenity of the local strain distribution. Indeed, S can have a very high value ~1.3 meV/nm for a dot far from the fiber (R~ 230 nm in Fig. 5b) as well as a very low value <0.3 meV/nm at a more central position (R<50 nm, Fig. 5a). Using the hydrostatic pressure coefficient of 8 meV/Kbar obtained for InAs QDs [6] we determined that the pressure is 27 kbar for z=70 nm and tip diameter 300 nm.

We found an increase of the pressure with increasing aperture diameter. Indeed, the maximum S value (3 meV/nm) was observed for the 300 nm tip (insert in Fig. 5b), while the minimum was 0.65 meV/nm for the 50 nm tip (QD1 in Fig. 4a). The ratio of pressure coefficient to the aperture diameter (effective pressure coefficient A=S/d) is approximately constant for the particular dot. Calculations of local deformation fields using numerical finite element methods coincide well with our experiments.

In conclusion we studied the effect of the tip-induced pressure on position, halfwidth and intensity of QD emission lines and demonstrated the use of near-field spectroscopy to probe their elastic properties. The pressure induced shift of the QD emission line increases with increasing aperture size and depends on QD stiffness. We observed an order of magnitude increase of single dot emission intensity. Our results show that the optical properties of single QDs can be tuned by using local strain. This work was supported by the DARPA/ONR grant N00014-01-1-0658, the W. M. Keck Foundation, and MACRO-DARPA Focus Center for Interconnects for Gigascale integration.

References

[1] Robinson H D Miller M G Goldberg B B and Merz J L 1998 Appl. Phys. Lett. 72, 2081-2083
[2] Chavez-Pirson A Temmyo J and Ando H 2000 Physica E 7 367-372
[3] Tokranov V Yakimov M Katsnelson A Dovodenko K Todt R and Oktyabrsky S Progress of Semiconductor Materials for Optoelectronic Applications 2002 (Proceedings of the Materials Research Society Symposium) 692 135-140
[4] Mintairov A M Vlasov A. S Merz J L Kovalenkov O V. Vinokurov D A Morphological and Compositional Evolution of Heteroepiaxial Semiconductor Thin Films 2000 (Proceedings of the Materials Research Society Symposium) 618 207-212
[5] Mintairov A M Blagnov P A. Kosel T. Merz J L. Ustinov V M. Vlasov A S. Cook R E 2001 Phys. Rev. Lett 87 277401-277404
[6] Inskevich I E Lyapin S G Troyan I A Klipstein P C Eaves L Martin P C and Henini M Phys.Rev. B 1998 58 R4250-R4253.

Magnetic properties of (Ga,Mn)N grown directly on 4H-SiC(0001) by molecular-beam epitaxy

S. Dhar, O. Brandt, A. Trampert, K. J. Friedland, and K. H. Ploog
Paul-Drude-Institut für Festkörperelektronik, Hausvogteiplatz 5–7, D-10117 Berlin, Germany

Abstract. We report on the growth, structural as well as magnetic characterization of (Ga,Mn)N epitaxial layers grown directly on 4H-SiC(0001) by reactive molecular-beam epitaxy. We focus on two layers grown under identical conditions except for the Mn/Ga flux ratio. Structural characterization reveals that the sample with the lower Mn-content is a uniform alloy, while for the layer with the higher Mn-content, Mn-rich clusters are found to be embedded in the (Ga,Mn)N alloy matrix. Although the magnetic behavior of both the samples is similar at low temperatures, showing anti-ferromagnetic characteristics with a spin-glass transition, the sample with higher Mn-content additionally exhibits ferromagnetic properties at and above room temperature. This ferromagnetism most likely originates from the Mn-rich clusters in this sample.

1. Introduction

Recently, the issue of achieving room-temperature ferromagnetism in diluted magnetic semiconductors (DMS) is gaining a lot of attention because of their importance for developing future 'spintronic' devices. Mn-doped III-V DMS are currently in the focus of interest, with (Ga,Mn)As being the most extensively studied compound of this class. The highest Curie temperature T_C reported for this material is 110 K. Dietl et al.[1] have calculated T_C for various Mn doped (5%) III-V semiconductors with very high (3.5×10^{20} cm^{-3}) hole concentration using the Zener model of carrier-induced ferromagnetism. This model predicts T_C to be above 300 K in $Ga_{0.95}Mn_{0.05}N$ as compared to 120 K for $Ga_{0.95}Mn_{0.05}As$. Several groups have thus initiated the growth and investigation of this material. The results, however, show significant discrepancies, particularly regarding the magnetic properties of the layers. For example, while some researchers have reported anti-ferromagnetic behavior for this material,[2,3] others observed ferromagnetism with various different values of T_C ranging from 20 K to 940 K.[3-7] All values of T_C above room temperature stem from n-type or even highly resistive samples.[4-6] The origin of the ferromagnetism observed is thus far from being understood. If it were an intrinsic property of insulating (Ga,Mn)N, it is clear that Dietl et al.'s [1] model would not apply to this case. However, it is important to note that there exist several Ga-Mn and Mn-N phases, which are ferro-, ferri- or anti-ferromagnetic in nature. Furthermore, some of these phases are ferromagnetic up to very high temperatures (for example, MnGa: ferromagnetic, $T_C > 600$ K; [8] Mn$_4$N: ferrimagnetic, $T_C = 738$ K [9]). The formation of such phases as precipitates during growth when exceeding the solubility limit (which is unknown at present) can dominate the magnetic properties of the material. Here, we present a systematic study of growth and characterization of (Ga,Mn)N layers in order to get a better understanding of this material system. (Ga,Mn)N layers are grown directly on 4H-SiC substrates using reactive molecular-beam epitaxy (RMBE). We focus here on two samples grown under identical conditions except for the Ga/Mn flux ratio, which is changed in order to adjust the Mn-content.

2. Experimental

Samples are grown in a custom designed two-chamber MBE system equipped with conventional effusion cells and an unheated NH$_3$ gas injector. A commercial filter purifies NH$_3$ and a mass-flow controller adjusts its flow into the growth chamber. 250 nm thick (Ga,Mn)N layers are grown directly on semi-insulating Si-face 4H-SiC(0001) substrates at a substrate temperature of 710°C (100°C lower than the temperature normally used for GaN growth). The NH$_3$ flux is controlled to keep the chamber pressure at $4-5\times10^{-5}$ Torr during growth. The Mn/Ga flux ratio was changed in order to adjust the Mn-content in the layers. Nucleation and growth is monitored *in situ* by reflection high-energy electron diffraction (RHEED). Structural properties of the layers are investigated by X-ray diffraction (XRD), transmission electron microscopy (TEM) and secondary ion mass spectroscopy (SIMS). Symmetric high resolution triple crystal x-ray ω-2θ scans are taken with a Bede D3 diffractometer utilizing Cu $K\alpha_1$ radiation and equipped with a Bartels-type Ge(002) monochromator and a Si(111) analyzer. TEM is performed using a JEOL3010 microscope operating at 300 kV. The magnetization measurements are done in a Quantum Design superconducting quantum interference device (SQUID) setup. Magnetization loops are recorded at various temperatures for magnetic fields between ±5000 Oe. Prior to measuring the temperature dependence of the magnetization, the sample is first cooled from room temperature to 2 K either under a saturation field of 10000 Oe (Field cooled: FC) or at zero field (zero field cooled: ZFC). The magnetic field is applied parallel to the surface, i. e., perpendicular to the c-axis, in all measurements. All data presented below are corrected for the diamagnetic contribution of the substrate. Both the samples are found to be electrically highly resistive ($\rho\sim1$ MΩcm) even at room temperature.

3. Results and discussion

During nucleation of the layers, a spotty (1×1) RHEED pattern is initially observed reflecting a purely three-dimensional (3D) growth mode. Upon the deposition of 10 ML (Ga,Mn)N, the RHEED pattern becomes streaky, reflecting two-dimensional (2D) growth. For sample A, the RHEED pattern remains entirely streaky throughout growth. For sample B, the RHEED pattern remains streaky for the first 100 nm of deposition, after which a superimposed 2D/3D pattern evolves.

The Mn-content as measured by SIMS is 7.6 and 13.7% for sample A and B, respectively, and is constant over the entire depth ruling out any accumulation of Mn on the surface during growth. XRD of samples with various Mn-content reveals a linear decrease of the *c* lattice constant with increasing Mn-content, indicating a substitutional incorporation of Mn. The x-ray rocking curve width in these samples is typically 300″, similar to the values we observe for equally thin pure GaN layers grown under these

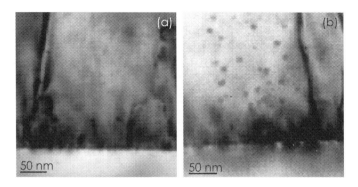

Figure 1. Bright field TEM micrographs of (a) sample A and (b) sample B. Note the presence of nm-scale clusters in sample B.

conditions. All these results suggest the formation of a uniform (Ga,Mn)N alloy with a good crystalline quality.

Figure 1 shows the TEM micrographs of both samples. Sample A is seen [Fig. 1(a)] to be a homogeneous layer without any evidence for a secondary phase. For sample B, in contrast, nm-size clusters are observed in the micrograph [Fig. 1(b)]. High-resolution TEM (not shown here) reveals that these clusters have wurtzite structure and are coherent to the surrounding matrix. We note that we were unable to detect the presence of these clusters by XRD, presumably because of their minuscule size resulting in a significant broadening of the reflection.

Figure 2 shows the temperature dependence of magnetization for samples A and B at a magnetic field of 100 Oe. Clearly, the low temperature behavior of both samples is similar, which is characterized by a FC-ZFC irreversibility and a sharp cusp in the ZFC curves. These two features are fingerprints for spin-glass systems. We have investigated the low temperature behavior in great detail using frequency and field dependent AC susceptibility measurements,[10] which demonstrate that the material indeed represents a Heisenberg spin-glass below 6 K.

At higher temperatures, sample B is ferromagnetic as seen from the separation of the FC and ZFC curves above 10 K. The inset of Fig. 2 (a) and (b) shows the magnetization

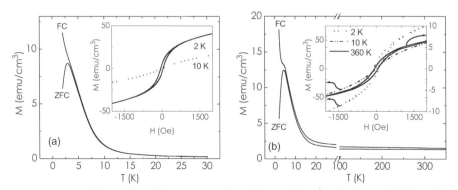

Figure 2. Temperature dependence of magnetization at field-cooled (FC) and zero-field-cooled (ZFC) conditions for (a) sample A and (b) sample B. Hysteresis loops obtained at various temperatures in these samples are shown in the insets of the respective figures.

loops for sample A and B, respectively. Both samples show a hysteresis below 6 K, as expected from the FC and ZFC curves. Sample A is paramagnetic at higher temperatures, whereas sample B is hysteretic again at temperatures higher than 20 K in agreement with the pronounced FC and ZFC separation at this temperature.

In the absence of carriers in the crystal, the Mn-Mn interaction is expected to be anti-ferromagnetic. Only the presence of an adequate density of carriers can turn the anti-ferromagnetic interaction into ferromagnetic, as has been observed in (Ga,Mn)As. In Fig. 3, the inverse of the DC susceptibility χ is plotted versus temperature for sample A. A fit of the Curie-Weiss law to the high-temperature data returns a value of $\theta = -10.4$ K, confirming the anti-ferromagnetic Mn-Mn interaction in insulating (Ga,Mn)N.

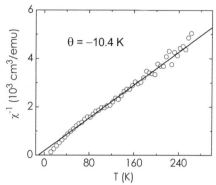

Figure 3. Inverse DC susceptibility measured at 200 Oe for sample A as a function of temperature (open circles). The line shows the fit to the data with the Curie-Weiss law [$\chi = C/(T - \theta)$].

Conclusion

The homogenous alloy (Ga,Mn)N exhibits anti-ferromagnetic Mn-Mn interaction and undergoes a spin-glass transition at temperatures around 5 K. Ferromagnetism is observed at room-temperature and above for insulating (Ga,Mn)N with a Mn-content of 14%. This ferromagnetism is not an intrinsic property of (Ga,Mn)N but originates from nm-scale Mn-rich clusters formed during growth.

References

[1] Dietl T, Ohno H, Matsukura F, Cibert J and Ferrand D 2000 Science 287 1019–1022
[2] Zajac M, Gosk J, Kamińska M, Twardowski A, Szyszko T and Podsiałdo 2001 Appl. Phys. Lett. 79 2432–2434
[3] Kuwabara S, Kondo T, Chikyow T, Ahmet P and Munekata H 2001 Jpn. J. Appl. Phys. 40 L724–L727
[4] Sasaki T, Sonoda S, Yamamoto Y, Suga K, Shimizu S, Kindo K and Hori H 2002 J. Appl. Phys. 91 7911–7913
[5] Reed M L, El-Masry A, Stadelmaier H H, Ritums M K, Reed M J, Parker C A, Roberts J C and Bedair S M 2001 Appl. Phys. Lett. 79 3473–3475
[6] Thaler G T, Overberg M E, Gila B, Frazier R, Abernathy C R, Pearton S J, Lee J S, Park Y D, Khim Z G, Kim J and Ren F 2002 Appl. Phys. Lett. 80 3964–3966
[7] Overberg M E, Abernathy C R, Pearton S J, Theodoropoudou N A, Mc-Carthy K T and Hebard A F 2001 Appl. Phys. Lett. 79 1312–1314
[8] Tanaka M, Harbison J P, DeBoeck J, Philips T, Cheeks T L and Keramidas V G 1993 Appl. Phys. Lett. 62 1565–1567
[9] Yang H, Al-Brithen H, Trifan E, Ingan D C, Smith A R 2002 J. Appl. Phys. 91 1053–1059
[10] Dhar S et al unpublished

ary, *Inst. Phys. Conf. Ser. No 174: Section 3*
Paper presented at 29th Int. Symp. Compound Semiconductors, Lausanne, Switzerland, 7–10 October 2002
©2003 IOP Publishing Ltd

Selective MBE Growth of GaAs Ridge Quantum Wire Arrays on Patterned (001) Substrates and Its Growth Mechanism

Taketomo Sato, Isao Tamai and Hideki Hasegawa

Research Center for Integrated Quantum Electronics (RCIQE) and
Graduate School of Electronics and Information Engineering,
Hokkaido University, North 13, West 8, Sapporo, 060-8628, Japan

Abstract. Selective MBE growth of an array of <-110>-oriented GaAs ridge QWRs was attempted on pre-patterned (001) substrates. Arrow-head shaped GaAs QWRs were selectively formed on the top (113)A facets of GaAs ridge structures. The underlying growth mechanism was clarified, and based on this, the wire width could be kinetically controlled by the growth process.

1. Introduction

Recently, intensive research efforts have been made on semiconductor quantum devices such as single electron transistors (SETs) and quantum wire transistors (QWR-Trs). For realization of such quantum devices, it is necessary to form high quality and highly uniform quantum structures in a size- and position-controlled fashion.

Among various approaches to from quantum structures, selective MBE/MOVPE growth on pre-patterned substrates is one of the most promising techniques for formation of position- and size-controlled arrays of III-V quantum wires (QWRs) and quantum dots (QDs) [1-5]. However, growth on non-planar substrates is complicated due to simultaneous involvement of various high-index facets and related kinetic processes. Proper understanding of the growth mechanism is, thus, inevitable for precise control of the feature size.

In this paper, <-110>-oriented GaAs/AlGaAs ridge QWR arrays were successfully fabricated on (001) patterned substrate by selective MBE growth technique for the first time. A new understanding of the formation mechanism has been achieved, and this has realized precise control of the wire size.

2. Experimental

As the templates for selective MBE growth of GaAs ridge QWRs, a patterned substrate shown in **Fig. 1(a)** was used. Here, an array of <-110>-orientated mesa stripes with (111)A side facets was formed on semi-insulating (001) GaAs substrates by electron-beam (EB) lithography and wet chemical etching.

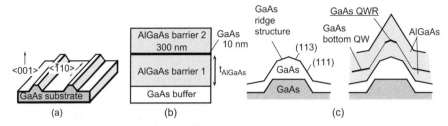

Figure 1. (a) Substrate, (b) material supply and (c) growth sequence.

As pre-growth treatments, organic cleaning and light wet chemical etching were applied in the atmosphere. Then, thermal cleaning under As pressure was done in the MBE chamber just before the growth in order to remove native oxides.

A typical material supply and the growth sequence are shown in **Fig. 1(b)** and **(c)**, respectively. After thermal cleaning in the MBE chamber, GaAs buffer layer was grown on the patterned substrate first. This led to formation of GaAs ridge structures defined by two (113) facets. Then, attempts of self-organized QWR growth were made by supplying $Al_{0.3}Ga_{0.7}As/GaAs/Al_{0.3}Ga_{0.7}As$ on the ridge structures. This approach is similar to that for growth of InP-based ridge QWRs developed by our group [3]. The growth rate of the AlGaAs layer and the V/III flux ratio were kept to be 500 nm/hour and 30, respectively, in terms of the values for growth on a planar (001) surface. Samples were grown at various substrate temperatures within the range from 600°C to 680°C.

3. Results and discussion

3.1. Growth of <-110>-oriented GaAs ridge QWRs

Figure 2 (a) shows a cross-sectional SEM image of the cleaved surface of a sample after stain etching using an alkali solution where GaAs was selectively dissolved into a solution. Before etching, the surface was featureless, but after etching, self-organized formation of a GaAs QWR became clearly visible. The wire structure is very different from that for the <110>-oriented case [2]. Namely, supply of AlGaAs/GaAs/AlGaAs materials resulted in self-organized formation of arrow-head shaped nano-wire on the top (113) facets of the GaAs ridge structure, as schematically shown in **Fig. 2(b)**. Furthermore, two boundary planes separating (111)/(113) growth region became also visible within the AlGaAs layer after etching.

Results of photo-luminescence (PL) and cathodo-luminescence (CL) measurements on GaAs ridge QWR are shown in **Fig. 3(a)** and **(b)**, respectively. The wire width of the measured QWR sample was 40 nm. As shown in **Fig. 3(a)**, two sharp peaks were observed at 1.52 eV (peak 1) and 1.56 eV (peak 2). From the results of spatially resolved monochromatic CL measurements shown

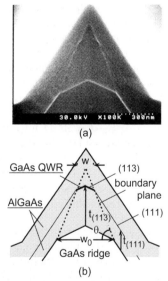

Figure 2. (a) Cross-sectional SEM image of GaAs QWR and (b) its schematic representation.

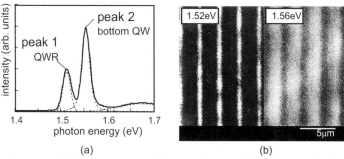

Figure 3. (a) PL spectra and (b) monochromatic CL images of top surface obtained from <-110> GaAs QWR at 30K.

in **Fig. 3(b)**, it was found that peak 1 and peak 2 came, respectively, from the top QWR and the bottom quantum well (QW) regions. The quantum wire peak 1 had a narrow full width at half-maximum (FWHM) of 20 meV. Thus, the results of PL and CL measurements have indicated that the spatially uniform GaAs wires can be formed on the pre-patterned substrate by the present selective MBE growth technique.

3.2. Formation mechanism and size controllability of QWR

For precise control of the QWR size, detailed understanding of the growth mechanism is necessary. The angle of the boundary planes formed within AlGaAs layer is one of the important parameters to control the QWR size, because the wire width is defined by the two boundary planes, as schematically shown in **Fig. 2(b)**. From the SEM observation shown in **Fig. 2(a)**, two boundary planes kept a constant angle, θ, with respect to (001) plane throughout the entire growth.

Figure 4 shows a plot of the measured boundary angle, θ, as a function of the growth temperature, T_{sub}. Rather surprisingly, any attempts to correlate these boundary planes with a particular high index facet failed. In fact, the value of θ was found to depend strongly on the growth temperature, T_{sub}, as shown in **Fig. 4**.

Growth rates on two (113) and (111) facets of the GaAs ridge structure were also investigated for various substrate temperatures. **Figure 5** plot the measured vertical growth thicknesses defined as $t_{(113)}$ and $t_{(111)}$ in **Fig. 2(b)** vs. the growth temperature, T_{sub}. The data were normalized by the thickness on the (001) plane, $t_{(001)}$. As the temperature increased, the growth thickness on (113) facets increased, while that on (111) facets decreased. This is due to the difference in migration and atom incorporation rates between on (113) and (111) facets. Thus, the growth selectivity on (113) facets with respect to (111) facets was enhanced at higher growth temperatures, as schematically illustrated in the insets of **Fig. 5**.

Furthermore, temperature-dependent differences in the migration rate and the atom incorporation rate lead to a Ga-rich thicker region on (113) facet, and a less Ga-rich thinner region on (111) facet at higher temperatures. Due to slight composition difference, the boundary

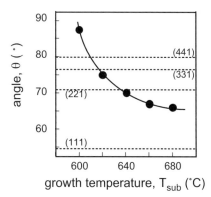

Figure 4. θ vs. growth temperature, T_{sub}.

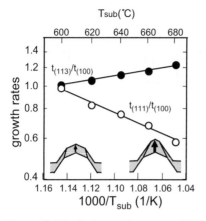
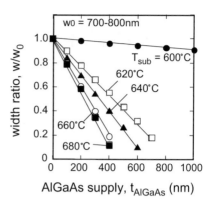

Figure 5. Vertical growth rates on (113) and (111) facets.

Figure 6. QWR width vs. AlGaAs supply thickness.

becomes visible by stain etching for cross-sectional SEM observations shown in **Fig. 2(a)**. Thus, it has been concluded that evolution of the boundary planes is the result of difference in the migration and the atom incorporation rates between (113) and (111) facets.

The measured variation of the normalized wire width, w/w_0, of the QWR is shown in **Fig. 6** as a function of the total AlGaAs supply thickness given before the start of the GaAs QWR growth. Here, w_0 denotes the initial GaAs ridge width prior to AlGaAs supply. As expected, the wire width, w, changed linearly with the AlGaAs supply thickness, t_{AlGaAs}. At high growth temperatures, the wire width rapidly decreased as the AlGaAs supply thickness increased, due to the reduction of the boundary angle. From these results, it was found that the size of the present QWR can be kinetically controlled precisely by the growth process.

4. Conclusion

In this study, the selective MBE growth of the <-110>-oriented GaAs ridge QWRs were made on pre-patterned (001) substrates for the first time.
(1) From SEM, PL and CL measurements, it was found that an array of arrow-head shaped GaAs wires was selectively formed on the top (113) facets of GaAs ridge structures with good size uniformity.
(2) The wire width of the present GaAs QWR is determined by boundary planes resulting from the difference in migration rate and the atom incorporation rate between top (113) facets and side (111) facets of the GaAs ridge structure. Thus, the wire width can be kinetically controlled precisely by growth conditions.

References

[1] Kapon E, Hwang D M and Bhat R: Phys. Rev. Lett., 63 (1989) 430.
[2] Koshiba S, Noge H, Ichinose H, Akiyama H, Nakamura Y, Inoshita T, Someya T, Wada K, Shimizu A and Sakaki H: Solid-State Electronics, 37 (1994) 729.
[3] Fujikura H and Hasegawa H: J. Electron. Mater., 25 (1996) 619.
[4] Wang X L, Ogura M and Matsuhata H: Appl. Phys. Lett., 66 (1995) 1506.
[5] Sato T, Tamai I and Hasegawa H: presented at ISCS 2001.

Large Transition Energy Separation at 1.3 μm Emission from InAs/GaAs Quantum Dots

Y.Q. Wei[a], S.M. Wang[a], F. Ferdos[a], Q.X. Zhao[b], J. Vukusic[a]
M. Sadeghi[c], and A. Larsson[a]

[a]*Photonics Laboratory, Department of Microelectronics,*
[b]*Physical Electronics and Photonics, Department of Microelectronics and Nanoscience,*
[c]*Microtechnology Centre at Chalmers (MC2),*

Abstract:

By capping InAs quantum dots (QDs) with a thin intermediate layer of InAlAs instead of GaAs, the radiative transition wavelengths are red-shifted. Surface morphology studies confirm that the redshift is due to a better preserved QD height as compared with capping by GaAs only. We also observe the energy separation between the ground and the first excited state increases significantly. Based on this, we demonstrate InAs QDs with a record energy separation of 108 meV and ground state emission at 1.3 μm.

1. Introduction

In the last few years, rapid progress has been made to realize GaAs-based long-wavelength lasers, which can be used as low-cost high-performance light emitters for telecommunication applications [1-7]. Using a self-organized quantum-dot (QD) structure as the active gain medium, the emission wavelength can be tuned to the telecommunication range. Up to now, high performance 1.3 μm QD lasers have been demonstrated with some favourable device performance parameters such as a threshold current density of 19 A/cm^2 with a threshold current of 1.2 mA [2], and a small-signal modulation bandwidth of more than 5 GHz [3]. However, recent experimental results show that, with increasing temperature, either the characteristic temperature T_0 abruptly decreases at or slightly above room temperature or the lasing wavelength switches to higher energy lasing states due to the thermal excitation of electrons and holes into the higher energy levels [2,4,5]. In particular, the growth approach of InAs QDs embedded in an InGaAs well (DWELL) has been widely used to tune the QD emission wavelength to 1.3 μm. However, this technique results in a decrease of the potential barrier and consequently reduces the separation between the discrete energy levels of the InAs QDs. In these reports [6,7], poor temperature stability of the devices has been shown, with a T_0 ranging from 50 to 60 K. Using GaAs directly instead of InGaAs as the confinement potential barrier, larger transition energy separations of up to 104 meV have been reported, which also leads to an improved high T_0 at room temperature. Unfortunately, the QD emission wavelength is only 1.24 μm. Recently, studies on energy level control for the InAs QDs with a thin AlAs (or InAlAs) cap layer has been carried out and the results are promising [8]. The authors attribute these merits to the suppression of In-Ga intermixing between the InAs QDs and the overgrown barrier layer, but the experimental evidence for this is poor.

In this paper, by using an AlGaAs barrier layer, instead of a GaAs barrier layer, with an InAlAs cap layer, we demonstrate an approach to simultaneously tune the InAs QD ground state emission wavelength to 1.3 μm and increase the separation between the QD ground and first excited state transition energies. We also show the distinct difference in surface morphology between InAs QDs capped with a thin AlAs layer and with a thin GaAs layer to aid in the interpretation of the experimental results. With a proper design, a large transition energy separation of 108 meV with ground state emission at 1.3 μm has been realized.

2. Experiment

Samples were grown in an EPI 930 solid-source molecular beam epitaxy system on GaAs (001) substrates. The growth temperature was 510°C for growth of the InAs and the composite InAlAs/InGaAs capping layers and 580°C for the rest of the layers. A set of samples, denoted samples A, B and C, was prepared. 3 ML InAs was deposited and then capped with 0, 4, and 9 MLs of $In_{0.2}Al_{0.8}As$, followed by 3 nm of $In_{0.2}Ga_{0.8}As$. All three samples were embedded in an $Al_{0.2}Ga_{0.8}As$ matrix. Finally, a 200 nm thick GaAs cap layer was grown for photoluminescence (PL) studies. Growth rates for all samples were 1 μm/h for GaAs, 0.1 μm/h for InAs, 0.5 μm/h for InGa(Al)As cap layers, and 1.25 μm/h for the $Al_{0.2}Ga_{0.8}As$ matrix. The samples were structurally characterized using *ex situ* atomic force microscopy (AFM). Room temperature PL measurements were performed using an argon-ion laser as the excitation source.

3. Results and discussion

Fig. 1 shows PL spectra for samples A, B and C. Two intensity peaks were observed for all samples in the PL spectra.

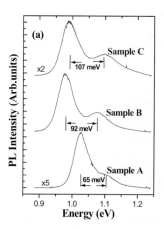

Fig. 1. Room-temperature (T=293K) PL spectra for samples with different thickness of $In_{0.2}Al_{0.8}As$ cap layers at an excitation power density of 60 W/cm^2.

We assign the two peaks to the ground state and the first excited state transitions. For sample A, the ground state transition is observed at 1.026 eV with a full width at half maximum (FWHM) of approximately 38 meV. After capping with 4 ML $In_{0.2}Al_{0.8}As$ (sample B), the ground state transition is red shifted to 0.981 eV, while the first excited state transition is red shifted from 1.091 eV to 1.073 eV. The energy separation between the ground and the first excited state transitions dramatically increases from 65 meV for sample A to 92 meV for sample B. By further increasing the thickness of the $In_{0.2}Al_{0.8}As$ capping layer to 9 ML, both the ground state and the first excited state transition energies is blue shifted to 0.99 eV and 1.097eV, respectively. Accordingly, an increased energy separation of 107 meV is observed.

To understand the puzzling results, we therefore grew another three samples for AFM measurement and the results are shown in Fig. 2. Fig. 2 (a) shows a 1×1 µm² AFM image of the uncapped InAs QDs. The QD density is $3.2\pm0.2\times10^{10}$ cm^{-2} while the average height and diameter are 6.2 and 45 nm, respectively. When InAs QDs were capped with 4 ML GaAs (Fig. 2 (b)), both the QD height and density decrease to about 4.9 nm and 0.5×10^{10} cm^{-2}, respectively. In contrast, we can clearly see from Fig. 2(c) that the surface morphology of the sample with a 4 ML AlAs cap is similar to that seen in Fig. 2(a). The average height and diameter are about 5.9 and 40 nm, respectively, while the QDs density increases to $4.5\pm0.3\times10^{10}$ cm^{-2}. For InAs/GaAs, we believe that the Ga and In atoms in the top region of the InAs QDs diffuse to the wetting layer upon capping, resulting in a reduction of the QD height. When the QD volume is below a critical value, the whole QD will collapse and therefore the density decreases. For InAs/AlAs, however, Al atoms are immobile at 510°C and the Al atoms on the InAs QDs are unable to diffuse to the wetting layer. The Al accumulation on the InAs QDs prevents the In atoms from diffusing to the wetting layer. Therefore, the QD decomposition is reduced and the QD height is larger than that in the case with GaAs as a cap layer. As the ground state transition energy in QDs is more sensitive to their height than their diameter, the larger QD height, when comparing sample B to A, results in a red shift that outweighs the blue shift due to the higher InAlAs barrier.

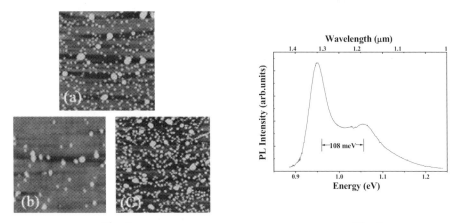

Fig. 2

Fig. 3

Fig.2. 1.0×1.0 µm² AFM images of the InAs QDs (a) without capping; (b) with a 4 ML GaAs cap layer; (c) with a 4 ML AlAs cap layer.

Fig. 3. Room-temperature (T=293K) PL spectrum of 3.5 ML InAs QDs emitting at 0.95 eV (1.31µm) at an excitation power density of 60 W/cm².

When using 3.5 ML InAs capped with 4 ML InAlAs and 3 nm InGaAs layers together with a $Al_{0.2}Ga_{0.8}As$ barrier layer, we can tune the ground state transition of the InAs QDs to 1.3 µm. The PL spectrum is shown in Fig. 3. The energy separation between the ground state and the first excited state transition energies is as large as 108 meV, indicating that this approach can be used for improving the temperature performance of 1.3 µm QD lasers.

4. Conclusion

In conclusion, we have investigated the influence of a thin InAlAs cap layer on the optical properties of self-organized InAs QDs. A fast red shift of the transition energies has been observed. The red shift when capping with a thin InAlAs layer is due to a better preserved QD height as confirmed by AFM observations. Using this approach, we have demonstrated InAs QDs emitting at 1.3 µm with a record energy separation of 108 meV.

5. Acknowledgement

This work was supported by Ericsson AB and the Foundation for Strategic Research (SSF).

References:
[1] Y.Arakawa, H.Asakaki, Appl. Phys. Lett. 40, 939 (1982)
[2] G. Park, O. B. Shchekin, D. L. huffaker, and D. G. Deppe, Photon. Technol. Lett. 13, 230 (2000)
[3] R. Krebs, F. Klopf, S. Rennon, J. P. Reithmaier, and A. Forchel, Electron. Lett. 37, 1223 (2001)
[4] X. Huang, A. Stintz, C.P. Hains, G. T. Liu, J. Chen, and K. J. Malloy, Photon. Technol. Lett. 12, 227 (2000)
[5] V. M. Ustinow and A. E. Zhukov, Semi. Sci. Technol. 15, R41 (2000)
[6] G. T. Liu, A. Stinz, H. Li, K. L. Malloy, and L.F. Lester, Electron. Lett. 35, 1163 (1999)
[7] X. Huang, A. Stinz, C. P. Hains, G. T. Liu, J. Cheng and K. J. Malloy, Electron. Lett. 36, 41 (1999)
[8] R. Jia, D. S. Jiang, H. Y. Liu, Y. Q. Wei, B. Xu, and Z. G. Wang, J. Crys. Growth. 236, 499 (2002)

Optical spectra of quantum dot aggregates in sub-wetting layer region

K Král and P Zdeněk

Institute of Physics, Academy of Sciences, Prague

Abstract. In polar semiconductor samples of self-organized quantum dots, grown by the Stranski-Krastanow growth method, the lowest-energy extended states of the motion of the electronic excitations, are assumed to be the wetting-layer states. The coupling between these extended states and the electronic states localized in the individual quantum dots, may influence the optical spectra of such samples in the sub-wetting layer region of energy. This effect is studied assuming Fröhlich's coupling between electrons and polar optical phonons. The contribution of this interaction to the appearance of the sub-wetting layer continuum in the optical spectra, and to the level broadening of the localized states, pointed out in some experiments, is estimated.

1. Introduction

As zero-dimensional structures with electronic motion confined in all three dimensions, the individual semiconductor quantum dots have very narrow spectral widths of optical lines [1,2] of the electronic bound states. Photoluminescence excitation (PLE) experiments [3,4], performed near $T=0$ K on the self-assembled quantum dot samples (SAQD), in the spectral region below the edge of the wetting-layer (WL) excited states of the charge carriers, often show a broadening of the lines corresponding to electronic states bound in quantum dots, and show also a presence of a continuum PLE background. These effects seem to be met at large densities of WL excitations. In this work a theoretical analysis of this effect is presented, basing on a simplified model. In this model the electrons are assumed to be scattered between the wetting layer states and the states bound in the individual quantum dots, emitting or absorbing an optical phonon (energy E_{LO}), via Fröhlich's interaction (see e. g. paper [2] for further references). The schematic picture of the electronic energy levels in the self-assembled quantum dot sample is shown in Fig. 1.

2. The model and the calculations

The self-assembled quantum dot system under consideration is rather complex. With the aim of obtaining simple estimates, we have to develop the theory within a model which is sufficiently simplified. In the case of real samples, the wetting layer is a quasi-two dimensional structure. In the presently used model, we assume that the wetting layer is simply three-dimensional, with the quantum dots distributed in the sample with the

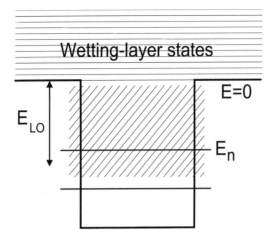

Figure 1. Schematic picture of the electronic energy structure of the single-electron states in the self-assembled quantum dot (SAQD) sample. The full line is the potential profile, which the electrons feel in a single quantum dot of the sample. The dashed horizontal lines denote the states bound in the quantum dots. The horizontally hatched area denotes the extended electronic states in the wetting-layer. The oblique hatching marks the sub-wetting layer area of electronic energies. E_n is the energy of a selected bound state in the sub-wetting layer energy region, E_{LO} is the optical phonon energy.

density of $8 \times 10^{21} m^{-3}$. The effect of holes in the valence band states is neglected, together with the exciton effect and other influence of the electronic correlation interaction. In the model, the electronic eigenstates corresponding to the electronic states localized in the individual dots are approximated by the electron wave functions taken from the infinitely deep cubic quantum dot with the lateral size of 20 nm. The wetting-layer electronic states are approximated by the three dimensional plane waves with the wave vector k, with the onset of the wetting layer electron energy levels at $E=0$. The electronic basis set defined in this way is not orthogonal. In order to avoid the non-orthogonality effect, and in order to assume a simple form of the electronic distribution function in the wetting layer states, we confine the wetting layer states to those having the wavelength not smaller than $3d/2$, d being the lateral size of our cubic quantum dot. These wetting layer states are assumed to be then homogeneously populated ($\overline{N_k}$) by electrons, independent on the electronic wave vector k. For simplicity, each quantum dot contains only a single bound state, with the wave function chosen to be equal to one of the lowest energy excited states of an electron in the infinitely deep isolated cubic quantum dot. The energy of this state is assumed to be positioned at $E_2 = -E_{LO} + \eta$, inside the sub-wetting layer region of energy. In this paper, the sub-wetting layer region of electron energy is understood to be the energy interval $(-E_{LO}, 0)$. The magnitude of the parameter η is chosen to be 10 meV. This choice of the position of the bound state energy, with respect to the bottom of the wetting layer states, allows us to demonstrate the broadening of the bound state spectral line already in the lowest order (in Fröhlich's coupling) of the approximation to the electronic self-energy. We neglect the polaron effect of electrons in WL states and in the quantum dots. The effect under consideration appears to be due to those terms of the electronic self-energy in the Born approximation, which are, in the limit of zero temperature of the lattice, proportional to the WL population $\overline{N_k}$. The photoluminescence excitation signal, or the optical absorption signal, as it is observed in experiment, is assumed to be approximately proportional to the electronic spectral densities of the electronic excitations of the sample. In this work, the electronic spectral density σ is calculated

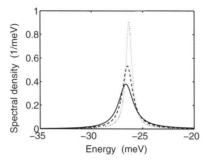

Figure 2. Energy (E) dependence of the spectral density $\sigma_n(E)$ of the selected bound state (with energy E_n) of a single quantum dot interacting with the wetting-layer states via Fröhlich's coupling at the WL population: $\overline{N_k}$ =0.25 (dotted line), 0.5 (dashed) and 0.75 (full).

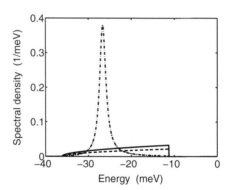

Figure 3. The full and the dash-dot lines give, respectively, the energy (E) dependence of the spectral densities of the phonon satellite $\sigma_{tot}(E)$ of the wetting-layer states, related to a single quantum dot, and the spectral density $\sigma_n(E)$ of the bound state (n) of a single dot, both computed for the wetting-layer population $\overline{N_k}$ =0.75. The dashed line is the spectral density of the phonon satellite of the WL states $\sigma_{tot}(E)$ computed for the WL states population $\overline{N_k}$ =0.5.

from the corresponding Green's function and from the electronic self-energy. The numerical calculations are performed near the zero temperature of the lattice. The system of the polar optical phonons is assumed to be identical with the optical phonon system of the bulk crystal of the polar semiconductor, ignoring in this way the effects which the assumed semiconductor heterostructure might have on the system of the optical phonons. Using material parameters of the gallium arsenide crystal, the results of the numerical calculations are presented in Figs. 2 and 3. In Fig. 2 we demonstrate the broadening of the bound state spectral line with increasing the WL population. Fig. 3 shows the low-energy LO phonon satellite of the WL populated states, in the sub-wetting layer region of energy. The intensity of this continuous background appears to be proportional to the first power of the WL electronic population. As the population of the wetting layer $\overline{N_k}$ increases, the relative intensity of the discrete lines, with respect to the continuous background, appears to weaken, which trend seems to be in accord with the experimental findings of the paper [3]. The details of the theoretical argumentation will be presented elsewhere.

3. Conclusions

In conclusion, the theoretical results, obtained with the present simplified model, appear to have plausible qualitative properties. Although we do not present in this work a detailed comparison with the experiment, e. g. with [3,4], the preliminary quantitative estimates allow us to express our expectation: basing on the Fröhlich's interaction in the SAQD samples, in which the population of the WL states is set up in the experiment to be large enough, there may be two separate phenomena, to which this electron-phonon coupling may contribute considerably. One is the broadening of the bound state spectral peaks, the magnitude of which should be proportional to the wetting layer population $\overline{N_k}$. The other effect may be the presence of the low-energy LO phonon satellite of the populated WL states, manifested in the PLE spectra as a continuous background in the sub-wetting layer region.

Acknowledgements

The work was supported by the grants IAA1010113, OCP5.20, RN19982003014 and by the project AVOZ1-010-914.

References

[1] Yoffe A D 2001 Adv. Phys. 50 1-208
[2] Král K and Zdeněk P 1999 Physica B 272 15-17
[3] Nakaema M K K, Brasil M J S S, Iikawa F, Ribeiro E, Heinzel T, Ensslin K, Medeiros-Ribeiro G, Petroff P M and Brum J A 2002 Physica E 12 872-875
[4] Toda Y, Moriwaki O, Nishioka M and Arakawa Y 1999 Phys. Rev. Lett. 82 4114-4117

Self-organized growth of InAs quantum dots and reduction of dot density by *in-situ* annealing

Y. Matsuzaki, T. Kobuse, R. Ohashi and M. Konagai

Department of Physical Electronics, Tokyo Institute of Technology,
2-12-1 O-okayama, Meguro-ku, Tokyo 152-8552, Japan

A. Yamada

Research Center for Quantum Effect Electronics, Tokyo Institute of Technology, 2-12-1 O-okayama, Meguro-ku, Tokyo 152-8552, Japan

Abstract. We have successfully grown InAs quantum dots (QDs) by the molecular beam epitaxy (MBE) method and we proposed the *in-situ* annealing to control the density of QDs. It was realized that the density of QDs could be reduced by using this annealing process and it became clear that QDs unisotropically migrated on the surface and tended to gather at the step edge.

1. Introduction

The growth process, Stranski-Krastanow (SK) mode [1], is widely used for the formation of self-organized quantum dots (QDs) because nano-scale structures can be fabricated without a need of lithography. In the mid-1980s, several groups studied on the InGaAs/GaAs islands growth via the SK mode [2, 3]. However, the early works did not attract great attention, at least as it applied to the fabrication of quantum dots. The research activity has rapidly increased since InAs islands on GaAs were found to work as quantum dots [4]. The early works were focused on their optical properties [5, 6]. Although the QD fabrication methods through self-organizing phenomena are useful, density and site control of dots has not been sufficient for the application of these QDs to electronic devices, such as single electron transistors.

In the paper, we will report the growth of InAs QDs using SK mode and demonstrate a method for the reduction of the QD density by *in-situ* annealing.

2. Experimental

InAs QDs were fabricated by molecular beam epitaxy with elemental sources, such as In, As and Ga. The surface reconstruction and roughness were monitored by the reflection high-energy electron diffraction (RHEED) during the growth.

N-type GaAs (001) substrates were used. Before growth, they were thermally cleaned at about 620°C for 10s under an arsenic pressure of 2.3×10^{-6} Torr. After the cleaning, a GaAs buffer layer with a thickness of 150 nm was grown at 570°C, then the substrate temperature was cooled down to 480°C which was a growth temperature of InAs QDs. During this process, the RHEED pattern was changed from As-stabilized (2x4) to c(4x4) reconstruction. When the substrate temperature settled at 480°C, the growth of InAs was started. The beam equivalent pressures of In and As were 6.0×10^{-8} and 2.3×10^{-6} Torr, respectively. The duration for the growth was changed from 20 to 60s. In our experiment, it took approximately 14s to grow one monolayer (ML) InAs.

3. Results and Discussions

3.1. Growth of InAs quantum dots

The RHEED pattern gives us important information on the surface state. When the layer-by-layer growth proceeds, RHEED shows streak patterns. The pattern transits to spots when three-dimensional islands start to grow. The RHEED intensity during the growth is represented in Fig. 1. The RHEED intensity increased after opening the In shutter, then it showed the maximum. In this duration, the RHEED showed a streaky pattern. We observed a kink in the RHEED intensity when the growth duration was about 22.5s (1.6 ML). At the time, RHEED pattern changed from the streaky pattern to a spotty pattern, indicating that a two-dimensional layer growth transited to a three-dimensional island growth. Then we again observed a streaky pattern. At the duration of 30s (2.1 ML), the streaky pattern was suddenly changed to a spot pattern with diffractions from the facet.

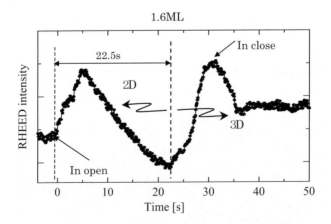

Figure 1. Time evolution of the RHEED intensity

The corresponding dynamic force microscope (DFM) images are shown in Figs. 2 (A)-(C). Fig. 2(A) shows the DFM image of the sample grown for 25s, 2(B) for 30s and 2(C) for 60s. These images refer to different epilayers grown under the same conditions. The scanned area is 1x1μm². We observe terrace and small dots in Fig. 2(A), indicating the start of the island growth as seen in the RHEED pattern. Then the island density rapidly increases and many dots are observed in Fig. 2(B). The average diameter and dot density of InAs QDs were about 40 nm and 5×10^{10} cm^{-2}, respectively. When the growth duration exceeded to 60s, the dots coalesced. The result well corresponded to the RHEED observations. For the application of InAs QDs to the single electron transistor, we have to reduce the density of InAs QDs. However, as shown in the DFM images, the QDs density rapidly increased when the layer thickness exceeded the wetting layer thickness which was about 1.6MLs. Therefore, we attempted the *in-situ* annealing of the QDs.

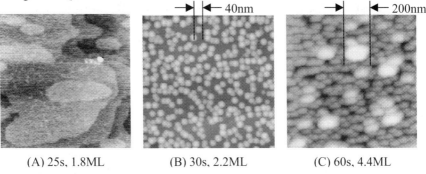

(A) 25s, 1.8ML   (B) 30s, 2.2ML   (C) 60s, 4.4ML

Figure 2. DFM images of InAs QDs

3.2. In-situ annealing

The annealing process was as follows: The temperature was raised up to 530°C with a rate of 1.5 °C/s just after the growth, then the substrate temperature was cooled down. During this process, the As beam with a pressure of 2.3×10^{-6} Torr was continuously irradiated on the surface. The DFM images of the sample after the annealing are shown in Fig. 3(A) with a scanned area of 5x5 μm². The square area (1x1 μm²) indicated in Fig. 3(A) is shown in Fig. 3(B).

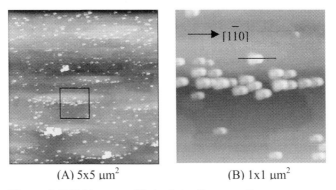

(A) 5x5 μm²   (B) 1x1 μm²

Figure 3. DFM images of InAs dots after annealing.

It is found from the images that the InAs QDs still remain after the annealing and that we have successfully reduced the density of InAs QDs. The density of QDs after the annealing was 3×10^9 cm^{-2}. Furthermore, it can be seen that QDs align at step edges. The phenomenon is more obvious in Fig. 3(B), that is, many dots gather around the step edge. Figure 4 is a cross-sectional view of the annealed InAs QD which is marked by a line in Fig. 3(B). The QD consisted of two dots whose height were about 4-6 nm and they aligned toward the [1$\bar{1}$0] direction. From these results it is concluded that the migration of InAs QDs during the annealing is not isotropic and that the QDs seem to gather at step edges.

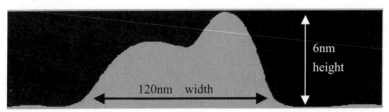

Figure 4. A cross-sectional view of InAs QDs after annealing process

4. Summary

We have successfully grown InAs QDs by MBE. However, it was found to be difficult to control the density of QDs by the intensity monitoring of RHEED due to the low reproducibility. Therefore, we proposed the *in-situ* annealing to control the density of QDs and it was realized that the density of QDs could be reduced by using this annealing process. Additionally, it became clear that QDs anisotropically migrated on the surface and tended to gather at the step edge.

Acknowledgements

This work partly supported by a Grant-in-Aid for Scientific Research (No. 13450124) from the Ministry of Education, Culture, Sports, Science and Technology.

References

[1] Stranski I. N. and Krastanow L. 1937 Akad. Wiss. Wien Math.-Natur. IIb 146, 797.

[2] Schaffer W. J., Lind M. D., Kowalczyk S. P. and Grant R. W. 1983 J. Vac. Sci. Technol. B1, 688.

[3] Lewis B. F., Lee T. C., Grunthaner F. J., Madhukar A., Fernandez R. and Masanian F. F. 1984 J. Vac. Sci. Technol. B2, 419.

[4] Tabuchi M., Noda S. and Sasaki A. 1992 Sci. & Tech. Mesoscopic Structures 379.

[5] Leonard D., Krishnamurthy M., Reaves C. M., Denbaars S. P. and Petroff P. M. 1993 Appl. Phys. Lett. 63, 3203.

[6] Mukai K., Ohtsuka N., Sugawara M. and Yamazaki S. 1994 Jpn. J. Appl. Phys. 33, L1710.

Narrow Size-Dispersion CdSe Quantum Dots Grown on ZnSe by Modified MEE Technique

I.V. Sedova[*], S.V. Sorokin, A.A. Sitnikova, O.V. Nekrutkina,
A.N. Reznitsky and S.V. Ivanov
Ioffe Physico-Technical Institute of RAS, St.-Petersburg 194021, Russia.

Abstract. The paper reports on a modification of the migration-enhanced epitaxy (MEE) technique applied to the growth of CdSe/ZnSe quantum dot (QD) structures. The CdSe insertions are formed at a growth temperature of 280°C in a multi-cycle MEE regime with a ~0.3 monolayer (ML) CdSe deposition per cycle. A one minute growth interruption introduced after each either Se or Cd deposition pulses instead of 10s in a reference structure results in the dramatic variation of a CdSe morphology – from bimodal to single-mode size distribution of CdSe QDs as observed by cross-section and plan-view transmission electron microscopy. It is accompanied by a ~50meV red shift of the photoluminescence (PL) peak energy. It has been found that the enhanced growth interruption after each Se pulse results in the formation of smaller lateral size (~8.5 nm) coherent CdSe-based QDs with narrower size-dispersion and higher Cd content due to different growth kinetics on Se- and Cd-terminated surfaces. The structures also demonstrate brighter PL than CdSe/ZnSe nanostructures grown using a standard MEE technique.

1. Introduction.

Self-organized CdSe/ZnSe quantum dot (QD) nanostructures with a CdSe nominal thickness (w) below 3 ML, generally defined as a critical thickness, still have attracted great interest both for basic studies of wide gap QDs and optoelectronic device applications, because they exhibit dramatically enhanced photoluminescence (PL) intensity in the 2.3-2.6 eV. Application of a 2.6-2.8 ML CdSe active region has resulted in fabrication of ultra-low threshold (3-4 kW/cm^2 at RT) optically pumped lasers as well as room temperature cw laser diodes [1]. However, great potential of CdSe/ZnSe nanostructures seems to be not exhausted, since their intrinsic morphology produced by molecular beam epitaxy (MBE) in a standard or in MEE mode in the growth temperature range of 270-310°C generally exhibits a disordered superposition of 2D extended (>15 nm) CdSe-rich islands (~10^{10} cm^{-2}) and small-scale (<10 nm) QD-like composition fluctuations (10^{11} cm^{-2}), floating in the broadened ZnCdSe QW rather than individual CdSe QDs [2]. Recently different approaches, such as a thermally activated self-organization process [3], the use of CdS as a Cd solid source [4], and BeSe fractional monolayer (FM) precursor [5], were proposed to fabricate narrow size-dispersion QDs consisting of nearly pure CdSe, which is necessary for narrowing a gain contour and increasing a carrier localization energy.

In this paper we report on a modification of MEE technique, resulting in fabrication of nearly pure coherent CdSe QDs with uniform lateral sizes of 8.5±1.5 nm and high

---

[*] Corresponding author: irina@beam.ioffe.rssi.ru, fax: +7 (812) 2471017

enough sheet density $\sim 10^9$ cm^{-2}, demonstrating brighter PL than CdSe/ZnSe nanostructures grown using a standard MEE technique [6].

2. Experiment.

The growth was performed in a two-chamber MBE system, equipped with elemental sources of Zn, Cd, and Se in the II-IV chamber and Ga and As in III-V one. CdSe/ZnSe structures containing bottom and cap ZnSe layers of 50-70 nm and 10-15 nm, respectively, were grown by MBE pseudomorphically on a (001) GaAs buffer epilayer at a substrate temperature of $T_S=280°C$. Since it has been established that the maximum Cd surface coverage of ZnSe per one MEE cycle does not exceed 0.5ML [7], the CdSe FMs were deposited in a multi-cycle MEE mode (by 0.3 ML per cycle) with a total nominal thickness of CdSe FM as high as ~2.1ML. In a reference structure A, 10s growth interruptions after both Cd and Se deposition pulses were used. To enhance surface atom migration and redistribution of the CdSe deposit the growth interruption time was increased up to 1min after each Se or Cd deposition pulses in the structures B and C, respectively. Structure D differs from A by a high temperature annealing within several minutes after the last CdSe deposition cycle at $T_S=350°C$.

Structural properties of the samples were studied by TEM using a Philips EM-420 microscope at 100kV in both plan-view and cross-section geometry. The samples for TEM study were prepared by Ar$^+$-ion milling. An analysis of Cd distribution in the FM was carried out using a (002) dark-field images, because (002) reflections are chemically sensitive in materials with a face-centered cubic lattice. A He-Cd laser (325nm) was used to measure the low-temperature (2K) PL spectra as well as PL temperature dependencies.

3. Results and discussion.

The reference structure A is characterized by a bimodal size distribution of CdSe islands. One can observe in Fig. 1a that large 2D islands with lateral sizes of 15-25 nm are superimposed with small Cd composition fluctuations (QDs) having lateral sizes below 10 nm. Average broadening of the CdSe sheet is 2.5-3 nm. Contrary to that, the (002) dark-field micrograph of structure B (Fig. 1b) displays appearance of isolated bright QD islands with lateral sizes of 8.5±1.5 nm as well as the pronounced narrowing (down to ~1.5 nm) of ZnCdSe "wetting" quantum well (QW) in between. Additional information on the CdSe FM morphology can be derived from the plain-view TEM, as presented in Fig. 2 for structure B. The CdSe QD islands observed as white spots with typical lateral sizes and density of 7-10 nm and ~10^9 cm^{-2}, respectively, are surrounded by a four-leaf strain field typical for highly strained coherent inclusions in a cubic crystal lattice. It permits one to suppose an extremely high Cd content in such QDs. Preliminary high-resolution TEM characterization using DALI composition evaluation give for the short-pulse MEE samples like structure A the maximum CdSe content in

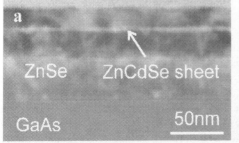
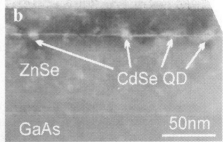

Fig. 1. Dark-field (002) TEM cross-section images of structure A (a) and structure B with a 1min growth interruption after each Se pulse (b).

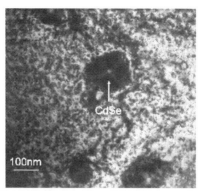

Fig. 2. Dark-field (002) plan-view TEM images of structure B with a CdSe white spots surrounded by four-leaf dark strain field. The relief between dots is artefact due to ion milling.

the islands as high as 85-90% [8]. It is approximately twice higher than that obtained in the samples grown earlier by MEE technique with a Cd pulse duration exceeding a 0.5ML surface coverage [2], which may be explained by large enough space for Cd surface migration, twice larger total time of CdSe redistribution between pulses and elimination of the parasitic influence on the intrinsic CdSe QD self-formation process of excessive (>0.5ML) Cd accumulation at growth surface imperfections [7].

The low-temperature PL spectrum of structure B (Fig. 3a) is characterized by enhancement of PL intensity and a 50 meV long-wavelength shift of the peak energy as compared to structure A. Simultaneously, the FWHM of the PL band decreases by more than 10meV down to 40meV. These observations confirm the efficient CdSe redistribution between large 2D islands and QDs during growth interruption at constant T_S, resulting in increase of Cd concentration in QDs and narrowing their lateral size dispersion, observed by TEM. The increase of Cd concentration estimated from the PL peak red shift, neglecting the lateral confinement variation, amounts to 5% with respect to structure A, which implies existence of nearly pure CdSe QD. Structure C with the 1min growth interruptions after each Cd pulse exhibits even more pronounced morphology transformation as compared to B. Plan-view TEM image reveals the

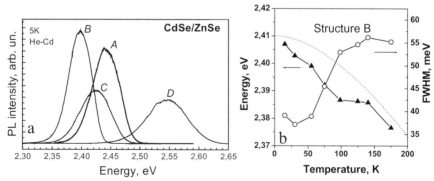

Fig. 3. (a) Low-temperature cw PL spectra of the CdSe/ZnSe nanostructures A (reference), B, C and D of the same CdSe deposition thickness, grown by different MEE techniques; (b) temperature dependencies of PL peak energy and FWHM for structure B.

CdSe-based islands observed as light circle-shaped objects with a factor of two lower density ($5\times10^8 \text{cm}^{-2}$) and larger lateral sizes (10-15 nm), in line with a cross-section TEM observation. However the PL spectrum of structure C demonstrates much smaller red shift of the peak energy and reduced intensity, perhaps due to defect occurrence, as well as line broadening up to 64 meV. Such smaller PL peak energy shift at larger island sizes evidences that the significant Cd re-evaporation takes place during growth interruption at the Cd-terminated surface in addition to faster Cd surface redistribution. Finally, structure D, demonstrating the effect of T_S rising on the structure A morphology shows more homogeneous Cd redistribution along the CdSe sheet, keeping its average broadening unchanged. The thinner TEM sample regions reveal high-density ($\sim 10^{11}$ cm^{-2}) small-scale QD islands with the lateral size ranging between 5 and 10 nm, which is reflected in dramatic broadening of the PL band width. Pronounced blue shift of the PL band indicates both an enhancement of the quantum confinement in the small QDs and, perhaps, an onset of Cd re-evaporation at high enough T_S.

The studies of PL temperature dependences (Fig.3b) indicates an complex structure of the excitonic states in the islands. Nonmonotonous behavior in the temperature dependence of PL band FWHM as well as pronounced red shift of the PL band maximum with temperature increase in comparison with the expected band gap temperature dependence - the features which are common for partly phase separated solutions - evidence that an appreciable part of the radiative states of excitons in islands are metastable at low temperature and can relax to deeper ground states at the temperature increase [9]. Data shown in Fig.3b reveal the existence of such metastable even in samples with highest CdSe composition, which can be treated as an evidence of the complex shape of the island. One should note, that the samples A and B demonstrate PL signal visible up to the room temperature, whereas PL of samples C and D decays at ~200K. This may be due to the extended defect occurrence in the latter two samples clearly revealed in TEM, which can serve as a non-radiative recombination centers.

In summary, using a 0.3ML mutli-cycle MEE technique for 2ML CdSe deposition on ZnSe with long growth interruptions after each Se pulse dramatically affects the CdSe redistribution on the surface, resulting in self-formation of coherent ~8.5 nm QDs with high Cd content and narrow size-dispersion. The structures demonstrate bright PL up to RT, which makes them promising candidates for the active region of ultra-low threshold green lasers.

Acknowledgements. The authors are thankful to A.A. Toropov and O.G. Lyublinskaya for useful discussion and D.Litvinov, A.Rosenauer and D.Gerthsen for preliminary information on HRTEM data. The work is supported in part by RFBR Grant 00-02-16999.

References
[1]. S.V. Ivanov, A.A. Toropov, S.V. Sorokin *et al.*, Appl. Phys. Lett. **74**, 498 (1999).
[2]. N. Peranio, A. Rosenauer, D. Gerthsen *et al.*, Phys. Rev. B **61** (23), 16015 (2000).
[3]. P.R. Kratzert, M. Rabe, and Henneberger, phys. stat. sol (b) **224**, 179 (2001).
[4]. E. Kurtz, M. Schmidt, D. Litvinov *et al.*, phys. stat. sol (b) **229**, 519 (2001).
[5]. M. Keim, M. Korn, A. Waag, *et al.*, J. Appl. Phys. **88**, 7051 (2000).
[6]. S.V. Ivanov, A.A. Toropov, T.V. Shubina *et al.*, J. Appl. Phys. **83**, 3168 (1998).
[7]. S.V. Sorokin, A.A. Toropov, T.V. Shubina *et al.*, J. Cryst. Growth **201/202**, 461 (1999).
[8]. D. Litvinov, D. Gerthsen, private communication.
[9]. A.Reznitsky, A.Klochikhin, L.Tenishev *et.al.*, Proc. 9 Int. Symp. "Nanostructures: Physics and Technology" p.538 (2001).

Spectroscopy of high-density assemblage of InAs/GaAs quantum dots

Z Ya Zhuchenko[1], J W Tomm[2], H Kissel[3], Yu I Mazur[4], G G Tarasov[1], W T Masselink[1]

[1]Department of Physics, Humboldt-Universität zu Berlin, 10115 Berlin, Germany

[2]Max-Born Institut für Nichtlineare optik und Kurzzeitspektroskopie, 12489 Berlin, Germany

[3]Ferdinand-Braun-Institut für Höchstfrequenztechnik, D-12489 Berlin, Germany

[4]Department of Physics, University of Arkansas, Fayetteville, Arkansas 72701, USA

Abstract. High-density arrays of InAs/GaAs quantum dots (QDs) have been studied by means of steady-state and time-resolved photoluminescence (PL) within a wide range of laser power. The ground state tunnelling between neighbouring QDs is suggested to be an important relaxation channel defining the PL spectral shape both at very low excitation density and at the temperature elevation. This channel becomes blocked when the QDs ground states population reaches the saturation at high excitation density.

1. Introduction

High-density self-assembled QDs are of interest for potential optoelectronic applications. However recently produced high-density QDs arrays possess essentially strong scattering both geometrical and relaxation parameters. Under the QDs areal density D_a comparatively high, $D_a \geq 10^{10}$ cm^{-2}, the behaviour of PL response, both steady state and dynamic, differs substantially from that of low-density QDs ensembles [1]. This difference reflects the intrinsic peculiarities of carrier transfer in the system of dense QDs arrays, which are still beyond the clarity.

Here we report a detailed investigation of carrier transfer in dense InAs/GaAs QDs arrays. A range of carrier densities in which inter-dot carrier transfer saturates is identified. This transfer involves transitions from higher lying ground states of small size QDs into lower-lying states of larger QDs.

2. Experimental details

GaAs superlattice structures containing thin layers of InAs were grown on GaAs (001) substrates by gas-source molecular beam epitaxy. Chosen growth conditions allowed to get the dense QDs arrays possessing nevertheless comparatively broad distribution by sizes. The AFM images of a sample with d_{InAs} = 2.46 ML and with substrate temperature T_G = 420 °C during the InAs growth display a co-existence of, at least, four QDs families with the average base length of b = 6, 8, 12, and 14 nm and the QDs density ratio of approximately 4:8:4:1, respectively. Transient PL measurements were performed with a self mode-locking Ti:sapphire laser (732 nm, 80 fs, 82 MHz) with excitation density between 10^9 and 10^{13} photons/(pulse×cm^2), allowing predominant excitation of the InAs QDs through the GaAs matrix. The steady-state PL was excited by the 514.5 nm line of Ar$^+$ laser with excitation densities in the range of 1 mW/cm^2 to 20 W/cm^2.

3. Results and Discussion

The low-temperature (T = 10 K) steady-state PL spectrum of the sample with d_{InAs} = 2.46 ML is depicted in Fig.1, representing a broad spectral band (FWHM ≈ 200 meV). The line-shape analysis shows that the PL signal is well reproduced by a convolution of four Gaussian-shaped peaks. Comparing the size distribution and areal densities derived from the AFM measurement and the results of PL analysis we assume that the peaks at 1.329 eV, 1293 eV, 1.224 eV, and 1.156 eV could be originated from the QDs families with the average bases b = 6, 8, 12, and 14 nm, respectively. The temperature elevation up to

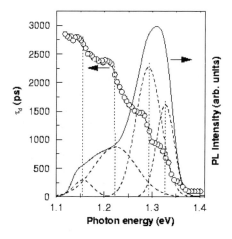

Figure 1. PL spectrum and staircase spectral dependence of $\tau_d(E)$ for InAs/GaAs QD's sample, T = 10 K, excitation density 2×10^{10} photons/(pulse×cm^2). Gaussians are shown by dash lines.

T = 240 K results in the complete decay of high-energy bands, and the low-energy band persists only. Thus the high temperature spectrum becomes effectively shrunken, being totally shifted towards the red side [2]. Time-resolved PL measurements were performed for different detection energies within the broad PL spectrum (Fig.1) and in a wide range of excitation intensities. Up to the excitation densities of 10^{11} photons/(pulse×cm^2), the PL emission at the different detection energies rises within the time resolution of our experiment of 10 ps. It displays a mono-exponential decay characterised by a decay time τ_d. In Fig.1, τ_d -data determined from a least-square fit of the transients for the excitation density 2×10^{10} photons/(pulse×cm^2) are summarised. The pronounced step-like behaviour of τ_d spectral dependence is observed. The τ_d step architecture completely reflects the PL band structure. The parts of the τ_d spectrum quickest slope correspond to the maximums of the PL components. As a function of sample temperature, the steps in τ_d shift by the same energy as the cw PL spectrum, providing independent evidence for the correlation between transient and steady-state PL features. This correlation is maintained up to the room temperature. The τ_d dependence allows revealing the various QD mode contributions even if these contributions could hardly be resolved spectroscopically in the conventional PL spectrum, thus providing a powerful experimental technique for analysis of very dense arrays of coupled QDs.

The τ_d spectral dependence is consistent with the model of inter-dot carrier transfer that considers the ground state relaxation in a system of coupled QDs. In our model, carriers in the ground state of a QD can relax by radiative recombination giving rise to PL and – in addition - carriers populating the ground states of smaller QDs can be transferred into the levels of larger QDs being even lower in energy. For weak excitation density of the QD ensemble and under the assumption that the carrier relaxation rate from an energy level in a QD is proportional to the number of vacant lower energy levels in adjacent QDs, the rate equation for a particular quantum level E_i can be written as

$$\frac{dn_i}{dt} = -\frac{n_i}{\tau_0^i} - \sum_{j<i} \frac{n_i(N_j - n_j)D_j}{\tau_t^{ij}} + \sum_{i<j} \frac{n_j(N_i - n_i)D_i}{\tau_t^{ji}}. \tag{1}$$

where τ_0^i is the total ground-state recombination lifetime in the i-th ground state, τ_t^{ij} is the inter-dot carrier transfer time between E_i and E_j states, D_j is the density of the (final) E_j states, N_j is the number of available QD ground states in the j-th dot distribution. If one neglects the non-linear terms in Eqn.(1) it can be solved analytically for the particular D_j distribution. The PL decay time τ_d is determined as $\tau_d = -n_i(t)/(dn_i(t)/dt)$. In fact, the proposed rate equations model of Eqn.(1) holds for a dense QD system independent of the number of modes of its size distribution. The presence of more than one distinct dot size distribution within a large QD ensemble can be taken into account simply by introducing additional Gaussians D_j into Eqn.(1), resulting in several step-like variations of τ_d with emission energy. The calculated transients $n_2(t)$ for the lower energy level of a

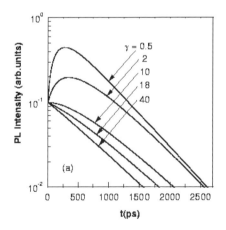
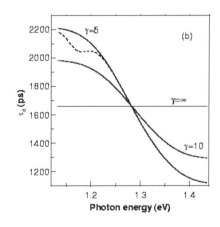

Figure 2. (a) Calculated transients and (b) spectral dependence of $\tau_d(E)$ for different for different ratios $\gamma = \tau/\tau_0$.

simplified QD model system with only two distinct dot sizes are shown in Fig.2a. The parameter $\gamma = \tau_t/\tau_0$ is a measure for the inter-dot coupling and describes the additional population of the lower level 2 by carriers from level 1. Small γ-values indicate strong coupling, i.e. efficient population of level 2. The spectral dependence of τ_d for different γ values is depicted for a single Gaussian (solid curves) and two Gaussians (dashed line) in Fig.2b. The spectral steps are clearly seen.

A realistic description of carrier dynamics in dense QD arrays has to include the case of saturation, i.e. situations in which the number of non-equilibrium carriers n_i reaches the number of available QD ground states N_j in the j-th dot distribution (see Eqn. (1)). The equations of set (1) are nonlinear differential equations, and do not have an analytical solution. Therefore the set was solved numerically with the natural assumption of a single time τ_0 for the recombination lifetime in the i-th ground state and a single time τ_t for the interdot carrier transfer. Applying this rate equation model to the actual experimental situation we can explain the details of QD kinetics. Indeed, the carrier transfer from smaller to larger QDs results in a delayed build-up of population in the latter, a behaviour reflected in the PL kinetics. With increasing excitation flux, saturation of population in such low-energy states occurs, leading to a faster rise of PL. This prediction of the rate equation models is in full agreement with our experimental results. For strong excitation, saturation leads to much faster population increase in the emitting QDs and a subsequent monoexponential decay of PL intensity. Applying the simple model for the low-density case and $\tau_0 = 300$ ps we derive a value of $\gamma = \tau_t/\tau_0 = 1.3$. It is clear that saturation of QD ground states could result in a lineshape variation of the cw PL at the smallest power densities due to saturation of inter-dot carrier transfer channel. Thorough PL study at the extremely low excitation density (less than 1 W/cm^2) displays such modification. It is observed the increase of the low-energy part of PL spectrum with the excitation density that reaches the saturation. We find that this saturation occurs already at 1W/cm^2, because the further power increase does not change the lineshape of this spectral region. The high energy part of the PL spectrum grows with the intensity of excitation, giving hint for the possible contribution of excited states which is expected to develop into pronounced structure under the very high excitation densities. It is also clear that the decrease of τ_d observed experimentally at high pumping levels can arise from many-body effects, which become more important in a dense QD system due to the fact that carriers can readily scatter out of the saturated states. Within the empirical model this depopulation of the lowest energy states could be treated by adding two similar nonlinear terms to the rate equations (1) but with opposite signs and a different, shorter scattering time $\tau_d(P_{exc})$ (which becomes a function of laser power for high excitation). This would lead to an effective decrease in the PL decay time, as experimentally observed in the spectral range of the QD emission regardless of QD size. This has little to do with the recombination times for higher energy states (or of smaller dots) being shorter.

In summary, we reported a study of steady-state and transient PL of dense InAs/GaAs quantum dot arrays. Our data are consistent with the model of carrier transfer from small to large QDs within the ensemble directly influencing the PL kinetics.

References
[1] Mazur Yu I, Tomm J W, Petrov V, Tarasov G G, Kissel H, Walther C, Zhuchenko Z Ya and Masselink W T 2001 Appl. Phys. Lett. 78 3214-6
[2] Tarasov G G, Mazur Yu I, Zhuchenko Z Ya, Maassdorf A, Nickel D, Tomm J W, Kissel H, Walther C and Masselink W T 2000 J. Appl. Phys. 88 7162-70

Properties of InGaAs Coupled Quantum Wire Structures Grown on Vicinal (111)B GaAs with Quasi-Periodic Corrugation

T. Noda, N. Kondo, Y. Akiyama, T. Kawazu, and H. Sakaki
Institute of Industrial Science, University of Tokyo,
4-6-1 Komaba, Meguro-ku, Tokyo 153-8505, Japan

Abstract. Transport properties of InGaAs coupled quantum wire (C-QWR) structures have been studied. Samples were formed by depositing a selectively doped InGaAs quantum well (QW) or an InAs/GaAs digital alloy channel layer on corrugated vicinal (111)B GaAs. Their capacitance-voltage (C-V) characteristics are found to depend on the channel direction and exhibit unique structures near the threshold, where the cross-corrugation transport is suppressed. Electron mobilities ($\mu_{//}$) in digital alloy channel for the current along the corrugation are by a factor ~3 larger than that $\mu_\perp$ across it. These features are interpreted in terms of C-QWR states induced by the local accumulation of In and also the large electron barriers induced by inhomogeneities of the corrugation.

1. Introduction

The epitaxial growth of GaAs on vicinal (111)B substrates is known to produce 10-30 nm scale periodic corrugation resulting from bunched multi-atomic steps. If pseudomorphic InGaAs quantum wells (QWs) are prepared on such corrugation, the in-plane potential is laterally modulated, inducing possibly lateral supperlattice (LSL) or coupled quantum wire (C-QWR) states. We showed earlier that these corrugations induce not only periodic but also random potentials, giving rise to anisotropic mobilities [1]-[3] and unique magneto-resistance characteristics [4].

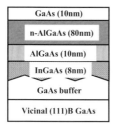 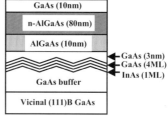

(a)   (b)

Fig.1 Schematic illustrations of samples A and B with InGaAs QW channel (a) and InAs/GaAs digital alloy channel (b).

In these systems electrons in neighboring wires are often coupled. But if one enhances the potential modulation or reduces the electron concentration Ns, the inter-wire coupling will be reduced and the one dimensional (1 D) nature enhanced. This situation appears when the gate voltage is set close to the threshold V_{th}. However, in this region the randomness of corrugation could start to dominate the electrical properties. To clarify these issues, we investigate two type QWs on such corrugated (111)B GaAs plane by performing and analysing capacitance-voltage (C-V) and Hall-effect measurements.

2. Sample fabrication

Two samples were grown by molecular beam epitaxy (MBE) on vicinal (111)B GaAs substrates with misorientation angle of 8.5° toward the <-10-1> and their structures are shown in Fig.1. They have selectively-doped channel structures with a 10 nm-thick AlGaAs spacer layer and 80 nm-thick n-AlGaAs layer. Sample A has an InGaAs C-QWR channel with average thickness of 8 nm and its In content is 11 %. Sample B has an almost identical structure, except the channel region being composed of an 8.7 nm-thick digital alloy layer, which consists of 4 periods of 1 monolayer (ML)-thick InAs and 4 ML-thick GaAs and a 3 nm-thick GaAs layer.

The growth rate of GaAs during the formation of the corrugated surface was 0.27 µm/hr and the As_4/Ga ratio was 12~35. Atomic force microscope (AFM) studies show that multiatomic steps formed on the vicinal (111)B GaAs run along the <10-1> direction and are about 20-30 nm in period and 1-2 nm in height. We fabricated FETs with Hall-bar geometry for transport and capacitance-voltage (C-V) studies. In particular, to clarify anisotropic transport we fabricated two identical FETs with the channel parallel (hereafter referred to para-FET) or perpendicular (perp-FET) to the corrugation. We also grew the same structures on nominally flat (001) GaAs as reference wafers with which two FETs with the channel along the <110> (2x-FET) and <-110> (4x-FET) directions were fabricated.

In sample B, In atoms may accumulate in the region of groove bottom and induce a strong in-plain potential modulation. To confirm the selective incorporation of In, we grew non-doped QWs with GaAs barrier layers instead of an AlGaAs barrier layer and performed photoluminescence (PL) measurements. The PL peak position of sample B is by 30 meV lower than that of the reference. This suggests that In atoms accumulate locally. In addition, AFM study of the QW surface showed that periodic corrugation is smeared, suggesting that In tends to be incorporated probably along the groove of the corrugation.

3. Experiment

We first study sample A. To clarify electronic states we carried out C-V measurements between the gate and the drain/source electrodes of FET samples at 4.2 K at a frequency of 1 KHz. Figure 2(a) shows the result for sample A. The data for reference sample is shown in Fig. 2(b). Note some specific differences between the two. C-V data for reference sample show the rapid rise, indicative of 2D electronic states. No clear structures were seen except for minor structures in 2x-FET, which may be related to localized states. In addition, the shift of the onset of C-V curves is due presumably to a slight difference of the sample structures such as channel width.

On the other hand, C-V data of sample A depend on the channel direction of FETs. The capacitance of perp-FET increases gradually until the drain current starts to flow at V_g ~ -0.35 V in this sample. For V_g beyond this region, the data coincide with that of para-FET. Note that the C-V curve in the para-FET has step-like structure, which can be

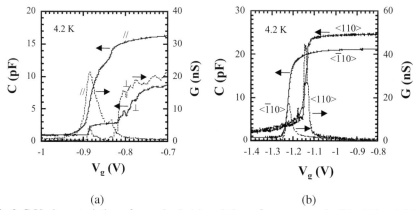

Fig.2 C-V characteristics of sample A (a) and the reference sample (b) at T = 4.2 K. Solid and dashed lines show capacitance C and loss G, respectively.

seen for other para-FETs (e.g., see Fig.4). This sharp rise in capacitance suggests that electrons have 2D character gas because the onset of N_s in N_s-V_g characteristics [3] extrapolate to $V_g \sim -0.8$ V. In the perp-FET complicated structures are seen. Note that the capacitance is less than that of the para-FET, indicating that electrons occupy only a fraction of the channel. Since no drain current flows until $V_g \sim -0.35$ V, localized states may be formed by inhomogeneities of the corrugations.

Next we study transport properties of sample B. Figure 3 shows its mobilities μ at T= 4.2 K for FETs with the channel parallel (plotted by closed circles) and perpendicular (open circles) to the corrugation as functions of electron concentration N_s. Data of the reference sample are also plotted by closed (for the current J along <110>) and open (J along <-110>) triangles.

Although the transport is anisotropic in both samples, sample B exhibits pronounced anisotropy. This large anisotropy suggests that In atoms deposit locally, which is consistent with PL and AFM data. To analyze the transport properties we calculated the

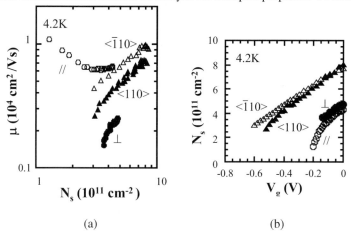

Fig.3 μ-N_s (a) and N_s-V_g (b) relations for sample B at T = 4.2 K. Closed and open circles show data measured across and along the corrugation, respectively. Mobilities of reference sample are also plotted by closed and open triangles for two current directions.

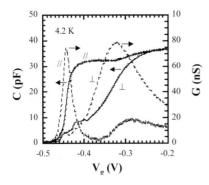

Fig.4 C-V characteristics of sample B at T = 4.2 K. Solid and dashed lines show capacitance C and loss G, respectively.

mobility of sample B. Here we used the same scattering model, which fairly well explain data of sample A [3]. However, an agreement between theory and experiment was poor, in particular, at low Ns, suggesting that 2D-based model may need to be modified. Note also that mobilities ($\mu_{//}$) along the corrugation tend to increase as Ns is reduced and exceeds μ of reference sample. The mechanism is not clear at present, but formation of wire-like structure may contribute to the increase of μ along the corrugation.

Figure 4 shows C-V data for sample B at 4.2 K. The sample for this study is the same FETs used for transport studies. Measurements were done under the same conditions as for the sample A. C-V curves qualitatively agree with those of sample A except that the second peak (V_g ~ -0.3 V) of the loss G in para-FET shifts at higher gate voltage, probably due to a larger potential modulation in this sample.

4. Conclusions

We have fabricated coupled quantum wire (C-QWR) structures on vicinal (111)B corrugated GaAs and studied their C-V and transport properties. We have found that C-V characteristics qualitatively have common features for both sample A and B. For the channel along the corrugation, the capacitance C increases step-like and the loss G has two peaks around the steep rise points of capacitance. On the other hand, for the channel across the corrugation, C-V curves have complicated structures. These features can be explained by regular and inhomogeneous corrugation. We have also found that in sample B, μ of para-FET increases as Ns is reduced and those exceed μ of reference sample. This result indicates formation of wire-like states. We believe that this system offer unique properties once the uniformity of corrugation is improved.

This work is supported by the COE project.

References

[1] Y. Nakamura, S. Koshiba, and H. Sakaki, Appl. Phys. Lett. 69 (1996) 4093
[2] Y. Nakamura, S. Koshiba, and H. Sakaki, J. Crst. Growth, 175/176 (1997) 1092.
[3] T. Noda, Y. Nagamune, Y. Nakamura, and H. Sakaki, Physica E 13, (2002) 333.
[4] T. Noda and H. Sakaki, to appear in Proc. of 28th Int. Symp. Comp. Semicond. (ISCS), Tokyo (2001).

One-dimensional free exciton in CdTe/Cd$_{0.74}$Mg$_{0.26}$Te quantum wires

S. Nagahara[1], T. Kita[1], O. Wada[1], L. Marsal[2], and H. Mariette[2]

[1]Department of Electrical and Electronics Engineering, Faculty of Engineering, Kobe University, 1-1 Rokkodai, Nada, Kobe 657-8501, Japan

[2]Laboratoire de Spectrometrie Physique, Universite J. Fourier, Grenoble I CNRS(UMR 5588), BP87, F-38402 Saint Martin d'Heres Cedex, France

Abstract. The magneto-optical properties in CdTe/CdMgTe quantum wires (QWRs) grown on a vicinal substrate were investigated. With increasing magnetic field, the exciton photoluminescence (PL) shows a slight energy shift. The energy shift of the exciton-PL peak can be described by the exciton Zeeman splitting and the diamagnetic shift. The effective g factor obtained from the Zeeman splitting for the QWRs were found to be larger than for CdTe/CdMgTe quantum well, indicating size dependence to the effective g factor in the QWRs. Furthermore, it was found that the biexciton formation is suppressed by magnetic field, and the suppression behaviour is dependent on the direction of magnetic field to the QWRs.

1. Introduction

Organized growth of semiconductors on a vicinal substrate is an attractive way to realize one-dimensional (1D) quantum wires (QWRs) in a simple technical step. [1] The repeated deposition of a fractional monolayer (ML) m of material A followed by a fractional ML n of material B results in a high-density QWRs, or a A$_m$B$_n$ tilted superlattice. At m and n = 0.5, the cross-section of the QWRs becomes square. The high-density QWRs, which are grown by a simple technical step, are promising for device applications.

Property of the 1D-free exciton (X) in the QWRs has been presented theoretically by Citrin [2]. The evaluated important results for the 1D X are a long intrinsic radiative lifetime due to a finite spatial coherence in the lateral direction and the $T^{1/2}$ dependence of the effective lifetime due to thermalization effects. Recently, we have demonstrated a clear 1D X behaving with the above-mentioned features in CdTe/Cd$_{0.74}$Mg$_{0.26}$Te QWRs structure grown on a vicinal substrate.[3,4] Furthermore, we confirmed biexciton (XX) photoluminescence (PL) of the QWRs for the first time [5]. In this study, we focus on magneto-optical properties of 1D X and XX in the QWRs. We observed the Zeeman splitting in the 1D system for the first time. In magnetic field, the XX-PL intensity decreases, which depends on the direction of the magnetic field to the QWRs.

2. Experiments

The growth of lateral superlattice on a vicinal substrate is based on the step-flow growth of fractional MLs over a very regular array of monomolecular steps. The steps align along

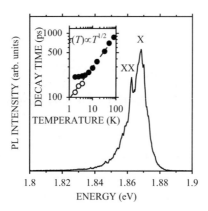

Figure 1. PL spectrum of the QWRs at 1.9 K. The inset shows radiative lifetimes of X in the QWRs as a function of temperature.

the [010] direction. In this experiment, the array of CdTe/Cd$_{0.74}$Mg$_{0.26}$Te QWRs, with m and $n \approx 0.5$ was grown on a Cd$_{0.96}$Zn$_{0.04}$Te (001) substrate misoriented 1° toward the [100] direction by molecular beam epitaxy. A repeated deposition of a fractional ML m of CdTe followed by a fractional ML n of Cd$_{0.74}$Mg$_{0.26}$Te resulted in CdTe/Cd$_{0.74}$Mg$_{0.26}$Te QWRs. This cycle was repeated 30 times. As a consequence, the 1° misoriented substrate gave rise to approximately square wire of 9.3×9.7 nm^2 [3]. After growing the QWRs, a 25-ML CdTe QW was grown. Magneto-PL measurement was performed at 1.9 K in an optical cryostat with a superconducting split-coil magnet up to 7 T. The laser wavelength was 387 nm.

3. Results

3.1. *Magneto-optical properties of 1D X in the QWRs*

Figure 1 shows the PL spectrum of the QWRs at 1.9 K. The PL at 1.867 eV is attributed to the X recombination. The temperature dependence of the radiative lifetime of the 1D X in the QWRs was measured using time-resolved spectroscopy. The inset of Fig. 1 shows radiative lifetimes of X in the QWRs as a function of temperature. The observed radiative lifetime in the QWRs (closed circles) obeys $T^{1/2}$, which clearly demonstrates the 1D-X nature [4]. At low temperature <10K, however, the radiative lifetime deviate from the $T^{1/2}$ dependence. At this temperature region, a new PL appears at the lower-energy side of X,

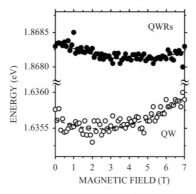

Figure 2. PL peak energies of the X in the QWRs and QW as a function of the magnetic field.

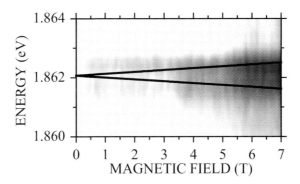

Figure 3. Counter plot for the intensity difference of XX PL by applying magnetic field.

which is due to the radiative recombination of XX [5]. By taking account of the XX generation, it was demonstrated that the radiative lifetime of the X (open circles) have been confirmed to obey the $T^{1/2}$ dependence in the wide temperature range.

We have investigated the magneto-optical properties of 1D X in the QWRs. Figure 2 shows the peak energies of the X PL in the QWRs and QW as a function of the magnetic field. The magnetic field dependence of the X-PL peak energy can be described well by the sum of the zero-field energy, the diamagnetic shift and the Zeeman splitting. The effective g factors of the QWRs and the QW obtained from the Zeeman splitting are –4.2 and –3.6, respectively. The X reduced masses for the QWRs and the QW obtained from the diamagnetic shift are $0.095m_0$ and $0.084m_0$, respectively. It is considered that the lateral confinement in the QWRs increases the effective g factor and X reduced masses.

Figure 3 shows a counter plot for the intensity difference of the XX PL in the magnetic field. The dark region indicates reduction of the XX-PL intensity. The full-width-at-half-maximum (FWHM) of the XX PL is increasing with increasing the magnetic field. Although the XX is a spin-singlet state, which can not be split by the magnetic field, the final state X shows the Zeeman splitting. The solid lines indicate the linear Zeeman splitting, in which we used the effective g factor of –4.2. The good agreement indicates that the effective g factor is almost linear in our measured field range.

3.2. Biexciton formation in magnetic field

The cross-section of the QWRs is comparable with the bulk-X-Bohr radius that is about 6.5 nm. Therefore, it is expected that the XX tends to align along the QWRs. The XX-PL intensity decreases by applying magnetic field as shown in Fig. 3. Time-resolved PL measurements reveal that the XX-radiative lifetime does not depend on magnitude of the magnetic field. This result reveals that the XX formation is suppressed by the magnetic field. We have investigated the XX-formation process in the magnetic field. The PL-intensity difference, $\Delta I/I$, caused by the magnetic field is defined as $\Delta I/I=(I(B)-I(0))/(I(B)+I(0))$, where $I(B)$ and $I(0)$ are PL intensity in magnetic field B and zero field, respectively. Solid and open circles in Fig. 4 (a) plot $\Delta I/I$ for the XX PL measured in magnetic field parallel and perpendicular to the QWRs, respectively. With increasing the magnetic field perpendicular to the QWRs, $\Delta I/I$ decreases monotonically. In magnetic field parallel to the QWRs, on the other hand, a different behaviour was observed. The magnetic field dependence of $\Delta I/I$ is quite small up to 3.5 T, and beyond this field it starts to decrease steeply. When the magnetic field perpendicular to the QWRs is applied, since the X spin tends to align perpendicular to the QWRs, the XX formation is suppressed with increasing the magnetic field. In the case of the magnetic field parallel to the QWRs, on the other hand, since the X spin tends to align parallel to the QWRs, the Xs can attract each other. After forming the X pair, the XX state may transform into a more stable spin-singlet state (J=0). Therefore, $\Delta I/I$ is almost independent of the magnetic field up to

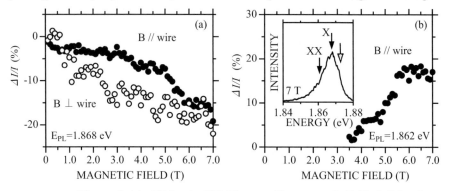

Figure 4. (a) $\Delta I/I$ for the XX PL caused by magnetic field. Solid and open circles plot the results in magnetic field parallel and perpendicular to the QWRs, respectively. (b) $\Delta I/I$ for the new PL appeared at the high-energy side of the X PL. The inset shows the PL spectrum at 7 T.

3.5 T. In addition to the steep decrease of $\Delta I/I$ above 3.5 T, a new PL appears at the high-energy side of the X PL as shown by an open arrow in the inset of Fig. 4 (b). $\Delta I/I$ for the new PL is shown in Fig. 4 (b). $\Delta I/I$ for the new PL increases with the magnetic field beyond the same magnetic field (3.5 T). The new PL is likely attributed to the spin-triplet state (J=2).

4. Conclusion

We have investigated the magneto-optical properties in CdTe/CdMgTe QWRs. With increasing magnetic field, the X PL shows a slight energy shift. The energy shift of the X-PL peak can be described by the Zeeman splitting and the diamagnetic shift. The Zeeman splitting was confirmed directly by observing split XX-PL signal in the magnetic field. The effective g factors of the QWRs and QW obtained from the Zeeman splitting are –4.2 and –3.6, respectively. The effective g factor for the QWRs was found to be larger than for the QW, indicating size dependence on the effective g factor of the QWRs.

The decrease of the XX-PL intensity is observed in the magnetic field. The radiative lifetime of XX does not depend on magnitude of the magnetic field. This result reveals that the XX formation is suppressed by the magnetic field. In the case of the magnetic field perpendicular to the QWRs, since the X spin tends to align perpendicular to the QWRs, the XX formation is suppressed. In the case of the magnetic field parallel to the QWRs, on the other hand, the Xs can attract each other, thus form the X pair, or XX. This XX state transforms into a more stable spin-singlet state below 3.5 T. In addition to the steep decrease of $\Delta I/I$ above 3.5 T, a new PL appears at the high-energy side of the X PL, which is likely attributed to the spin-triplet state.

References

[1] P.M. Petroff, A. C. Gossard, R. A. Logan, and W. Wiegmann, Appl. Phys. Lett. **41**, 635 (1982).
[2] D. S. Citrin, Phys. Rev. Lett. **69**, 3393 (1992).
[3] L. Marsal, A. Wasiela, G. Fishman, H. Mariette, F. Michelini, S. Nagahara, and T. Kita, Phys. Rev. B **63**, 165304 (2001).
[4] S. Nagahara, T. Kita, O. Wada, L. Marsal, and H. Mariette, Phys. Stat. Sol. (a) **190**, 699 (2002).
[5] S. Nagahara, T. Kita, O. Wada, L. Marsal, and H. Mariette, Inst. Phys. Conf. Ser. **170**, 519 (2002).

Analyses of Self-Assembled GaN Nanorods on Si (111) Substrate

L. W. Tu[1,a], C. L. Hsiao[1], T. W. Chi[1], J. F. Wu[1], I. Lo[1], K. Y. Hsieh[2,3], T. T. Sheng[3], and C. F. Hu[3]

[1]Department of Physics, [2]Institute of Materials Science and Engineering, National Sun Yat-Sen University, Kaohsiung, Taiwan 80424, Republic of China [3]Macronix International Co., Ltd., Hsinchu, Taiwan 30055, Republic of China

Abstract. With the techniques of molecular beam epitaxy and self-assembly, GaN pillars with a dimension of nanoscale are demonstrated. No catalysis or extra nanostructure are used. These nanorods are hexagonal phase GaN and grow in a vertical direction to the substrate along the c-axis. Analyses are done with scanning electron microscopy, transmission electron microscopy, x-ray diffraction, photoluminescence, and micro-Raman spectroscopy.

1. Introduction

Among the compound semiconductors, direct-bandgap III-V semiconductors GaN and related N-based compounds possess many unique properties [1-7]. They are good light emitters covering a broad electromagnetic wave spectrum from IR (InGaAsN) to deep UV (AlGaN). They have high mechanical hardness, thermal stability, and chemical inertness. They are good candidates in high-power and high-speed devices and in applications under harsh environments. Although devices made of N-based semiconductors, like light emitting diodes and laser diodes are already on the market for some years, the defect density, especially the dislocation density, is very high as compared with other semiconductors, like Si, Ge, GaAs, ZnS, and etc. To further improve the quality, many techniques have been studied including dimensionality reduction.

Two important implications accompanied with dimensionality reduction are achievement of crystal perfection and observation of size-related effects [8-11]. In this report, GaN vertical pillars in the scale of nanometer are grown by plasma assisted molecular beam epitaxy (PAMBE). These nearly one-dimensional structures do not contain dislocation defects and show their crystal perfection in the micrographs of transmission electron microscopy (TEM). Nanorods related luminescence emissions are observed and growth mechanism is discussed.

---

[a] To whom correspondence should be addressed. E-mail: lwtu@mail.nsysu.edu.tw

2. Experiments

The crystal growth facility is a Veeco-Applied Epi 930 system using N_2 as the plasma source. The nitrogen gas is 6N pure and further purified through an N_2 purifier. The plasma cell has a Uni-Bulb design with a high generation efficiency of active nitrogen species. Ga source is 7N5 pure Ga metal in a Knudsen cell. Substrates are Si (111) wafers. Standard degrease procedures and HF light etching are applied to the Si substrate before loading into the load-lock chamber. A (7x7) reflection high-energy electron diffraction pattern ensures the good starting surface of the Si substrate. Low temperature GaN buffer layer is deposited first at 500 – 600 °C. Then, the temperature is raised to 720 °C before the growth of the GaN layer. The GaN pillars are formed naturally protruding above the surrounding GaN film.

3. Results and discussion

Vertical GaN nanorods are formed through self-assembly with a lateral size δ 10 nm – 200 nm. Nanorods are shown in the image of field emission scanning electron microscopy (FESEM) as in Fig. 1. The vertical growth rate of the nanorods is ~20% higher than the film. With a film thickness of 1.02 μm, the GaN nanorods have a nominal height of ~170 nm above the film surface. TEM samples are prepared by focused ion beam and a GaN nanorod is shown in Fig. 2., which shows its perfection without dislocations.

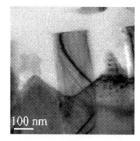

Figure 1. SEM images of GaN nanorods.   **Figure 2**. TEM image of nanorods.

Diffraction pattern of TEM shows that the nanorods are wurtzite GaN with a direction of <0001> along the rod length, i.e., along the Si <111> direction. A diffraction pattern of the nanorod is shown in Fig. 3 along zone axis <1-210>. This is consistent with the results of x-ray θ/2θ scan, which gives a single peak at 2θ = 34.6° of GaN (0002) Bragg diffraction. Strong strain fields in the GaN/Si interface are clearly seen in the TEM as in Fig. 4.

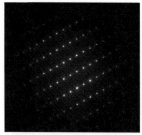

Figure 3. TEM diffraction pattern of one GaN nanorod.   **Figure 4**. TEM image of GaN/Si Interface.

Temperature dependent photoluminescence (PL) spectra have been measured on a sample with nanorods structure as shown in Fig. 5. There is a single peak at 364 nm at room temperature. An extra peak emerges at lower temperatures and the peak positions are at 358 nm and 362 nm at 66 K. This extra peak is associated with nanorods structure. Samples without nanorods do not show this extra peak.

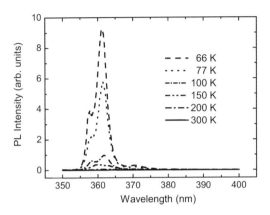

Figure 5. Temperature dependent PL.

Micro-Raman spectroscopy (Jobin Yvon T64000) at room temperature in a backscattering configuration is performed. The results show Stokes scattering of a hexagonal GaN with lines at 566.5 cm^{-1} (E_2 mode, high) and 142.5 cm^{-1} (E_2 mode, low) with 532 nm laser focused on the nanorod and at 564.8 cm^{-1} (E_2 mode, high) and 142.5 cm^{-1} (E_2 mode, low) focused on the film from the top surface as shown in Fig. 6(a) and (b), respectively.

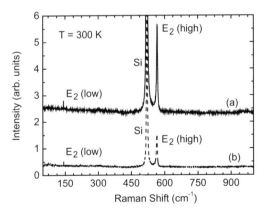

Figure 6. μ-Raman spectroscopy with laser beam focused on (a) the nanorod and (b) the film.

From the large strain revealed on the images of TEM results (Fig. 4), a possible strain-releasing growth mechanism is proposed. With a large difference of the lattice constant and the coefficient of thermal expansion between the Si substrate and the GaN layer, the GaN crystal grown over the Si substrate experiences large stress. The role of the buffer layer could serve as a stress-releasing layer, which absorbs a major portion of the stress and leaves nanoscale open windows. Nanorods are formed within these open windows and maintain a ~20% higher growth rate than the surrounding film in a nearly three-dimensional growth mode.

4. Conclusions

Vertical GaN nanorods are demonstrated and analysed. SEM image shows the well-formed geometry of these nanorods. TEM data reveals that these GaN nanopillars have a dislocation-free crystal quality and a wurtzite phase. An extra peak is observed in the PL spectra at low temperatures, which is ascribed to the nanorods structure. The origin of the growth of the nanorods is possibly due to the stress between the GaN and the Si substrate.

5. Acknowledgments

We acknowledge the support from Center for Nanoscience and Nanotechnology of NSYSU and the National Science Council of the Republic of China.

6. References

[1] Pankove J. I. and Moustakas T. D., ed., Gallium Nitride (GaN) II (Academic Press, New York, 1999).
[2] Nakamura S., Pearton S., and Fasol G., The Blue Laser Diode (Springer, New York, 2000).
[3] Morkoç H., Nitride Semiconductors and Devices (Springer, New York, 1999).
[4] Tu L. W., Lee Y. C., Chen S. J., Lo I., Stocker D., and Schubert E. F., Appl. Phys. Lett., 73, 2802 (1998).
[5] Tu L. W., Lee Y. C., Stocker D., and Schubert E. F., Phys. Rev. B, 58, 10696 (1998).
[6] Tu L. W., Kuo W. C., Lee K. H., Tsao P. H., Lai C. M., Chu A. K., and Sheu J. K., Appl. Phys. Lett., 77, 3788 (2000).
[7] Tu L. W., Tsao P. H., Lee K. H., Lo I., Bai S. J., Wu C. C., Hsieh K. Y., and Sheu J. K., Appl. Phys. Lett., 79, 4589 (2001).
[8] Calleja E., Sánchez-García M. A., Sánchez F. J., Calle F., Naranjo F. B., Muñoz E., Jahn U., and Ploog K., Phys. Rev. B, 62, 16826 (2000).
[9] Kusakabe K., Kikuchi A., and Kishino K., Jpn. J. Appl. Phys., 40, L192 (2001).
[10] Yoshizawa M., Kikuchi A., Mori M., Fujita N., and Kishino K., Jpn. J. Appl. Phys., 36, L459 (1997).
[11] Mamutin V. V., Cherkashin N. A., Vekshin V. A., Zhmerik V. N., and Ivanov S. V., Phys. Solid State, 43, 151 (2001).

Electroluminescence of asymmetric coupled GaAs/AlGaAs V-groove quantum wires

K F Karlsson[*], H Weman[*], M –A Dupertuis, K Leifer, A Rudra, and E Kapon

Laboratory of Physics of Nanostructures, Institute of Quantum Electronics and Photonics, Swiss Federal Institute of Technology Lausanne, Switzerland

Abstract. We have studied systems of asymmetric coupled GaAs V-groove quantum wires, in which the electrons and holes are injected into different wires. Our experimental results indicate efficient electron and hole tunneling, despite a 7 nm thick AlGaAs tunnel barrier. Temperature dependent electroluminescence exhibit clear effects of tunneling up to room temperature but cannot distinguish electron/hole tunneling from exciton tunneling.

1. Introduction

Carrier tunneling has been extensively studied in two-dimensional quantum wells (QWs) due to the interesting physical effects and their potential use in device applications [1, 2]. Studies concerning coupling and tunneling in one-dimensional quantum wire structures (QWRs) have so far mainly been focused on transport properties [3]. Recently we reported a photoluminecence (PL) study on electronic coupling and tunneling between two asymmetric, coupled GaAs V-groove QWRs [4], but without beeing able to separate the effect of electron and hole tunneling. In this paper, we report on an electroluminescence (EL) study of tunneling in asymmetric coupled GaAs/AlGaAs V-groove QWR light-emitting diode (LED) structures, with thicker tunneling barrier than previously used. In contrast to the PL experiments, the present use of electrical carrier injection offers the possibility to populate the two coupled QWRs with different charge type. Two samples with inverted doping sequence around the QWRs were fabricated in order to separate electron and hole tunneling.

2. Samples and Experimental Setup

The samples studied consist of $GaAs/Al_{0.32}Ga_{0.68}As$ double-QWR structures (DQWRs) introduced in the intrinsic region of p-i-n-diodes. Two samples, A and B, were produced with an inverted doping sequence, resulting in opposite direction of carrier injection. Sample A (B) was grown by low-pressure organometallic chemical vapor deposition on n^+- (p^+-) doped (100)-GaAs substrates patterned with 0.5-µm pitch [$01\bar{1}$]-oriented V-grooves. After the growth, the samples where processed into mesa diodes (62×62 µm^2). Details of the fabrication of the LEDs and V-groove QWRs can be found elsewhere [4, 5]

Crescent-shaped GaAs QWRs are formed at the tip of the V-groove while QWs are formed on the sidewalls of the V-grooves as well as on the top planar part between the V-

---

[*] Also at Materials Science, Dept. of Physics, Linköping University, Sweden

grooves. Furthermore, a self-ordered AlGaAs vertical QW (VQW) is formed between the doped top- and bottom regions, intersecting the two differently sized QWRs, serving as an efficient selective injection channel for the charge carriers from the doped regions directly into the QWRs [5]. A dark-field TEM micrograph of the QWR region of sample B is shown in Fig. 1. The thickness of the narrow QWR (n-QWR), the tunnel barrier and the wide QWR (w-QWR) is 5.5 nm, 7 nm and 7.5 nm, respectively. Samples A and B differ by the way the electrons and holes are injected according to the schematic band diagrams shown as insets in Fig 4; in sample A (B) the electrons are injected into the n-QWR (w-QWR) while the holes are injected into the w-QWR (n-QWR).

The samples were placed in a helium cryostat at a temperature in the range from 10 K to 300 K. A constant current source (0.65 A/cm^2) or a tuneable Ti:Sapphire laser (50 W/cm^2) was used to excite the sample for EL or photoluminescence excitation (PLE) measurements, respectively. The luminescence was dispersed by a monochromator and detected by a GaAs-photomultiplier. The PLE measurements were performed at flat-band conditions with a forward bias just below the threshold voltage for the onset of the EL.

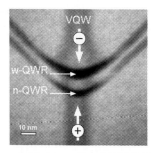
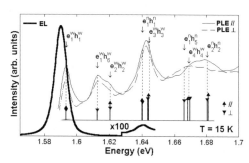

Figure 1. Cross-sectional TEM micrograph of the QWR region in sample B.

Figure 2. EL and polarized PLE spectra of sample B compared with calculations.

3. Results and Discussion

The QWRs were designed to have the first electron subband of the n-QWR (e_1^n) coinciding with the third electron subband of the w-QWR (e_3^w) in order to obtain resonant electron tunneling between the two QWRs. With this design, we previously achieved also resonance between the first hole subband of the n-QWR (h_1^n) and the fourth hole subband of the w-QWR (h_4^w) [4]. With a tunnel barrier thickness of 7 nm we are in a weak-coupling regime, with the wavefunctions mainly localized either in the n-QWR or the w-QWR [4]. Since the electrons and holes are injected into different QWRs, carrier transfer between them is necessary for the EL to arise.

The low-temperature (T = 15 K) EL spectrum of sample B has a strong peak at 1.590 eV with a weak satellite peak at 49 meV higher energy, as shown in Fig. 2. These energies agree well with the energy positions of the exciton ground states in previously studied isolated QWRs of comparable size and shape [6]. The peaks are therefore directly attributed to the recombination of excitons formed from electrons and holes in the ground subbands of the w-QWR ($e_1^w h_1^w$) and the n-QWR ($e_1^n h_1^n$), respectively. At such low temperature, the electrons (holes) are effectively relaxed to the ground subband in the w- and n-QWR, making the tunneling possible merely in one direction: from the n-QWR to the w-QWR. Consequently, EL only from the w-QWR is expected at low temperatures, consistent with the experimental data shown in Fig. 2, where the intensity from the w-

QWR is three orders of magnitude larger than from the n-QWR. The finite contribution from the n-QWR might be due to a small leakage of electrons directly into the n-QWR.

Information about the electronic structure related to the main peak $e_1^w h_1^w$ is revealed by the polarized PLE measurements presented in Fig. 2. The excitonic resonances exhibited in the PLE spectra correspond to the ground state, $e_1^w h_1^w$, and excitons associated with higher subbands $e_i^{w(n)} h_j^{w(n)}$ ($i,j > 1$) in the DQWR.

The PLE spectra in Fig. 2 are compared with a theoretical calculation of the optical interband transitions for the DQWR based on the solution of the 2D single-particle Schrödinger equation using the full 4 × 4 **k·p** Luttinger Hamiltionan for the valence bands. The shape of the confining potential was extracted from the cross-sectional TEM micrograph, and the material parameters used was equivalent to those used for modeling of isolated QWRs. The calculated energy positions and the polarization dependent oscillator strengths are represented by the arrows shown below the PLE spectra, horizontally translated in order to make the calculated and measured ground states coincide. The agreement with the experiment is very good, and the model can be used as a guide to identify the resonances in the PLE spectra. Furthermore, the calculated energy separation between e_3^w and e_1^n (h_4^w and h_1^n) is 8.1 meV (-0.3 meV), corresponding to non-resonant (near-resonant) electron (hole) tunneling, for this specific sample location on which the calculations are based.

For later comparison, the experimental energy separation between $e_1^w h_1^w$ and $e_1^w h_6^w$ is determined to be 20 meV and between $e_1^w h_1^w$ and $e_1^n h_1^n$ ($=e_3^w h_3^w$) it is 49 meV. The PLE resonance related to the $e_1^n h_1^n$ and $e_3^w h_3^w$ peak is abnormally strong compared to analogous measurements of isolated QWRs [6] due to an increased density-of-states originating from the coupling of the two QWRs at this energy. This suggests also the possibility of resonant coupling between these two excitonic levels.

The low-temperature EL and PLE measurements performed on sample A gave similar results as sample B. However, the intensity of the $e_1^n h_1^n$ transition was about four (instead of three) orders of magnitude smaller than the intensity of $e_1^w h_1^w$, (indicating that hole leakage is smaller than electron leakage into the n-QWR), and the subband separation was slightly increased (the energy separation between $e_1^w h_1^w$ and $e_1^w h_6^w$ was 21 meV and the separation between $e_1^w h_1^w$ and $e_1^n h_1^n$ ($=e_3^w h_3^w$) was 52 meV). The difference in subband separation is probably due to small fluctuations in the growth rate.

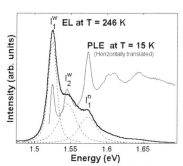

Figure 3. EL (with peak fitting) and PLE spectra, for sample B.

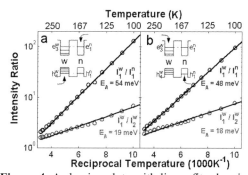

Figure 4. Arrhenius plots with linear fits showing the temperature evolution of a) sample A and b) sample B. Insets show carrier injection in A and B.

The above-mentioned experimental results demonstrate that both electrons and holes are effectively transferred from the n-QWR to the w-QWR at low temperatures. By increasing the sample temperature, the reversed transfer becomes possible since the thermally populated higher subbands of the w-QWR are coupled to the n-QWR. A thermal equilibrium of carriers populating both the w-QWR and the n-QWR is therefore expected. The EL spectrum of sample B at $T = 246$ K is presented in Fig. 3 together with the low-temperature PLE spectrum. In contrast to the low-temperature case, contributions from higher subbands is clearly seen in this EL spectrum. Three Voigt peaks, fitted to reproduce the shape of the spectrum, are shown below the spectrum. Hereafter these peak intensities will be referred to as I_1^w, I_2^w, and I_1^n, as indicated in Fig. 3. The two closely spaced transitions $e_2^w h_2^w$ and $e_1^w h_6^w$ are thus approximated as one single peak I_2^w. The ratio I_1^w / I_2^w is consistent with what we have observed at the same temperature for isolated QWRs, while the ratio I_1^w / I_1^n is about a factor two larger in the present DQWR due to the coupling.

To further investigate the temperature dependence of the relative intensites, the ratio I_1^w / I_1^n is shown as Arrhenius plots in Figs. 4a and 4b for samples A and B, respectively. For comparison, the corresponding temperature evolution of the excited subbands, I_1^w / I_2^w, are plotted as well for both samples. Qualitatively, both samples behave in the same way; a weak temperature dependence of the relative intensities up to about 50 K (not shown) and thereafter a linear decrease in the Arrhenius plot indicating a well-defined activation energy. The activation energies for sample A (B) obtained with a least-square fit for T $\geq$ 100 K is 19 meV (18 meV) for I_1^w / I_2^w and 54 meV (48 meV) for I_1^w / I_1^n. These values are close to the actual energy separations of 21 meV (20 meV) and 52 meV (49 meV) obtained from the PLE spectrum. These observations can be explained with a model analogous to the one proposed by Lee et al. [2], but applied to QWRs. The theoretical temperature dependence of the intensity ratio I_1^w / I_1^n then yields an activation energy that is equal to the energy separation between the exciton peaks, independent of if electrons and holes tunnel separately or together as excitons. This theoretical result is not general but is the result of the electronic structure of the present DQWR structure. Therefore, no conclusion of the exact nature of the tunneling process can be made at present.

4. Conclusions

We have in the present contribution extended our studies on DQWR structures to include electrical carrier injection and a thicker tunnel barrier (7 nm). We still observe effective tunneling between the QWRs for both electrons and holes. No difference in electron and hole tunneling was observed indicating fast hole tunneling in the DQWR structures, probably due to the attained resonance between hole levels in the two QWRs. The EL spectra exhibit a clear effect of thermally activated tunneling up to room temperature. The thermal activation energy corresponds to the energy separation between the ground states of the two QWRs, indicating that the coupled states are in thermal equilibrium. However, with the present DQWR structure this does not allow us to distinguish whether electrons and holes tunnel together as excitons or separately.

References
[1] R. Ferreira et al., Rep. Prog. Phys. **60**, 345 (1997).
[2] D.H. Lee et al., J. Appl. Phys. **74**, 3475 (1993).
[3] O. M. Auslaender et al., Science **295**, 825 (2002);
    Y. Wang et al., Appl. Phys. Lett. **74**, 1412 (1999).
[4] H. Weman et al., Phys. Rev. B **58**, 1150 (1998).
[5] H. Weman et al., Appl. Phys. Lett. **73** 2959 (1998).
[6] F. Vouilloz et al., Phys. Rev. B **57**, 12378 (1998).

Effects of Nitrogen Incorporation in In(Ga)As/GaAs Quantum Dots

K H Park and W G Jeong
Sungkyunkwan University, Suwon, Korea

J W Jang
NanoEpi Technologies Corporation, Suwon, Korea

Y D Jang, N J Kim, and D Lee
Chungnam National University, Daejeon, Korea

Abstract It is observed that the incorporation of nitrogen into InAs quantum dots (QD) during growth has an effect of shifting the photoluminescence (PL) peak to longer wavelength and reducing the FWHM of the PL peak. The room temperature PL peak shifts from ~1160 nm to ~1300 nm when the InAs QDs are grown with dimethylhydrazine (DMHY)-to-AsH3 ratio as low as 0.027 as compared to the cases when the dots are grown without nitrogen. The FWHM of the room temperature PL peak is measured to be as low as 33 meV as compared with ~65 meV measured from the InAs/GaAs and InGaAs/GaAs QDs grown under the same condition except for the DMHy supply. The PL yield of the 1300 nm peak from InAsN/GaAs remains at 8% of the value at 10 K that is more than three times as large as those from InAs and InGaAs dots. The carrier decay times of the 1300 nm luminescence at 10 K and room temperature are measured to be similar at ~600 ps. All these characteristics suggest that the number of crystal defects is smaller and the uniformity in the size and composition is enhanced when InAs dots are grown with nitrogen supplied as compared with the cases without it.

1. Introduction

The growth of In(Ga)As quantum dots (QD) on GaAs substrates utilizing self-assembling characteristics has attracted much attention recently due to high crystal quality obtainable and the possiblity of utilizing them to enhance the performance of laser diodes, infra-red photodiodes, and so on by taking advantage of three dimensional quantization effects. Especially, the growth of high quality QDs emitting at 1300 nm on GaAs substrates is of intense interests in an effort to realize GaAs based telecom lasers.

Even though there have been reports that 1300 nm QD LDs have been realized using In(Ga)As/GaAs QDs[1-3], it is generally not easy to grow QDs with the luminescence wavelength beyond 1300 nm with a linewidth comparable to those of quantum wells due to non-uniformity in the shape and size of the QDs. In an efforts to achieve luminescence wavelength beyond 1300 nm, several approaches have been tried. In and Ga reactants have been supplied in an alternating mode to grow InGaAs[4,5], the InGaAs has been used as a cover layer over InAs QDs to reduce the strain[6], the InAs QDs were grown in an InGaAs quantum well[7], to name a few.

In this letter, we report a new simple method to grow QDs on a GaAs substrate with a strong photoluminescence (PL) in the 1300 nm region. A small amount of nitrogen in a form of dimethylhydrazine (DMHy) as low as 2.7% of AsH_3 supplied during the growth of InAs QDs has an effect of generating a PL peak at 1300 nm and reducing the FWHM of the peak as narrow as 33 meV at room temperature. The luminescence characteristics such as temperature and power dependence and carrier decay behavior as compared with InAs and InGaAs QDs grown without nitrogen are discussed.

2. Experiments

The QDs were grown in a low pressure horizontal Metal-Organic Chemical Vapor Deposition reactor operating at 76 torr. Trimethylindium (TMIn), trimethylgallium (TMGa), 20% AsH_3 diluted in H_2, and DMHy were used as reactants. Five stacks of QDs were grown with 20 nm thick GaAs spacer layers at around 540°C with As/III ratio of 60~75 on (100) GaAs substrates. When nitrogen is incorporated, DMHy was flown in with the other reactants for the growth of QDs.

3. Results and discussions

A typical Atomic Force Microscope (AFM) image of InGaAs/GaAs QDs (Q1584) grown at 545°C with TMIn/TMGa = 5 and As/III = 61 is shown in Fig. 1. Relatively uniform InGaAs QDs with a diameter of ~20 nm and a height of ~2 nm are seen. These InGaAs/GaAs QDs show a room temperature PL peak at 1155 nm with a FWHM of ~65 meV as in Fig. 2 (a). The InAs/GaAs QDs (Q1506) grown at 540°C show a room temperautre PL peak at 1160 nm as in Fig. 2 (b). But when DMHy is supplied with AsH_3 with a ratio of $DMHy/AsH_3$ = 0.027 as in the sample Q1524, a strong peak at ~1300 nm is seen together with a much less intense peak at ~1180 nm as in Fig. 2 (c). This effect of generating a strong luminescence at ~ 1300 nm as in Q1524 has been observed for a wide range of $DMHy/AsH_3$ ratios up to 0.3 which is the highest tried in this experiments.

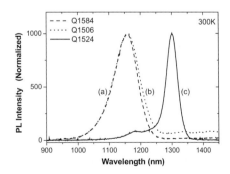

Figure 1. AFM image of InGaAs/GaAs QDs (Q1584)

Figure 2. Room temperature PL spectra of three types of QDs.

In Fig. 3, PL spectra from the InAsN/GaAs QDs (Q1524) taken at various temperatures are shown. At all temperatures, the strong and narrow PL peak at long wavelength side is dominant. At low temperatures the relatively broad peak at short wavelength side is significant but it diminishes as the temperature rises. The incorpoartion of nitrogen would reduce the bandgap energy of the QD material[8,9] and reduce the lattice mismatch with the GaAs substrate that would lead to larger QDs. Therefore, it is understood that the reduction in the bandgap energy and the increase in the QD size have shifted the luminescence peak to 1300 nm in Q1524. In addition, a narrow peak with a FWHM of 33 meV of the dominant peak indicates that the grown QDs are relatively uniform in size and composition. This contrasts to the FWHM of ~65 meV of the PL peaks of the QDs grown without nitrogen. This contrast suggests that the incorporation of nitrogen has an additional effect of controlling the growth mechanism in a way to make the sizes of the QDs more uniform. Furthermore, no change in the shape of the peaks and the relative intensity between two peaks is observed when the excitation power is changed. This suggests that there is virtually no carrier transfer between two groups of the dots and there are almost no extrinsic localization centers.

The integrated intensities of the PL peaks of three samples are shown in Fig. 4. The integrated intensity of Q1524 at room temperature remains at 8.2% of the value at 10 K. On the contrary, those of Q1506 and Q1584 drop to 1.7% and 2.8% of the values at low temperature, respectively. The difference could have been originated from the deeper potential for 1300 nm luminescence in Q1524. However, the relatively smaller drop in the intensity in InAsN/GaAs QDs also suggests that larger number of carriers thermalized to the GaAs barrier are recaptured into the QDs without being lost by recombining in the cyrstal defects around the QDs and the GaAs barrier. We believe that this strongly evidences that the grown QDs are of high crystal quality.

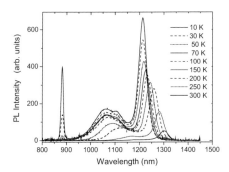
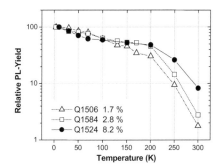

Figure 3. PL spetra of InAsN/GaAs QDs (Q1524)

Figure 4. Integrated intensities of QDs at various temperatures

Decay characteristics of carriers in Q1524 are shown in Fig. 5. The decay time of the dominant PL peak is measured to be around 600 ps both at 10 K and at room temperature. Meanwhile, the decay times for Q1506 are 610 and 330 ps at 10 K and room temperature, respectively, and those for Q1584 are 650 and 470 ps. The comparable decay time at 300 K as at 10 K in Q1524 once again may have been partly caused by a deep potential . However the long decay time of 600 ps at room temperature suggests that most of the carriers thermalized to the GaAs barrier are recapture to the dots without being recombined in the crystal defects. This once again indicates that the QDs in Q1524 are of high crystal quality..

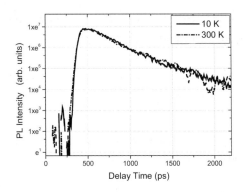

Figure 5. Carrier decay characteristics in InAsN/GaAs QD (Q1524)

4. Conclusion

A strong and narrow 1300 nm PL peak has been obtained by incorporating a small amount of nitrogen in InAs/GaAs QDs. The narrower PL peak width of 33 meV, the smaller deterioration of the PL intensity and no change in the carrier decay time at room temperature as compared to at low temperatures that the InAsN/GaAs QDs show suggest that the incorporation of nitrogen into InAs QDs is an effective way to grow QDs with highly uniform size and composition and of high crystal quality. We believe that the simple growth method explained in this report could be refined further and lead to a way to grow high quality 1300 nm QDs on GaAs that can be used as an active medium for high performance optoelectronic devices.

Acknowledgement

This work has been supported in part by Korean Science and Engineering Foundation (Grant R01-2000-00037) and in part by the Ministry of Commerce, Industry and Energy in 2001.

References

[1]   Huffaker D L et al. Appl. Phys. Lett., 73, 2564 (1998)
[2]   Mukai S et al., IEEE Photon. Technol. Lett. 11, 1205 (1999)
[3]   Park G et al. Appl. Phys. Lett. 75, 326 (1999)
[4]   Mirin R P et al. Appl. Phys. Lett. 67, 3795 (1995)
[5]   Huffaker D L and Deppe D G, Appl. Phys. Lett. 73, 520 (1998)
[6]   Nishi K et al. Appl. Phys. Lett. 74, 1111 (1999)
[7]   Ustinov V M et al. Appl. Phys. Lett. 74, 2815 (1999)
[8]   Weyers M et al. Jpn. J. Appl. Phys. Part 2 31, L853 (1992)
[9]   Kondow M et al. J. Crystal Growth 188, 255 (1998)

Photoconductivity of GaAs/AlGaAs quantum wires measured along the wires direction

V. Donchev[1], M. Saraydarov[1], K. Germanova[1], Xue-Lun Wang[2,3], Seong-Jin Kim[4], Mutsuo Ogura[2,3]

[1]Faculty of Physics, Sofia University, 5. blvd. J.Bourchier, 1164 Sofia, Bulgaria

[2]Photonics Research Institute, National Institute of Advanced Industrial Science andTechnology (AIST), Tsukuba Central 2, Tsukuba 305-8568, Japan

[3]CREST, Japan Science and Technology Corporation (JST), 4-1-8 Honcho, Kawaguchi 332-0012, Japan

[4]Electronics and Telecommunications Research Institute (ETRI), 161Kajong-Dong, Yusong-Gu, Taejon 305-350, South Korea

Abstract. A new photoconductivity (PC) study of undoped GaAs/AlGaAs quantum wires (QWRs) is carried out measuring the PC along the wires direction. The PC spectrum reveals several peak structures, related to the QWRs. This suggestion is confirmed by the observed dependence of the spectrum on the exciting light polarisation and by photoluminescence and photoluminescence excitation measurements on a similar sample. A long pre-illumination of the sample with infrared light (hν = 1.18 eV) is found to reduce considerably the substrate related background in the PC spectrum, which makes the QWR structures better prnounced.

1. Introduction

GaAs/AlGaAs quantum wires (QWRs) grown on V-grooved substrates are very promising for the fabrication modern optoelectronic devices such as high efficiency semiconductor lasers [1]. It has been shown that the photoconductivity (PC) characterisation of QWRs can give information equivalent to the absorption spectroscopy one. Up to now PC measurements have been carried out perpendicularly to the QWRs [2]. In this paper we present the first study, in which the PC is measured along the wire direction (longitudinal PC).

2. Experimental

The samples are fabricated by the MOCVD technique on 4 μm pitch V-grooved semi-insulating GaAs substrates. The layer structure consists of a 300 nm GaAs buffer layer, a 1000 nm $Al_{0.4}Ga_{0.6}As$ lower barrier layer, a 9 nm GaAs QWR grown by flow rate modulation epitaxy [3], a 100 nm $Al_{0.4}Ga_{0.6}As$ upper barrier layer and a 10 nm GaAs cap layer. The lateral width of the crescent is about 60 nm. Interdigitated finger contacts are made on the top surface by metal evaporation, the finger spacing being 2 μm.

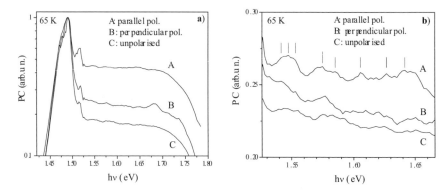

Figure 1. Normalised PC spectra of GaAs/AlGaAs QWRs (a). In (b) only the QWR spectral region is shown and the curves are shifted vertically to get closer. The vertical bars indicate the optical transitions (see Table 1).

The fingers cross perpendicularly 24 wires. PC measurements have been carried out at 65K, scanning from high toward low wavelengths in the range 900 - 710 nm (1.37-1.75 eV) and keeping the photon flux constant (~ $10^{15} cm^{-2} s^{-1}$) at each wavelength. Spectra are measured with unpolarised light as well as with light polarised perpendicular and parallel to the wire direction taking advantage of our sample geometry.

3. Results and discussion

Figure 1 shows PC spectra measured with different polarisation of the exciting light. At photon energies higher than the GaAs band-gap (hv ≥ 1.51 eV) the PC spectrum reveals several small structures some of which are connected with the QWRs. This suggestion is confirmed by the dependence of the PC spectrum on the exciting light polarisation. The near-band-edge energy region of the parallel polarisation PC spectrum (curve A) reveals three well resolved peak structures located at 1.542, 1.547 and 1.553 eV, respectively. In the unpolarised light spectrum (curve C) these structures are not so well pronounced. The perpendicular light polarisation (curve B) suppresses the first two peaks, but the third one (1.553eV) is well seen. In the high-energy region of the PC spectra one observes five peak structures, which are also polarisation dependent.

Photoluminescence (PL) and photoluminescence excitation (PLE) measurements performed at 4.5K on a similar sample (Fig.2) also confirm the QWR origin of the above-discussed structures. The assignment of the PL and PLE spectral structures in Fig.2 is made after a comparison with theoretical calculations of the transition energies [4]. In Table 1 we compare the energies of the spectral structures observed in the PC and the PLE spectra. A correction for the temperature dependence of the GaAs band-gap [5] is made by subtracting 8 meV from the PLE energies. As seen from Table 1, the PC and PLE spectra reveal similar structures, keeping the same energy spacing between them. The shift of about 10 meV between the two sets of values could be due to the fact that the PLE is measured in a special sample designed to achieve a zero Stokes shift.

Taking into account the light polarisation behaviour of the PC structures and the comparison between PC and PLE we can conclude about the optical transitions corresponding to the observed PC structures. They are given in the last column of Table 1. The first two peaks (1.542 and 1.547 eV) are due to excitonic transitions between the

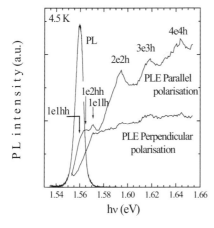
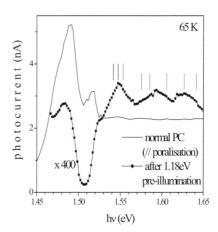

Figure 2. PL and PLE spectra of GaAs/AlGaAs QWRs.

Figure 3. Effect of the 1.18eV pre-illumination on the PC spectrum of GaAs/AlGaAs QWRs. The vertical bars form Fig.1 are also shown.

lowest electron state and the lowest and the second heavy hole states, respectively. The second transition is allowed with an appreciable amount of oscillator strength in V-grooved QWRs, because of the low symmetry of these QWRs [6]. The third peak (1.553 eV) has a light-hole-like behaviour and is attributed to excitonic transitions between the lowest electron and light hole states. In accordance with the PLE results the high energy peaks at 1.577, 1.606 and 1.627 eV can be ascribed to excitonic transitions with the same electron and hole quantum numbers, i.e. 2e-2h, 3e-3h, and 4e-4h. Some of the PC structures are not seen in the PLE spectrum (see Table 1). They could be related to transitions with other combinations of the electron and hole quantum numbers.

Table 1. Energies (in eV) of the QWR related spectral structures, observed in the PC and the PLE spectra and their identification.

PC	PLE	Optical transition
1.542	1.552	1e-1hh
1.547	1.557	1e-2hh
1.553	1.562	1e-1lh
1.575	1.587	2e-2h
1.585	-	
1.606	1.611	3e-3h
1.627	1.635	4e-4h
1.642	-	

The GaAs substrate and buffer layer are believed to have a large contribution to the smooth background under the QWR structures. At $h\nu \approx 1.49$ eV a large peak is observed in the PC spectrum, which is ascribed to electron transitions in the GaAs substrate between residual shallow acceptors, compensated by the deep donor EL2, and the conduction band. It has been found that a long pre-illumination of the sample with infrared light (hv = 1.18 eV) suppresses this peak and reduces considerably the background under the QWR structures (Fig.3). In the same time the dark current remains below the detection limit (< 10 fA) as in the case without pre-illumination.

This observation has been explained as follows. We believe the pre-illumination transforms the major part of the EL2 centres in the GaAs substrate into the metastable state EL2*, which is optically and electrically inactive [7]. The EL2* centres do not take part in the compensation mechanism of the material. As a result the major part of the shallow acceptors become empty after the pre-illumination and therefore act as trapping and recombination centres, which reduces the PC about 400 times. The effect of the EL2 quenching is much less important in the QWRs, because of the much lower EL2 concentration there. Therefore the pre-illumination with 1.18eV is not expected to affect the PC of the QWRs but only that of the GaAs substrate. Thus, the contribution of the substrate to the PC is strongly reduced. This is confirmed by the fact that the band-gap peak of bulk GaAs at 1.517eV is transformed into a shoulder whose inflection point has the same energy. As a result the QWRs structures become much more pronounced, keeping approximately the same energy positions as in the normal (without pre-illumination) PC spectrum.

4. Conclusion

We have presented a new PC study of GaAs/AlGaAs QWRs in which the PC is measured along the wires direction. Peak structures related to optical transitions in the QWRs have been observed in the above bandgap region of the PC spectra. They are identified taking advantage of their dependence on the exciting light polarisation and also comparing with PLE results. The observation of well-resolved QWR structures in the PC spectrum, which are similar to the structures in the PLE one, demonstrates the possibilities of the PC characterisation technique. An original approach is applied to reduce the background contribution of the GaAs substrate to the PC, consisting in EL2 quenching by infra-red pre-illumination [7], which results in an enhanced recombination via shallow acceptors in the substrate.

5. Acknowledgements

We acknowledge financial support from the Bulgarian National Research Council, the Sofia University Research Fund and the AvH Foundation (Germany).

References

[1] Kim T G, Wang X L, Suzuki Y, Komori K and Ogura M. 2000 IEEE J. Selected Topics in Quantum Electron. **6** 516-521.
[2] Hamoudi A, Ogura M and Wang X L 1997 J.Appl. Phys. **81** 6229-6233.
[3] Wang Xue–Lun, Ogura M and Matsuhata H 1995 Appl.Phys.Lett. **66** 1506-1508
[4] Wang Xue-Lun, Liu Xing-Quan, Ogura M, Guillet T, Voliotis V and Grousson R. (unpublished data).
[5] Madelung O (ed.) 1991 Semiconductors-Basic Data (Springer, Berlin) pp.75-152
[6] Dupertuis M A, Martinet E, Oberli D Y and Kapon E 2000 Europhys. Lett. **52** 420.
[7] Martin G M and Makram-Ebeid S 1986 in: Deep centers in semiconductors, edited by S T Pantelides. (Gordon and Breach, New York) Chap.6 "The mid-gap donor level EL2 in gallium arsenide", pp.399-488.

GaAs HBTs with reduced collector capacitance for high-speed ICs and microwave power applications

K Mochizuki
Central Research Laboratory, Hitachi, Ltd.

Abstract. This paper presents two new technologies for reducing collector capacitance and the offset voltage $V_{CE,sat}$. One is a very-high-resistance poly-GaAs planar isolation of emitter-up HBTs (heterojunction bipolar transistors), and the other is collector-up HBTs with a thin wide bandgap tunnel barrier that blocks hole injection. The former technology is used to improve microwave performance. The latter effectively achieved area- and temperature-independent zero-$V_{CE,sat}$ characteristics, which is suitable for high-efficiency high power amplifiers.

1. Introduction

GaAs heterojunction bipolar transistors (HBTs) have great potential for high-speed ICs and microwave power amplifiers. Enhancing their performance requires increasing the cutoff frequency f_T and the maximum oscillation frequency f_{max}, as well as reducing the collector capacitance C_{TC}. For power applications, reducing the offset voltage $V_{CE,sat}$ (the value of collector-emitter voltage V_{CE} when collector current I_C is 0) is also important. To reduce C_{TC}, previous studies applied a) deep-ion implantation [1], b) elimination of the subcollector under the base electrode [2] [3], or c) collector-up (C-up) configuration [4]. However, a) and b) require a high-energy ion-implantation apparatus or a complex regrowth process, and c) was not suited for power applications because of a large $V_{CE,sat}$ [5].

This paper presents two new technologies for GaAs HBTs: polycrystalline GaAs (poly-GaAs) isolation to reduce C_{TC} of emitter-up (E-up) HBTs and a tunnel-collector structure to reduce $V_{CE,sat}$ of C-up HBTs.

1.1. Poly-GaAs isolation technology

Poly-GaAs should have as great an impact on GaAs HBTs as poly-Si had on Si bipolar technology. However, applying poly-GaAs has been limited to isolating laser diodes [6], Schottky diodes [7], and field-effect transistors [8]. Applying high-resistance poly-GaAs to HBTs has been difficult because its resistivity decreases rapidly when the n-type doping level (N_D) exceeds 4×10^{18} cm^{-3} (Fig. 1), while $N_D > 5 \times 10^{18}$ cm^{-3} in the subcollector layer is required to achieve a low collector resistance.

We have developed an isolation technology by using very-high-resistance poly-GaAs to fabricate GaAs HBT-ICs. The high resistivity is maintained at a higher N_D by decreasing the poly-GaAs grain size L_g. To decrease L_g, we use low-temperature gas-source molecular beam epitaxy (LT-GSMBE). This technology provides planar isolation and reduces C_{TC}. The planar isolation prevents wire disconnection and can drastically increase the number of transistors in an IC.

Figure 1 shows the dependence of the resistivity of poly-GaAs on N_D. Compared to a previous experiment [8], in which L_g is large (0.3-0.5 μm) due to a relatively high growth temperature T_g (480°C-500°C), we use a fairly low T_g of 430°C to reduce L_g to as small as 0.15 μm and achieve a high resistivity of 2 MΩcm. Compared to previous technologies [1]-[3], ours can drastically reduce C_{TC} by simply burying carrier-depleted poly-GaAs under the base electrode.

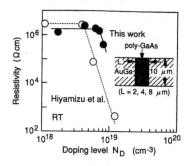

Figure 1. Dependence of the resistivity of poly-GaAs at room temperature on the *n*-type doping level. The inset shows the sample we used for mesurements.

1.2. Tunnel collector structure technology

Improving the efficiency of high-power amplifiers, particularly at a low power-supply voltage, requires reducing the knee voltage V_k (the minimum value of V_{CE} at a given operating I_C). The value of V_k is determined by $V_{CE,sat}$, an emitter, and a collector resisitance, and by voltage drops associated with any internal barriers. In conventional E-up GaAs HBTs, $V_{CE,sat}$ must be reduced to obtain a low V_k because $V_{CE,sat}$ rapidly increases with a reduction in the emitter area S_E and with an increase in substrate temperature T_s [9].

The value of $V_{CE,sat}$ is related to the difference between the base-emitter voltage V_{BE} associated with a given I_C, and the base-collector voltage V_{BC} associated with the same amount of forward current of the base-collector (BC) junction, I_{BCF}. As shown in Fig. 2a, I_{BCF} of p^+n GaAs BC junction in E-up HBTs consists of (i) an area-dependent diffusion current and a bulk-recombination current of holes (indicated by hb) and electrons (eb), and (ii) a perimeter-dependent surface recombination current of holes (hs) and electrons (es). When S_E is reduced, I_C proportionally decreases, while I_{BCF} gradually decreases, owing to the perimeter-dependent current component. When T_s is increased, the increase in surface recombination currents (hs and es) in I_{BCF} is larger than the increase in I_C. These phenomena result in S_E- and T_s-dependent $V_{CE,sat}$ of E-up HBTs.

To suppress surface recombination of the BC junction, the use of a C-up tunneling-collector HBT (TC-HBT) structure should be effective. As shown in Fig. 2b, the InGaP tunnel barrier inserted at the BC junction suppresses hole injection from base to collector (hb' and hs') and decreases surface recombination on the *n*-GaAs collector mesa. Since the InGaP tunnel barrier allows electrons to flow from collector to base, the surface recombination current caused by the electron injection (es') still exists. However, in a p^+n GaAs junction, the current indicated by es' in Fig. 2b will be much smaller than the current indicated by hs in Fig. 2a. In C-up TC-HBTs, the size and temperature dependences of I_{BCF} are thus expected to be similar to those of I_C. Moreover, C-up configuration is suited for decreasing the thickness of the tunnel barrier to minimize V_k in the InGaP/GaAs system because the conduction band discontinuity ΔE_c at the interface of InGaP on GaAs is lower than ΔE_c at the interface of GaAs on InGaP [10].

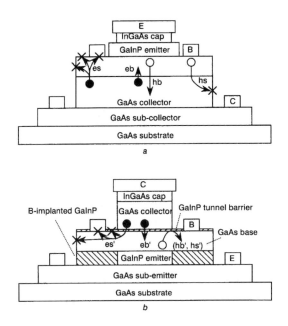

Figure 2. Schematic illustration of forward current of base-collector junction of (a) E-up HBT and (b) C-up TC-HBT. The solid and open circles repressent electrons and holes, respectively.

2. Device fabrication and performance

2.1. E-up HBTs with buried poly-GaAs

2.1.1. *Device fabrication.* The E-up HBT layer structure was grown by LT-GSMBE at T_g = 430°C on semi-insulating GaAs (100) substrates patterned with a 15-nm-thick photo-CVD (chemical vapor deposition) SiO_2 film. To fabricate conventional HBTs without poly-GaAs on the same wafer, the SiO_2 film was removed from one half of a 3-in wafer. The HBT structure consists of a 500-nm-thick GaAs subcollector layer (Si: 7 x 10^{18} cm^{-3}), a 200-nm-thick undoped GaAs collector layer, a 30-nm-thick GaAs base layer (C: 1 x 10^{20} cm^{-3}), a 100-nm-thick $In_{0.5}Ga_{0.5}P$ emitter layer (Si: 3 x 10^{17} cm^{-3}), 50-nm-thick $In_{0.5}Ga_{0.5}P$ (Si: 8 x 10^{18} cm^{-3}), 100-nm-thick GaAs (Si: 8 x 10^{18} cm^{-3}), and 50-nm-thick InGaAs (Si: 4 x 10^{19} cm^{-3}) emitter-cap layers.

Figure 3 illustrates the process for fabricating polycrystal-isolated InGaP/GaAs E-up HBTs. First, WSi was deposited on the InGaAs emitter-cap layer by RF sputtering and patterned into a non-alloyed emitter electrode by reactive ion etching (RIE) (Fig. 3a). This emitter electrode was used as a mask for wet-etching the InGaAs and GaAs emitter-cap layers and the InGaP emitter layer (Fig. 3b). Subsequently, a Au/Pt/Ti/Mo/Ti/Pt base electrode was formed on the base layer by using a lift-off technique (Fig. 3c). This electrode was also used as a mask for wet-etching the base mesa. Then, the Au/Ni/W/AuGe collector electrode was formed by using a lift-off technique

(Fig. 3d). Finally, transistors without poly-GaAs were isolated by wet chemical etching. Note that transistors with poly-GaAs required no isolation etching because of automatic isolation provided in the growth stage.

2.1.2. *Microwave characteristics.* We measured C_{TC} of the fabricated E-up HBTs with an emitter width W_E of 1.5 μm and an emitter length L_E of 5-20 μm at 3 GHz with zero bias. Poly-GaAs was buried under the base electrode in the 3.1 x 2.5-μm area of each transistor. Assuming a relative dielectric constant of 13 for GaAs [11] and complete carrier depletion in the buried poly-GaAs, we expected C_{TC} to reduce by 4.5 fF. Since the experimentally obtained reduction of 4.8 fF (Fig. 4) agrees fairly well with this expected reduction, we believe that the capacitance in the area where the poly-GaAs is buried is completely eliminated.

Figure 5 shows the frequency dependence of the common-emitter current gain $|h_{21}|$, Mason's unilateral gain U, and the maximum stable gain (*MSG*) of three parallel 0.7 x 8.5-μm E-up HBTs. The measurements were carried out at I_c = 13 mA and V_{CE} = 1.7 V. The f_T and f_{max} of the HBT with poly-GaAs reached 120 and 230GHz, while those of the HBT without poly-GaAs reached 110 and 200 GHz. The *MSG* of the HBT with poly-GaAs was 1.2 dB higher than the *MSG* of the HBT without poly-GaAs. We attribute this increase in *MSG* to the reduced C_{TC} due to complete carrier depletion in the poly-GaAs.

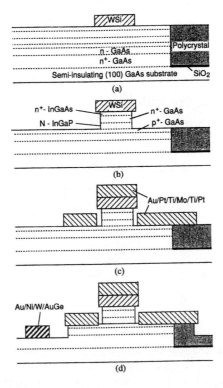

Figure 3. Steps for fabricating polycrystal-isolated InGaP/GaAs E-up HBTs.

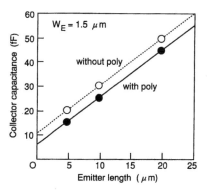

Figure 4. Dependence of emitter length on collector capacitance at zero bias for E-up HBTs with a 1.5-µm emitter width. Solid and open circles show the results for transistors with and without poly-GaAs, respectively.

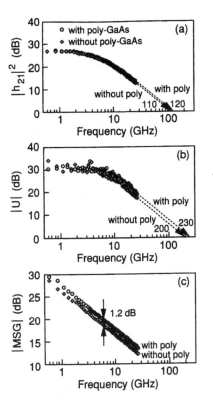

Figure 5. Frequency dependence of (a) the common-emitter current gain $|h_{21}|$, Mason's unilateral gain U, and (c) the maximum stable gain MSG of three parallel 0.7 x 8.5-µm E-up HBTs. The collector current is 13 mA and the collector-emitter bias 1.7 V.

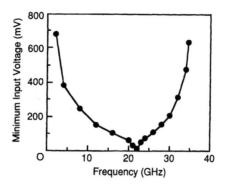

Figure 6. Input sensitivity of the 1/8 static frequency divider.

2.1.3. *Circuit performance.* As a circuit application of the polycrystal-isolated InGaP/GaAs E-up HBTs, we fabricated a 1/8 static frequency-divider circuit. In this circuit, we used 173 HBTs with an S_E of 0.8 x 5 µm². The size of the chip was 0.9 x 1.8 mm. The divider consists of an input buffer composed of emitter followers with 50-Ω on-chip resistors, three stages of divide-by-two master-slave T-type flip-flops (TFF) with an internal buffer in series, and an output buffer consisting of a differential amplifier. Each TFF was constructed with series-gated emitter-coupled logic. The supply voltage was –6.5 V.

The frequency divider operated up to 34.6 GHz driven by a single-ended input signal. The measured minimum input voltage versus the input frequency of the divider is plotted in Fig. 6. The free-running frequency is about 22 GHz.

2.2. C-up TC-HBTs

2.2.1 *Device fabrication.* The C-up TC-HBT layer structure was grown on semi-insulating GaAs (100) substrates by metalorganic chemical vapor deposition. The structure consists of 1-µm-thick GaAs (Si: 5 x 10^{18} cm⁻³) and 50-nm-thick $In_{0.5}Ga_{0.5}P$ (Si: 2 x 10^{18} cm⁻³) sub-emitter layers, a 150-nm-thick $In_{0.5}Ga_{0.5}P$ emitter layer (Si: 5 x 10^{17} cm⁻³), a 5-nm-thick undoped GaAs spacer layer, a 70-nm-thick GaAs base layer (C: 3 x 10^{19} cm⁻³), a 20-nm-thick undoped GaAs spacer layer, a 5-nm-thick $In_{0.5}Ga_{0.5}P$ tunnel barrier (Si: 5 x 10^{17} cm⁻³), an 800-nm-thick GaAs collector layer (Si: 3 x 10^{16} cm⁻³), as well as 50-nm-thick GaAs (Si: 5 x 10^{18} cm⁻³) and 100-nm-thick InGaAs (Si: 1 x 10^{19} cm⁻³) collector-cap layers.

Transistors with a collector area S_C ranging from 4 x 4 µm² to 50 x 50 µm² were fabricated by using a non-self-aligned process. First, WSi was deposited on the InGaAs collector-cap layer by RF sputtering and patterned into a non-alloyed collector electrode by RIE. This collector electrode was used as a mask, while the InGaAs and GaAs collector-cap layers were etched by wet chemical etching. The GaAs collector layer was then etched by RIE and wet chemical etching. To make the extrinsic emitter highly resisitive, boron-ion-implantation (dose: 2 x 10^{12} cm⁻²) was carried out at 50 keV (e. g. [5]). Subsequently, a Au/Pt/Ti/Mo/Ti/Pt base electrode was formed on the InGaP tunnel barrier by using a lift-off technique. The base mesa was then formed by photolithography and wet chemical etching of the InGaP tunnel barrier, undoped GaAs spacer, GaAs base, undoped GaAs spacer, InGaP emitter, and InGaP sub-emitter layers. Finally, a Au/Ni/W/AuGe emitter electrode was formed by using a lift-off technique.

2.2.2 *Results and discussion.* Figure 7 shows the common-emitter current-voltage characteristics measured at $T_S = 20°C$ and $50°C$ by using an HP 4145B curvetracer. The characteristics of typical InGaP/GaAs E-up HBTs with the same base and collector layers are also shown for comparison. When S_C (or S_E) = 50 x 50 µm², $V_{CE,sat}$ is 10 mV in C-up TC-HBTs and 80 mV in E-up HBTs at $T_S = 20°C$ (Fig. 7a). This difference in $V_{CE,sat}$ suggests that the bulk-recombination current (hb and eb in Fig. 2) in E-up HBTs is larger than that in C-up TC-HBTs since the surface recombination is considered to be small in these large transistors. We attribute the larger bulk-recombination current of E-up HBTs to the 80% larger area of the BC junction.

At $T_S = 20°C$ (Fig. 7a and b), $V_{CE,sat}$ of E-up HBTs increases rapidly as S_E decreases: $V_{CE,sat} = 200$ mV when $S_E = 4 \times 4$ µm. This rapid increase is due to the large dependence of I_{BCF} in E-up HBTs on the perimeter of the BC junction. Conversely, $V_{CE,sat}$ of C-up TC-HBTs depends little on S_C: $V_{CE,sat} = 10\text{-}14$ mV at $T_S = 20°C$. This low dependence is attributed to suppressed surface recombination on the *n*-GaAs collector mesa owing to reduced hole-injection current over the InGaP tunnel barrier.

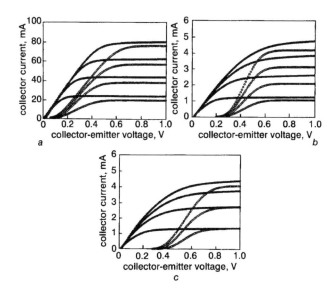

Figure 7. Dependence of collector current on collector-emitter voltage in InGaP/GaAs C-up TC-HBTs (solid circles) and E-up HBTs (open circles).

(a) Transistor size: 50 x 50 µm²; base current: 2-8 mA with 2 mA step
 (C-up TC-HBTs), 0.1-0.4 mA with 0.1 mA step (E-up HBTs)

(b) Transistor size: 4 x 4 µm²; base current: 0.5-2 mA with 0.5 mA step
 (C-up TC-HBTs), 5-20 µA with 5 µA step (E-up HBTs)

(c) Transistor size: 4 x 4 µm²; base current: 0.5-2 mA with 0.5 mA step
 (C-up TC-HBTs), 10-30 µA with 0.1 µA step (E-up HBTs)

When T_s is increased to 150°C, $V_{CE,sat}$ of the 4 x 4-µm E-up HBT increases to 280 mV, while $V_{CE,sat}$ of the 4 x 4-µm C-up TC-HBT stays constant (14 mV) (Fig. 7c). This result indicates that in C-up TC-HBTs, I_{BCF} is dominated by an electron-diffusion current indicated by eb' in Fig. 2b, so it has a similar temperature dependence to that of I_C.

3. Conclusions

We have developed two new technologies for GaAs HBTs: poly-GaAs isolation to reduce C_{TC} and a tunnel-collector structure to reduce $V_{CE,sat}$. We used the former technology to fabricate 0.8 x 8.5-µm InGaP/GaAs HBTs with poly-GaAs buried under the base electrode and attained an f_T of 120 GHz and an f_{max} of 230 GHz. Compared to the *MSG* of conventional HBTs without poly-GaAs, *MSG* of our HBTs was improved by 1.2 dB, which is due to complete carrier depletion in the buried poly-GaAs. We also fabricated a 1/8 static frequency-divider circuit using 173 of these transistors and demonstrated its operation at 34.6 GHz. The latter technology was used, together with C-up configuration, to fabricate InGaP/GaAs TC-HBTs. In these transistors, a 5-nm-thick InGaP layer was employed as a tunnel barrier at the BC junction to suppress hole injection into the collector. We found that the $V_{CE,sat}$ was almost zero (10-14 mV) and was independent of transistor size and temperature.

These results show that the present E-up HBTs with buried poly-GaAs and C-up TC-HBTs are attractive candidates for high-speed digital and microwave power applications.

Acknowledgments

The author thanks Prof. P. M. Asbeck and Dr. R. J. Welty of the University of California, San Diego, USA, and Prof. T. Nakamura of Hosei University, Japan for their helpful discussions. He is also grateful to Drs. T. Oka, T. Tanue, T. Mishima, H. Uchiyama, A. Terano, K. Ouchi, T. Taniguchi, and T. Shiota of Hitachi, Ltd., and Dr. K. Hirata of Hitachi ULSI Engineering Corp. for their help with processing and measurements.

References

[1] Ho M C, Johnson R A, Ho W J, Chang M F, and Asbeck P M 1995 IEEE Electron Device Lett. 16 512
[2] Frei M R, Hayes J R, Song J I, Caneau C, Bhat R, and Cox H 1992 Abst. 50th Device Research Conf. IVA-5
[3] Yang Y F, Hsu C C, Yang E S, and Ou H J 1996 IEEE Electron Device Lett. 17 531
[4] Yamahata S, Matsuoka Y, and Ishibashi T 1992 IEEE Electron Device Lett. 14 173
[5] Giradot A, Henkel A, Delage S L, Diforte-Poisson M A, Chartier E, Floriot D, Cassette S, and Rolland P A 1999 Electron. Lett. 35 670
[6] Lee T P and Cho A Y 1976 Appl. Phys. Lett. 29 164
[7] Ballamy B C and Cho A Y 1976 IEEE Trans. Electron Devices 23, 481
[8] Hiyamizu S, Nanbu K, Fujii T, Sakurai T, Hashimoto H, and Ryuzan O 1980 J. Electrochem. Soc. 127 1562
[9] Bovolon N, Schulthesis R, Mueller J R, Zwicknagl P, and Zanoni E 1999 IEEE Trans. Electron Devices 46 622
[10] Lee T W, Houston P A, Kumar R, Hill G, and Hopkinson M 1992 Semicond. Sci. Technol. 7 425
[11] Sze S M 1981 Physics of Semiconductor Devices 2nd Ed. (Wiley, New York) p 850

InP/GaAsSb/InP Heterojunction Bipolar Transistors

C R Bolognesi, †M W Dvorak, S P Watkins
Simon Fraser University, Compound Semiconductor Device Lab. (CSDL)
8888 University Drive, Burnaby BC, Canada, V5A 1S6
E-mail: colombo@ieee.org

†Current Address: Agilent Technologies Inc., Santa Rosa CA.

Abstract. InP/GaAsSb/InP double heterojunction bipolar transistors (DHBTs) have demonstrated record performances with current gain cutoff and maximum oscillation frequencies simultaneously exceeding 300 GHz while maintaining breakdown voltages $BV_{CEO} > 6$ V. This material system is particularly appealing for DHBTs because excellent device figures of merit are achievable with relatively simple structures involving abrupt junctions and uniform doping levels and compositions: this is a tremendous manufacturability advantage in comparison to GaInAs-based DHBT alternatives when high transistor count HBT circuits need to be implemented with aggressively scaled high-performance devices. A description of device operation, and a discussion of cutoff frequency and breakdown voltage limits are given.

1. Introduction

InP/GaAsSb/InP double heterojunction bipolar transistors (DHBTs) have demonstrated record performances despite the relative scarcity of available information on the $GaAs_{1-x}Sb_x$ ternary alloy near lattice-matching conditions to InP ($x_{Sb} = 0.49$): we previously reported current gain cutoff and maximum oscillation frequencies simultaneously exceeding 300 GHz while maintaining breakdown voltages $BV_{CEO} > 6$ V (Dvorak et al., 2001). We anticipate that $f_T = 400$ GHz should be possible with a breakdown voltage of ~ 4.0 V based on a 1000-1500 Å InP collector. The staggered band lineup and the absence of collector blocking effect in abrupt junction InP/GaAsSb/InP DHBTs enable a very high current drivability (as high as 7 mA/μm² even with relatively lightly doped emitter layers) that makes these devices attractive for the ultrahigh speed digital circuits needed for fiber network and instrumentation applications at data rates of 40 Gb/s and beyond. InP/GaAsSb/InP DHBTs feature a very low turn-on voltage $V_{BE,ON}$ ~ 0.4 V at $J_C = 1$ A/cm² (Bolognesi et al., 1999a) which also makes them attractive for long talk-time wireless applications operating with low battery voltage requirements. These DHBTs are particularly appealing because excellent device figures-of-merit are achievable with simple structures that only involve abrupt junctions and uniform doping levels and compositions (see Fig. 1): the applicability of such a technology outside a research laboratory setting has been validated at *Agilent* with the realisation of various circuit blocks based on conservative device structures: performances are already competitive with published state-of-the-art results for both GaInAs SHBTs and DHBTs (Moll, 2002). The benefits of simple epitaxial structures allowing superior device performance simply cannot be overstated: InP/GaAsSb DHBTs can be fabricated using selective etching techniques, and this is a critical manufacturability advantage for the realisation of high transistor count circuits. This is particularly true when faced with the implementation of ultrahigh-speed HBTs with sub- 300 Å base layers in a production setting. In addition, C-doped GaAsSb can be grown by MBE and MOCVD, but the MOCVD growth proves advantageous because of its widely appreciated compatibility with InP growth, and because it can be done with hydrogen as the carrier gas and in conjunction with organo-metallic sources without running into H-passivation problems like C-doped GaInAs does. In this fashion, C-doped GaAsSb layers with hole concentrations as high as 3×10^{20} cm^{-3} have been produced at SFU (Watkins et al., 2000). In general, SIMS analysis performed on our material has shown that MOCVD-grown C-doped GaAsSb epilayers feature a much lower [H]/[C] concentration ratio than found in GaInAs, a finding of apparent interest from a burn-in / reliability point of view if parallels can be drawn between past experience with GaAs HBTs and present InP/GaAsSb/InP DHBTs.

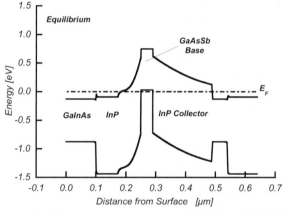

Figure 1. Equilibrium band diagram for an InP/GaAsSb/InP DHBT. The electron injection into the base is thermal, and there is no collector blocking effect, even under zero field conditions at the B/C junction. A "type-I" AlInAs emitter is also clearly a valid option.

The situation with InP/GaAsSb/InP DHBTs is in marked contrast to GaInAs -based DHBTs: the latter require sophisticated grading schemes at both the emitter and the collector (unless an InP emitter is used): by now it is well understood that even a single extraneous/missing layer in a superlattice grading can have dire consequences on both the static and dynamic characteristics of the device (such as negative differential conductance, depressed current carrying capabilities, low current gains, large offset and knee voltages, and low cutoff frequencies). With

InP/GaAsSb/InP DHBTs the E/B and B/C junction characteristics are determined by the doping levels, heterojunction band offsets, and the bandgaps, not by the effectiveness —*or lack thereof*— of various grading schemes intended to mask the natural blocking band offset between GaInAs and InP. From this point of view, the use of a staggered (type-II) base collector heterojunction makes nature work *for* the device designer: in the alternative, one must overcome type-I blocking heterojunctions with clever designs which are characterized by limited ranges of applicability (Tiwari and Frank, 1989).

One can expect a high degree of device reliability with InP/GaAsSb/InP DHBTs for a number of reasons: *a*) carbon is known to show a low diffusivity in comparison to other acceptors such as Be and Zn, and C –doped InP has been reported to be n-type (Cohen et al., 1996): even if C were to shift from the base to the emitter under bias, the E/B junction would remain self-aligned to the N-p heterojunction; *b*) the GaAsSb base is characterized by a low recombination velocity and an energy gap of ~0.72 eV (Hu et al., 1998) which provides little energy to drive the type of recombination enhanced impurity diffusion processes at play in GaAs; and finally, *c*) hydrogen passivation is not an issue with C-doped InP/GaAsSb/InP DHBTs. The expectations of device reliability are supported by the long life of transistors fabricated without any passivation layers and operated at high current densities (≥ 3 mA/μm^2) for prolonged test periods. We have recently presented some data from our collaborators at *Agilent Technologies* [IPRM'02 presentation in Stockholm, Sweden] which demonstrate stable device operation at junction temperatures as high as 300°C. At this point in time it does appear that InP/GaAsSb/InP DHBTs will offer sufficient reliability to fulfil very demanding test & measurement and telecom applications.

2. Staggered lineup base/collector and high-current operation of DHBTs
2.1. Emitter/Base Considerations

The equilibrium band alignment (Hu et al., 1998, Peter et al., 1999) for an InP/GaAsSb/InP DHBT is shown in Fig. 1: the advantages from a collector blocking point of view are immediately apparent in comparison to the GaInAs-based DHBT case —because the p+ base conduction band edge sits above the InP collector CB edge, electrons are injected into the collector even under zero electric field conditions at the B/C heterojunction. Electrons do experience thermal injection from the InP emitter into the GaAsSb base when the junction is forward biased. Back injection of holes into the emitter is simply a non-issue because of the very large valence band discontinuity of $\Delta E_V \sim 0.78$ eV. This band lineup enables a very simple emitter structure consisting of a GaInAs contact layer followed by an abrupt junction to the InP emitter per se. On the other hand, one can also opt for a type-I $Al_{0.48}In_{0.52}As$ emitter that would launch electrons into the GaAsSb base with some ~0.1 eV of kinetic energy, based on our measurement of the band alignment between InP and $GaAs_{0.51}Sb_{0.49}$ (Hu et al., 1998), and on transitivity arguments with AlInAs. Care probably should be used in launching 'hot' electrons in GaAsSb because the ternary alloy may well have a low intervalley separation since both its binary constituents feature low Γ-L valley separations (GaAs: 0.29 eV and GaSb: 0.08 eV). Thus, even if feasible, hot electron injection beyond ~0.1 eV with an Al-rich $Al_{0.48+x}In_{0.52-x}As$ emitter is probably not very useful inasmuch as speeding up transport across base is concerned.

Once across the E/B junction and into the quasi-neutral GaAsSb base, minority carrier electrons diffuse toward the collector with an apparent mobility of 900-1000 cm^2/Vs according to our most recent estimates —a direct measurement remains to be performed. This notwithstanding, the possibility of grading the base exists if it ever becomes necessary, although doing so would complicate the implementation of InP/GaAsSb transistors. With an AlInAs emitter, nearly $4kT$ become available for bandgap grading across the base layer. It appears the most favorable grading approach would likely involve the formation of a Ga-rich (Al,Ga)As$_{0.51}$Sb$_{0.49}$ base layer because the GaAsSb alloy features a larger bandgap bowing coefficient than GaInAs does, and this would likely diminish the benefits of group-V grading across the base.

2.2. Base/Collector Considerations

The staggered band lineup at GaAsSb/InP heterojunctions reduces the collector design problem to a selection of the InP collector doping level and thickness that are necessary to meet a specification for a given peak f_T current density J_C. B/C grading design is not necessary, as the alloy potential effect first described by Tiwari is non-existent in this system (Tiwari, 1988). A simple uniformly doped InP collector of thickness W_C can thus be used with an abrupt heterojunction to the GaAsSb base. The resulting InP/GaAsSb/InP DHBT structure is thus nearly symmetric and about as simple as any transistor structure is going to get. Another advantage of the structure is that the GaAsSb base layer is sandwiched between (relatively) high thermal conductivity InP layers.

It has long been known that type-I GaInAs-based DHBTs with good device performances (i.e. sufficient for 40 Gb/s) can be achieved (Matsuoka et al., 1996), but it is interesting to consider the issues involved in such an undertaking: *a*) the blocking potential of 0.25 eV between GaInAs and InP results in a retarding quasi-electric field of $125/t$ kV/cm if a $200t$ Å grading length is used, and this sizable quasi-electric field intensity must be compensated by a combination of doping profile and applied junction reverse bias V_{CB}; *b*) the extent to which grading layers affect the electron velocity profile through the collector region; and *c*) the extent to which potential base dopant (Be/Zn, but not C) out-diffusion will interfere with the intended B/C band profile. Obviously, all these design issues can be addressed by a combination of less than mature numerical modeling tools (in comparison to the mature state of silicon device/process simulators) and brute force trial-and-error.

The staggered band lineup in InP/GaAsSb/InP DHBTs has interesting consequences inasmuch as charge storage is concerned during transistor operation. We previously contrasted the behavior of GaInAs SHBTs and DHBTs to that of GaAsSb based devices and found that the GaInAs DHBT features a dramatic increase of charge storage at high current densities because the alloy potential effect at the B/C junction results in a sharp increase of the minority electron storage in the base layer as shown in Fig. 2 (Bolognesi et al., 2001). The InP/GaAsSb/InP DHBT features an altogether different behavior which was first discovered by Tom MacElwee of *Nortel* while measuring devices fabricated at the SFU-CSDL: C_{BE} first dips with increasing current density and reaches a minimum value corresponding to the peak f_T bias occurring for the condition of zero electric field at the B/C junction. Further increases in current density reverse the electric field at the B/C junction resulting in the formation of a small field-induced thermionic electron barrier at the B/C junction. Despite its smallness, the induced barrier E_B is seen to have a tremendous effect on the base charge storage because it can be expected to reduce the effective base exit velocity by a factor $\sim\exp(-E_B/kT)$, resulting in a sharp rise in C_{BE} and a drop in $f_T(J_C)$. An indication of the smallness E_B is confirmed by measurements of $f_T(J_C)$ characteristics as a function of temperature: the f_T roll-off with increasing current at higher temperatures becomes less abrupt as the chuck temperature increases because more base electrons can thermally overcome the small field-induced barrier at the B/C junction (Bolognesi et al., 2001).

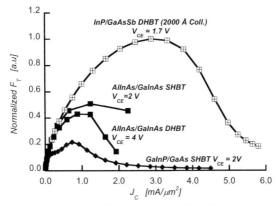

Figure 2. Normalised $f_T(J_C)$ dependencies for various technologies, with a normalisation to 300 GHz (Bolognesi et al., 2001). Note the range of V_{CE} biases and the clear manifestation of alloy potential blocking in the AlInAs/GaInAs DHBT despite the high collector voltage. In contrast, GaInAs and GaAs SHBTs feature a slow f_T rolloff due to the Kirk effect.

3. Potential improvements in high-speed performance

The realisation of transistors with cutoff frequencies simultaneously exceeding 300 GHz through optical contact lithography and all-wet etching has far exceeded our early expectations for InP/GaAsSb DHBTs (Bolognesi et al., 1999b). In fact, such performances will probably be difficult to better by relying on contact photolithography. However, modern stepper-based lithography tools offer much better scaling capabilities in comparison to contact aligners. We may thus anticipate the type of performance likely to be achieved in a reasonably optimised stepper-based process.

For this optimised process we assume contact resistances of 10^{-7} Ωcm^2, a 0.5 μm emitter metal with a 0.05 μm undercut, a base doping level of 10^{20} cm^{-3} with a hole mobility of 40 cm^2/Vs, and 0.4 μm wide self-aligned base Ohmic contacts. Assuming a conservative working current density of $J_C = 3$ mA/μm^2 and a two-region InP collector velocity profile adjusted to match our experimental data for a range of devices, we obtain the $f_{MAX}-f_T$ dependence shown in Fig. 3 for a variety of base and collector thickness combinations. The model suggests it should be possible to achieve $f_T = 400$ GHz and $f_{MAX} = 600$ GHz with a 150 Å base and a 1500 Å InP collector. These estimates —like most estimates— should not be taken too literally, but they certainly do indicate that there exists significant room for performance improvements with proper scaling in InP/GaAsSb/InP DHBTs. It must be emphasized that such high performances are predicted for structures relying on thermal injection and grown without the use of any kind of grading scheme.

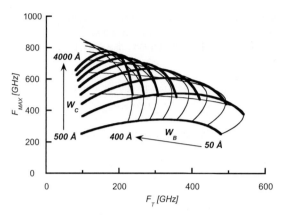

Figure 3. Anticipated performance for a 0.5 μm stepper-based DHBT process with a base doping level of 10^{20} cm^{-3} and a base hole mobility of 40 cm^2/Vs. The base Ohmic contacts are 0.4 μm wide and self-aligned to the emitter stripe. The distributed nature of the B/C network is taken into account for f_{MAX}.

4. Impact ionization and breakdown voltages in InP collectors

The realisation of aggressively scaled transistors with cutoff frequencies exceeding 300 GHz will involve the vertical scaling of the InP collector region in order to minimize the collector signal delay and enable operation at even higher current densities. Our devices have shown $f_T \times BV_{CEO}$ products in excess of 1800 GHz·V (Dvorak et al., 2001), and it becomes necessary to understand the implications of the InP collector scaling on impact ionisation related effects.

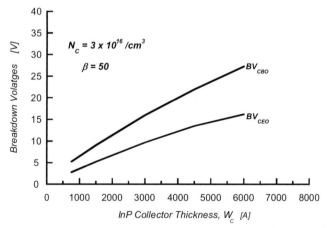

Figure 4. Calculated breakdown voltages for InP collector DHBTs with a collector doping of 3×10^{16} cm^{-3}.

We have calculated the three-terminal common-emitter breakdown voltage BV_{CEO} (with open-circuited base terminal) and the collector-base breakdown voltage BV_{CBO} (with open-circuited emitter) dependencies on the InP collector thickness for a collector doping level $N_C = 3 \times 10^{16}$ cm^{-3} and a transistor current gain $\beta = 50$. Our model takes into account dead space effects (inasmuch as electron impact ionisation is concerned) as well as ionisation events due to secondary impact ionisation-generated holes. Fig. 4 shows that a BV_{CEO} of 3.5 (5) V is predicted for a 1000 (1500) Å InP collector. This is an interesting result in that, taken with Fig. 3, it suggests InP DHBTs with cutoff frequencies of 300-400 GHz should display breakdown voltages of > 3.5 V. Such performance currently seems out of reach for SiGe/Si HBTs which have nevertheless reached the 200 GHz domain with breakdown voltages $BV_{CEO} \sim 1.7$ V (which can however be exceeded by a factor of up to 2 in actual circuit operation). This suggests that if raw device speed and high breakdown voltages are required, InP will remain the material of choice for ultrahigh-speed electronic applications. It must be acknowledged that factors other than pure device performance may dictate other choices in the end, at least as far as commercial applications are concerned.

Acknowledgments

This work was funded by NSERC, *Agilent Laboratories*, and *Nortel Networks*. The authors are indebted to Nick Moll (*Agilent*) and Tom MacElwee (*Nortel*) for their support.

References

Bolognesi, C. R., Dvorak, M. W., Pitts, O., Watkins, S. P. and MacElwee, T. W. (2001) In International Electron Device Meeting Tech. Digest, pp. 768-771.

Bolognesi, C. R., Matine, N., Dvorak, M. W., Xu, X. G., Hu, J. and Watkins, S. P. (1999a) IEEE Electron Dev. Lett., **20,** 155-157.

Bolognesi, C. R., Matine, N., Dvorak, M. W., Xu, X. G. and Watkins, S. P. (1999b) IEEE GaAs IC Symposium Tech. Digest, pp. 63-66.

Cohen, G. M., Benchimol, J. L., Roux, G. L., Legay, P. and Sapriel, J. (1996) J. Appl. Phys., **68,** 3793-3795.

Dvorak, M. W., Bolognesi, C. R., Pitts, O. J. and Watkins, S. P. (2001) IEEE Electron Dev. Lett., **22,** 361-363.

Hu, J., Xu, X. G., Stotz, J. A. H., Watkins, S. P., Curzon, A. E., Thewalt, M. L. W., Matine, N. and Bolognesi, C. R. (1998) Appl. Phys. Lett., **73,** 2799-2801.

Matsuoka, Y., Yamahata, S., Kurishima, K. and Ito, H. (1996) Jpn. J. Appl. Phys., **35 Part 1,** 5646-5654.

Moll, N. (2002) , Personal communication.

Peter, M., Herres, N., Fuchs, F., Winkler, K., Bachem, K.-H. and Wagner, J. (1999) Appl. Phys. Lett., **74,** 410-412.

Tiwari, S. (1988) IEEE Electron Dev. Lett., **9,** 142-144.

Tiwari, S. and Frank, D. J. (1989) IEEE Trans. Electron Devices, **36,** 2105-2121.

Watkins, S. P., Pitts, O. J., Dale, C., Xu, X. G., Dvorak, M. W., Matine, N. and Bolognesi, C. R. (2000) J. Cryst. Growth, **221,** 59-65.

Industrial Aspects of III/V Electronics: GaAs IC Manufacturing in Taiwan for Wireless Communications

P. C. Chao

www.winfoundry.com
WIN Semiconductors Corp., No. 69, Technology 7th Rd., Hwaya Technology Park, Taoyuan, Taiwan

Abstract. With the explosive growth in broadband applications and the increasing demand in the wireless communication area (including cellular/PCS handsets, WLAN and systems), GaAs process technology has been accepted as a superior one to produce high frequency, high power and low noise products for these applications. In Taiwan, the GaAs IC market is widely considered the fastest growing opportunity of semiconductor technology in the past several years. The success in supplying Si foundry services to the leading IDM and chip design companies provides a strong foundation for the rapid development of GaAs IC foundry Fabs in Taiwan. There are four pure-play GaAs foundries in Taiwan - two are 6" and two are 4" ready. These foundries offer devices and MMICs based on GaAs MESFET, HBT and pHEMT and mHEMT technologies with huge capacity. Despite the presence of high technical barrier to entry in GaAs processing, and the pressures and challenges associated with low cost production, the investors and engineering teams are optimistic about the high growth wireless market. There has been a strong confidence that Taiwan will follow the successful Si foundry model and be the world leader in providing a low cost, high quality GaAs foundry service to serve the worldwide wireless market needs.

1. GaAs Foundry Model

Taiwan has become a major center for the manufacture of laptop computers. As the emerging wireless market evolves, and is driven toward high-volume, low-cost manufacturing, Taiwan is also becoming a major center for ODM and/or OEM of handsets and WLAN for the worldwide wireless market. The increasing demand for wireless products stimulates a demand for key electronic components such as power amplifiers (PA) as well as other RF front-end components fabricated with GaAs technology. In Taiwan, the GaAs IC market is widely considered the fastest growing opportunity of semiconductor technology in the past several years. This change in focus is driven, in part by the rapid growth in wireless sales in China.

The success of Taiwan Semiconductor Manufacturing Corporation (TSMC) and United Microelectronics Corporation (UMC) in supplying Si foundry services to the leading IDM and chip design companies provides a strong foundation for the rapid development of 4" and 6" GaAs IC foundry fabs in Taiwan [1, 2]. The foundry business model is based on the concept that outsourcing is the only option for the long run because someone else can do something better than you can - due to economies of scale, capital funding and specific technological prowess. In the Si world, IDMs are now putting ~25% of their work into foundries [3].

Following the Si business model, most of the GaAs IC foundry players in Taiwan are seeking to leverage their low cost manufacturing capability, high production yield, and efficient management in an effort to convince IDMs to forego new plant construction or expansion in favor of outsourcing. The trend to outsourcing for GaAs is real, as has been demonstrated in the Si world. Whether driven by opportunity or desperation, it has been found that increasing GaAs companies - both fabless and IDMs - are looking to improve their overall competitiveness by making smart business alliances on both strategic and tactical levels with foundries [4].

2. GaAs Foundries in Taiwan

In 1991, Hexawave Photonic Systems became the first GaAs company on the island, opening its facility in Hsinchu's Science-based Industrial Park. Seven years later, Advanced Wireless Semiconductor Company (AWSC) opened its facility in the southern city of Tainan. As shown in Table I, today, there are four pure-play GaAs foundries in Taiwan - two are 6" and two are 4" capable, representing an investment in excess of $250M in this business opportunity. These foundries offer devices and MMICs based on GaAs MESFET, HBT, pHEMT and mHEMT (metamorphic HEMT) technologies with huge capacity. Some even offer optical devices, SAW, AWG and MEMS technologies. Most companies claim to provide "one-stop shopping" turnkey solution, including design, testing and packaging, to customers.

Table I Summary of GaAs Pure-Play Foundries in Taiwan

Founded Date	Company	Claimed Yearly Capacity, 2002	Employees	Capital	Supporting Technology
12/1998	AWSC	4" : 40,000	160	US$27M	2μm InGaP HBT
10/1999	WIN	6" : 20,000	270	US$85M	1, 2, 3μm InGaP HBTs 0.15, 0.35, 0.5μm pHEMTs 0.15μm mHEMT 0.5μm e-mode pHEMT 0.5μm pHEMT switch
07/2000	GCTC	6" : 30,000 4" : 48,000	120	US$50M	2μm InGaP HBT 0.35, 0.5μm pHEMTs SAW, MEMS (4")
08/2000	Suntek	4" : 60,000	240	US$100M	2μm InGaP HBT 0.5μm pHEMT 0.5μm MESFET Optical devices

AWSC is the first GaAs HBT foundry service company in Taiwan. It is also the smallest in scale among these four companies. AWSC offers a single product - 4" GaAs HBT. AWSC is partially invested by Conexant (now Skyworks) as the second source of HBT MMIC fabrication with a comparable technology.

WIN Semiconductors Corporation (WIN) - the world's first pure-play 6" GaAs foundry - was founded in October 1999 in recognition of the growing demand in manufacturing high-speed and high quality MMICs. WIN supplies InGaP HBT, pHEMT and mHEMT MMIC fabrication to the worldwide communication IC manufacturers using state-of-the-art automated 6" GaAs process technology (see Fig. 1). WIN's HBT and pHEMT technologies have been successfully qualified by major customers. WIN has ~270 employees and with the 4-shift operation, it provides the shortest process cycle time (4 weeks) among the foundries to serve its customers. In 2002, WIN formed a strategic alliance with Raytheon Company to manufacture Raytheon's 3μm InGaP HBT products at WIN.

 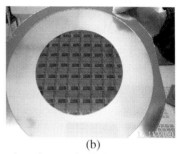
(a) (b)

Fig. 1 (a) Highly automated 6" foundry of WIN Semiconductors in Taiwan. (b) The first 6" GaAs HBT wafer was produced 2 months after the installation of equipment.

Following WIN, Global Communication Technology Corporation (GCTC) - funded by shareholders of US Global Communications Semiconductors (GCS) - is also a pure-play foundry service provider serving the needs of wireless and optical communications (OC) markets. It offers HBT, pHEMT and discrete devices as well as SAW filters and MEMS devices. It claims that its current fab has a capacity of 2,500 6" wafers per month for the HBT and pHEMT processes. The SAW capacity is 4,000 4" wafers per month. The broad product line and especially the mixing of 4" and 6" processes in the same fab are unique among the foundries.

Suntek Compound Semiconductor Corporation was established in August 2000 - nearly at the same time as GCTC - by Procomp Informatics. Suntek has formed significant partnerships with both Mitsubishi and Celeritek. Instead of developing the technology by itself, Suntek adopted the strategy of transferring 4μm HBT and 0.5μm MESFET/pHEMT processes from Mitsubishi and Celeritek. Suntek's current wafer process capacity is 5,000 4" wafers per month and, with expansion, the projected capacity will reach 20,000 wafers (6" equivalent) per month in 2004 [5].

3. Competence in Technology

To transform from a PC-based electronic industry into a center of mobile communications equipment manufacturing for the worldwide market, there are 9 GaAs epitaxy vendors, 4 GaAs fabs for pure-play foundry service, at least 4 or 5 RF testing service companies, and several RF module and packaging ODM/OEM vendors being established to support the 10 cellular cell phone ODM/OEM vendors.

Although it took very difficult years for TSMC to build up it technology since it is founded in 1987, the advanced technology that TSMC has developed on its own or jointly with international partners has convinced the leading IDM companies to focus on core design and marketing skills, while leaving the manufacturing function to foundry players. By the same token most of the GaAs IC foundry players are seeking to leverage their low cost manufacturing and efficient management for high production yield in an effort to convince IDMs to forego new plant construction or expansion in favor of outsourcing.

Foundries in Taiwan have developed or plan to establish more advanced production technology for heterojunction transistor ICs to persuade IDMs to abandon the costly development of high volume production for next generation technology and to rely instead on foundry players for cutting-edge capacity. The device and IC technologies range from GaAs MESFET, HFET, pHEMT, mHEMT, InGaP HBT to InP HBT, SAW filter, MEMS and optical components such as VCSEL, PIN and LD. For example, WIN has a very complete GaAs technology roadmap - in addition to the conventional InGaP HBT and pHEMT technologies, it has developed very advanced processes for high-performance, low-cost MMICs. As shown in Fig. 2., the advanced technologies include 1μm HBT, 0.15μm pHEMT, 0.15μm mHEMT, 0.5μm e-mode (enhancement-mode) pHEMT and 0.5μm pHEMT switch - all on 6" wafers.

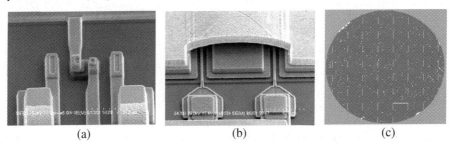

(a) (b) (c)

Fig. 2 (a) 1μm InGaP HBT produced at WIN for WLAN and high integration digital applications. (b) Very high-performance 0.15μm GaAs pHEMT was successfully demonstrated, for the first time, on 6" wafer for very low-cost microwave/MMW and OC applications. (c) 99% PA die yield shown in 6" GaAs HBT wafer mapping.

4. Technical and Business Challenges

The build up in GaAs IC fabrication facilities in Taiwan is going forward at a rapid pace despite the worldwide economic slowdown. Foundry service providers, however, have faced several technical and business challenges. Technical challenges include shortage of experienced manpower in the GaAs area, yield enhancement and ramping up for low-cost volume manufacturing. The business challenges include: 1) cash pressure due to long product validation period (12-18 months) and economy downturn, 2) a lack of technical development on the part of local, RF module packaging vendors, and 3) meeting the challenges from competition, such as SiGe, for low frequency RF components from a strong CMOS based IC industry which has a strong tendency toward SOC approaches.

Most of the technical teams had either 3" or 4" GaAs wafer process experience before joining the foundries. With limited time allowed for process development in new start-up fabs, they relied heavily on equipment vendors to assist in process development on 6" GaAs lines. The new process on 6" line, however, requires significant tweaking in most steps to make process reproducible. The on-line technicians, although equipped with solid background from CMOS production experience, did not have the knowledge of GaAs process; often resulting in mis-operations and therefore impacting on the device/IC yield and cycle time. Extensive training and online supervision of operators by highly skillful engineers are essential in handling the technical challenging in the GaAs manufacturing area.

Since there is in general a lack of experienced GaAs technologists, RF designers and microwave engineers locally, the competition is not only on the production cost, but also on the technology edge for a vertical integration virtually through strategic alliance (e.g., WIN/Raytheon, Suntek/Mitsubishi/Celeritek, AWSC/Skyworks). Despite the engineer resource limitation, the 6" GaAs IC lines still produced very respectable device results which are comparable to regular 3" and 4" lines.

Although high performance GaAs HBTs and HEMTs have been fabricated from successful pilot runs with 4" and 6" GaAs wafers, there are few technical barriers need to be overcome; (1) reliable suppliers of consistent and high-quality of 6" epitaxial wafers, (2) the backside process, including wafer mounting/thinning/demounting, is still highly manual and remains as the production challenge in throughput and yield, and (3) lack of experts and product chain infrastructures as driving forces to quickly establish IC qualification, reliability assessment, and failure analysis. The shortage of process, RF testing, product and reliability engineers also creates a great challenge for local industry and to the high education programs in colleges and universities.

5. Future of GaAs Foundries in Taiwan

The cost of operating a fab is high. With low utilization rate, some fabs may find it is simply more efficient and profitable to mothball their own facilities and outsource to foundries. During the recession, when the outside competition is at its highest and the internal investment is next to impossible, the pressure of outsourcing will be even stronger. Equipped with the latest process equipment and skilled engineers, GaAs foundries in Taiwan are quite experienced in producing high-performance and low-cost working circuits, with fast turn around, from the design stage. The rapidly developed fabrication capacity for 4" and 6" GaAs wafers, however, relies heavily on external design capability and product applications.

Despite the presence of high technical barriers to entry in large wafer size GaAs processing, and the pressures and challenges associated with low cost production, the investors and engineering teams are optimistic about the high growth wireless market, particularly in China. It is well understood that GaAs foundries require a long time to establish and sufficient capital support over a long period is necessary to guarantee the success of the venture. Before the fabless design houses are established, most foundries recognize that an alliance with IDM houses is essential for success - at least in the short run. Overall, there has been a strong confidence that Taiwan will follow the successful Si foundry model and be the world leader in providing a low cost, high quality GaAs foundry service to serve the worldwide wireless market needs.

References

[1] Burggraaf P 2002 Solid State Tech. S10
[2] Bavin A 2001 Solid State Tech. S18-S20
[3] Buehler S 2002 Semiconductor Magazine 10-18
[4] Williams D 2002 Compound Semiconductor 35-39
[5] Lin J Compound Semiconductor Outlook 2002, Tai

Silicon germanium technologies for high-speed digital and analog applications

H. Knapp, J. Böck, M. Wurzer, K. Aufinger, T. F. Meister

Infineon Technologies AG, Otto-Hahn-Ring 6, 81730 Munich, Germany

Abstract. This paper gives an overview of an advanced silicon germanium bipolar process and presents circuits that demonstrate the performance of this technology. Transistors manufactured in this technology achieve a cut-off frequency f_T of 106 GHz and a maximum oscillation frequency f_{max} of 145 GHz. Three different circuits are presented to evaluate the suitability of this process for analog and digital high-speed applications. A single-stage broadband amplifier has a 3-dB bandwidth of over 26 GHz. A dual modulus prescaler with divide ratios of 256 and 257 operates up to 41 GHz at a power consumption of 118 mW. Finally, a regenerative frequency divider with a maximum operating frequency of 92.5 GHz is presented.

1. Introduction

Silicon germanium technologies have found wide-spread interest for applications ranging from digital communications systems, such as 10 Gb/s and 40 Gb/s fiber-optic links, to microwave analog circuits. They offer excellent RF performance with cut-off frequencies beyond 100 GHz [1-5].

This paper gives an overview of a pre-production SiGe bipolar technology developed at Infineon [1] and presents circuits manufactured in this technology which demonstrate its potential for analog and digital high-speed circuits.

2. Device fabrication

The silicon germanium transistors are based on the double-polysilicon self-aligned transistor configuration. The most important feature is the selective epitaxial growth of the silicon germanium base. The germanium content is linearly graded across the base up to a maximum germanium content of 20%. This generates a drift field which reduces the base transit time. A boron spike in the neutral base provides a low base sheet resistance. To prevent undesirable boron diffusion during further transistor fabrication the boron spike is surrounded by carbon.

The minimum lithographic feature size is 0.35 μm which results in an effective emitter width of only 0.18 μm and therefore a small intrinsic base resistance (figure 1). The extrinsic base resistance is reduced by salicided base electrodes. The metallisation consists of four layers of aluminium. The transistors offer well-balanced performance with low parasitic capacitances, low base resistance, a cut-off frequency f_T of 106 GHz (figure 2) and a maximum oscillation frequency f_{max} of 145 GHz.

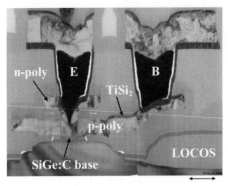

Figure 1. TEM cross section of the emitter base complex of an npn transistor.

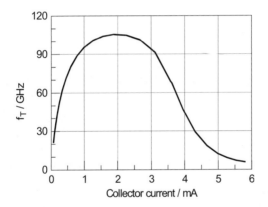

Figure 2. Cut-off frequency f_T versus collector current of a transistor with an effective emitter area of 0.18×2.8 μm^2.

3. Demonstrator circuits

To evaluate the performance of the SiGe transistors we have designed various analog and digital circuits. This section describes three circuits which illustrate the suitability of SiGe bipolar technologies for a wide range of applications. The first circuit is a two-transistor broadband amplifier. A dual-modulus prescaler serves as example for a larger digital circuit where operating speed and low power consumption are important design goals. The last circuit is a regenerative frequency divider which illustrates the potential of SiGe bipolar technologies for operating frequencies approaching 100 GHz. Apart from these three circuits presented here a wide range of other circuits have been realised. These include low-noise amplifiers, microwave mixers, and static frequency dividers [1, 6, 7].

3.1. Broadband amplifier

A broadband amplifier was designed to evaluate the performance of the SiGe technology in wideband analog applications. The amplifier consists of a single Darlington stage with emitter degeneration and shunt feedback (fig. 3). An on-chip spiral inductor provides a further increase of the bandwidth. A high quality factor Q of the inductor is not required because it is connected in series to the feedback resistor. To achieve a high self-resonant frequency only the top metal layer is used for the inductor.

The measured gain versus frequency can be seen in figure 4. The low-frequency gain is 11.3 dB and the 3 dB bandwidth is higher than 26 GHz. This on-wafer measurement shows a slight gain peaking at frequencies around 15 GHz. This gain peaking was intended to counteract the gain roll-off caused by bond wires when the chip is mounted. Large ground pads allow the use of multiple bond wires to keep the parasitic inductance small. Figure 3 shows a chip photograph of the amplifier. The chip size of 450 x 450 μm^2 was dictated by the shared reticle used for the fabrication of the chip. The amplifier requires only a small fraction of this area.

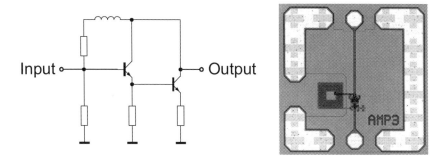

Figure 3. Schematic diagram and chip photograph of the broadband amplifier.

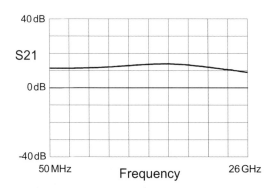

Figure 4. Measured S21 of the broadband amplifier.

3.2. Dual-modulus prescaler

Dual-modulus prescalers are frequency dividers with two selectable divide ratios. They are widely used in frequency synthesisers to extend the frequency range of programmable frequency dividers. Figure 5 shows the block diagram of the prescaler. It consists of a synchronous divide-by-four/divide-by-five input stage which is followed by a six-stage asynchronous divider [8]. The overall divide ratio is 256 or 257, depending on the level at the modulus control input.

The synchronous counter at the prescaler input determines the maximum operating frequency and is responsible for most of the power consumption of the circuit. The flip-flops in the asynchronous divider operate at lower frequencies and require less tail current. All flip-flops and gates are realised in emitter-coupled logic (ECL) and use differential signals with a voltage swing of 2 x 200 mV_{pp}.

The input sensitivity of the prescaler, measured on wafer, is shown in figure 6. The circuit operates over a wide frequency range up to a maximum input frequency of 41 GHz. The response at low frequencies is limited by the slew rate of the sinusoidal input signal. The power consumption of the prescaler is only 118 mW at a supply voltage of 3.3 V.

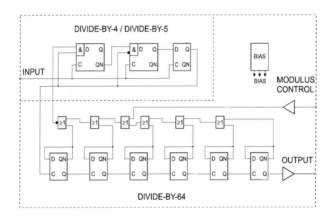

Figure 5. Block diagram of the dual-modulus prescaler.

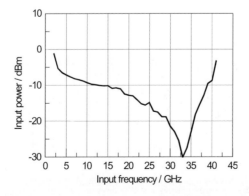

Figure 6. Measured input sensitivity of the dual-modulus prescaler.

3.3. Regenerative frequency divider

Static frequency dividers use flip-flops for frequency division. Their maximum operating frequency is limited by the gate delay τ_D to a value of approximately $1/(2\,\tau_D)$. Higher operating frequencies can be achieved by dynamic frequency dividers. However, unlike static frequency dividers, they do not operate at arbitrarily low input frequencies. In many high-speed applications this poses no problem because operation is only required in a limited frequency band.

We have designed a dynamic frequency divider which uses regenerative frequency division [9]. The circuit is based on an active double-balanced mixer consisting of a Gilbert cell. The input signal is applied to one of the two mixer inputs. The mixer output signal is fed back to the second mixer input via three emitter follower stages (fig. 7). The regenerative frequency divider is followed by an output buffer which contains two limiter stages. Input and output of the circuit are terminated by on-chip 50 Ω resistors.

Measurements of the regenerative frequency divider were performed on wafer. The circuit operates with a supply voltage of 6.5 V and consumes 85 mA. A single-ended input signal was used and the second divider input was left unconnected. The maximum operating frequency of the circuit is 92.5 GHz. Figure 8 shows the output signal at this input frequency.

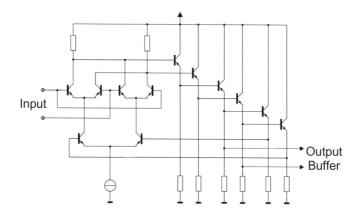

Figure 7. Schematic diagram of the regenerative frequency divider.

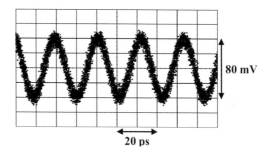

Figure 8. Single-ended output signal of the regenerative frequency divider (input frequency 92.5 GHz).

4. Conclusion

We have presented analog and digital high-speed circuits manufactured in a pre-production SiGe bipolar technology. A dual-modulus prescaler combines high operating speed with low power consumption. It operates up to 41 GHz with a power consumption of only 118 mW. A regenerative frequency divider reaches a maximum operating frequency of 92.5 GHz which is among the highest values reported for any semiconductor technology.

These results show that SiGe bipolar technology is a promising candidate for analog and digital applications in the millimeter-wave range.

References

[1] Böck J., Schäfer H., Knapp H., Zöschg D., Aufinger K., Wurzer M., Boguth S., Stengl R., Schreiter R., Meister T.F. 2001 IEDM Technical Digest 344-347

[2] Jagannathan B., Khater M., Pagette F., Rieh J.-S., Angell D., Chen H., Florkey J., Golan F., Greenberg D.R., Groves R., Jeng S.J., Johnson J., Mengistu E., Schonenberg K.T., Schnabel C.M., Smith P., Stricker, A., Ahlgren D., Freeman G., Stein K., Subbanna S. 2001 IEEE Electron Device Letters. 23(5) 258-260

[3] Oda K., Ohue E., Suzumura I., Hayami R., Kodama A., Shimamoto H., Washio K. 2001 IEDM Technical Digest 332-335

[4] Racanelli M., Schuegraf K., Kalburge A., Kar-Roy A., Shen B., Hu C., Chapek D., Howard D., Quon D., Wang F., U'ren, G., Lao L., Tu H., Zheng J., Zhang J., Bell K., Yin K., Joshi P., Akhtar S., Vo S., Lee T., Shi W., Kempf P. 2001 IEDM Technical Digest 336-339

[5] Heinemann B., Knoll D., Barth R., Bolze D., Blum K., Drews J., Ehwald K.-E., Fischer G.G., Köpke K., Krüger, D., Kurps R., Rücker H., Schley P., Winkler W., Wulf H.-E. 2001 IEDM Technical Digest 348-351

[6] Hackl S., Böck J., Wurzer M., Scholtz, A.L. 2002 International Microwave Symposium 1241-1244

[7] Wurzer M., Böck J., Knapp H., Aufinger K., Meister T.F. 2002 Bipolar/BiCMOS Circuits and Technology Meeting

[8] Knapp H., Wurzer M., Böck J., Meister T.F., Ritzberger G., Aufinger K. 2002 IEEE RFIC Symposium 239-242

[9] Knapp H., Meister T.F., Wurzer M., Zöschg D., Aufinger K., Treitinger L. 2000 International Solid-State Circuits Conference 208-209

AlGaN/GaN HEMTs grown by Molecular Beam Epitaxy on sapphire, 6H-SiC, and HVPE-GaN templates

N G Weimann, M J Manfra, J W P Hsu, K Baldwin, L N Pfeiffer, and K W West

Lucent Technologies Bell Laboratories, Murray Hill, NJ, USA

S N G Chu, D V Lang

Agere Systems, Murray Hill, NJ, USA

R J Molnar

MIT Lincoln Labs, Lexington, MA, USA

Abstract. We report on a systematic study of the growth of AlGaN/GaN HEMT structures by plasma-assisted Molecular Beam Epitaxy (MBE) on different substrates, including sapphire, 6H-SiC, and GaN templates prepared by Hydride Vapour Phase Epitaxy (HVPE). We obtained record low-temperature mobilities of 75,000 cm^2/Vs from HEMT structures with low sheet charge ($\sim 10^{12}$ cm^{-2}) grown on HVPE GaN templates, and demonstrated functioning AlGaN/GaN HEMT devices on HVPE templates grown by MBE for the first time. HEMT devices of equal geometries were fabricated on all three substrates, allowing for a direct comparison between the different substrate technologies for GaN MBE growth. Our best devices on semi-insulating 6H-SiC substrates delivered a saturated power density of 6.2 W/mm at 2 GHz, and 4.9 W/mm at 7 GHz in continuous wave class A mode. A record drain current density of 1798 mA/mm was recorded from a submicron HEMT on SiC.

1. Introduction

AlGaN/GaN HEMT devices have delivered remarkable RF power performance at microwave frequency, enabled by both the wide bandgap of GaN and the electron transport properties in two-dimensional electron gases in AlGaN/GaN heterostructures [1, 2]. Recently, Molecular Beam Epitaxy of GaN and related alloys is becoming a rival to the more established Metal Organic Vapour Phase Epitaxy [3]. Excellent control of impurity, interface abruptness, and in-situ monitoring of the growth are driving the increase in quality of MBE epilayers. We have developed plasma-source MBE nucleation schemes on three types of substrates, consisting of sapphire, semi-insulating (SI-) SiC, and HVPE SI-GaN templates on sapphire. While sapphire and SI-SiC are established substrates for the growth of AlGaN/GaN HEMT epilayers, HVPE GaN templates may provide a path to low-cost and large-diameter substrates for electronic devices. We fabricated HEMT devices on our samples to optimize the MBE growth parameters. The optimization goals were the minimization of the 2DEG sheet resistance, as measured by the Hall effect and by TLM, and the RF output power density of finished *unpassivated* HEMT devices. Both short and long-gate devices were manufactured to assess the power performance at S- and X-band.

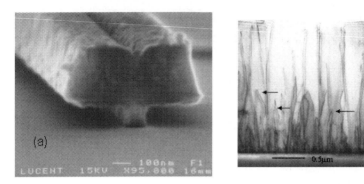

Figure 1. T-gate metal structure (a) fabricated with the bilayer resist electron beam lithography process. Cross-sectional TEM image (b) of the two-step buffer on sapphire

2. MBE growth of AlGaN/GaN HEMT device layers

We developed nucleation schemes on the foreign substrates sapphire and SiC. In order to accommodate the ~13% lattice mismatch between GaN and sapphire, and to ensure Ga-face polarity of the MBE layer, a thin aluminium-rich AlN nucleation layer is first deposited. The nucleation layer is followed by a 2 µm thick undoped GaN buffer layer. The barrier consists of 250Å of $Al_{0.30}Ga_{0.70}N$, followed by a 50Å thick GaN cap. Part of the barrier is doped n-type with silicon to decrease the resistance of our ohmic contacts. The growth rate is 0.5 µm/hr at a growth temperature of 745°C. The buffer layer is insulating with a residual carrier concentration of less than 10^{15} cm^{-2}, as measured by room-temperature CV. The quality of the 2DEG depends largely on the conditions during growth of the buffer. We devised a two-step growth method for the buffer that leads to defect reduction, while maintaining a smooth surface at the GaN-AlGaN interface [4]. The first half of the buffer is grown under slightly nitrogen stable condition. In this case the surface of the growth front is relatively rough, the observed RHEED pattern is indicative of a 3D growth mode. The 3D growth mode appears to increase dislocation interactions, and thus, reduces the number of dislocations that propagate to the surface. In the second phase of growth of the insulating GaN buffer, the Ga flux is increased to produce metal stable growth. The RMS roughness of the surface is less than 1 nm, as measured by AFM over a 2x2 µm^2 area. During the second phase of the growth, no additional defect interaction occurs, as can be seen from the TEM image shown in figure 1b. On sapphire, we were able to reproducibly yield 2" wafers with sheet resistances around 400 Ω/□. On 6H-SiC substrates, we obtained repeatedly sheet resistances of 300-350 Ω/□. For the MBE growth of AlGaN/GaN HEMT layers on HVPE GaN templates, no additional nucleation layer is required. The 20 µm thick HVPE GaN films already have the required polarity. These low-defect (~10^8 cm^{-2}) quasi-substrates enable AlGaN-GaN 2DEGs with very high low-field mobility due to the reduced defect scattering. We measured low-field mobilities up to 75,000 cm^2/Vs with a density of ~10^{12} cm^{-2} at 4K [5]. In order to realise HEMT devices on HVPE templates, the residual conductivity attributed to silicon and oxygen incorporation during the HVPE growth required compensation doping. Zinc was used as a compensating acceptor, a concentration of 10^{17} cm^{-3} allowed for the compensation of the residual impurities.

3. HEMT device technology

We developed a robust process for HEMT devices on AlGaN/GaN heterostructures. Initially, the MBE material was optimised with 2 μm gate-length HEMTs using contact aligner lithography, ensuring quick turn-around. Recently, we added electron beam lithography to the process to realise 200 nm long gates between source and drain spaced 2 μm apart. A novel bilayer T-gate lithography process was adapted to AlGaN-GaN HEMT structures, reducing writing time and improving the yield by simplifying the development scheme as compared to our established PMMA-based trilayer resist process [6]. A typical T-gate structure is shown in figure 1a. The ohmic contacts consist of Ti/Al/Ni/Au, alloyed at 850°C for 30 sec in nitrogen atmosphere. Ni/Au is used as a gate Schottky contact. The devices are mesa-isolated in a chlorine ICP etch step. The devices are unpassivated prior to measurement. The periphery of the HEMTs varied between 50 and 200 μm.

4. HEMT device measurements

HEMT devices in a coplanar configuration were tested on-wafer. From small-signal S-parameters, we extracted an f_t of 43 GHz and an f_{max} of 109 GHz for a 0.2 x 50 μm^2 HEMT on SiC. A record DC drain current density of 1798 mA/mm was recorded for the same device (figure 2a). Our power devices with 2 μm long gates, targeted for S-band operation, showed an output power density of up to 6.3 W/mm at 2 GHz on SiC. Over four consecutive growth runs, our average devices delivered 4.5 W/mm at 2 GHz on SiC. At 7 GHz, our 200 nm gate-length devices delivered a saturated power output density of 4.9 W/mm, P$_{-3dB}$ and P$_{-1dB}$ were 4.3 and 2.8 W/mm, respectively (figure 2b). At P$_{-1dB}$, a gain of 11 dB was available. All results given are in cw class A operation, all power data are for devices with 200 μm periphery. On sapphire, the RF output power density is limited to 2.2 W/mm by the thermal impedance of this substrate.

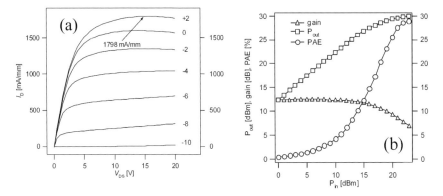

Figure 2. IV curve (a) of HEMT with 0.2 x 50 μm^2 gate. cw load-pull data (b) measured at 7 GHz in class A from HEMT with 0.2 x 200 μm^2 gate

The thermal management on HVPE GaN templates is expected to lie between sapphire substrates and SiC, with a thermal conductivity of 2 W/cmK for GaN compared to 5 W/cmK for SiC and 0.3 W/cmK for sapphire. However, we only observed an RF output power density of 1 W/mm at 2 GHz on HVPE templates. The Zn compensators in the HVPE GaN may act as traps, which can be charged and then effectively back-gate the channel under high RF voltage swing, unable to follow the microwave excitation due to their relatively long relaxation time. In the past, this effect has been observed in Cr-compensated GaAs substrates and, more recently, in V-compensated SiC substrates. Other possible trap states may be associated with the AlGaN barrier and the device surface. The trapping behaviour on HVPE GaN substrates is currently under further investigation. We measure the drain current dispersion in our devices by plotting the DC drain current versus the RF input power during the load-pull power sweep. HEMTs with 2 μm gate length on SiC have shown as little as 7% drop of the drain current in saturation as compared to the small-signal regime. HEMTs with 0.2 μm long gates from the same wafer show a drop in drain current of 23%, indicating a field-dependent mechanism of trapping in these devices.

5. Conclusion

We have developed plasma-assisted MBE growth processes of AlGaN/GaN heterostructures on different substrates, and demonstrated repeatedly HEMT devices fabricated from these layers. On semi-insulating 6H-SiC substrates, we recorded the highest power density of 6.2 W/mm at 2 GHz, and 4.9 W/mm at 7 GHz, along with a record drain current density of 1798 mA/mm. All results are from *unpassivated* devices.

6. References

[1] J. R. Shealy, V. Kaper, V. Tilak, T. Prunty, J. A. Smart, B. Green, and L. F. Eastman, "An AlGaN/GaN high-electron-mobility transistor with an AlN sub-buffer layer", *Journal of Physics: Condensed Matter*, vol. 14, pp. 3499-509, 2002.

[2] H. Xing, S. Keller, Y. F. Wu, L. McCarthy, I. P. Smorchkova, D. Buttari, R. Coffie, D. S. Green, G. Parish, S. Heikman, L. Shen, N. Zhang, J. J. Xu, B. P. Keller, S. P. DenBaars, and U. K. Mishra, "Gallium nitride based transistors", *Journal of Physics: Condensed Matter*, vol. 13, pp. 7139-57, 2001.

[3] M. Micovic, J. S. Moon, T. Hussain, P. Hashimoto, W. S. Wong, and L. McCray, "GaN HFETs on SiC substrates grown by nitrogen plasma MBE", *Physica Status Solidi A*, vol. 188, pp. 31-5, 2001.

[4] M. J. Manfra, N. G. Weimann, J. W. P. Hsu, L. N. Pfeiffer, K. W. West, and S. N. G. Chu, "Dislocation and morphology control during molecular-beam epitaxy of AlGaN/GaN heterostructures directly on sapphire substrates", *Applied Physics Letters*, vol. 81, pp. 1456-1458, 2002.

[5] M. J. Manfra, N. G. Weimann, J. W. P. Hsu, L. N. Pfeiffer, K. W. West, S. Syed, H. L. Stormer, W. Pan, D. V. Lang, S. N. G. Chu, G. Kowach, A. M. Sergent, J. Caissie, K. M. Molvar, L. J. Mahoney, and R. J. Molnar, "High mobility AlGaN/GaN heterostructures grown by plasma-assisted molecular beam epitaxy on semi-insulating GaN templates prepared by hydride vapor phase epitaxy", *Journal of Applied Physics*, vol. 92, pp. 338-345, 2002.

[6] L. E. Ocola, D. M. Tennant, and P. D. Ye, "Bilayer process for T-gates and Gamma-gates using 100 kV e-beam Lithography," presented at International Conference on Micro- and Nano-Engineering, Lugano, Switzerland, 2002.

High Power AlGaN/GaN HEMT's

Lester F. Eastman, Vinayak Tilak, Richard Thompson, Bruce Green, Valery Kaper, Tom Prunty, Richard Shealy, Joseph Smart, and Hyungtak Kim

Cornell University, Electrical and Computer Engineering and CNF
Ithaca, NY 14853-5401, U.S.A.
e-mail: lfe@iiiv.tn.cornell.edu

Abstract – The efficiency impacts of loading, channel width, and the phase difference between channels are presented for passivated AlGaN/GaN HEMT's that have undoped, polarization-induced 2DEG. For 1.5 mm periphery HEMT's it is shown that 10 channels yield the best high power performance in class B operation at 10 GHz.

1. Introduction

Undoped $Al_xGa_{1-x}N/GaN/AlN/SiC$ HEMT devices have high electron sheet density (> 1 x $10^{13}/cm^2$) for x > .30. Bulk GaN has a breakdown electric field strength of 3 MV/cm, allowing 10 GHz devices with .3 µm gate length to have instantaneous drain-source voltages up to 80 V. Together these properties allow high-power microwave amplification [1]. The HEMT performance dependence on design, and fabrication technology will be covered here.

2. Approach

The polarization difference [2] between pseudomorphic $Al_xGa_{1-x}N$ and GaN is ((-) .0593 x (-) .0492 x^2)C/m^2. 200 Å of $Al_{.33}Ga_{.67}N$ is grown on the Ga-face of GaN to form a positive, fixed polarization charge of 1.56 x $10^{13}/cm^2$ equivalent electric charges at the heterojunction. A 2DEG is induced in the GaN channel, and a Ni Schottky barrier with 1.66 V potential, depletes .36 x $10^{13}/cm^2$ electrons, leaving 1.2 x $10^{13}/cm^2$ under the gate for V_{gs} = 0V. A chlorine-based ECR etch is used to form the mesas, and ohmic contacts are formed using Ti (200 Å)/Al (1000 Å)/Ti (450 Å)/Au (550 Å) annealed for 60 seconds at 800°C in nitrogen. Tri-level electron-beam resist is used to form a mushroom-cross-section gate with ~ 120 Ω/mm static resistance. Small periphery layouts include a single gate, center-fed, and a pair of gates separated by 50 µm. Large periphery manifold layouts have a set of gates with various widths, with 50 µm pitch between adjacent gates. In all cases a passivation layer of PECVD Si_3N_4 is deposited at 300°C to suppress current slump, sometimes called dispersion [3], caused by transient surface state charges. Using short-pulse gate bias, over a range of drain bias, the gate-lag I(V) characteristics show minimal current slump after this passivation [4]. Gate leakage currents are reduced by using a 2000 Å AlN layer under the GaN channel/buffer layer [5].

HEMT's with two parallel .3 μm gates, of various widths, are laid out with gate-feed pads at both ends, in order to measure the microwave voltage drop along the gate. HEMT's with the same gate widths, but with a gate-feed at only one end, are tested to correlate efficiency changes with the voltage changes. In addition, manifold layouts are laid out with 6, 8, 10, and 12 channels, all with a total periphery of 1.5 mm, to determine the effect of the incremental phase difference between channels. These large periphery devices are tested using Maury^{TM} automatic tuners on the input and output.

3. Experimental results

Small periphery devices with center-fed, .30 x 100 μm gates have yielded 11.2 W/mm in CW class B operation at 10 GHz [5]. Based on heat-flow simulations, the maximum channel temperature is determined to be at ~ 450 °K. The measured channel resistance rises [5] in proportion to (T channel/300°K)$^{1.8}$, thus doubling its value. This in turn raises the minimum voltage during RF voltage swing, lowering the efficiency. With the 4 Ω-mm R_{SD} being made up of 3.2 Ω-mm channel resistance and .8 Ω-mm for the sum of the two contact transfer resistances, a 50 Ω-mm load resistance efficiency factor is lowered, for example, from .86 times its ideal value to .78 times its ideal value.

The attenuation factor for the RF electric field along the gate depends on the real part of $\sqrt{j\omega C_g R_{gate}}$, where ω is the operating radian frequency, C_g is the sum of the normalized C_{gs} and C_{ds} capacitances, and R_{gate} is the normalized gate resistance. The static normalized resistance of the gate, with .3 μm footprint, is 120 Ω/mm. At 10 GHz it is ≥ 175 Ω/mm, raised by the skin depth effect. This gives rise to the voltage dropping as $e^{-2.5\,l}$, where l is the distance in mm, along the gate from its drive point. At the end of the gate there is the reflection. For W_g gate width in mm, the ratio of the RF drive voltage to the RF voltage at the end of the gate is then cosh (2.5 W_g). The modulated channel current drops in proportion, impacting the power-added efficiency. The experimental drop in power-added efficiency with the gate width leads to an efficiency factor η_W, that depends on gate width as:

$$\eta_W = 1 - .5 \left(1 - \frac{1}{\cosh^2(2.5W_g)}\right) \quad (1)$$

Experimental variations in power-added efficiency were determined for a set of manifold layout designs with 6 x 250 µm, 8 x 187.5 µm, 10 x 150 µm, and 12 x 125 µm peripheries. The power-added efficiencies are shown in the chart below.

Class B Power-added efficiency for 1.5 µm HEMT, with 50 µm pitch with different layouts, for .3 µm gates, at 10 GHz

Layout	CW η_{PA} (10 V_{ds})	CW η_{PA} (20 V_{ds})	Pulsed η_{PA} (35 V_{ds})
6 x 250 µm	.57	.54	.37
8 x 187.5 µm	.62	.62	.45
10 x 150 µm*	.57	.58	.49
12 x 125 µm	.53	.55	.44
	optimum load	28 Ω load	40 Ω load

*10 W CW at 40% η_{PA} at 30 V

At 10 V bias, CW self-heating is low, and at higher bias, these values change, and the layout with 10 channels yields the best result. The narrower channels have less RF voltage drop along the channel, and also less temperature rise, together more than making up for the cumulative impact of the phase difference among the increased number of channels.

After the impacts of loading and gate width on the efficiency have been taken into account, the impact of the incremental phase difference between adjacent channels remains. There is both a gate drive, and a drain collection, component involved. Using 60 Ω as the characteristic impedance of the lines connecting the gate and drain drive points, and using the total ($C_{gs} + C_{gd}$) input capacitance, and the C_{ds} capacitance, the magnitude of this phase difference is, approximately

$$\Delta\theta = 2\pi f \left[\sqrt{L(C_{gs} + C_{gd})} + \sqrt{LC_{ds}}\right], \quad (2)$$

where L is the inductance of the 50 µm length of the 60 Ω transmission line. At 10 GHz this relation gives .192 radian, or 11° phase difference between adjacent channels. A 15° phase difference more closely fits the experimental data, possibly due to mutual inductance between these gate and drain feed lines, or due to a source inductance effect.

Using .12 x 75 µm gate, center-fed, performance at 35 GHz was measured by Dr. R. Quay at the Fraunhofer IAF in Freiburg, Germany. He obtained 2.32 W/mm at 22% power-added efficiency in the 12-15 V_{ds} bias range. Preliminary data on a four-section cascode, traveling-wave amplifier has also yielded 31 GHz band width, with 6 db maximum gain, for .3 µm gate length.

4. Summary

The limits of power and efficiency for .3 µm gate AlGaN/GaN HEMT's on SiC has been determined for various layouts, including 1.5 mm periphery, where 10 W CW has been obtained at 40% power-added efficiency at 10 GHz. The impacts of loading, channel width, and phase difference between channels have been covered.

5. References

[1] Eastman, L.F., Tilak, V., Smart, J., Green, B.M., Chumbes, E.M., Dimitrov, R., Kim, H., Ambacher, O.S., Weimann, N.G., Prunty, T., Murphy, M., Schaff, W.J., and Shealy, J.R., *IEEE Trans. on Elect. Dev.*, **48** (3), pp. 479-484, March 2001

[2] Ambacher, O., Majewski, J., Miskys, C., Link, A., Hermann, M., Eickhoff, M., Stutzmann, M., Bernardini, F., Fiorentini, V., Tilak, V., Schaff, W.J., and Eastman, L.F., *Condens. Matter,* **14,** pp. 3399-3434, 2002

[3] Green, B.M., Chu, K.K., Chumbes, E.M., Smart, J.A., Shealy, J.R., and Eastman, L.F., *IEEE Elect. Dev. Lett,.* **21** (6), pp. 268-270, 2000

[4] Tilak, V., Kaper, V., Thompson, R., Prunty, T., Kim, H., Smart, J., Shealy, J.R., and Eastman, L.F., *ISCS 2002*, paper TU-P-13

[5] Shealy, J.R., Kaper, V., Tilak, V., Prunty, T., Smart, J.A., Green, B.M., and Eastman, L.F., *J. Phys: Condens. Matt.* **14** (13), pp. 3499-3509, 2002

High breakdown electric field for Npn-type AlGaN/InGaN/GaN heterojunction bipolar transistors

Toshiki Makimoto, Kazuhide Kumakura, and Naoki Kobayashi
NTT Basic Research Laboratories, NTT Corporation
3-1 Morinosato Wakamiya, Atsugi-shi, Kanagawa 243-0198, Japan

Abstract. Npn-type AlGaN/InGaN/GaN heterojunction bipolar transistors have been fabricated for high power application. Their common-emitter current-voltage characteristics showed a high breakdown voltage of 120 V, corresponding to the breakdown electric field of 2.3 MV/cm in the collector. The cross-section transmission electron microscopy image showed that the V-shape defects in InGaN were filled with wider bandgap AlGaN layers during the emitter growth. This is considered to prevent the leakage current from the base to the collector, resulting the high breakdown electric field.

1. Introduction

Group-III nitride heterojunction bipolar transistors (HBTs) are promising electronic devices that require high power and/or operate under high temperatures. These nitride HBTs have some advantages such as high breakdown voltage, good linearity performance, high current density, low phase noise, single power supply, and good threshold voltage uniformity. For AlGaN/GaN HBTs using the p-GaN base, however, there are two major problems. One problem is that the hole concentration of the p-type GaN base is lower than 1×10^{18} cm^{-3} at room temperature, resulting in the high sheet resistance. The other problem is severe etching damage for p-GaN. Recently, we have reported that p-type InGaN layers show high hole concentrations at room temperature [1,2] and that their sheet resistance was reduced down to one tenth compared with p-GaN [3]. We have also reported that In atoms doped in p-type GaN reduce the etching damage to improve p-type ohmic characteristics on the etched surface [4]. These two characteristics of p-type InGaN layers are preferable for the base of nitride HBTs, so we have fabricated an InGaN/GaN double heterojunction bipolar transistor (DHBT) with a p-type InGaN base to show a high common-emitter current gain of 20 [5-7]. While relatively high current gains were obtained for these nitride HBTs, their breakdown voltages were, however, still lower than that expected for the wide bandgap of the GaN collector. In this paper, we report the greatly improved breakdown electric field in the collector for Npn-type AlGaN/InGaN/GaN HBTs and discuss the growth mechanism of the AlGaN/InGaN structure.

2. Experimental Procedure

The structures were grown by low-pressure metalorganic vapor phase epitaxy (MOVPE) and consist of the N-type AlGaN emitter, the p-type InGaN base, and the n-type GaN collector. Trimethylgallium (TMG) and NH$_3$ were used for the growth at 1000 °C, while triethylgallium (TEG), trimethylindium (TMIn), trimethylaluminium (TMA) and NH_3 were used for the growth at 780 °C. The p-type and n-type doping gases were bis-cyclopentadienyl-magnesium (Cp_2Mg) and silane (SiH_4), respectively. First, an AlN buffer was deposited at 1100 °C on a SiC substrate using TMA. Next, the Si-doped GaN sub-collector and collector layers were grown at 1000 °C. Then, Si-doped graded InGaN, the Mg-doped InGaN base, and the Si-doped AlGaN emitter were grown at 780 °C. The base mesa was defined by electron cyclotron resonance (ECR) plasma using Cl_2. Pd/Au and Al/Au metals were used for p- and n-type ohmic contacts, respectively. The In mole fraction in the InGaN base and the Al mole fraction in the AlGaN emitter were determined using X-ray diffraction measurements in separate experiments.

3. Results and discussions

3.1. Common-emitter current-voltage characteristics

Figure 1 shows the common-emitter current-voltage (I-V) characteristics of the Npn-type AlGaN/InGaN/GaN HBT at room temperature. The AlGaN/InGaN/GaN HBT had an emitter size of 50 μm x 30 μm. The maximum common-emitter current gain was 4. Current gains reported for nitride HBTs are not high mainly due to high dislocation densities above 10^9 cm^{-3}. These current gains will be improved by reducing dislocation densities. The common-emitter I-V characteristics were observed up to a collector-emitter voltage of 70 V and a collector current of 14.5 mA, corresponding to a power of 1 W. Figure 2 shows the breakdown voltage in the common-emitter I-V characteristics of the AlGaN/InGaN/GaN HBT. The breakdown voltage is more than 120 V. A very high breakdown voltage of 330 V was reported for an AlGaN/GaN HBT using a relatively thick collector of 8 μm [8]. The corresponding breakdown electric field in the collector was 0.41 MV/cm for this AlGaN/GaN HBT. In contrast, the breakdown electric field obtained in this work is as high as 2.3 MV/cm, which is comparable to the expected value for GaN (3.3 MV/cm) [9]. There are some reports on the breakdown voltages of HBTs using other material systems such as Si/Ge [10], InP/InGaAs [11], and InGaP/GaAs systems [12]. From the reported breakdown voltages and their structures, the breakdown electric fields in the collector are calculated to be 0.25, 0.33, and 0.42 MV/cm for Si/Ge, InP/InGaAs, and InGaP/GaAs DHBTs, respectively. Therefore, the breakdown electric field obtained in this work is much higher than those reported values for the other material systems, indicating that nitride HBTs are promising electronic devices for high power application.

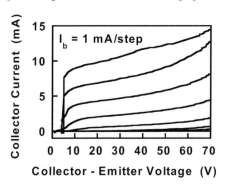

Figure 1. Common-emitter I-V characteristics of an Npn-type AlGaN/InGaN/GaN HBT.

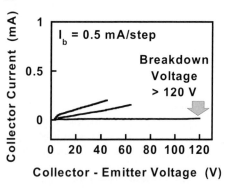

Figure 2. Breakdown voltage in the common-emitter I-V characteristics.

3.2. Growth Mechanism of the AlGaN/InGaN structure

It has been reported that the defects such as V-pits or V-defects are often observed in the InGaN layer [13-16]. These V-shape defects are considered to become the origin of leakage current paths in HBTs using p-InGaN bases, so we have observed cross-section transmission electron microscopy (TEM) images of AlGaN/InGaN/GaN HBT structures. Figure 3 shows a typical cross-section TEM image of a V-shape defect in the AlGaN/InGaN/GaN HBT structure and its schematic depiction. This cross-section image shows that the V-shape defect was formed on the threading dislocation and this defect was filled with AlGaN during the emitter growth. The defect density was about 10^9 cm^{-2}. Compared with In and Ga atoms, Al atoms migrate slowly on the growing surface during MOVPE growth, so the V-shape defect was filled with AlGaN during the emitter growth. The Al mole fraction of this AlGaN layer is considered to become high due to slower migration of Al atoms. Next, the HBT structure was etched to expose the base surface and the Pd/Au metal was deposited on this surface, in a similar way to the HBT fabrication process. Then, we observed a cross-section TEM image. Figure 4 shows a typical cross-section TEM image of the base/collector junction and its schematic depiction. The tip of the V-shape defect was sharp before etching as shown in Fig. 3, similar to the

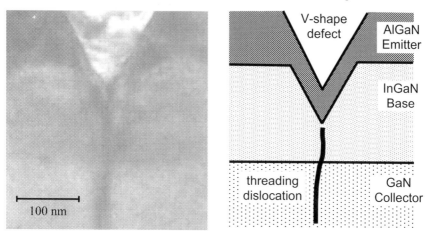

Figure 3. Typical cross-section TEM image of a V-shape defect in the AlGaN/InGaN/GaN HBT structure and its schematic depiction.

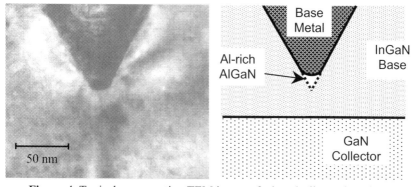

Figure 4. Typical cross-section TEM image of a base/collector junction and its schematic depiction.

previous reports [13-16]. In contrast, it became rounded after etching, as shown in Fig. 4. Such a rounded tip was observed for all V-shape defects after etching, meaning that the etching-rate was low around the rounded tip due to Al-rich AlGaN and that the AlGaN layer are expected to still remain just below the tip of the V-shape defect. This wider bandgap AlGaN layer below the tip, shown in the schematic depiction of Fig. 4 using dotted lines, is considered to prevent the leakage current from the base to the collector. This is the reason for the high breakdown electric field obtained in this work along with the wide bandgap of the n-GaN collector and the less etching damage of the p-InGaN base.

4. Conclusions

Npn-type AlGaN/InGaN/GaN HBTs have been fabricated for high power application. Their common-emitter I-V characteristics showed a high breakdown voltage of 120 V, corresponding to the breakdown electric field of 2.3 MV/cm in the collector. This high electric field is much higher than those reported for HBTs composed of other material systems such as Si/Ge, InP/InGaAs, and InGaP/GaAs systems. This indicates that nitride HBTs are promising electronic devices for high power application. The cross-section TEM image showed that the V-shape defects in InGaN were filled with wider bandgap AlGaN layers during the emitter growth. This is considered to prevent the leakage current from the base to the collector. As a result, the high breakdown electric field was obtained in this work, which is also ascribed to the wide bandgap of the n-GaN collector and the less etching damage of the p-InGaN base.

References

[1] K. Kumakura, T. Makimoto, and N. Kobayashi, Jpn. J. Appl. Phys. **39**, L337 (2000).
[2] K. Kumakura, T. Makimoto, and N. Kobayashi, J. Cryst. Growth **221**, 267 (2000).
[3] T. Makimoto, K. Kumakura, and N. Kobayashi, physical status solidi (a) **188**, No.1, 183 (2001).
[4] T. Makimoto, K. Kumakura, and N. Kobayashi, J. Cryst. Growth **221**, 350 (2000).
[5] T. Makimoto, K. Kumakura, and N. Kobayashi, Appl. Phys. Lett. **79**, 380 (2001).
[6] T. Makimoto, K. Kumakura, and N. Kobayashi, *28th International Symposium on Compound Semiconductors* (ISCS 2001), ThM-4, Tokyo (2001).
[7] T. Makimoto, K. Kumakura, and N. Kobayashi, Inst. Phys. Conf. Ser. **170**: Chapter 1, 33 (2002).
[8] H. Xing, P. Chavarkar, R. Sharma, J. G. Champlain, S. Keller, U. K. Mishra, and S. P. DenBaars, *28th International Symposium on Compound Semiconductors* (ISCS 2001), ThM-7, Tokyo (2001).
[9] J. C. Zolper, Solid-State Electron., **42**, 2153 (1998).
[10] T. Tatsumi, H. Hirayama, and N. Aizaki, Appl. Phys. Lett. **52**, 895 (1988).
[11] K. Kurishima, H. Nakajima, T. Kobayashi, Y. Matsuoka, and T. Ishibashi, IEEE Trans. Electron Devices, **41**, 1319 (1994).
[12] J. -I. Song, C. Caneau, W. -P. Hong, and K. B. Chough, Electron. Lett. **29**, 1881 (1993).
[13] Z. Lillental-Weber, Y. Chen, S. Ruvimov, and J. Washburn, Phys. Rev. Lett. **79**, 2835 (1997).
[14] P. Vennegues, B. Beaumont, M. Vaille, and P. Gibart, Appl. Phys. Lett. **70**, 2434 (1997).
[15] Y. Chen, T. Takeuchi, H. Amano, I. Akasaki, N. Yamada, Y. Kaneko, and S. Y. Wang, Appl.Phys.Lett.**72**,710 (1998).
[16] X. H. Wu, C. R. Elsaaa, A. Abare, M. Mack, S. Keller, P. M. Petroff, S. P. DenBaars, and J. S. Speck, Appl. Phys. Lett. **72**, 692 (1998).

Multiwafer Epitaxy of GaN/AlGaN Heterostructures for Power Applications

K Köhler, S Müller, N Rollbühler, R Kiefer, R Quay and G Weimann

Fraunhofer-Institut für Angewandte Festkörperphysik,
Tullastraße 72, 79108 Freiburg, Germany

Abstract. Heterostructures of GaN/AlGaN for power applications are grown by metal organic vapor phase epitaxy in a multiwafer reactor on sapphire and SiC substrates. Electrical properties of the two-dimensional electron gas are discussed. The sheet carrier concentration was varied from 4×10^{12} cm^{-2} to 1.5×10^{13} cm^{-2} by modulation doping. From electrical and material properties we show the excellent uniformity of the wafers. The influence of the Al content on the sheet carrier concentration is demonstrated. A maximum electron mobility of 1500 cm^2/Vs is achieved for an undoped sample. Device fabrication was done on full 2" wafers by electron-beam and contact lithography. The quality of the grown epi layers and the technology is confirmed by CW load pull measurements at 2 GHz and 10 GHz

1. Introduction

High electron mobility transistors (HEMTs) based on GaN/AlGaN heterostructures are well suited for microwave high power amplifiers covering the 2 – 40 GHz frequency range [1-3]. The main advantages of GaN over Si and GaAs stem from its wide bandgap, hence high breakdown field and high electron saturation velocity. This material offers large potential to surpass existing device limitations in RF output power, operation voltage and operation temperature. The HEMT structures can be grown on sapphire or SiC substrates. Sapphire is available at significantly lower costs and in larger sizes than SiC and is successfully used for opto-electronic devices. However, to develop its full RF performance, the dissipated power has to be removed as effectively as possible. Therefore SiC is the substrate of choice providing an excellent thermal conductivity of 3.5 W/cm K, which is an order of magnitude higher than that of sapphire.

2. Epitaxy and characterization

The HEMT structures are grown by metal organic vapor phase epitaxy (MOVPE) in an AIXTRON 2000 G3 HT 6×2" multiwafer reactor. On SiC a 500 nm thick AlGaN layer is grown followed by a 2.75 μm thick highly insulating GaN buffer layer. The spontaneous and piezoelectric polarization of a 25 nm thick Al$_{0.25}$Ga$_{0.75}$N layer on top of the buffer forms a two-dimensional electron gas (2DEG) at the GaN/AlGaN interface (Figure 1) [4]. In Si-doped HEMT-structures the Si-doped AlGaN supply layer is

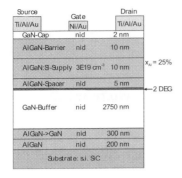
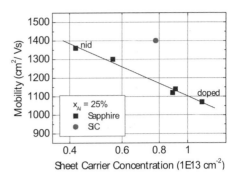

Figure 1: Epitaxial layer structure of the GaN/AlGaN HEMT grown by MOCVD on SiC substrates.

Figure 2: Comparison of the mobility and sheet carrier concentration between the same HEMT layer structures grown on sapphire and SiC substrates

sandwiched between the spacer and the barrier layer The structure is capped with a 2 nm thick GaN layer. A similar layer structure is used for the growth on sapphire. The Al-content of the actual HEMT-structures is determined frequently by spectral ellipsometry (SE). Using a 200 nm AlGaN layer we verified the Al-content determined by our standard SE (24.7 %) with energy dispersive X-ray analysis to (26.2 ± 0.7) % across a 2" wafer.

The 2DEG concentration for a HEMT-structure with an Al-content of 25 % can be increased from 4×10^{12} cm^{-2} to 1.5×10^{13} cm^{-2} by modulation doping, where the nominal Si-doping level varies from undoped to 3×10^{19} cm^{-3}, respectively. Results of Hall measurements (Figure 2) are from structures on sapphire (squares) and on SiC (circles). The slight decrease in mobility (squares) might be due to Coulomb scattering of the electrons caused by the increasing number of Si ions in the supply layer. The mobility of HEMT-structures on SiC is slightly higher for the same carrier concentration. From Hall measurements typical sheet carrier concentrations and electron mobilities of 8×10^{12} cm^{-2} and 1400 cm^2/Vs were obtained.

Sheet resistance (R_s) mappings across the wafers and from wafer to wafer proved an excellent uniformity. In Figure 3 the R_s across a 2" SiC wafer is shown with an average value of 483 Ω and a standard deviation of ± 0.6 %. The wafer to wafer uniformity of one epitaxial run on SiC substrates is demonstrated in Figure 4 with an average R_s of 474 Ω and a standard deviation of 3 %. The standard deviation across the wafers is given for each satellite in the figure. Device structures were grown reproducibly over a 1.5 year period with a standard deviation in sheet resistance of 3 - 4 %.

There is a trend towards devices with higher Al content in HEMT structures as this results in a higher sheet carrier density in the channel due to an increase in the spontaneous and piezoelectric polarization [4]. In addition the higher bandgap of the AlGaN layer promises a higher composite breakdown field. The influence of the increasing Al content from 25 % to 35 % is demonstrated in Figure 5. We observe an increasing sheet carrier density and almost constant mobility. Each point triplet represents wafers with different doping levels. However, the highest mobility of 1500 cm^2/Vs is achieved for an undoped structure with an Al-content of 30 %.

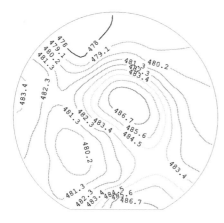 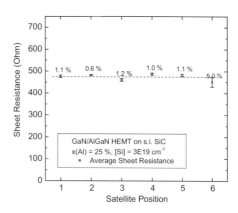

Figure 3: Distribution of the R_s of a GaN/AlGaN HEMT-structure with x(Al) = 25 % on a 2" s.i. SiC substrate. Average = 483 Ω ± 0.6 %.

Figure 4: Wafer-to-wafer R_s of HEMT-structures on SiC in the 6 x 2" MOCVD reactor. Numbers in the figure: Standard deviation for each wafer.

3. Process technology and device performance

Device fabrication was done on full 2" wafers by contact lithography. Device isolation was accomplished by a 180 nm deep mesa dry etch into the GaN buffer by chemically assisted ion beam etching using chlorine. Ti/Al/Au alloyed ohmic contacts were used obtaining an average contact resistance of 0.6 Ωmm. The T-shaped Ni/Au-gates were defined by electron-beam lithography with a minimum gatelength of 0.15 µm. A scanning electron microscope image of the T-gate in the source drain area is shown in Figure 6. All devices were passivated by a 100 nm thick SiN layer and multi-finger transistors were fabricated using first level metal interconnects and galvanic Au air bridges.

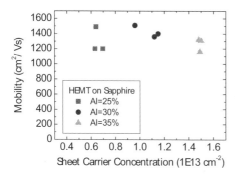 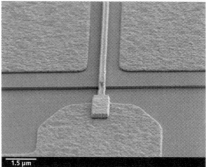

Figure 5: Mobility and sheet carrier concentration of GaN/AlGaN HEMT structures on sapphire as a function of the Al content in the AlGaN layer. Each point triplet represents wafers with different doping levels.

Figure 6: Scanning electron microscope image of a HEMT with a gatelength lg = 150 nm.

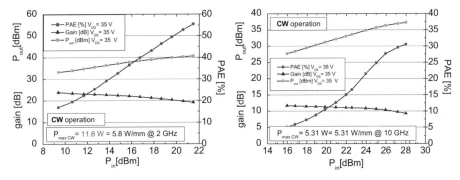

Figure 7: Output power, PAE and gain at 2 GHz for a 300 nm GaN/AlGaN HEMT on SiC with a gatewidth of 10×200 µm.

Figure 8: Output power, PAE and gain at 10 GHz for a 150 nm GaN/AlGaN HEMT on SiC with a gatewidth of 8×125 µm

The quality of the grown epitaxial layers and the technology is confirmed by CW load pull measurements at 2 GHz and 10 GHz (Figure 7 and 8, respectively). Results from a device with 2 mm gatewidth and a gatelength of 300 nm are shown in Figure 7. A linear gain of 23 dB, a maximum output power of 11.6 W yielding 5.8 W/mm, and a power added efficiency beyond 55 % are obtained under CW operating conditions. For a 150 nm AlGaN/GaN HEMT with a gatewidth of 1 mm (Figure 8), a linear gain beyond 11 dB, a maximum output power of 5.31 W yielding 5.31 W/mm, and a maximum PAE beyond 30 % are achieved. The results represent state of the art values. Furthermore the process is shown to be also suitable for K_a-band applications [5,6].

4. Conclusion

We have demonstrated the growth of GaN/AlGaN heterostructures by MOVPE in a multiwafer reactor for power applications. The electrical properties of the 2DEG as well as the excellent uniformity of the wafers are good conditions for device fabrication. The quality of both epitaxy and technology is confirmed by CW load pull measurements at 2 GHz and 10 GHz.

References

[1] Wu Y-F, Keller B P, Keller S, Kapolnek P, Kozodoy P, DenBaars S P and Mishra U K 1997 Solid State Electron. 41 1569-1574
[2] Shur M S 1998 Solid State Electron. 42 2131-2138
[3] Morkoc H, Di Carlo A and Cingolani R 2002 Solid State Electron. 46 157-202
[4] Ambacher O, Smart J, Shealy J R, Weimann N G, Chu K, Murphy M, Dimitrov R, Wittmer L, Stutzmann M, Rieger W and Hilsenbeck J 1999 J. Appl. Phys. 85 3222-3233
[5] Kiefer R and Quay R 2002 Workshop Notes: Wide Band Gap Materials, Microwave Theory Techniques Symposium, Seattle
[6] Kiefer R, Quay R, Müller S, Köhler K, van Raay F, Raynor B, Pletschen W, Massler H, Ramberger S, Mikulla M and Weimann G 2002 Abstract Book: IEEE Lester Eastman Conference on High Performance Devices, Newark 78-79

Fabrication and Electrical Performance of Oscillators in GaInP/GaAs-HBT MMIC Technology up to 40 GHz

J. Hilsenbeck, F. Brunner, F. Lenk and J. Würfl

Ferdinand-Braun-Institut für Höchstfrequenztechnik, Albert-Einstein-Straße 11, 12489 Berlin, Germany

Abstract. The fabrication and electrical performance of monolithic coplanar 19- and 38-GHz oscillators in GaInP/GaAs-HBT MMIC technology are presented. Both fixed frequency and Voltage Controlled Oscillators (VCOs) have been realized. The fixed frequency oscillators show very low phase noise (PN), in case of the 19-GHz oscillator PN is -96 dBc/Hz, the 38-GHz oscillator reaches -89 dBc/Hz at 100 kHz offset.

1. Introduction

Low phase noise oscillators are one of the major building blocks both in sensor application and wireless communication systems. In order to reduce cost, monolithically integrated circuits (MMIC) are well suited. However, the active devices in MMICs have to show very low 1/f noise, because the quality factors of on-chip resonators are fairly low. Principally Heterojunction Bipolar Transistors (here: GaInP/GaAs-HBTs) show lower 1/f noise than Fieldeffect Transistors (e.g. AlGaAs/GaAs-HEMTs) because the current flow is perpendicular to semiconductor interfaces and thus the influence of interface and surface states on the output signal is drastically reduced.

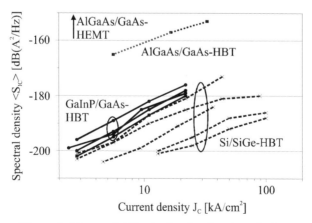

Figure 1. Measured 1/f noise levels of various devices vs. current density J_C in $dB(A^2/Hz)$ at a frequency of 100 kHz (input resistance $R_S = 10\ \Omega$).

Although SiGe-HBTs demonstrate the lowest 1/f noise of all devices suitable for MMIC oscillators (see Fig. 1), these devices suffer from their frequency limitations and can hardly cover the emerging market for Ka- and W-band operation. GaInP/GaAs-HBTs fulfill both requirements and thus are well suited for low noise oscillators in the 20 - 77 GHz range.

2. Experimental

The epitaxial layer structures were grown by Metalorganic Vapor-Phase Epitaxy (MOVPE) on semi-insulating LEC-GaAs substrate. For the epitaxial HBT design a 55 nm n^+-InGaAs graded emitter-contact-layer, a 100 nm thick n-GaAs layer, a 30 nm n-InGaP emitter and a 100 nm uniformly doped p-GaAs base layer was used. Finally, the HBT structure was completed with a 1 µm n-GaAs collector, a 20 nm GaInP etch stop layer and a 700 nm thick n^+-GaAs subcollector.

The HBT MMICs were fabricated in-house using an industry-compatible 4" process line with stepper lithography [1]. In total 14 lithography levels were used.

From simulation experiments [2] two sources mainly affect the low frequency noise of HBTs: base current and 1/f noise of emitter resistance. In order to achieve low 1/f noise and high frequency levels especially the emitter-base-mesa fabrication is one of the most critical steps: With stronger under-etching of the emitter metal (used as etch mask) the distance from emitter to base-metalization increases. This leads to higher base-resistance R_B and directly lower f_{max}. Similarly the cross-section area of the emitter decreases leading to a higher emitter resistance. Therefore a special double dry etch process has been developed combining anisotropic and isotropic etching. Since radiation damage strongly increases R_B and causes surface recombination currents the anisotropic step must not reach the GaInP-emitter. Thus, in the first step 80 % of the InGaAs/GaAs emitter constact layer is etched anisotropic, in the second step the remaining GaAs is etched isotropic using the GaInP-emitter as etch stop layer (Fig. 2 a). To control the depth of each etch step an in situ interferometrical measuring unit (NanoMES®) was employed [3].

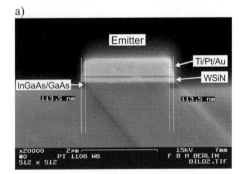 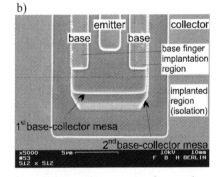

Figure 2. a) SEM cross section of the emitter area after emitter metalization and etching (typical: 90 - 120 nm); b) SEM photograph of an HBT after collector metalization.

Ledge technology was applied to suppress emitter-base-surface-leakage and further reduces 1/f noise. He^+-implantation was used for isolation between devices. Base-collector mesa was etched in two steps: in the first step dry etching was used to control the mesa width. In the second step wet chemical etching was applied in order to achieve under-etching and thus to reduce the parasitic base-collector capacitance C_{BC} (see Fig. 2 b). An additional He^+-implantation in the outer region of the base fingers was introduced. In this way, the underlying base layer and the upper part of the collector (approx. 600 nm) are isolated and thus further reduction of C_{BC} is achievable. For the

MMIC's passive elements MIM-capacitors (PECVD-SiN$_x$ as dielectric material), thin-film-resistors (sputtered NiCr), spiral-inductors, coplanar waveguides and air bridges (electroplated gold) were used.

3. Results

From process control monitoring (PCM) the gain β_{max} is 120. With a corresponding intrinsic base sheet resistance R_{SBI} of 215 Ω/sq the β_{max}/R_{SBI}-ratio is greater than 0.5 $(\Omega/sq)^{-1}$ indicating both excellent epitaxial material and device processing. Passive elements show very good uniformity over 4-inch wafer. In case of NiCr thin-film-resistivity the standard deviation σ is 3.5 %. Using 110 nm thick SiN$_x$ for MIM-capacitors the breakdown voltage is around 70 V (E_{Br} > 6 MV/cm, σ < 2 %) indicating the high quality PECVD-SiN$_x$. From S-parameter measurements f_T and f_{max} of our standard HBT (emitter-base-distance d_{EB} = 1.3 µm, emitter area 1x3x30 µm^2) were determined to be 40 and 100 GHz, respectively. These values are quite sufficient for oscillators up to 40 GHz, but for future 77-GHz-oscillators f_{max} has to be increased. Since

$$f_{max} = \sqrt{\frac{f_T}{8\pi R_B C_{BC}}}$$

drastic improvements can be achieved if the base resistance R_B and the base-collector capacitance C_{BC} are reduced. According to $R_B = R_{CB} + R_{SBE}*d_{EB}$ (R_{CB}: base contact resistance, R_{SBE}: extrinsic base sheet resistance) the base resistance drops down from 0.5 Ωmm to 0,25 Ωmm by reducing d_{EB} from 1.3 to 0.5 µm. As can be seen from Fig. 3 in this way f_{max} could be further increased to 150 GHz. By introduction of a shallow He$^+$-implantation in the outer region of the base fingers C_{BC} could be further reduced leading to an f_{max} of 170 GHz.

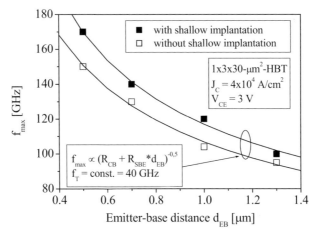

Figure 3. Maximum frequency of oscillation f_{max} vs. emitter-base-separation d_{EB} with and without shallow basefinger implantation.

As can be seen from Fig. 1 our HBTs show very low 1/f noise levels comparable to those of SiGe-HBT. It has to be mentioned that oscillators based on SiGe-HBTs typically run at higher current densities.

Fig. 4 illustrates our oscillator results. Details about oscillator simulations and measurement techniques were published in [4-6]. With single ended 19 GHz oscillator design the phase noise PN at 100 kHz offset exhibits very low levels and excellent uniformity over 4-inch (σ < 2%, Fig. 4 a). Further improvements could be achieved with the push-push oscillator concept. Here we measured -96 dBc/Hz at the first harmonic of 19 GHz and -89 dBc/Hz at the second harmonic (Fig. 4 b). For the VCO version the PN is somewhat higher. Here the oscillator exhibits -86 dBc/Hz at the fundamental and -81 dBc/Hz at the second harmonic at 100 kHz offset.

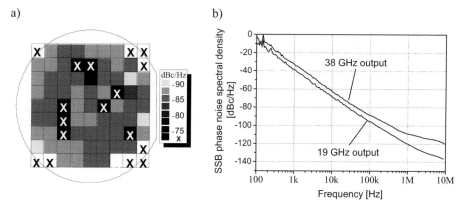

Figure 4. a) Phase noise map of a 19-GHz-oscillator (single ended, PN @ 100 kHz = -87.3± 1.7 dBc/Hz); b) phase noise of a 19-GHz differential oscillator (push-push design).

4. Summary

Fixed frequency oscillators and VCOs with low phase noise levels in the –90 dBc range were fabricated demonstrating the suitability of the GaInP/GaAs-HBT as the active device. By reducing parasitic elements of the HBT the maximum frequency of oscillation could be increased from 100 to 170 GHz. Conclusively, these devices combine very good 1/f noise properties with the high frequency potential of GaAs-based transistors.

5. Acknowledgments

The authors wish to thank D. Rentner, W. John, A. Klein, M. Mai for their contributions in processing the MMICs, P. Heymann, S. Schulz, W. Köhler for performing microwave, phase noise and 1/f-noise measurements and M. Rudolph, M. Schott, H. Kuhnert for device and MMIC simulations. Moreover, financial support by the German Ministry of Research and Technology (BMBF) under grant 01BM050 is gratefully acknowledged.

6. References

[1] M. Achouche, T. Spitzbart, P. Kurpas, F. Brunner, J. Würfl, G. Tränkle, *Electr. Letters*, Vol. 36, No. 12, 2000, pp. 1073-1075.

[2] P. Heymann, M. Rudolph, R. Doerner, and F. Lenk, *MTT-S Int. Microwave Symp. Dig.*, 2001, Vol. 3, pp. 1967-1970.

[3] W. John, L. Weixelbaum, H. Wittrich, G. Frankowski, J. Würfl, *GaAs MANTECH Dig.*, 2000, pp. 21-24.

[4] H. Kuhnert, F. Lenk, J. Hilsenbeck, J. Würfl, and W. Heinrich, *IEEE MTT-S Int. Microwave Symp. Dig.*, 2001, Vol. 3, pp. 1551-1554.

[5] M. Schott, H. Kuhnert, F. Lenk, J. Hilsenbeck, J. Würfl, and W. Heinrich, *IEEE MTT-S Int. Microwave Symp. Dig.*, 2002, Vol. 2, pp. 839-842.

[6] F. Lenk, M. Schott, and W. Heinrich, *2001 IEEE EuMC 31st European Microwave Conference Digest*, 2001, Vol. 1, pp. 181-184.

Experimental Demonstration of a Resonant Tunneling Delta-Sigma Modulator for High-Speed, High-Resolution Analog-to-Digital Converter

Yuji Yokoyama, Yutaka Ohno, Shigeru Kishimoto, Koichi Maezawa and Takashi Mizutani

Graduate School of Engineering, Nagoya University, Furo-cho, Chikusa-ku, Nagoya 464-8603, Japan

Abstract. A resonant tunneling delta-sigma ($\Delta\Sigma$) modulator was fabricated on an InP substrate. The circuit is extremely simple due to the high performance comparator based on the monostable-bistable transition logic element (MOBILE). Basic operation of this $\Delta\Sigma$ modulator was demonstrated.

1. Introduction

A high-speed and high-resolution analog-to-digital converter (ADC) is a key component for software defined radio (SDR) for microwave communication, and considerable efforts have been made to realize such ADCs [1-3]. Recently, we proposed a novel high-speed delta-sigma ($\Delta\Sigma$) ADC using resonant tunneling diodes (RTDs) [4]. This type of ADC has a significant advantage; higher resolution can be easily obtained by increasing the sampling rate. Furthermore, it does not require high-accuracy analog components to achieve high resolution. The $\Delta\Sigma$ ADC using RTDs is promising for high-speed and high-resolution application including SDR. In this paper, we demonstrate the basic operation of the resonant tunneling $\Delta\Sigma$ modulator, which is a key component of the $\Delta\Sigma$ ADC (as shown in Fig. 1). The $\Delta\Sigma$ modulator converts input analog signal to the pulse density signal.

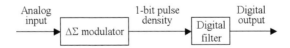

Figure 1. Block diagram of a $\Delta\Sigma$ ADC.

2. Circuit configuration and operation principle

The circuit configuration of the $\Delta\Sigma$ modulator is shown in Fig. 2. This is an extremely simple circuit consisting of a capacitor, two RTDs, and six high electron mobility transistors (HEMTs). The capacitor C and the input HEMT Tr1 function as an integrator, and the two RTDs form monostable-bistable transition logic element (MOBILE) [5], which functions as a high performance comparator.

Here, we will briefly describe the operating principle of the proposed $\Delta\Sigma$ modulator. The current I_1 flows depending on the input voltage V_{in}, so that the charge stored at the capacitor gradually decreases. When the capacitor voltage, V_C, decreases to the threshold voltage of the MOBILE, it outputs a pulse. This output pulse charges the capacitor C through Tr2 (I_2), and it inhibits output of the pulse (negative feedback loop). This process repeats according to the clock, and the period of the process depends on the discharge rate of the capacitor C through Tr1. Consequently, the input voltage V_{in} is converted to the output pulse density ($\Delta\Sigma$ modulation). The advantage of this circuit is that it requires no extra clock cycle in the feedback loop, thus this circuit can be operated at an extremely high sampling rate.

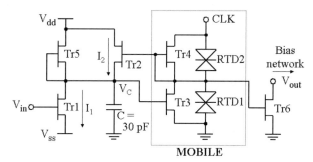

Figure 2. Circuit diagram of the resonant tunneling $\Delta\Sigma$ modulator circuit.

The circuit fabricated here is a modified version of the previously-proposed circuit. First, we added Tr4 to fabricate the circuit with depletion mode HEMTs, which are easy to fabricate. The Tr4 compensates the drain current of Tr3 at $V_g = 0$ V, and makes the design of the RTD areas easy. Second, we added Tr5 to improve linearity. The previous circuit configuration [4] has a serious problem that the operating region of the input voltage V_{in} is close to the threshold voltage of Tr1, which degrades the linearity. The current through the Tr5 pulls up the operating region of V_{in}. Fig. 3(a) and (b) show simulation results of output pulse density as a function of the input voltage for the circuits with and without Tr5, respectively. These figures demonstrate the improved linearity of the pulse density modulation when employing the Tr5. The areas of the RTD1 and RTD2 are 27 μm^2 and 28.5 μm^2, respectively. The gate widths of HEMTs are designed to be 10 (Tr1, Tr3, Tr4, and Tr5), 5 (Tr2), and 15 (Tr6) μm.

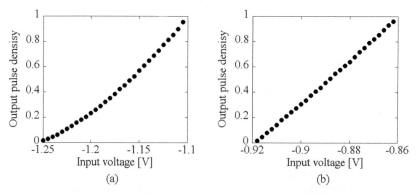

Figure 3. Simulation results of output pulse density as a function of the input voltage. (a) the previous circuit (without Tr5). (b) the improved circuit (with Tr5).

We fabricated the circuit by using an InP-based RTD/HEMT integration technology [6][7]. The fabricated HEMT has the gate length of 1.2 μm, a maximum transconductance of 450 mS/mm, and a threshold voltage of –0.2 V. The peak voltage, the peak current density, and the peak-to-valley current ratio of the fabricated RTD are 0.35 V, 6.0×10^4 A/cm^2, and 6 at room temperature, respectively. A chip microphotograph of the circuit is shown in Fig. 4. The chip size is 750 x 750 μm^2.

3. Experimental results

The circuit was tested using an on-wafer measurement system. A synthesised sweeper was used to supply sinusoidal clock signal. Dc voltages were supplied to clock offset and output terminal (V_{out}) through the bias network. Figure 5(a)-(d) shows measured output waveforms of the operation with the different dc input voltage. The clock frequency of

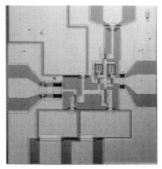

Figure 4. Microphotograph of fabricated circuit.

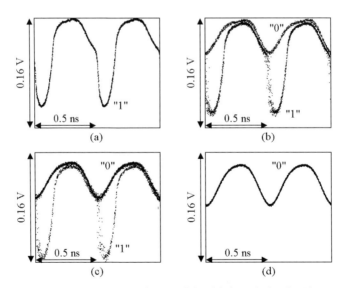

Figure 5. Output waveforms of the fabricated circuit. The output signals are long period pulse sequences, and the pulse density is visible in the plot density. (a) V_{in}=-1.4 V. The output signal is all "1". (b) V_{in}=-1.45 V. The output signal is a mixture of "0" and "1". (c) V_{in}=-1.55 V. The output signal is a mixture of "0" and "1". The density of the pulse "1" is smaller than that in (b). (d) V_{in}=-1.6 V. The output signal is all "0".

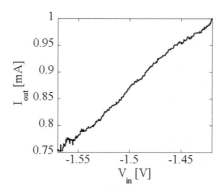

Figure 6. The dc component of output current, which corresponds to the output pulse density, as a function of the dc input voltage.

MOBILE was 2 GHz. V_{dd} and V_{ss} are 1.4 V and -1.3 V, respectively. The waveforms are inverted by the open-drain buffer Tr6. The output signals are random pulse sequences, and the pulse density is visible in the plot density of the traces. The density of the "1" pulses decreased, as the input voltage decreased from (a) to (d) in order. Figure 6 shows the measured dc component of the output current, which corresponds to output pulse density, as a function of the input voltage. This figure clearly indicates that the circuit converted the input voltage to the output pulse density. An extremely high clock frequency higher than 100 GHz should be possible by reducing the gate length of the HEMTs to 0.1 µm and optimizing the circuit parameters.

4. Conclusion

A resonant tunneling delta-sigma ($\Delta\Sigma$) modulator was designed and fabricated on an InP substrate. The circuit was extremely simple owing to the high performance comparator based on the MOBILE. Basic operation of this $\Delta\Sigma$ modulator was demonstrated at 2 GHz.

References

[1] T. Itoh, T. Waho, K. Maezawa and M. Yamamoto: IEICE Trans. Inf. & Syst. E82-D (1999) 949.
[2] T. P. E. Broekaert, B. Brar, J. P. A. van der Wagt, A. C. Seabaugh, F. J. Morris, T. S. Moise, E. A. Beam, III and G. A. Frazier: IEEE J. Solid-State Circuits 33 (1998) 1342.
[3] T. P. E. Broekaert, B. Brar, F. Morris, A. C. Seabaugh and G. A. Frazier: IEEE Proc. Ninth Great Lakes Symp. VLSI (IEEE Press, Michigan, 1999), pp.123-126.
[4] Y. Yokoyama, Y. Ohno, S. Kishimoto, K. Maezawa and T. Mizutani: Jpn. J. Appl. Phys., 40 No. 10A, (2001) L1005-L1007.
[5] K. Maezawa, H. Matsuki, M. Yamamoto and T. Otsuji: IEEE Electron Device Lett. 19 (1998) 80.
[6] K. J. Chen, K. Mawzawa, T. Waho and M. Yamamoto: IEICE Trans. Electron., E79-C (1996) 1515-1524.
[7] K. Maezawa, J. Osaka, H. Yokoyama and M. Yamamoto: Jpn. J. Appl. Phys., 37 No. 10, (1998) 5500-5502

InAs/AlGaSb heterostructure displacement sensors

H Yamaguchi,[1] S Miyashita,[2] and Y Hirayama[1,3]

[1] NTT Basic Research Laboratories, NTT Corporation, 3-1, Morinosato Wakamiya, Atsugi, Kanagawa 243-0198, Japan

[2] NTT Advanced Technology, 3-1, Morinosato Wakamiya, Atsugi, Kanagawa 243-0198, Japan

[3] CREST-JST, 4-1-8 Honmachi, Kawaguchi, Saitama 331-0012, Japan

Abstract. We have successfully fabricated a novel self-sensing mechanical displacement sensor with a surface InAs conductive layer of nanometer-scale thickness based on MBE-grown InAs/AlGaSb heterostructures. Sub-angstrom cantilever displacement is detectable and the sensitivity is increased with decreasing thickness. Tapping mode AFM characterization clarified the spatially resolved frequency response of this device, showing the lowest mode resonance frequency of about 300kHz.

1. Introduction

Micro- and nano-electromechanical systems (MEMS/NEMS) have the potential to bring about a revolution in the application of semiconductor fine-structure devices, such as high-resolution actuators and sensors, high-frequency signal processing components, and medical diagnostic devices [1]. Compound semiconductor MEMS/NEMS have several additional advantages. Integration with low-dimensional heterostructures provides additional functionalities, such as optical and quantum mechanical operations, on MEMS/NEMS devices [2-5]. The precise controllability in semiconductor nanostructure fabrication using ultra-thin film growth techniques, such as molecular beam epitaxy (MBE) and vapour phase epitaxy (VPE), enables us to fabricate ultrasmall mechanical structures [6,7]. The stress controllability for lattice-mismatched systems also makes it possible to fabricate novel three-dimensional structures [6].

InAs-based mechanical structures have the additional feature [7-9] that the surface Fermi level pinning in the conduction band allows us to fabricate conductive structures that are much smaller than those possible with commonly used materials systems, such as Si/SiO$_2$ [1,10,11] and GaAs/AlGaAs-based heterostructures [2-4]. We have successfully fabricated novel self-sensing mechanical displacement sensors with a surface InAs conductive layer of nanometer-scale thickness based on MBE-grown InAs/AlGaSb

Fig. 1 SEM image of a fabricated InAs/AlGaSb displacement sensor

heterostructures. The device has a sub-angstrom detection limit. We confirmed about five times enhanced sensitivity for the samples with reduced InAs thickness. The size of this self-sensing device can be reduced to nanometer scale and it is expected to be a key component in future NEMS technologies.

2. Experimental

InAs/AlGaSb heterosturcutres totally 300-nm-thick were grown on GaAs substrate. The use of (111)A-oriented substrate allowed the growth of a high-quality AlGaSb layer directly on the GaAs substrate despite the 7% lattice mismatch [12]. The GaAs sacrificial layer was etched away by NH_4OH solution to form an InAs/AlGaSb cantilever after defining the lateral dimension of the structures by the standard photolithographic technique. Figure 1 shows a SEM image of a fabricated displacement sensor. The InAs thickness was varied from 8 to 20 nm in order to study the surface and interface effect on the

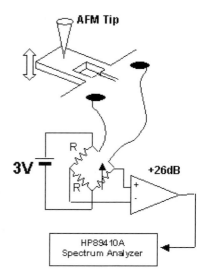

Fig. 2 AFM characterization setup used to measure the piezo-resistivity

piezoresistivity. Without intentional doping, all the cantilevers showed conductivity due to the native electron surface accumulation in the InAs layer.

Because the InAs layer is formed on one side of the cantilever, its deflection induces a two-dimensional (2D) resistivity change in the InAs film. We measured this *piezoresistivity* by contact-mode atomic-force microscopy (AFM). The device was mounted in a commercial AFM apparatus and the mechanical motion was actuated by z-scanner modulation as the AFM tip approached the sensor cantilever. The deflection-induced resistance change was measured through a standard dc-biased Wheatstone bridge (Fig. 2)

3. Results and Discussions

The measured sensor sensitivity, $\delta R/R/\delta z$, where δR is the resistance change, R the device resistance, and δz the sensor cantilever deflection, at the frequency of 1 kHz is summarized in Table. 1. The sensitivity shows an increase with decreasing InAs thickness and about five times increase in the sensitivity was observed for 8nm sample compared to the 20nm one. Because the bulk piezoresistivity is thickness independent, the increase is surface and interface effects. The sensitivity of 6.3×10^{-5} nm^{-1} for the 8-nm-thick sample is roughly ten times higher than that reported for Si piezoresistive cantilevers [10,11]. The detection limit Δ is, however, nearly two orders of magnitude below that of Si cantilevers because of the much higher 1/f noise for our devices in this frequency range. The sensor detection limit is largely improved for higher frequency regions. The thermal vibration of the similarly fabricated cantilever with the amplitude of 10^{-2} nm was electrically detected in vacuum at 300 kHz, where the 1/f noise is sufficiently suppressed [13].

The resonance characteristics of the sensor cantilever were also characterized by tapping-mode AFM. The cantilever was periodically tapped by the AFM tip and the induced resistance change was measured in the frequency domain with a spectrum analyser. Figure 3 shows typical spectra obtained for the 15-nm-thick-InAs sample with the tapping frequency F_{ex} of 113.9 kHz. The signals at the tapping frequency (F_{ex}) and its higher harmonics ($2F_{ex}$ and $3F_{ex}$) are confirmed, although tapping-force-independent signal is mixed in with the F_{ex} peak. This signal originates from the capacitance coupling between the device and the piezoelectric-bimorph used for the AFM cantilever actuation. Higher harmonic signals originate only from the mechanical motion of cantilever. The tap induces pulse-shaped time-dependent force on the cantilever, which includes higher harmonic components.

d (nm)	$n_{s,Hall}$ (cm^{-2})	μ (cm^2/Vs)	$\delta R/R/\delta z$ (nm^{-1})	Δ (nm$_{rms}$/Hz$^{1/2}$)
20	2.3×10^{12}	6,400	1.1×10^{-5}	0.076
15	2.4×10^{12}	5,040	3.8×10^{-5}	0.045
10	1.8×10^{12}	3,820	5.7×10^{-5}	0.05
8	1.1×10^{12}	1,870	6.3×10^{-5}	0.0625

Table. 1 The measured piezoresistance per unit displacement, $\delta R/R/\delta z$, sheet carrier concentration n_s and the mobility μ for various InAs thicknesses d. The displacement detection limit at 1 kHz, Δ, i.e. the measured noise density divided by $\delta R/R/\delta z$, is also shown. All the data were obtained at room temperature. The Hall measurements were performed for the unprocessed samples with van der Pauw geometry.

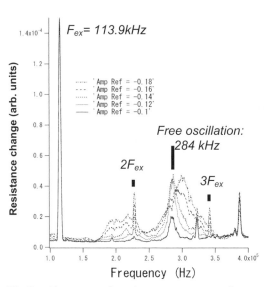

Fig.3 Frequency-domain spectrum of resistance change induced by the AFM tapping at 113.9kHz. "Amp Ref" corresponds to the tapping force and has a larger negative value for stronger tapping

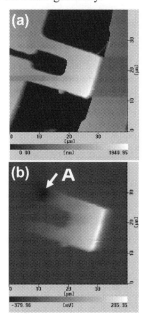

Fig.4 (a) Tapping mode AFM topographic image and (b) the map of simultaneously obtained F_{ex} components in output voltage.

The free mechanical vibration of the sensor cantilever after it was tapped by the AFM tip was also detected at 284 kHz. This frequency shows good agreement with the laser-interferometer measurement for the same device. The peak height increased with increasing tapping force and finally shifted to the high-frequency side with significant peak broadening. This could be due to the non-linear and dissipative interaction between the AFM tip and the sensor cantilever surface.

Figure 4 shows the topographic image (a) and the map of the simultaneously measured F_{ex} component of resistance change (b) obtained from the tapping-mode AFM characterization. The latter successfully images the cantilever deflection induced by the taps by the AFM tip and clearly shows that larger deflection is induced at the cantilever apex. The dark contrast obtained at A can be observed only at the side of the contact for voltage measurement and is probably due to the electrostatic fields induced by the AFM tip. This novel characterization method is expected to be used for nanoscale mechanical structures and will be a powerful tool for the characterization of nanoscale electromechanical devices.

4. Conclusion

In conclusion, we have fabricated InAs/AlGaSb heterostructure microelectromechanical displacement sensors, where the nanometer-thick surface InAs layer could electrically sense the sub-angstrom scale displacement of the cantilever. The resonance frequency at about 300 kHz was measured by tapping-mode AFM characterization. The results promise fabrication of high-speed nanomechanical displacement sensors with length of 200 nm and thickness of 20 nm, for which a simple scaling estimation predicts the resonance frequency of 500 MHz.

Acknowledgements

The authors are grateful to Dr. Sunao Ishihara and Dr. Takaaki Mukai for their continuous encouragement throughout this study. This work was partly supported by the NEDO International Joint Research Program "*Nano-elasticity*".

References

[1] Craighead H G 2000, Science 290 1532-1535.
[2] Uenishi Y, Tanaka H, and Ukita H 1995, IEICE Trans. Electron. E78-C 139
[3] Beck R G, Eriksson M A, Westervelt R M, Campman K L, and Gossard A C 1996 Appl. Phys. Lett. 68 3763-3765
[4] Beck R G, Eriksson M A, Topinka M A, Westervelt R M, Maranowski K D, and Gossard A C 1998 Appl. Phys. Lett. 73 1149-1151
[5] Harris J G E, Knobel R, Maranowski K D, Gossard A C, Samarth N, and Awschalom D D 2001 Phys. Rev. Lett. 86, 4644
[6] Prinz V Ya, Seleznev V A, Gutakovsky A K 1999 *Proc. 24th Int. Conf. on Physics of Semiconductors* ed D Gershoni (Singapore: World Scientific)
[7] Yamaguchi H and Hirayama Y 2002 Appl. Phys. Lett. 80 4428
[8] Yamaguchi H, Dreyfus R, Hirayama Y, and Miyashita S 2001 Appl. Phys. Lett. 78 2372-2374
[9] Yamaguchi H, Dreyfus R, Hirayama Y, and Miyashita S 2002 Jpn. J. Appl. Phys. 41 2519-2521
[10] Tortonese M, Barrett M R C, and Quate C F 1993 Appl. Phys. Lett. 62 834-836
[11] Yuan C W, Batalla E, Zacher M, de Lozanne A L, Kirk M D, and Tortonese M 1994 Appl. Phys. Lett. 65 1308-1310
[12] Yamaguchi H and Hirayama Y 1998 Jpn. J. Appl. Phys. 37 1599-1602
[13] Yamaguchi H, Miyashita S, and Hirayama Y 2002 to be published

Charge balanced Ga$_2$O-GaAs interface and application to self-aligned GaAs p-channel enhancement mode MOS heterostructure field-effect transistor

M Passlack, J K Abrokwah, R Droopad, Z Yu, and C Overgaard

Motorola Inc., Tempe, AZ 85284 USA

S I Yi, M Hale, J Sexton, and A C Kummel

Department of Chemistry, University of California, San Diego, La Jolla, CA 92093 USA

Abstract. The surface structure formed upon deposition of Ga$_2$O onto the GaAs(001) surface has been investigated by scanning tunneling microscopy and spectroscopy. A charged balanced (2x2) surface order that is electronically unpinned was found. Further, self-aligned GaAs enhancement mode MOS-HFETs employing a Ga$_2$O$_3$ gate oxide have been manufactured. The p-channel devices with a gate length of 0.6 μm exhibit a maximum dc transconductance g_m of 51 mS/mm which is an improvement of more than two orders of magnitude over previously reported results. With the demonstration of a complete process flow and 66% of theoretical performance, GaAs MOS technology has moved into the realm of reality.

1. Introduction

III-V semiconductors have become an enabling technology, in particular in wireless and fiber optic communications. However, the lack of a III-V MOSFET technology has limited functionality, scalability, and performance; consequently, market acceptance of III-V technologies has been restricted. Therefore, III-V gate oxide research has been an active field for decades. Conclusive evidence of accumulation and inversion in n- and p-type capacitors manufactured by deposition of a Ga$_2$O$_3$ layer on GaAs was provided in 1996 [1]. However, the mechanism of Fermi level unpinning was not understood at the time and later published transistor data using a Ga$_2$O$_3$(Gd$_2$O$_3$) layer as gate oxide [2]- [5] were inconsistent with the low interface state density D_{it} reported in [1].

The scope of this paper is to discuss the mechanism of Fermi level unpinning at the Ga$_2$O-GaAs interface on an atomic level and to subsequently present data on the first self-aligned GaAs enhancement mode MOS heterostructure field effect transistors (MOS-HFET) using a Ga$_2$O$_3$-GaAs interface.

2. Ga$_2$O-GaAs surface structure and Ga$_2$O-GaAs interface

The surface structures formed upon deposition of Ga$_2$O onto the technologically important As-rich GaAs(001)-(2x4) surface have been studied in an ultra-high vacuum chamber using scanning tunneling microscopy (STM) and spectroscopy (STS).

An atomically clean GaAs(001)-(2x4) surface has been obtained by thermally desorbing a capping layer of As. After an atomically ordered (2x4) surface was confirmed by STM, the surface has been exposed to Ga$_2$O at an elevated surface temperature for 5 minutes. STM images confirm that a saturated monolayer of Ga$_2$O is formed; this monolayer is a two dimensional wetting layer of Ga$_2$O. At monolayer coverage of Ga$_2$O, the surface shows a new quasi (2x2) order. The 2x order reflects the dimerization of the As atoms along [$\bar{1}$10] while the x2 order along [110] shows that the troughs consisting of 3rd layer As dimer and 4th layer Ga atoms have been filled. Fig. 1 shows a ball and stick model for the proposed structure of the (2x2) order. The surface is completely As terminated with Ga$_2$O occupying all available sites between the As dimers.

The dI/dV (density of states) spectra for a monolayer of Ga$_2$O adsorbed onto both n and p-type GaAs is shown in Fig. 2. The Ga$_2$O curves for both n and p-type materials are indicative of clean, doped samples with a measured band gap at the expected 1.4 eV with the Fermi level positioned at its bulk level close to the conduction and valence band, respectively. In contrast, when only 2-3% of a monolayer of O$_2$ adsorbs onto both n and p-type GaAs, enough states are induced into the band gap to pin the Fermi level at midgap. This case is shown for n-type GaAs in Fig. 2 for comparison purposes.

Density functional calculations have been carried out to gain insight into the local electronic environment. The model calculations have shown that the surface arsenic atoms rehybridize when Ga$_2$O adsorbs on the clean cluster and that their charges tend toward more bulk-like values. The gallium atoms from the Ga$_2$O molecule also attain a more bulk-like charge, as well as the 2nd layer gallium atoms. This restoration of charge to bulk values causes the Fermi level to remain unpinned when Ga$_2$O is adsorbed onto the surface. Further details concerning Ga$_2$O surface structure and Ga$_2$O-GaAs interface properties will be published in [6].

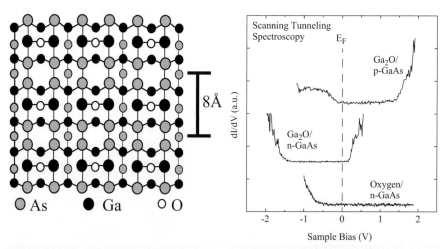

Figure 1. Ball and stick diagram for adsorption sites of Ga$_2$O.

Figure 2. dI/dV for Ga$_2$O/p-GaAs, Ga$_2$O/n-GaAs, and oxygen/n-GaAs.

3. Self-aligned GaAs p-channel enhancement mode MOS heterostructure field-effect transistor

Fig. 3 shows the cross section of a self-aligned p-channel enhancement mode MOS-HFET. The epitaxial layer structure is grown by molecular beam epitaxy on 3" semi-insulating GaAs substrates and consists of a 0.2 µm undoped GaAs buffer layer, a 15 nm undoped $In_{0.2}Ga_{0.8}As$ channel layer, a 15 nm undoped $Al_{0.75}Ga_{0.45}As$ barrier layer, and a 2 monolayer thick GaAs cap layer. A silicon δ-doping with an areal density of 3.3×10^{11} cm^{-2} is positioned 3 nm below the channel. Subsequent to completion of epitaxial layer growth, the wafer platen is transferred under ultra-high vacuum to a second oxide deposition chamber. Finally, the 9 nm thick Ga_2O_3 gate oxide layer is deposited by thermal evaporation of Ga_2O_3 from an effusion cell. Details of the Ga_2O_3 deposition process are described elsewhere [7]. Rutherford backscattering spectroscopy proved the oxide films to be stoichiometric with 60 and 40 at.% of oxygen and gallium, respectively. High resolution cross-sectional TEM of the completed device layer structure shows an atomically abrupt interface between the single crystal semiconductor surface and the amorphous Ga_2O_3 layer.

The 3" wafers were processed using Motorola's standard CGaAs^{TM} process [8]. A refractory gate metal (TiWN) is used with a 20 nm W diffusion barrier. The source/drain implants are implemented self-aligned to the gate and activated by RTA at 700 °C for 10 sec. The device dimensions L_G, L_{GS}, and L_{GD} are 0.6 µm, 1.2 µm, and 1.2 µm, respectively.

Fig. 4 shows the measured dc output characteristics of a 0.6 µm GaAs p-channel MOS-HFET. The transistor characteristics have been measured using a HP4145 semiconductor parameter analyzer; repeated curve scanning has always exactly reproduced the previously measured data and no hysteresis could be observed. Simulated data (not shown) have been obtained using the two-dimensional simulator PISCES [9] under the assumption of $D_{it} = 0$. The measured and simulated maximum dc transconductance g_m and threshold voltage V_{th} are 51 and 77 mS/mm, and -0.93 V and - 0.76 V, respectively. Al-

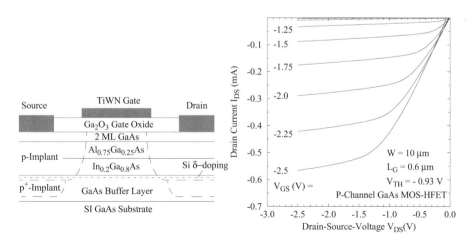

Figure 3. Cross section of a self-aligned GaAs enhancement mode MOS heterostructure field effect transistors.

Figure 4. Measured dc output characteristics of a 0.6 µm GaAs p-channel MOS-HFET.

though reduced g_m and shifted V_{th} indicate that interface states still affect the devices to a certain extent, the demonstrated performance of 66% of theoretical dc transconductance is a clear breakthrough. The device statistics for a 3" wafer are g_m = 46.7 ± 3.9 mS/mm and V_{th} = -0.93 ± 0.1 V.

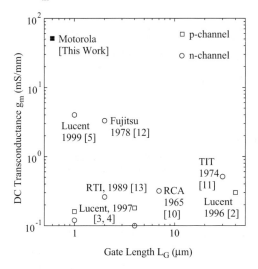

Figure 5. Comparison of dc transconductance data of p-channel (squares) and n-channel (circles) enhancement mode GaAs MOSFETs.

Fig. 5 shows a comparison of published dc transconductance data of enhancement mode GaAs MOSFETs. For p-channel devices, the highest dc g_m previously reported was 0.3 mS/mm [2]. The dc g_m of our MOS-HFETs is higher by a factor of 170. All previously reported dc g_m data for both n- and p-channel GaAs enhancement mode MOSFETs [2]-[5], [10]-[13] were at least two orders of magnitude lower than those of competing GaAs technologies such as PHEMT and CGaAs^{TM}. A further comparison of previously published data for GaAs n-channel enhancement mode MOSFETs reveals that essentially no progress had been made since 1978 when a dc g_m of 3.3 mS/mm was reported for a 2 µm GaAs MOSFET.

References

[1] Passlack M, Hong M Mannaerts J P 1996 Appl. Phys. Lett. 68 1099-1101
[2] Ren F Hong M W Hobson W S Kuo J M Lothian J R Mannaerts J P Kwo J Chen Y K and Cho A Y 1996 IEDM Tech. Dig. 943-945
[3] Ren F Hong M Kuo J M Hobson W S Lothian J R, Tsai H S Lin J Mannaerts J P Kwo J Chu S N G Chen Y K and Cho A Y 1997 Proc. IEEE GaAs IC Symp. 18-21
[4] Ren F Hong M Hobson W S Kuo J M Lothian J R Mannaerts J P Kwo J Chu S N G Chen Y K and Cho A Y 1997 Solid-State Electron. 41 1751-1753
[5] Wang Y C, Hong M, Kuo J M Mannaerts J P Kwo J Tsai H S Krajewski J J Weiner J S Chen Y K and Cho A Y 1999 Proc. Mat. Res. Soc. Symp. 573 219-225
[6] Yi S I Hale M Sexton J Kummel A C and Passlack M submitted to Phys. Rev. Lett.
[7] Yu Z Passlack M Bowers B Overgaard C D Droopad R Abrokwah J K 2000 US Patent 6,159,834
[8] Abrokwah J K Huang J H Ooms W Shurboff C Hallmark J A Lucero R Gilbert J Bernhardt B and Hansell G 1993 Proc. IEEE GaAs IC Symp. 127
[9] G-PISCES-IIB Manual Gateway Modeling Inc. 1998
[10] Becke H Hall R and White J 1965 Solid-State Electron. 8 813-823
[11] Ito T and Sakai Y 1974 Solid-State Electron. 17 751-759
[12] Mimura T Odani K Yokoyama N Nakayama Y and Fukuta M 1978 IEEE Trans. Electron Devices 25 573-579
[13] Fountain G G Rudder R A Hattangady S V Markunas R J and Hutchby J A 1989 IEDM Tech. Dig. 887-889

Investigation of Quantum Transport Phenomena in Resonant Tunneling Structures by Simulations with a Novel Quantum Hydrodynamic Transport Model

J. Höntschel, W. Klix, R. Stenzel

University of Applied Sciences Dresden, Department of Electrical Engineering, Friedrich-List-Platz 1, D-01069 Dresden, GERMANY

Abstract This paper describes the simulations of resonant tunneling structures with a novel quantum hydrodynamic transport model. For the simulation the device simulator SIMBA is used, which is capable to handle complex device geometries as well as several physical models represented by certain sets of partial differential equations. As a new feature the involvement of a quantum potential is implemented to include quantum mechanical transport phenomena in different quantum size devices. The coupled solution of this quantum correction potential with a hydrodynamic transport model allows to model resonant tunneling of electrons through potential barriers and particle build up in potential wells. The experimental results of a resonant tunneling structure are compared with the simulated data of the device.

1. Introduction

One intention of modern semiconductor technology is the reduction of the device length and width. With the realization of these nanometer structures several quantum mechanical effects appear. The advancing miniaturizations in the semiconductor technology make it necessary to model quantum effects like resonant tunneling of electrons. The quantum hydrodynamic simulation, which is based of a quantum fluid dynamical model [1], offers expanding possibilities for the understanding as well as the design of novel quantum sized semiconductor devices. The derivation of the full three-dimensional quantum hydrodynamic model (QHD-model), based on the Wigner-Boltzmann equation by a moment expansion, delivers the same conservation laws for the classical hydrodynamic equations, but the energy density and the stress tensor have additional quantum terms. The consequence of these additional quantum terms is the extension of the classical drift diffusion transport equation by an accessory expression, which describes a quantum correction potential. The equation for the energy flux density and the energy balance equation are likewise extended to include quantum effects by the same quantum correction potential. With these quantum terms the resonant tunneling of particles through potential barriers and the accumulation in potential wells can be calculated. The advantage of this model is the macroscopic character, because a description without knowledge of quantum mechanical details like initial wave function is obtained.

2. Simulation model

The Poisson equation

$$\nabla(\varepsilon_s\varepsilon_0\nabla(\phi)) = -q \cdot (n - p - N_D^+ + N_A^- + \rho_{ADD}), \qquad (1)$$

(N_D^+, N_A^- ionized donor and acceptor density, ρ_{ADD} additionally fixed charge),
the continuity equations

$$\nabla \cdot \mathbf{J}_p = -q \cdot \left(R - G + \frac{\partial p}{\partial t}\right), \qquad (2)$$

$$\nabla \cdot \mathbf{J}_n = q \cdot \left(R - G + \frac{\partial n}{\partial t}\right), \qquad (3)$$

and the transport equations

$$\mathbf{J}_p = -qp\mu_p \nabla(\phi - \lambda_p - \Theta_p) - D_p q \nabla(p) - k_B p \mu_p \nabla(T_p), \qquad (4)$$

$$\mathbf{J}_n = -qn\mu_n \nabla(\phi + \lambda_n + \Theta_n) + D_n q \nabla(n) + k_B n \mu_n \nabla(T_n), \qquad (5)$$

both for electrons and holes, are solved self-consistently in the Gummel algorithm to get the device characteristics at different bias conditions. R and G are the recombination and generation rate, Θ_p and Θ_n are so called band parameter for holes and electrons respectively, which make it possible to simulate heterostructures. In the drift gradient of the transport equations an accessory expression is included, which describes a quantum potential. Most conveniently the additional force term from the transport equation is personated as the quantum correction potential for electrons and holes

$$\lambda_p = -2 \cdot \frac{\gamma_p \cdot \hbar^2}{12 \cdot m_p \cdot q} \cdot \frac{\nabla^2 \sqrt{p}}{\sqrt{p}}, \qquad (6)$$

$$\lambda_n = 2 \cdot \frac{\gamma_n \cdot \hbar^2}{12 \cdot m_n \cdot q} \cdot \frac{\nabla^2 \sqrt{n}}{\sqrt{n}}. \qquad (7)$$

For a semiconductor with multiple conduction band minima and anisotropic effective mass the values for m_p and m_n are not clear. In this simulation model m_p and m_n are the constant effective masses and the problem of the anisotropic effective mass is handled by the fitting factors γ_p and γ_n. Non-equilibrium device phenomena, like short-channel and overshoot behavior are taken into account by the energy balance equation and the energy flux density (HD transport model), which are included as additional equations in the self-consistent Gummel-algorithm. The energy flux density equations

$$\mathbf{S}_p = -\kappa_p \cdot \nabla(T_p) - \frac{5}{2} \cdot \frac{k_B}{q} \cdot T_p \cdot \mathbf{J}_p + \frac{3}{2} \lambda_p \cdot \mathbf{J}_p, \qquad (8)$$

$$\mathbf{S}_n = -\kappa_n \cdot \nabla(T_n) + \frac{5}{2} \cdot \frac{k_B}{q} \cdot T_n \cdot \mathbf{J}_n + \frac{3}{2} \lambda_n \cdot \mathbf{J}_n, \qquad (9)$$

and the energy balance equations

$$\nabla \cdot \mathbf{S}_p = \mathbf{J}_p \cdot \mathbf{E}^* - \frac{3}{2}k_B p \frac{(T_p - T_L)}{\tau_{wp}} - \frac{3}{2}k_B \frac{\partial}{\partial t}(pT_p) - \frac{3}{2}k_B T_p(R-G) - \frac{1}{2}q\lambda_p\left(\frac{p}{\tau_{wp}} - (G-R)\right) - \frac{1}{2}q\frac{\partial}{\partial t}(p\lambda_p), \qquad (10)$$

$$\nabla \cdot \mathbf{S}_n = \mathbf{J}_n \cdot \mathbf{E}^* - \frac{3}{2}k_B n \frac{(T_n - T_L)}{\tau_{wn}} - \frac{3}{2}k_B \frac{\partial}{\partial t}(nT_n) - \frac{3}{2}k_B T_n(R-G) + \frac{1}{2}q\lambda_n\left(\frac{n}{\tau_{wn}} - (G-R)\right) + \frac{1}{2}q\frac{\partial}{\partial t}(n\lambda_n), \qquad (11)$$

are likewise extended to include quantum effects by the same quantum correction potential both for electrons and holes respectively. τ_{wp} and τ_{wn} are the energy relaxations times, which can be calculated by a modified Baccarani-Wordeman model. With these extensions it is possible to consider quantum mechanical effects, like resonant tunneling of electrons and holes through potential barriers and the accumulation of carriers in potential wells.

3. Results

The device structure of the resonant tunneling diode, which was used for the simulation, is represented in Fig. 1 together with the n-type doping densities. This structure was experimentally investigated in [2]. Fig. 2 shows the calculated current-voltage characteristic of a resonant tunneling diode by the QHD-model. Additionally the results of a simulation by a transfer matrix method (TM-method) and measurement data [2] are inserted. The calculated characteristics agree quantitatively at T = 300 K with the experimental values and show a negative differential resistance (NDR). The peak-to-valley current ratio (PVCR) for the QHD-model amounts 2.5, the TM-method results in a PVCR = 5.5 and the experimental results delivers a ratio of 3.0. Fig. 3 illustrates the electron density distribution at different voltages in the range of the NDR and shows charge build up in the quantum well. The electron temperature for the same voltages is represented in Fig. 4. A high electron temperature behind the barriers can be detected. In this range a strong increase of the temperature gradient can be indicated, which is included in the transport equations.

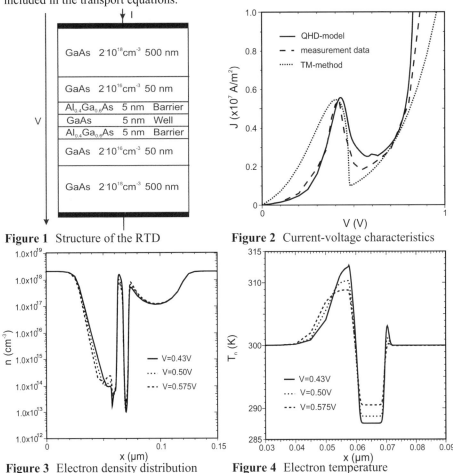

Figure 1 Structure of the RTD

Figure 2 Current-voltage characteristics

Figure 3 Electron density distribution

Figure 4 Electron temperature

Further investigations of some structure modifications have been carried out. Especially the thickness of the quantum well is varied between $l_w = 3$ nm to $l_w = 7$ nm. Fig. 5 shows the dependence of the first sub-band energy and the separation to the second sub-band energy on the thickness of the quantum well in the thermal equilibrium, which is calculated by a self-consistent solution of Schrödinger and Poisson equation. With a decrease of thickness an increase of the first sub band energy and a stronger separation to the second sub band energy occurs. This behavior results in a shift of the tunneling maximum to higher bias voltages and larger PVCRs. Fig. 6 illustrates the simulated current-voltage characteristics of a parallel connection of 3 RTDs with different thickness ($l_w = 3$ nm, $l_w = 4$ nm and $l_w = 7$ nm). Additionally the separated current-voltage characteristics of the three different RTDs and the sum of these characteristics are inserted. A good agreement between the local maximums of the separated characteristics and whose simulated results of the parallel connection can be detected. The differences between the sum of the characteristics and the simulated data of the real calculated quantum size device results from the corners at the interface of the different RTDs with varied thickness of the quantum well.

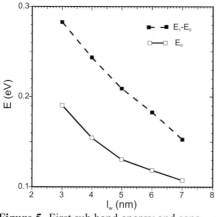

Figure 5 First sub band energy and separation to the second sub band energy as a function of the quantum well thickness

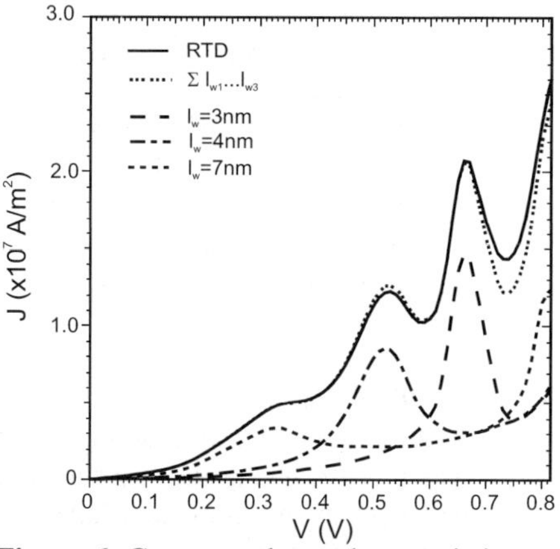

Figure 6 Current-voltage characteristics for a parallel connection of 3 RTDs with different quantum well thickness

4. Conclusions

Numerical 2D-simulations with a novel quantum hydrodynamic transport model of resonant tunneling structures have been carried out. For the calculations, a solution of a quantum correction potential is used, which is included in the hydrodynamic transport model. The simulated RTD structure shows a good agreement with measurement data. The structure variations exhibit a shift of the tunneling maximum to higher bias voltages at a decrease of the quantum well thickness. With the novel quantum hydrodynamic model it is possible to take into account quantum mechanical effects in all directions of a complex nanometer structure.

References

[1] C.L. Gardner: The Quantum Hydrodynamic Model for Semiconductor Devices. SIAM J. Appl. Math., vol. 54, no 2, pp. 409-427, (1994).

[2] A.M.P.J. Hendriks, W. Magnus, and T.G. van de Roer: Accurate Modeling of the Accumulation Region of a Double Barrier Resonant Tunneling Diode. Solid-State Electronics, 39, pp. 703-712, (1996).

Correlation between channel temperature and negative resistance in AlGaN/GaN HEMTs

Naoteru Shigekawa and Kenji Shiojima

Photonics Laboratories, NTT Corporation, 3-1 Morinosato-Wakamiya, Atsugi, Kanagawa, 243-0198 Japan

Abstract. We investigated the relationship between the channel temperature and the negative resistance for AlGaN/GaN HEMTs by means of μ-PL spectroscopy. We found that the negative-resistance characteristics correlate with the channel temperature, irrespective of gate-bias voltages and gate widths of devices. We also analysed the relationship between the channel temperature and the effective gate-bias voltage, and found that the negative resistance is assumed to be due to a decrease in the electron velocity.

1. Introduction

AlGaN/GaN high-electron-mobility transistors (HEMTs) have been intensively investigated as promising devices for handling high-frequency and high-power signals, and their excellent performances have been realised [1]. However, several drawbacks related to self-heating effects, such as negative resistance, or negative slope in drain-current-to-drain-bias-voltage (I_D-V_{DS}) characteristics [2], and the variation of I_D due to change in pulse width or duty cycle of bias-voltage pulses [3] have emerged.

For fully understanding the self-heating effects, we need to know the channel temperature, or lattice temperature of devices in operation. Actually its bias-voltage dependence and spatial gradient were examined by analysing micro-Raman [4] or micro-photoluminescence (μ-PL) [5] spectra of AlGaN/GaN HEMTs. In this paper, we discuss the relationship between the channel temperature and the device characteristics of AlGaN/GaN HEMTs fabricated on sapphire substrates, with emphasis on their negative resistance.

2. Experiment

2.1. Sample characteristics and setup of measurement

AlGaN/GaN HEMTs with 2-μm-long 10- or 50-μm-wide gates, which had been fabricated on a 430-μm-thick sapphire substrate, were employed. Their typical I_D-V_{DS} characteristics are shown in Fig. 1. Their threshold voltage V_{th} and maximum transconductance were typically –6 V and 80 mS/mm, respectively. In the characteristics of the 50-μm-wide HEMT, the negative resistance is definitely seen, and it is more pronounced at larger (smaller in absolute value) gate-bias voltage V_{GS}. In the characteristics of the 10-μm-wide HEMT, in contrast, the negative resistance is not clearly observed.

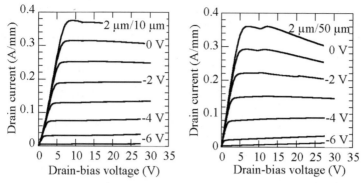

Figure 1. Typical I_D-V_{DS} characteristics of employed HEMTs with 10- and 50-μm-wide gates.

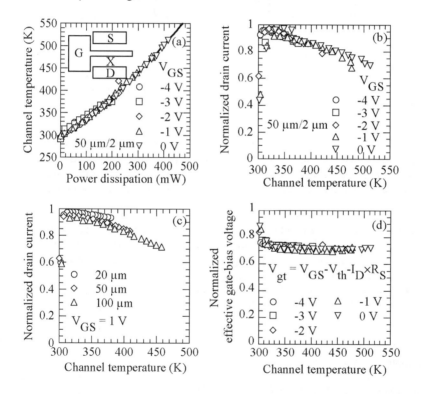

Figure 2. (a) Relationships between the channel temperature at the centre of the gate-to-drain opening in a 50-μm-wide AlGaN/GaN HEMT and P_{dis} at different V_{GS}'s. The position of measurements and a curve visualising the relationships are also shown. (b) Relationships between $\hat{I}_D$ and the channel temperature in a 50-μm-wide HEMT for different V_{GS}'s. (c) Relationships between $\hat{I}_D$ and the channel temperature for HEMTs with different gate widths. (d) Relationships between $\hat{V}_{gt}$ and the channel temperature in a 50-μm-wide HEMT at different V_{GS}'s.

Using the previously reported setup [5], we measured μ-PL spectra at the centre of gate-to-drain opening [For position, see the inset of Fig. 2(a).] in DC-biased AlGaN/GaN HEMTs. The highest V_{DS} value was 34 V, and V_{GS} was varied between -4 and 0 V. We deduced the channel temperature from each spectrum with reference to the ambient-temperature dependence of the PL-peak energy, which had been measured without applying bias voltages [5]. We also measured I_D during the μ-PL measurement.

2.2. Channel temperature and negative resistance

The channel temperature is shown in Fig. 2(a) as a function of the power dissipation P_{dis} ($\equiv I_D \times V_{DS}$). It is found that the channel temperature monotonically rises when P_{dis} increases and all the data lie on a single curve, regardless of V_{GS} values.

We divided I_D by its zero-drain-bias-voltage limit for each V_{GS} and defined the normalised drain current $\hat{I}_D$. ($\hat{I}_D \equiv I_D / I_D(V_{DS} \to 0 \text{ V})$.) Note that $\hat{I}_D$ is a measure of the negative-resistance characteristics. The relationship between $\hat{I}_D$ and the channel temperature is shown in Fig. 2(b). For each V_{GS}, $\hat{I}_D$ decreases when the channel temperature rises beyond 310 or 330 K. For channel temperatures with a decrease of $\hat{I}_D$, the variation in $\hat{I}_D$ due to V_{GS} change is negligibly small.

We also measured the channel temperature of 20-, 50-, and 100-μm-wide HEMTs fabricated on another wafer at $V_{GS} = 1$ V. Their gate length was commonly 2 μm. The relationship between $\hat{I}_D$ and the channel temperature is shown in Fig. 2(c). We found that $\hat{I}_D$ is slightly larger in narrower devices. The observed variation in $\hat{I}_D$ is, however, as small as ≈ 8% at maximum.

2.3. Effective gate-bias voltage

We measured the ambient-temperature dependencies of the contact and sheet resistances. By combining these ambient-temperature dependencies and the bias-voltage dependence of the channel temperature at the centre of the source-to-gate opening (not depicted), we estimated the source resistance R_S and effective gate-bias voltage V_{gt}, which is defined as

$$V_{gt} \equiv V_{GS} - V_{th} - I_D \times R_S.$$

We employed a constant for V_{th} in this estimation because we had observed no significant variation in V_{th} when the ambient temperature was raised up to 490 K. For each V_{GS}, furthermore, we divided V_{gt} by its zero-drain-bias-voltage limit and obtained $\hat{V}_{gt}$, or normalised effective gate-bias voltage.

The relation between the channel temperature and $\hat{V}_{gt}$ is shown in Fig. 2(d). In contrast to the behaviour of $\hat{I}_D$, $\hat{V}_{gt}$ remains almost unchanged for channel temperatures higher than 330 K.

3. Discussion

The relationships between the channel temperature and P_{dis} for different V_{GS} values are almost the same with one another. Since the heat-loss profile inside the channel is sensitive to V_{GS}, this finding means that the channel temperature is not influenced by the heat-loss profile and is likely to be a good measure of the self-heating.

It is notable that the relationships between $\hat{I}_D$ and the channel temperature for different V_{GS} values and device widths quantitatively agree with one another. This means that the negative-resistance characteristics are uniquely determined by the channel temperature. This result, more importantly, suggests that a "universal model" of the negative-resistance characteristics, or a model applicable for describing the negative-resistance characteristics of the HEMTs irrespective of their bias voltages and device geometry, can be established by further analysing their relationship to the channel temperature.

The electron concentration in the channel is likely to be proportional to V_{gt}, on the assumption that the gate-source capacitance is independent of the bias voltage and the channel temperature. [We had found in a preparatory study that the capacitance slightly increases when the ambient temperature rises from the room temperature to 200 °C. The increase was less than 10%.] Consequently, the difference between the channel-temperature dependencies of $\hat{I}_D$ and of $\hat{V}_{gt}$ suggests that the negative resistance is mainly due to a decrease in the electron velocity caused by rise in the channel temperature, not to the reduction of the electron concentration in the channel.

4. Conclusion

We investigated the relationship between the channel temperature at the centre of the gate-to-drain opening and the negative-resistance characteristics of AlGaN/GaN HEMTs by means of μ-PL spectroscopy. We found that the channel-temperature-to-power-dissipation characteristics are on a single curve, irrespective of the gate-bias voltages. We also found that the normalised drain current definitely correlates with the channel temperature. These results suggest that the negative-resistance characteristics can be modelled by proceeding with the analysis of their relationship to the channel temperature. Furthermore, we examined the channel-temperature dependence of the effective gate-bias voltage, and found that the negative resistance is assumed to be due to a decrease in the electron velocity.

Acknowledgement

The authors are grateful to Tomofumi Furuta and Suehiro Sugitani for their advice in μ-PL and gate-source capacitance measurements and to Takatomo Enoki and Masahiro Muraguchi for their encouragement.

References

[1] See, for example, Keller S Wu Y-F Parish G Ziang N Xu J J Keller B P DenBaars S P and Mishra U K 2001 IEEE Trans. Electron Devices, 48 552-559
[2] Gaska R Osinsky A Yang J W and Shur M S 1998 IEEE Electron Device Lett. 19 89-91
[3] Nuttinck S Gebara E Laskar J and Harris H M 2001 IEEE Trans. Microwave Theory Tech. 49 2413-2420
[4] Kuball M Hayes J M Uren M J Martin T Birbeck J C H Balmer R S and Hughes B T 2002 IEEE Electron Device Lett. 23 7-9
[5] Shigekawa N Shiojima K and Suemitsu T 2002 J. Appl. Phys. 92 531-535

Effect of temperature on the avalanche properties of sub-micron structures

C N Harrison, J P R David, C Groves, M Hopkinson and G J Rees

Department of Electrical and Electronic Engineering, University of Sheffield, S1 3JD, UK

Abstract. Photomultiplication measured in a series of $Al_{0.6}Ga_{0.4}As$ p^+-i-n^+ diodes is observed to fall with decreasing temperature. The effect is reduced with decreasing i-region thickness and increasing electric field. It is argued that this results from reduced phonon scattering in the thinner avalanche regions.

1. Introduction

For successful design of high power HBTs and HFETs the avalanche multiplication process must be understood and quantified accurately. This is conventionally done in terms of the impact ionization coefficients for electrons, α, and for holes, β, which represent the inverse mean distance between ionization events. In high field regions shorter than ~1μm the dead space, the distance over which α and β reach equilibrium with the local electric field, becomes an important factor in avalanche multiplication [1,2].

The ionization process is also known to depend strongly on temperature. The phonon population and the associated carrier-phonon scattering rate increase with temperature, cooling the electron (hole) gas heated in the electric field and suppressing impact ionization and multiplication. To date investigations of the temperature dependence of impact ionization have been performed only in bulk structures and the effects of temperature on dead space have not been studied [3-5]. We report measurements of the effect of temperature on avalanche multiplication in both thick and thin p^+-i-n^+ diodes where dead space effects become important.

2. Experimental Technique

A series of $Al_{0.6}Ga_{0.4}As$ p^+-i-n^+ structures with nominal i-region thicknesses varying from $w = 0.05$ to 0.8μm were grown epitaxially by MBE. $Al_{0.6}Ga_{0.4}As$ was chosen because of its large energy gap and resulting low dark current, I_{dark}, and the fact that α/β is close to unity, simplifying interpretation of the results. Circular mesa diodes with diameters ranging from 400 to 100μm and with annular top contacts for optical access were fabricated using chemical wet etching. Both dark current and capacitance-voltage measurements were made and modelling of the latter was performed to improve our estimates of the i-region thicknesses. The results were supported by SIMS measurements on a selection of samples.

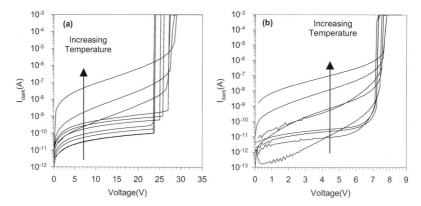

Figure 1. Variation of dark current with bias at temperatures ranging from 13K to 450K, a) for a $w = 0.51\mu m$ device, b) for a $w = 0.046\mu m$ device.

Measurements of pure electron initiated photomultiplication, M_e, versus bias were made at various temperatures and optical intensities, using optical excitation at 457nm. The optical excitation was chopped mechanically and the resulting ac signal was measured using a lock-in amplifier to distinguish the ac photocurrent from dc leakage. When the dark current was three orders of magnitude smaller than the photocurrent the latter was measured directly. The slight linear increase with bias in primary photocurrent before the onset of photomultiplication is due to the improved collection efficiency associated with the widening depletion layer and was corrected for using the method described by Woods et al. [6]. Assuming that α and β are equal and that the electric field in the multiplication region is uniform the effective ionisation coefficients can be determined from measurements of M_e alone using a local model [7], which ignores dead space effects, to give,

$$\alpha (=\beta) = \frac{M_e - 1}{wM_e} \qquad (1)$$

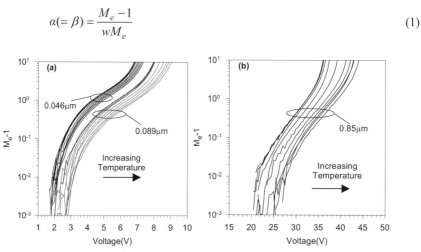

Figure 2. Dependence of multiplication curves on temperature in the range 13K to 450K for a) $w = 0.046\mu m$ and $0.089\mu m$ devices and b) a $w = 0.85\mu m$ device.

3. Experimental results

Fig.1(a) shows measurements of dark current for the $w = 0.51\mu m$ device at temperatures in the range T = 13K-450K. Very low dark currents, typically < 0.9nA close to breakdown, are measured below 200K but begin to increase dramatically at higher temperatures. By contrast fig.1(b) shows the softer breakdown and higher dark currents in the $w = 0.046\mu m$ device at all temperatures. Increasing temperature reduces multiplication in both thick and thin structures, and so increases the breakdown voltage, V_{bd} (fig. 2). However, it can be seen that this temperature dependence is significantly reduced for the thinner devices (fig.2a). Fig. 3(a) also shows that, in addition to the increase in V_{bd} with the avalanche width, its temperature coefficient also increases. Fig. 3(b) shows that the effective ionization coefficients in the $w = 0.85\mu m$ structure fall with increasing temperature, as expected, and converge to values common to all device thicknesses at high fields. However, in the thinner structures and at lower fields, where dead space effects are more significant, multiplication is suppressed and the values of effective ionization coefficient lie below those obtained from the thicker devices, as seen elsewhere [8]. Increasing temperature from 13K to 450K produces a greater effect on the effective ionization coefficients in the thicker devices, particularly at low fields.

4. Discussion

The dependence of phonon population on temperature is given by

$$n = \left[\exp\left(\frac{\hbar\omega}{kT}\right) - 1 \right]^{-1} \qquad (2)$$

where $\hbar\omega$ is the phonon energy and k is Boltzmann's constant. Increasing temperature increases the phonon population and hence carrier-phonon scattering, cooling the carrier gas and leading to the observed decrease in ionization coefficients in bulk devices. Thinner devices achieve the same multiplication at increased electric fields and carriers scatter off fewer phonons before ionizing over shorter path lengths, reducing temperature sensitivity.

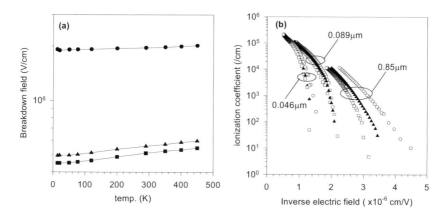

Figure 3 a) Breakdown field of $w = 0.046\mu m$ (●), $0.51\mu m$ (▲), and $0.85\mu m$ (■) devices as a function of temperature, b) ionization coefficients at temperatures 13K (○), 290K (▲) and 450K (□) for these three devices thicknesses.

Monte Carlo simulations [9] show that at high fields the dead space can be estimated approximately as the distance, d travelled by a carrier in the electric field, ξ before acquiring the ionization threshold energy, E_{th}, given by

$$d = \frac{E_{th}}{q\xi}, \qquad (3)$$

where q is the electron charge. This simple approximation holds because, although the carrier may scatter from phonons many times before covering this distance, the mean energy exchanged with the lattice is relatively small since $\hbar\omega \ll E_{th}$. Changing the temperature therefore has relatively little effect on this dead space, which represents a significant fraction of the ionization mean free path in short structures. It is therefore not surprising that avalanche multiplication is less sensitive to temperature in thinner devices.

5. Conclusions

Multiplication characteristics and avalanche breakdown voltages which are temperature sensitive in long multiplication regions show reduced sensitivity in short multiplication regions. This is because in shorter structures the ionization mean free paths are necessarily reduced at the higher operating fields needed to maintain the same level of multiplication. The number of carrier-phonon scattering events prior to ionization is correspondingly reduced, leading to a reduction in temperature sensitivity.

6. Acknowledgements

The authors wish to thank B. K. Ng and C. H. Tan for useful discussions. This work was funded by EPSRC (UK).

7. References

[1] Plimmer S A, David J P R, Herbert D C, Lee T W, Rees G J, Houston P A, Grey R, Robson P N, Higgs A W and Wight D R 1996 IEEE Trans. Elec. Dev. 43 1066-1072

[2] Plimmer S A, David J P R, Grey R, Rees G J 2000 IEEE Trans. Elec. Dev. 47 1089-1097

[3] Zheng X G, Yuan P, Sun X, Kinsey G S, Holmes A L, Streetman B G and Campbell J C 2000 IEEE J. Quant. Elect. 36 1168-1173

[4] Osaka F and Mikawa T 1985 J. Appl. Phys. 58 4426-4430

[5] Brozel M R and Stillman G E IEE Emis Datareviews series 16 ch. 4.9D

[6] Woods M H, Johnson W C and Lampert M A 1973 Solid-State Electron. 16 381-394

[7] Stillman G E and Wolfe C M 1977 Semiconductors and Semimetals (New York: Academic) 12 291-393

[8] Tan C H, David J P R, Plimmer S A, Rees G J, Tozer R C and Grey R 2001 IEEE Trans. Elect. Dev. 48 1310-1317

[9] Ong D S, Li K F, Rees G J, Dunn G M, David J P R and Robson P N 1998 IEEE Trans. Elect. Dev. 45 1804-1810

$In_{0.53}Ga_{0.47}As$ ionization coefficients deduced from photomultiplication measurements

J S Ng, M C Yee, J P R David, P A Houston G J Rees and G Hill
University of Sheffield, UK

Abstract. The ionization coefficients in $In_{0.53}Ga_{0.47}As$ are deduced from phase-sensitive photomultiplication measurements to give more accurate interpretation of measurement data. These results confirm the low field impact ionization observed in electrons and the conventional field dependence of holes but show a larger α/β ratio than previously reported. Low temperature results show decreasing breakdown voltages in $In_{0.53}Ga_{0.47}As$ with decreasing temperature, an observation in disagreement with previous HBT results.

1. Introduction

Impact ionization in $In_{0.53}Ga_{0.47}As$ has been linked to device breakdown of heterojunction bipolar transistors (HBTs) with $In_{0.53}Ga_{0.47}As$ collector and high electron mobility transistors (HEMTs) with $In_{0.53}Ga_{0.47}As$ channel region. While breakdown of a common-base configured HBT is determined by avalanche breakdown of the collector, most breakdown conditions of these devices require not avalanche breakdown of the $In_{0.53}Ga_{0.47}As$ layers, but some current multiplication caused by $In_{0.53}Ga_{0.47}As$ impact ionization. The ability to predict breakdowns in devices therefore relies on the accuracy of the $In_{0.53}Ga_{0.47}As$ ionization coefficients used to calculate multiplication factors.

Recent measurements of the electron and the hole ionization coefficients, α and β, of $In_{0.53}Ga_{0.47}As$ HBTs [1-4] cover a wider electric field range than the earlier results obtained from photomultiplication measurements [5-7] and show a significant low field α as well as the fact that it has a negative temperature dependence. However these HBT measurements cannot measure α and β independently and so their interpretations relied on the simplifying assumption that $\alpha = \beta$. Furthermore the HBT measurements employed DC technique so variations in current due to leakage current mechanisms cannot be differentiated easily from the impact ionization induced current. Therefore measurements of $In_{0.53}Ga_{0.47}As$ ionization coefficients using unambiguous multiplication factors from a series of layers are desirable.

In this work photomultiplication measurements were performed on a series of $In_{0.53}Ga_{0.47}As$ p-i-n layers employing phase-sensitive technique, which differentiates photocurrent from dark current, to improve the accuracy of ionization coefficients. Also the results were interpreted without assuming $\alpha = \beta$. The temperature dependence of the multiplication characteristics and breakdown voltage was also studied.

2. Structure details

The $In_{0.53}Ga_{0.47}As$ layers measured in this work consists of three heterojunction p-i-n diodes grown by MOVPE on n^+ InP substrates. Each layer has an $In_{0.53}Ga_{0.47}As$ i-region, of thickness w, with $0.5\mu m$ thick p^+ and n^+ InP cladding layers. Fitting measured capacitance-voltage characteristics of the layers using abrupt 3-region doping profile gives estimated w of $1.8\mu m$, $3.2\mu m$ and $4.8\mu m$.

3. Phase-sensitive photomultiplication measurements

The multiplication factor, M, was obtained by measuring multiplied photocurrent, $I(V)$, as a function of reverse bias, V. To facilitate phase-sensitive measurements, the light injected into the devices was chopped mechanically and a lock-in amplifier was used to detect the resultant ac photocurrent. The measurements are insensitive to the leakage current, consequently large multiplication values could be obtained, allowing an accurate determination of the breakdown voltage.

The corrections for multiplication factors of most photomultiplication measurements are due to diffusion of carriers created in undepleted cladding regions [8]. In our measurements by creating carriers only in the avalanche regions we achieved carrier injection conditions which did not require corrections for multiplication factors, thereby reducing the ambiguity of the data. Light with $\lambda = 1064nm$ was used to illuminate the devices from top and back to obtain mixed carrier multiplication factors, M_{mix1} and M_{mix2}. Since the InP claddings and substrate of the devices are transparent to the 1064nm light, the light was only absorbed in the i-$In_{0.53}Ga_{0.47}As$ region. Hence the measurements will not require correction and $M_{mix1}(V)$ as well as $M_{mix2}(V)$ are given by $I(V)/I(0)$. The breakdown voltages of these devices are in fact slightly higher than those of GaAs p-i-n diodes with corresponding w.

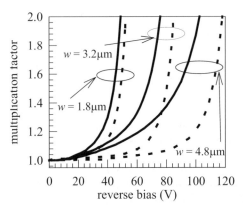

Figure 1. Experimental M_{mix1} (solid lines) and M_{mix2} (dotted lines) of the three layers.

4. Ionization coefficients

Calculations of ionization coefficients using measured multiplication factors are explained as follows. The carrier generation rate due to light absorption is $G(x) \propto \exp(\pm\gamma x)$ for M_{mix2} and M_{mix1}, where $\gamma = 2\times 10^4 cm^{-1}$ [9] is the absorption coefficient at 1064nm in $In_{0.53}Ga_{0.47}As$. The multiplication factor for an electron-hole pair injected at position x, $M(x)$ is a function of α, β and w. Integrating $M(x)$ over the multiplication region weighted by $G(x)$ gives M_{mix1} and M_{mix2}. The values of α and β were adjusted until the calculated values of M_{mix1} and M_{mix2} agreed with the experimental data within a tolerance of 10^{-4}.

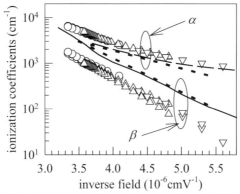

Figure 2. Comparison of the α (upper set) and β (lower set) of this work ($w = 1.8\mu m$ (O), $3.2\mu m$ ($\triangle$) and $4.8\mu m$ (∇)) with the results of Urquhart et al. (short dashed lines), Ritter et al. (long dashed line) and Buttari et al. (solid line).

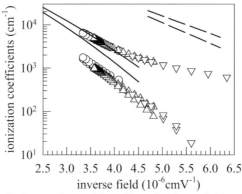

Figure 3. Comparison of $In_{0.53}Ga_{0.47}As$ data of this work (symbols) with the data of GaAs (solid lines) and Ge (dashed lines).

The α and β deduced from the photomultiplication measurements give good agreement within their common field range, as shown in Fig. 2 together with recently published data. These results show that even at high electric fields, α/β ratio is significantly higher than previously assumed for $In_{0.53}Ga_{0.47}As$. Our results of α, exhibiting low field impact ionization, are similar to those of Ritter et al. [1]. The values of β in this work is however lower than all previous results. The $In_{0.53}Ga_{0.47}As$ ionization coefficients are also found to be similar in magnitude to those of GaAs but significantly lower than those of Ge, which has a similar band-gap as shown in Fig. 3. This explains the slightly higher breakdown voltages observed in $In_{0.53}Ga_{0.47}As$ than those in GaAs.

5. Temperature dependence

Figure 4 shows the multiplication characteristics in the $1.8\mu m$ and $4.8\mu m$ structures as a function of temperature. Clearly, the decreasing temperature shifts the multiplication curves to lower voltages, suggesting a positive temperature coefficient of breakdown voltage unlike those reported recently. The magnitude of the change however is much smaller than has been reported in other materials such as GaAs or InP and may account for the discrepancies in the reported data.

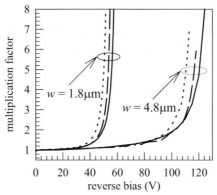

Figure 4. Multiplication factors of the $w = 1.8\mu m$ and $4.8\mu m$ structures at $T = 20K$ (dotted lines), 140K (dashed lines) and 300K (solid lines).

6. Conclusions

$In_{0.53}Ga_{0.47}As$ ionization coefficients have been measured and the results provide a larger α/β ratio than previously reported. Low field impact ionization was observed in α but not in β, consistent with the HBT results. Contrary to the negative temperature coefficient measured using HBTs, our low temperature measurement shows that breakdown voltages in $In_{0.53}Ga_{0.47}As$ increase with temperature. The small magnitude of change in breakdown voltage may account for the discrepancies in the reported data.

References

[1] Ritter D, Hamm R A, Feygenson A and Panish M B 1992 Appl. Phys. Lett. 60 3150-3152
[2] Canali C, Forzan C, Neviani A, Vendrame L, Zanoni E, Hamm R A, Malik R J, Capasso F, Chandrasekhar S 1995 Appl. Phys. Lett. 66 1095-1097
[3] Shamir N and Ritter D 2000 IEEE Electron Device Lett. 21 509-511
[4] Buttari D, Chini A, Meneghesso, Zanoni E, Sawdai D, Pavlidis D and Hsu S S H 2001 IEEE Electron Device Lett. 22 197-199
[5] Pearsall T P 1980 Appl. Phys. Lett. 36 218-210
[6] Osaka F, Mikawa T, Kaneda T 1985 IEEE J. Quan. Electron. QE-21 1326-1338
[7] Urquhart J, Robbins D J, Taylor R I and Moseley A J 1990 Semicond. Sci. Technol. 789-791
[8] Woods M H, Johnson W C and Lampert M A 1973 Solid State Electron. 16 381-394
[9] Bacher F R, Blakemore J S, Ebner J T and Arthur J R 1988 Phys. Rev. B 37 2551-2557

InGaP/InGaAs/GaAs Double Channel Pseudomorphic High Electron Mobility Transistor

H. M. Chuang, K. H. Yu, K. W. Lin[1], C. C. Cheng, J. Y. Chen, and W. C. Liu*

Institute of Microelectronics, Department of Electrical Engineering, National Cheng-Kung University, 1 University Road, Tainan, Taiwan 70101, Republic of China

[1]Department of Electrical Engineering, Chien Kuo Institute of Technology, Changhua, Taiwan 50045, Republic of China

*Corresponding author: E-Mail: wcliu@mail.ncku.edu.tw

Abstract. The device characteristics of a novel InGaP/InGaAs/GaAs double channel pseudomorphic high electron mobility transistor (PHEMT) are reported. The double InGaAs layers are used to increase the total channel thickness. Experimentally, a flat and wide transconductance and microwave operation regimes over 300 mA/mm are obtained. In addition, the compression of transconductance and frequency response are insignificant even operated at high forward gate-source voltage of +2.0V. Therefore, the studied device provides the promise for microwave circuit applications.

1. Introduction

Field-effect transistors (FETs) have been studied extensively for high-speed microwave and digital circuit applications [1, 2]. In addition, in GaAs-based heterostructure FETs, due to the higher mobility, higher peak electron velocity, and lower effective mass of InGaAs material, it is favorable to use an InGaAs to replace the GaAs as a channel layer [3, 4]. However, the lattice mismatch between the InGaAs and GaAs limits the use of thicker and higher In mole fraction of InGaAs layer in the device fabrication [5]. This restricts the device characteristics for high-performance applications. To overcome this disadvantage and improve the device performances, a new InGaP/InGaAs double channel pseudomorphic high electron mobility transistor (PHEMT) is presented in this work. The use of InGaAs double channel structure can effectively increase the total thickness of channel layer without decreasing the In mole fraction. In addition, the employed top, middle, and bottom triple δ-doped sheets as carrier suppliers cause the uniform distribution of carriers in InGaAs double channel layers [6]. Therefore, device with good linearity in transconductance and frequency behaviors can be obtained.

2. Experimental

The epitaxial structure of the studied device was grown on a (100) oriented semi-insulating (S.I.) GaAs substrate by a metal-organic chemical vapor deposition

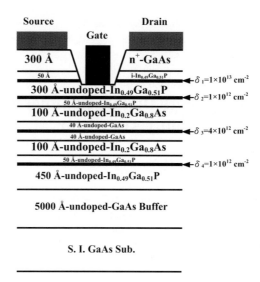

Figure 1. The schematic cross section of the studied device

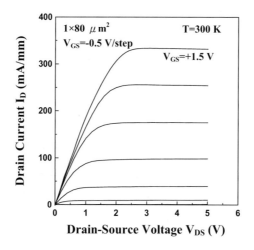

Figure 2. The common source I-V characteristics at room temperatures

(MOCVD) system. The schematic cross-section of the studied device is shown in Fig. 1. It consisted of a 0.5- μm GaAs buffer, a 450-Å $In_{0.49}Ga_{0.51}P$ buffer, a delta-doped sheet δ_4 $(n^+)=1 \times 10^{12}$ cm^{-2}, a 50-Å $In_{0.49}Ga_{0.51}P$ spacer, a 100-Å $In_{0.2}Ga_{0.8}As$ channel, a 40-Å GaAs spacer, a delta-doped sheet δ_3 $(n^+) = 4 \times 10^{12}$ cm^{-2}, a 40-Å GaAs spacer, a 100-Å $In_{0.2}Ga_{0.8}As$ channel, a 50-Å $In_{0.49}Ga_{0.51}P$ spacer, a delta-doped sheet $\delta_2(n^+)=1 \times 10^{12}$ cm^{-2}, a 300-Å $In_{0.49}Ga_{0.51}P$ gate "insulator", a delta-doped sheet δ_1 $(n^+)=1 \times 10^{13}$ cm^{-2}, a 50-Å $In_{0.49}Ga_{0.51}P$, and a 300-Å n^+-GaAs $(n^+=4 \times 10^{18}$ cm$^{-3})$ cap layers. The δ_1 (n^+) is used to reduce the source and drain Ohmic contact resistance and δ_2 (n^+), δ_3 (n^+), and δ_4 (n^+) are used for carrier supplier layers. After epitaxial growth, wet chemical etching, conventional vacuum evaporation, and lift-off techniques were used to fabricate mesa-type devices. Drain-source Ohmic contacts were formed on the n^+-GaAs cap layer by alloying evaporated AuGe/Ni/Au metals at 250 ℃ for 10s. The wet chemical etching process was used for device isolation. Finally, n^+-GaAs, $In_{0.49}Ga_{0.51}P$, and δ_1 (n^+) layers are removed and gate Schottky contacts with gate length of 1 μm were achieved by evaporating Au metals on the $In_{0.49}Ga_{0.51}P$ "insulator" layer.

3. Results and Discussion

For two-terminal gate-drain current-voltage (I-V) characteristics, the forward turn-on voltage V_{on}, defined at a gate leakage current of 1 mA/mm, is 1.46 V. The reverse gate leakage current I_G is only of 60 μm/mm at V_{GD}=-15 V. The high V_{on} increases the forward biased operation regimes and low I_G is suitable for power applications. Figure 2 shows the common-source I-V characteristics of the studied device at room temperature. The maximum applied gate-source voltage V_{GS} is +1.5 V. With the good carrier confinement in the InGaAs double channel layers, the studied device shows good pinch-off and saturation characteristics. The measured threshold voltages V_{th} are -1.51 V. The drain saturation current I_{DS} and transconductance g_m versus gate-source voltage V_{GS} at V_{DS}=3.5 V are shown in Fig. 3. The frequency responses of the studied device are shown in Fig. 4. Obviously, the transconductance and frequency responses of the studied device show good linearity behaviors. For a 1 μm gate length dimension, the maximum transconductance $g_{m,max}$ of 161 mS/mm and the maximum output current of 460 mA/mm with 320 mA/mm broad operation regime (>0.9 $g_{m,max}$) are achieved. The maximum unity current gain cut-off frequency f_T and maximum oscillation frequency f_{max} are 13 and 32 GHz, respectively, at V_{GS}=0.5 V and V_{DS}=3.5 V and maintains 90% of its f_T and f_{max} peak values over a large drain current range of 300 mA/mm. In addition, due to the high V_{on}, the compression of I_{DS} is insignificant and g_m, f_T, f_{max} still maintain high values of 135 mS/mm, 9.9 GHz, 28.4 GHz, respectively, even operated at high forward gate-source voltage of +2.0 V.

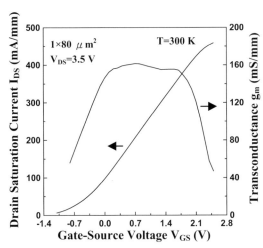

Figure 3. The drain saturation current I_{DS} and transconductance g_m as a function of the gate-source voltage V_{GS}

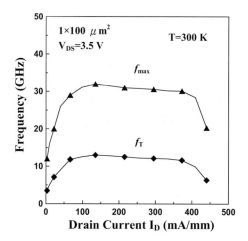

Figure 4. The measured frequency responses of the studied device

4. Conclusion

The device characteristics of a novel InGaP/InGaAs/GaAs double channel PHEMT are reported. The double InGaAs layers are used to increase the total channel thickness. Experimentally, a flat and wide transconductance and microwave operation regimes over 300 mA/mm are obtained. In addition, the compressions of transconductance and frequency response are insignificant even operated at high forward gate-source voltage of +2.0 V.

Acknowledgments - Part of this work was supported by the National Science Council of the Republic of China under Contract No. NSC-90-2215-E-006-021. The authors are also grateful to National Nano Device Laboratories (NDL) for RF measurements.

References

[1] Chertouk M, Dammann M, Kohler K and Weimann G 2000 IEEE Electron Device Lett. 21 97-99
[2] Tsai M K, Tan S W, Wu Y W, Lour W S and Yang Y J 2002 Semicond. Sci. Technol. 17 156-160
[3] Liu W C, Yu K H, Liu R C, Lin K W, Lin K P, Yen C H, Cheng C C and Thei K B 2001 IEEE Trans. Electron Devices 48 2677-2683
[4] Laih L W, Cheng S Y, Wang W C, Lin P H, Chen J Y, Liu W C and Lin W 1997 Electron. Lett. 33 98-99
[5] Matthews J W 1975 J. Vac. Sci. Technol. 12 126-133
[6] Hsu W C, Shieh H M, Wu C L and Wu T S 1994 IEEE Trans. Electron Devices 41 456-457

70-nm-gate PHEMT fabricated by a trilayer process of ZEP/P(MMA-MAA)/PMMA resist

S. C. Kim, B. O. Lim, H.S. Lee, S. K. Kim, H. C. Park, D-H Shin, and J. K. Rhee
Millimeter-wave Innovation Technology Research Center, Seoul, Korea

Abstract. We applied to a novel multilayer resist system that consists of commercially available positive resist, ZEP520 and P(MMA-MAA), for fabricating T-gate PHEMTs with gate lengths of 70 nm or less, and has manufactured a 70 nm T-gate using electron beam lithography. In this study, devices of 70 μm unit gate width and two gate fingers with a drain-source saturation current density of 357.14 mA/mm, extrinsic transconductance of 505.45 mS/mm, f_T of as high as 115 GHz and f_{max} of 175 GHz have been fabricated.

1. Introduction

For millimeter wave application, the T-shaped gates are required for the production of high performance HEMTs and MMICs to achieve very small gate length and hence low gate resistance [1]. Recent advances in electron-beam lithography have made possible the fabrication of PHEMTs with gate length well in the nanometer regime [2]. A number of multilayer resist processes have been proposed to fabricate T-gates [3].
In this paper, we reported a novel multilayer resist process with positive resist, ZEP520, P(MMA-MAA), and PMMA, for fabricating T-gate PHEMTs with gate lengths of 70 nm or less, because the ZEP 520 resists offers higher processing latitude than PMMA in both etch resistance and resist sensitivity without loss in resolution and the DC and RF characteristics for the 70 nm scale T-gate.

2. Fabrication process

We designed the epi structure of AlGaAs/InGaAs/GaAs with double heterojunction to manufacture PHEMT's. Devices were fabricated using 3500 Å mesa isolation using $H_2SO_4 : H_2O_2 : H_2O$(1 : 8 : 160) etchants and AuGe/Ni/Au(1150/280/1600 Å) ohmic metallization with a specific contact resistance of around $1 - 2\times10^{-7}$ $\Omega\cdot cm^2$. Ohmic alloying was performed by two-step rapid thermal annealing at 300 and 350 ^{0}C for 10 and 20 sec, respectively. 70 nm T-gate patterning was performed by electron beam lithography using a Leica Microsystems lithography electron pattern generator (EBPG-4HR). The ZEP520 : DCB(1 : 1.5)/P(MMA-MAA)/PMMA-950K : MCB(1 : 1) T-gate process is similar to the process for T-gate fabricating using PMMA/P(MMA-MAA)/PMMA [4]. Before achieving suitable resist profiles with an overhang structure for metal lift-off and a 70 nm T-gate with electron beam resists of a trilayer structure, a few dose tests and optimum resist thickness are required.

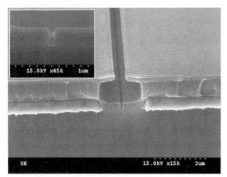

Fig. 1 A T-shaped resist profile .

Therefore, a thin resist thickness of 1000 Å, which will define the T-gate foot, by using a mixing liquid of ZEP520 and DCB(dichrobenzen)(1: 1.5) and a compatible coating speed of 2500 rpm were produced and we determined an appropriate dose of 550 μC/cm^2 by conducting exposure dose tests in the range from 500 to 700 μC/cm^2 in 50 μC/cm^2 increments. A methyl isobutyl ketone and isopropyl alcohol(MIBK : IPA = 1 : 3) solution was used to develop the patterned resist using both spray and immersion with the development of 4 min. We determined the optimum fabrication conditions with a development time of 4 min and a dose of 550 μC/cm^2 to obtain the 70 nm gate foot length. The top layers which have different sensitivity, will form a reliable overhang structure for metal lift-off. The linewidth of the top layers determines the cross-sectional width of the cap. Using the fact that the cross section of the head can easily be increased by an over-development process, we found an appropriate dose of 90 μC/cm^2. After exposure to a dose of 90 μC/cm^2, the top layer(PMMA : MCB = 1 : 1) was developed using MCB for 15 sec, and then the middle layer of P(MMA-MAA) was developed using a mixed solution of methanol and IPA(1:1) for 28 sec. Fig. 1 shows a patterned T-shape profile with a gate length of a 70 nm. After gate fabrication, the devices were fully passivated with a Si_3N_4 thickness of 800 Å deposited at low temperature (250 ^{0}C) using a plasma-enhanced chemical vapor deposition (PEVCD) system. Fig. 2a shows the cross section of T-gate with 70 nm gate length. However, in the case of T-gates with a 50 nm gate length, we obtained the collapse of the gate head as shown in Fig. 2b, because of the poor mechanical reliability in the narrow foot for such a heavy T-gate.

Fig. 2 SEM micrograph (a) showing 70 nm T-gate, (b) collapse of the gate head with a 50 nm gate length due to mechanical instability.

3. Device results

The DC and RF characteristics were evaluated by measuring the devices in a HP 4156A DC parameter analyzer and a HP 8510C vector network analyzer, respectively. The current-voltage characteristics of the 70 μm unit gate width with 2 gate fingers PHEMT's are given in Fig. 3. A good DC characteristic was achieved. We obtained a pinch-off property of $V_P = -1$ V and a drain-source saturation current (I_{dss}) of 50 mA. The maximum drain current density, defined as the saturation current density measured at a gate-to-source voltage(V_{gs}) of 0 V, is 357.14 mA/mm. The maximum extrinsic transconductance(g_m) is 505.45 mS/mm at $V_{ds} = 1.5$ V and $V_{gs} = -0.25$ V, as shown in Fig. 4.

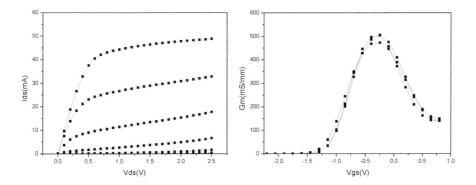

Fig. 3 I_{ds} vs. V_{ds} characteristics of PHEMT's

Fig. 4 Transconductance characteristics of PHEMT's

The RF measurements were performed in a frequency range of 1.0 ~ 50 GHz. The drain and gate voltages applied in the RF measurements were 1.5 and -0.25 V. From the measured S parameters, the small-signal equivalent circuit was extracted using a direct extraction technique and the corresponding f_T and f_{max} were estimated. Fig. 5 shows the extracted RF characteristics for a 70 nm T-gate PHEMT and shows a plot of the S_{21} gain, the current gain H_{21} and G_{ms} values at 50 GHz by an extrapolation of 6 dB/octave. At 50 GHz, a maximum stable gain(MSG) of 9.6 dB and a S_{21} gain 3.52 dB were obtained. A current gain cut-off frequency(f_T) of 115 GHz and a maximum frequency 175 GHz were achieved from the fabricated PHEMT's of 70 nm gate length.

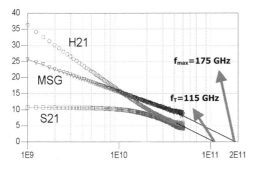

Fig. 5 RF characteristics of the fabricated PHEMT

4. Conclusions

A 70 nm foot print of the T-shaped gate is realized by using ZEP520/ P(MMA-MAA)/PMMA trilayer resists. The PHEMT's with 70 nm T-shaped Ti/Au gate are fabricated using this tri-layer process without a thin dielectric SiN_x supporting layer on the PHEMT's epi-layer. This PHEMT's show well-controlled profile and electrical properties.

Acknowledgement

This work was supported by KOSEF under ERC program through the Millimeter-wave INnovation Technology (MINT) research center at Dongguk University in Seoul, Korea.

References

[1] Matsumura M, Tsutsui K, and Naruke Y 1981 Electron Lett. 12 (12) 429
[2] Samoto N, Makino Y, Onda K, Mizuki E, and Itoh T 1990 J. Vac. Sci. Technol. B8 1335
[3] Ahmad M M and Ahmad H 1997 J. Vac. Tech. B15 no. 2, pp306-310
[4] Kim H S, Lim B O, Kim S C, Lee S D, Shin D-H, Rhee J K 2002 Microelectronic Engineering 63 417-431.

Studies on the Low-k BCB passivation of 0.1 μm gamma gate PHEMTs

Woo-Suk Sul, Hyo-Jong Han, Seong-Dae Lee, and Jin-Koo Rhee
Millimeter-wave INnovation Technology Research Center, Seoul, Korea

Abstract

In this paper, we compared Si_3N_4 passivation, which is mainly used in PHEMT, to BenzoCycloButene (BCB) passivation with various advantages. The DC and RF characteristics of PHEMT's passivated BCB are similar to those of PHEMT's passivated Si_3N_4. However, we obtained the results that the noise characteristics of the former were 3 times superior to those of the latter.

1. Introduction

Millimeter-waves are expected to be in high demand for wireless communication services because they can provide an extremely wide bandwidth with the frequency reuse characteristics. Therefore, the necessity of the development and the manufacturing of the high-speed device that can work in millimeter wave bandwidth are increased.
GaAs as III-V compound semiconductor have higher mobility and saturation velocity than Si, and GaAs based HEMT's has been proved to be the key active devices for various millimeter wave applications and high-speed circuits. However, GaAs-based PHEMT is very sensitive to the surface effect, since the channel layer for the electron transport can be strongly affected by the surface region exposed during the device fabrication. The devices of exposed surface can give rise to various problems such as the degradation in the electrical performance and physical damage through the oxidation, moisture, and the pollution by dust and chemical material. Hence, it is essential to perform the proper passivation process to ensure the device performances. In this paper, we fabricated PHEMT's passivated Si_3N_4 and PHEMT's passivated BCB and compared the noise, DC, and RF characteristics of the former with that of the latter. BCB has many advantages of the passivation of PHEMT's for various millimeter wave applications, i. e. relative dielectric constant (~ 2.7) regardless of the frequency, low dielectric loss tangent (~ 0.0008), low curing temperature, low moisture absorption, low cost, and simple manufacturing process [1]. The BCB can also alleviate the oxidation problem in Schottky barrier layer, such as AlGaAs layer in the PHEMT, due to its low oxygen impurity level.

2. Fabrication Technology

The PHEMT epitaxial structure was grown by molecular beam epitaxy (MBE) on a semi-insulating GaAs substrate and consists of the following layers: 1 μm undoped GaAs buffer, 120 Å undoped $In_{0.2}Ga_{0.8}As$ channel, 40 Å undoped $Al_{0.25}Ga_{0.75}As$ spacer layer, delta doping plane (5×10^{12} cm^{-2}), 250 Å undoped $Al_{0.25}Ga_{0.75}As$ Schottky contact layer, 300 Å n-type doped GaAs cap (5×10^{18} cm^{-3}). These epitaxial layers showed a 2 dimensional electron carrier density of 2.1×10^{12} cm^2 and Hall mobility of 6670 cm^2/V-s at room temperature. Devices were fabricated using 3500 Å mesa isolation by using the etchant of H$_2$SO$_4$/H$_2$O$_2$/H$_2$O (1:8:160) and AuGe/Ni/Au (1250/280/1600 Å) ohmic metallization with a specific contact resistance of around $1\sim2 \times 10^{-7}$ Ω·cm. Ohmic alloy was performed by the two-step rapid thermal annealing at 200 and 270 °C for 10 and 20 seconds, respectively. After ohmic metallization, we performed the gate recess by the wet-etching method using the NH$_4$OH/H$_2$O$_2$/H$_2$O solution (1:1:2000). Then, 0.1 μm offset Γ-gate patterning was done by electron beam lithography using a Leica Microsystems Ltd Electron Beam Pattern Generator (EBPG-4HR system) operating at an acceleration voltage of 50 keV, a beam size of 50 nm and a beam current of 1 nA. For Γ-shaped gate with reproducibility, we used tri-layer resist of PMMA 4%+MCB / P(MMA-MAA) / PMMA 4% and carried out double exposure, and then we evaporated gate metals of the Ti/Au (500/4500 Å). After gate fabrication, we fabricated the passivation using the two materials. The first one is fully passivated by 800 Å Si$_3$N$_4$ deposited at a RF plasma enhanced CVD (PECVD) system. The other one is fully passivated by curing for 10 min at 250 °C by the vacuum condition in RTP after coating the BCB with 3 μm thickness using the spin coater. The BCB thin film has the dielectric constant of 2.7 and the oxidation problem. A SEM micrograph of 0.1 μm offset Γ-gate after lift off and passivation with a Si$_3$N$_4$ film and BCB film are shown in Fig. 1 and Fig. 2, respectively.

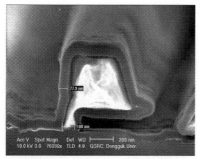
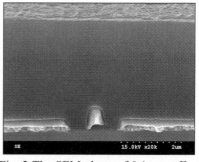

Fig. 1 The SEM photo of 0.1 μm offset Γ-gate by Si$_3$N$_4$ passivation

Fig. 2 The SEM photo of 0.1 μm offset Γ-gate by BCB passivation

3. Results and Discussion

We measured the fabricated PHEMT's with gate length of 0.1 μm, unit gate width of 70 μm and 2 gate fingers. The DC Characteristics of PHEMT's are as follows. A HP 4156A DC parameter analyzer examined the DC characteristics of fabricated PHEMT's. In the Si$_3$N$_4$ passivated PHEMT, we obtained a pinch-off property (V_P) is -1.3 V and the maximum drain current density, defined as the saturation current density measured at a gate-to-source voltage (V_{gs}) of 0 V, is 364.86 mA/mm, and those values are -1.3 V and 332.3 mA/mm for unpassivated device. The results increased about 9.8 %. The maximum extrinsic transconductance (g_m) is 333 mS/mm, and this value is 290 mS/mm for unpassivated device. The results increased about 14.6 % at $V_{ds} = 1$ V and $V_{gs} = -0.65$ V. The DC characteristics of the Si$_3$N$_4$ passivated device shown in Fig.3. In the BCB passivated PHEMT, we obtained a pinch-off property (V_P) is -1.3 V and the maximum drain current density, defined as the saturation current density measured at a gate-to-source voltage (V_{gs}) of 0 V, is 365.5 mA/mm, and those values are –1.3 V and 333.85 mA/mm for unpassivated device. The results increased about 9.5 %. The maximum extrinsic transconductance (g_m) is 318.12 mS/mm, and this value is 289.051 mS/mm for unpassivated device. The results increased about 9.6 % at V_{ds} = 1 V and V_{gs} = -0.65 V. The DC characteristics of the BCB passivated device shown in Fig. 4.

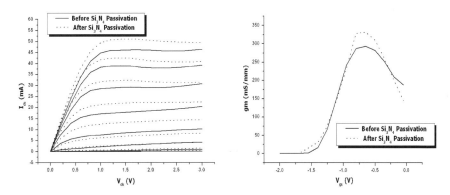

Fig. 3 The DC characteristics of Si_3N_4 passivated PHEMT's

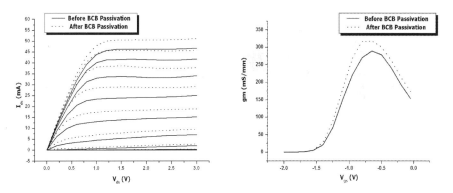

Fig. 4 The DC characteristics of BCB passivated PHEMT's

The RF measurements were performed in a frequency range of 0.5 GHz ~ 50 GHz by a HP 8510C network analyzer. In the Si_3N_4 passivated PHEMT, the S_{21} gain decreased about 3.17 dB and the H_{21} gain decreased about 5.44 dB at 50 GHz. The current gain cut-off frequency (f_T) is 76.56 GHz, and this value is 143.6 GHz for unpassivated device. The results decreased about 46.7 %. In the BCB passivated PHEMT, the S_{21} gain about 4.34 dB and the H_{21} gain decreased about 4.6 dB at 50 GHz. The current gain cut-off frequency (f_T) is 84.157 GHz, and this value is 142.58 GHz for unpassivated device. The results decreased about 40.9 %.

Finally, the noise measurements were performed in a frequency range of 52 ~ 62 GHz using a HP 8970B noise figure meter. The measured noise characteristics of the

PHEMT's with the BCB and the Si_3N_4 passivations are shown in Fig.5 (a) and (b), respectively. After the Si_3N_4 passivation, the PHEMT's showed a clear increase of ~ 2.6 dB in noise figure (NF) at 60 GHz. However, we observed almost no degradation in NF (~ 0.6 dB increase at 60 GHz after the passivation) in the case of BCB passivation. This effect of BCB passivation on the noise characteristics was not fully understood yet. However, it is supposed that the surface state between the passivation and the active region is less pronounced in the BCB film than in the Si_3N_4 because the BCB layers provides a simple fabrication process with lower film stress and less stress induced damages than the PECVD Si_3N_4 [3].

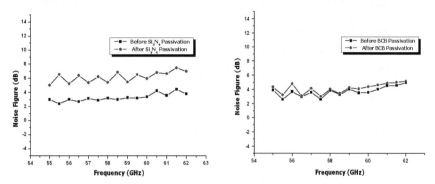

Fig. 5 The noise characteristics of Si_3N_4 passivation (a), BCB passivation (b)

4. Conclusions

We fabricated PHEMT's passivated Si_3N_4 and PHEMT's passivated BCB and compared the characteristic of the former with that of the latter. The characteristic of DC and RF of PHEMT's passivated BCB are similar to those of PHEMT's passivated Si_3N_4 and the similar result. However, we obtained the results that the noise characteristics of the former were 3 times superior to those of the latter. We consider the BCB passivation of PHEMT's has good characteristics and possible low noise application for millimeter-waves.

Acknowledgement
This work was supported by KOSEF under ERC program through the Millimeter-wave INnovation Technology (MINT) research center at Dongguk University in Seoul, Korea.

References
[1] P. Garrou, W. Rogers, D. Scheck, A. Standjord, Y. Ida, and K. Ohba, 1999, IEEE Trans. Advanced Packaging, vol. 22, no. 3
[2] R. Leoni, J. Bao, J. Bu, X. Du, M. Shirokov, and C. Hwang, 2000, IEEE Trans. Electron Devices, vol. 47, no. 3
[3] Hsien-Chin Chiu, Ming-Jyh Hwu, Shih-Cheng Yang, and Yi-Jen Chan, 2002, IEEE Electron Device Letters, Vol. 23, No. 5.

Design of Low Loss Transmission Lines on GaAs Substrates using the Surface Micromachining Methods

Young-Hoon Chun, Sung-Chan Kim, Byoung-Ok Lim, Han-Shin Lee,
Moon-Kyo Lee, Hae-Sung Kim, Dong-Hoon Shin, Soon-Koo Kim,
Hyun-Chang Park, Jin-Koo Rhee, *Sang-Won Yun

Millimeter-wave INnovation Technology Research Center (MINT),
Dongguk University, SEOUL, 100-715, KOREA
*Dept. of Electronic Engineering, Sogang University, SEOUL, 121-768, KOREA

Abstract. This paper describes a new GaAs-based micromachined microstrip line with elevated and overlaid signal line with ground metal. This new type of overlay microstrip line (OML) structure is developed using micromachining techniques to provide easy means of airbridge connection between the signal lines, as well as to achieve low losses over wide impedance ranges.

1. Introduction

In recent, the micromaching technology could improve the performance of MMIC's. Especially, in the mm-wave ranges, it is useful to make low-loss elements. As reported in [1]~[4], many authors made several attempts to get low loss transmission line, which could make low loss integrated passive circuits in microwave or mm-wave ranges.

This paper describes a new GaAs-based surface-micromachined microstrip line with elevated and overlaid signal line with ground metal. This new type of overlay microstrip line (OML) structure is developed using surface micromachining techniques to provide easy means of airbridge connection between the signal lines, as well as to achieve low losses over wide impedance ranges. The OML showed less than 2 dB/cm loss at 40 GHz over a wide impedance range from 20 to 85 Ω. It also offered low effective dielectric constant, and insensitivity to the substrate losses. Wide impedance ranges and simple process steps make OML a promising uniplanar transmission line medium for mm-wave monolithic applications. Additionally, comparing with other CPW line structures, this microstrip structure can have a flexibility to build some applications such as tight coupling elements, hybrids, baluns and transformers.

2. Design and Simulation

The low-loss transmission lines using MEMS technologies for monolithic application are reported several times, those are mostly fabricated with Si-substrates. Shielded microstrip lines, inverted microstrip lines, and elevated CPWs are those ones. These can be applicable to the essential components for wireless communications, such as resonators, matching networks, and baluns. Si-based MEMS chips on which passive components are fabricated would show good results at high frequency band. But, they should be interfaced with active

MMICs. At millimeter-wave frequency band, their interface problems become serious. Therefore, we need low-loss interfacing technologies for mm-wave applications.

In this paper, a new structure which is compatible with the conventional MMIC technologies is proposed. This compatibility will decrease the problems of interfaces and increase the degrees of freedom in RF front-end design. Using this result, we could integrate the passive MEMS components on the active GaAs MMIC, which can make the cost lower and the size smaller with good performances.

The schematic of the OML is shown in Fig. 1. The centered signal lines are elevated using surface micromachining techniques and overlapped with the finite ground plane. The signal line is over the finite ground metal to facilitate airbridge connections, making OML readily compatible with MMICs.

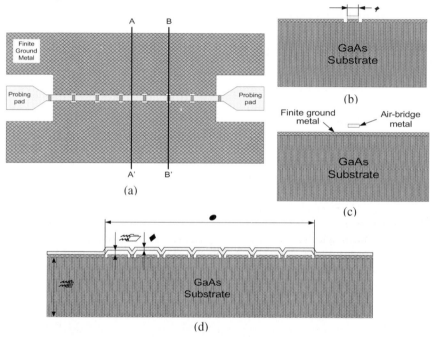

Fig. 1 Schematic diagram of OML (a), transverse cross sections; (b) A-A', (c) B-B', and longitudinal cross section (d).
(t = 1.0/10 μm, h1 = 3/10 μm, h2 = 600 μm, W = 10 ~ 100 μm)

Using FEM simulation tool, the full wave analyses are performed. The simulated results which show some variations as several parameters are changed are shown in Fig. 2. These results show that OML can be designed with low loss and low effective dielectric constant over wide impedance ranges above 30 GHz.

As in the fig. 2-(c), the higher the signal lines, the lower the losses. It is caused that the main loss factor of OML, when the height is quite low, is the dielectric loss which stems from the GaAs substrates. And, in the case of the h=10 μm, the conductor loss could be critical. Especially, as the characteristic impedance become higher, there is a wide difference between the insertion losses with different line width. So, in order to design a low loss OML, we should have high and thick elevated signal lines.

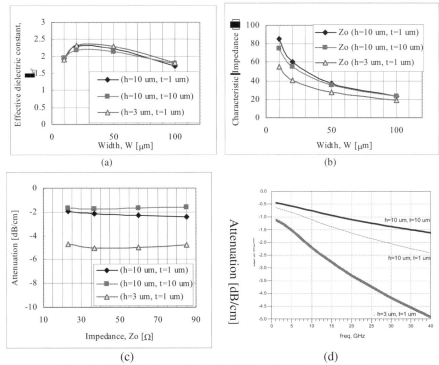

Fig. 2 Simulated results of OML; effective dielectric constants (a), characteristic impedances (b) and insertion losses (c) in the variation with line width, height and thickness at 40 GHz. (d) is the insertion loss of the OML with same width, W=10 μm, different height and thickness, which varies with the operating frequency.

3. Fabrication and Measurement

Several OMLs were characterized for line widths of 12, 18, 24, 30, 40, 50, 60, and 70 μm. For the elimination of parasitic of probing pads and transitions between CPW line and OML, the length of OML was fabricated long as possible. We have designed and fabricated OMLs 5 mm long. Fig. 3 shows its photograph and measured results. In these results, there are some differences with the simulated ones. We could analyze from the simulated results that it may occur when the signal line fabricated is lower and thinner than we expected. According to the calculated results, we can estimate that their height (h1) and thickness (t) are approximately 2.5 μm and 0.8 μm.

4. Conclusion

This paper has presented an in depth computational and experimental characterization of novel overlay microstrip lines (OMLs) on GaAs substrates. It has shown that the conductor loss for the OML is much greater than the dielectric loss in the case of the height is quite high. In this paper, we presented the insertion loss of OML can be designed below 2 dB/cm above 40 GHz, and we have fabricated some OMLs and shown that the measured results agreed well with the simulated ones.

After these results, we could have flexibility in the design of low loss microwave and mm-wave passive integrated circuits. And its compatibility of existing MMIC technology can make possible to build a high performance RF front-end ICs at the high frequency ranges.

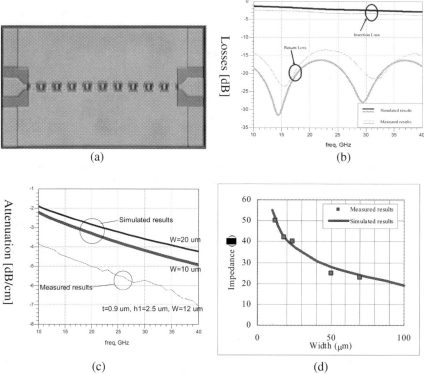

Fig. 3 The microphotograph of a fabricated OML (a), comparisons between simulated (t = 1.0 μm, h1 = 3 μm) and measured results (t = 0.8 μm, h1 = 2.5 μm), insertion and return losses for l=5 mm (b), normalized attenuations (c), and measured line impedances (d).

Acknowledgement

This work was supported by KOSEF under the ERC program through the Millimeter-wave INnovation Technology (MINT) Research Center at Dongguk University.

References

[1] U. Bhattacharya S.T. Allen and M.J.W. Rodwell 1995 IEEE Microwave and Guided Wave Letters 5-2 50-52
[2] F. Schnieder R. Doerner and W. Heinrich 1996 IEEE Microwave and Guided Wave Letters 6-3 117-119
[3] T.M. Weller 1998 Topical Meeting on Silicon Monolithic Integrated Circuits in RF Systems Digest of Papers 173-177
[4] Youngwoo Kwon Hong-Teuk Kim Jae-Hyoung Park and Yong-Kweon Kim, 2001 IEEE Microwave and Wireless Components Letters 11-2 59-61

Correlation of pulsed IV measurements and high power performance of AlGaN/GaN HEMTs

V.Tilak[1], V. Kaper, R. Thompson, T. Prunty, H. Kim, J. Smart, J. R. Shealy, L. Eastman

Cornell Nanofabrication Facility and School of Electrical and Computer Engineering, Cornell University, Ithaca, NY 14853-5401
e-mail: tilak@crd.ge.com

Abstract – The key to achieving record microwave high power densities [1] from AlGaN/GaN HEMT's is the use of optimized silicon nitride passivation [2]. Current and 10 GHz power are doubled by this method, and improved performance correlates with pulsed I-V curves on passivated HEMT's.

1. Introduction

The inherent advantages of gallium nitride like large band gap (3.4 eV), large breakdown fields and high electron peak velocity (2.8 X 10^7 cm/s) make it an ideal candidate for high power RF applications as they can combine high breakdown voltages with high frequency operation, AlGaN/GaN HEMTs with power densities of greater than 11 W/mm have been reported at 10 GHz [1]. A problem that initially plagued the field of microwave power transistors was that the RF power measured during a load pull measurement was not as high as expected from the measured static I-V characteristics. This phenomenon was due to current dispersion. A simple calculation of expected RF power at a drain bias of 15V, with a knee voltage of 4V will give us an output power of 3.25 W/mm ($P_{out} = \Delta V \Delta I/8$). However, typical power measured was around 1 W/mm. Vetury et al. showed that surface states play an important role in causing current dispersion [3] and Green et al. showed that PECVD silicon nitride passivates the surface and thereby mitigates the problem of current dispersion [2]. Pulsed IV measurements as described below are an easy way to determining the current dispersion seen in the device and this correlates very well with the RF performance of the device. Hence this measurement can be used to determine the high power performance potential of an AlGaN/GaN HEMT device.

2. Fabrication of transistors

AlGaN/GaN structures were grown on semi-insulating (SI) SiC substrates at 1050 °C. A layer of AlN sub-buffer was grown before the AlGaN/GaN epi layers were deposited. This

---

[1] Currently at G.E. G.R.C. Niskayuna, NY

is critical to the high power performance in this material system [1]. Further details on growth and characterization of the epitaxial layers can be found in [1]. The AlGaN/GaN layers were characterized by C-V to have a sheet charge density of 1.2×10^{13} cm^{-2}. Next, transistor structures were fabricated using our baseline process. Mesa isolation was achieved using a Chlorine based ECR etch. Ti(20nm) /Al(100nm)/Ti(45nm) /Au(55nm) annealed at 800 °C in N_2 atmosphere for 75s was used as the ohmic contact recipe. Mushroom gates were fabricated using e-beam lithography and Ni(25nm)/Au(375nm) were evaporated to form the gates. The samples were passivated using PECVD silicon nitride at 300 °C. The silicon nitride was monitored and optimized as described in [4] to ensure that the devices did not degrade after passivation.

3. Characterization of transistors

Static DC I-V curves were obtained using an Agilent™ HP-4142. Large signal load pull testing was done using a Maury™ loadpull sytem. Details of these test setups and testing conditions are described elsewhere [1]. The setup for pulsed I-V testing consists of HP™ 8114A pulser and a EGG 800 box car integrator on the gate circuit and a HP 3230 power supply on the drain circuit. The output and the input of the device is co-planar waveguide (CPW) probed and terminated into a parallel 50 Ω. This 50 Ω load helps in maintaining an overall 50 Ω environment as the device sees large changes in output conductance during its operation from the linear region to the saturation region. The CPW probes used were placed as close as physically possible to the drain circuit to minimize inductance which can cause oscillation and ringing at the onset of the pulse and can be detrimental to the device .The box car integrator is used to acquire the current and voltage pulse. A sampling window routine is used to sample a number of points in the pulse. This helps in averaging any undesired transients due to ringing. A 100ns pulse was applied to the gate with low duty cycle (1 %). The low duty cycle eliminates self heating effects. The drain current response is measured with a current probe with 5 ns transient response. Small device peripheries are used (150 μm or less) which result in a time constant of less than 10 ns due to parasitic pad capacitances that can interfere with the measurement of actual gate lag.

4. Results

To perform the pulsed I-V measurement, the drain was swept from 0 to 20V in 128 steps and the gate electrode is pulsed for each measurement from its quiescent value, typically -8V, to +1V in 1V steps. If the gate is pulsed from -8V to +1V,while holding the drain at 6V, the drain current pulse shows a clear lag. As the dispersion is seen due to gate pulsing this is known as gate lag. For plotting I-V curves, the drain current can be measured at

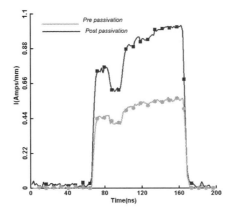

Figure 1 Drain Pulse before and after passivation, V_{drain} = 5V, V_{gate} pulsed from –8V to 0V

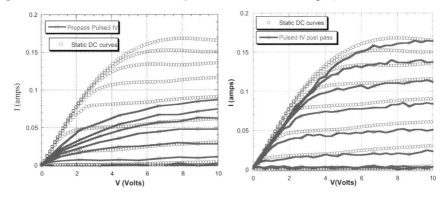

Figure 2. Comparison of static I-V curves to pulsed I-V curves, (a) Prepassivation and (b) Post passivation

different points on the pulse and plotted as a function of the drain voltage for a given gate voltage. For example, if we measure the drain current at say t = 92 ns (see figure 1), then we shall get a much lower current than at t = 162ns. This initial dip is seen on all devices and is an artifact associated with the ringing in the system. However, useful comparisons can be made, by measuring pre and post passivation pulsed I-V at one particular time step, say at t=162ns in figure 1. Figure 2(a) superposes the pulsed I-V characteristics on the static I-V characteristics measured on the *same* device before passivation.

We can clearly see that the full channel current measured with the pulsed I-V technique is less than 50% of that seen by the static I-V measurements. Further, the knee of the I-V curve is clearly defined in the static DC curves but is less clearly defined and significantly higher for the pulsed I-V curves. The pulsed I-V and the static I-V curves almost overlay each other after passivation.

A drain lag phenomenon has been observed in AlGaN/GaN devices by Binari [4]. This drain lag phenomenon results from the injection of hot electrons into trapping states outside of the channel during the pulse, partially depleting the 2DEG in the channel. The gate lag effect seen in these devices has been previously observed by Yeats [5] in GaAs MESFET technology. The explanation for this effect in the GaAs MESFET case was due to a slow transient behavior of the surface state existing in the ungated regions in the channel [6]. Current dispersion seen in this material system has been linked to the surface state phenomenon [3]. The gate lag measurement can be used to effectively probe the current dispersion as it is a surface phenomenon.

The current dispersion causes the reduction of saturated output power at RF frequencies as compared to the power expected from the static DC I-V curves. However, If we estimate the saturate RF power from the *pulsed I-V curves* instead of the static, we see excellent correlation. From the pulsed I-V curves, we measure the full channel current to be 0.6 A/mm and a knee voltage of 5V on the device before passivation. The before passivation RF power expected at 15V bias is 1.875 W/mm and 1.6 W/mm was measured on the device at 15V bias and gate bias of –5V at a PAE of 44%. Post passivation, the pulsed I-V full channel current is 1 A/mm and a knee voltage of 5V, so the expected RF power at 15V bias is 3.13 W/mm and the measured RF power was 3 W/mm.

5. Conclusions

Gate pulsed I-V measurements were performed on AlGaN/GaN HEMTs. The pulsed I-V characteristics were compared to the static characteristics, and significant current dispersion was observed for unpassivated devices. The pulsed I-V characteristics showed good correlation with the RF power measured both pre and post passivation. Thus, the pulsed I-V measurement technique can be used to determine whether there is significant current dispersion in the device and estimate the saturated output power more reliably than static I-V curves.

6. References

[1] Shealy, J. R., Kaper, V., Tilak, V., Prunty, T., Smart, J. A., Green, B., and Eastman, L.F., *J. Phys.: Condens. Matter* , **14** (13), pp. 3499-3509, 2002.
[2] Green, B.M., Chu, K.K., Chumbes, E.M., Smart, J.A., Shealy, J.R., Eastman, L.F., *IEEE Elect. Dev. Lett.* **21** (6), pp. 268-270, 2000.
[3] Vetury,R., Zhang, N.Q., Keller, S., Mishra, U.K., *IEEE Trans. Electron. Dev.* **48** (3), pp. 560-566, 2001.
[4] Binari, S. C., Ikossi-Anastasiou, K., Roussos, J. A., Park, D ., Koleske, D. D., Wickenden, A. E., and Henry, R. L., *Proc. Int. Conf. Nitride Semiconductors*, pp. 476-478, 1997
[5] Yeats, R., D'Avanzo, D. C., Chan, K., Fernandez, N., Taylor, T. W., and Vogel, C., *IEDM Tech. Dig.,* pp. 842-845, 1988.
[6] Lo, S-H., Lee, C-P., *IEEE Trans. Elect. Dev.,* **41** (9), pp. 1504-1511, 1994.

// 120 nm Gate Length E-Beam and Nanoimprint T-Gate GaAs pHEMTs Utilising Non-Annealed Ohmic Contacts

Euan Boyd, Dave Moran, Helen McLelland, Khaled Elgaid, Yifang Chen, Douglas Macintyre, Stephen Thoms, Colin Stanley, Iain Thayne

Ultrafast Systems Group, Nanoelectronics Research Centre,
University of Glasgow, Glasgow G12 8LT, Scotland, U.K.
Tel: +44 (0)141 330 6125   e.boyd@elec.gla.ac.uk

Abstract.

We report the fabrication, DC and RF performance of 120 nm gate length GaAs pHEMT's whose T-gates were defined using electron-beam lithography and nanoimprinting. All devices were realised using novel non-annealed Ohmic contact and selective succinic acid gate recess etching technologies. The 120 nm gate length nanoimprinted devices showed DC transconductance of up to 450 mS/mm whilst e-beam defined gate devices showed f_T and f_{max} of 135 GHz and 190 GHz respectively.

1. Introduction

In the fabrication of high speed GaAs pHEMT's the definition of the T-gate is critical. In a typical GaAs pHEMT process flow, the gates of a device are defined by electron beam lithography after the Ohmic contacts so that any high temperature processing associated with low resistance Ohmic contact formation does not affect the Schottky gate contact performance. When the gate resist is spun over the Ohmic contacts however, resist thickness variations occur which can lead to gate yield and uniformity issues, particularly for sub-100 nm geometries.

As electron beam lithography is a relatively slow and expensive technique, alternative methods of high-resolution pattern transfer are currently being investigated. Nanoimprinting is a particularly attractive technology as it has sub-50 nm resolution, and is relatively inexpensive[1]. The main challenge of incorporating nanoimprint technology into a GaAs pHEMT process flow is the issue of alignment of the gate to the Ohmic contacts which currently means the gate must be defined prior to the Ohmic contact.

Figure 1- Profile of an electron beam defined T-Gate showing optimised recess etch.

To address the above issues, we have developed a low resistance, non-annealed Ohmic contact technology which is defined *after* the gate lithography step – thereby offering improved yield and uniformity in sub-100 nm gate devices. In addition, the technology enables the use of nanoimprinting for the realisation of GaAs pHEMTs.

2. Fabrication

The GaAs pHEMT used in this study was grown in Glasgow by Molecular Beam Epitaxy (MBE) on SI GaAs substrates. The layer structure includes the following, a 20nm thick $In_{0.2}Al_{0.8}As$ channel, 5nm $Al_{0.3}Ga_{0.7}As$ spacer layer, 10nm $Al_{0.3}Ga_{0.7}As$ barrier layer containing a single Si δ-doping layer. The etch stop layer is a 5nm $Al_{0.3}Ga_{0.7}As$ layer. The cap structure, the key to the formation of the low resistance non-annealed Ohmic contact comprised of a 15nm $In_{0.2}Ga_{0.8}As$ heavily δ-doped upper cap above a 15nm $1\times10^{18}\,cm^{-3}$ n-doped GaAs layer. The room temperature transport properties measured using Van de Pauw structures gave a sheet density, n_{sh} of $1.18\times10^{13}\,cm^{-2}$ and a mobility, μ_H, of 3100 cm^2/Vs with the cap in place. After the cap layer was removed using a selective wet etch n_{sh} of $4.6\times10^{12}\,cm^{-2}$ and μ_H of 4100 cm^2/Vs were measured.

For the e-beam defined gates, lithography was carried out directly after the mesa isolation stage. This was performed using a Leica EBPG5-HR100 lithography tool operating at 50keV. The gate profile was formed in a trilayer PMMA/P(MMA/MAA) resist stack. A succinic acid based wet etch was used for the gate recess etch. This was found to etch both GaAs and InGaAs cap layers stopping on the 5nm $Al_{0.3}As_{0.7}As$ etch stop layer (Figure 1) The length of the recess offset was controlled by etch time to give the desired recess offset of 35nm. The Ti:Pd:Au Schottky gates was then metallised by electron beam evaporation.

The Ohmic contacts were subsequently defined either side of the gate with a source-drain separation of 1.5μm. The 100nm thick Ni:Ge:Au based Ohmic contacts were then formed by electron beam evaporation. The contact resistance of the unannealed contacts was measured using the transmission line method, the contact resistance was found to be 0.16Ωmm. The devices were completed with the addition of coplanar waveguides to enable on-wafer DC and RF characterisation.

293

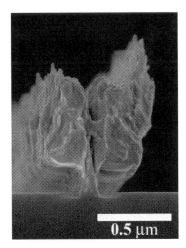

Figure 2 – Nanoimprinted T-gate with recess etch.

The nanoimprinted devices followed a similar process flow with the exception of the mesa isolation which was formed after the Ohmic contact definition. The nanoimprinted gates were realised using high resolution stamping tools fabricated by electron beam lithography and dry etching methods. The stamp tool patterns were imprinted into a resist bilayer [2] using an Obducat Nanoimprinter with a 60 mm substrate holder and computer control. The bilayer imprinting technique allowed controlled removal of post imprint residual resist by chemical dissolution and a short oxygen ash. This eliminated the need for a long dry etch step which can damage epitaxial layers on III-V substrates. The gate recess etch was performed using a selective wet etch and the gate was then metalised. The cross section of the nano-imprinted gate is shown in Figure 2.

3. Measurement

The devices were characterised on-wafer at DC and RF using an HP4155 parameter analyser and a Anristu 360B Vector Network Analyser with on-wafer Picoprobes from GGB. The devices were measured in the bias range of 0 to 1V drain bias and -0.3 to 0.2V gate bias. Figure 3 shows the DC transfer characteristics of the nanoimprint device. This

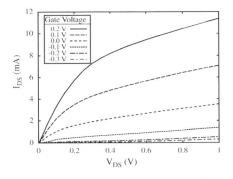 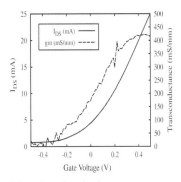

Figure 3 - DC Characteristics of the Nanoimprint device.

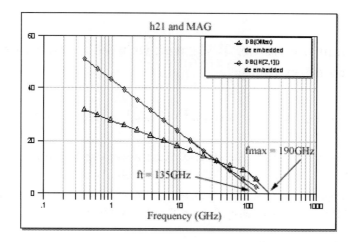

Figure 4 – RF Characterisation of e-beam devices showing f_t and f_{max}.

shows good DC behaviour with a pinch off voltage of -0.3V and a peak transconductance of 450mS/mm.

On wafer S-parameters in the frequency range 40MHz to 60GHz of the e-beam defined gate were obtained after calibration performed using the LRRM technique on a Cascade Microtech Impedence Standard Substrate (ISS). A small signal model was then extracted from measurement over a similar bias range to the DC measurements. This model was used to de-embed the effects of the input and output co-planar waveguide feedlines of the device. Figure 4 shows the plots of the de-embedded h_{21} and MSG/MAG as a function of frequency for the bias point Vds=1V, Vgs=-0.2V for a 2x50µm device. This bias point provides the optimum RF performance. Extrapolating h_{21} and MAG/MSG to unity gain gives f_T and f_{max} of 135GHz and 190GHz respectively.

3. Conclusions

We have successfully demonstrated the used of non-annealed Ohmic contacts in the fabrication of 120nm gate length GaAs pHEMT devices whose T-gates have been defined by both electron beam lithography and nano-imprint lithography. The performance is comparable with similar devices that have been fabricated using conventional process flows. This demonstrates the viability and the potential of these novel fabrication methods.

References

[1] Imprint of sub-25nm vias and trenches in polymers, S. Y. Chou, P. R. Krauss and P. J. Renstrom, Appl. Phys. Lett. 67 (21), 3114 (1995)

[2] The Fabrication of High Electron Mobility Transistors with T-Gates by Nanoimprint Lithography, Y. Chen et al., To be published in J. Vac. Sci. Technol.

New composite-emitter HBTs with reduced turn-on voltage and small offset voltage

M K Tsai[1], Y W Wu[2], S W Tan[2], Y J Yang[1], and W S Lour[2,a]

[1]Department of Electrical Engineering, National Taiwan University,
1 Sec. 4, Roosevelt Road, Taipei, Taiwan, Republic of China.
[2]Department of Electrical Engineering, National Taiwan Ocean University,
2 Peining Road, Keelung, Taiwan, Republic of China.

[a]corresponding author

Abstract. This paper reported a new composite-emitter heterojunction bipolar transistor (CEHBT) with a composite emitter formed of a 0.04-μm $In_{0.5}Ga_{0.5}P$ bulk layer and a 0.06-μm $Al_{0.45}Ga_{0.55}As$/GaAs digital graded superlattice (DGSL) layer. The CEHBT's exhibit a small collector-emitter offset voltage of 55 mV and a base-emitter turn-on voltage of 0.87 V, which is 0.4 V lower than that of 1.27 V of the InGaP/AlGaAs abrupt-emitter HBT. It is found that CEHBT's exhibits a current gain as high as 250 and is even enhanced to 385 when only a DGSL layer is used for passivation layer.

1. Introduction

The Npn heterojunction bipolar transistors (HBT's) have recently received intensive attention in both wireless and wired consumer products [1]. It has become one of the most important issues to reduce the base-emitter (B-E) turn-on voltage, $V_{ON(B-E)}$, of HBT's. InGaAs-based HBT's are first considered as candidates for the next-generation power-amplifier. However, InP technology is expensive and a large spike at the B-E junction severely limits the reduction of $V_{ON(B-E)}$ [2]. Other p-InGaAsN-based HBT's reported are double-heterojunction ones. The blocking effect at the B-C heterojunction results in a high knee voltage. Furthermore, the expected reduction of $V_{ON(B-E)}$ is usually not so significant due to increased ΔE_C [3].

In this work, we report the use of a composite emitter comprising an InGaP layer and a digital graded superlattice (DGSL) layer. The InGaP bulk layer functions as 1) hole confinement layer, 2) conduction-band transition layer between DGSL and narrow-gap GaAs cap layer, and 3) etching stop layer for well-controlled passivated devices with optimum passivation-layer thickness. Whereas, the DGSL layer forms a step-wise graded composition to smooth out the potential spike associated with the hetero-interface.

2. Device structures and fabrication

Three different types of device are compared in this study: the InGaP/DGSL-passivated CEHBT, the DGSL-passivated CEHBT and the AlGaAs-passivated InGaP/AlGaAs HBT. Both the CEHBT's employ the same device structure grown on a (100)-oriented GaAs substrate by MOCVD. As shown in Fig.1, the DGSL structure comprises four superlattice unit cells of four different barrier/well thicknesses (10/40, 20/30, 30/20 and 40/10 Å upon base in sequence). Each of the four superlattice unit cells has a combination of 3 periods of $Al_{0.45}Ga_{0.55}As/GaAs$ quantum wells. The device structure for the InGaP/AlGaAs HBT is the same except that the $Al_{0.45}Ga_{0.55}As/GaAs$ DGSL layer is replaced by an AlGaAs bulk layer.

The calculated results (by transfer matrix method) under various W:B conditions according to $Al_{0.45}Ga_{0.55}As/GaAs$ 3-period quantum wells shows the lowest mini-band energy for electrons is 60 meV for W:B=1:4, 113 meV for W:B=2:3, 180 meV for W:B=3:2, and 260 meV for W:B=4:1, respectively. It is evident that the DGSL layer forms a step-wise graded composition to smooth out the potential spike associated with the hetero-interface. The fabrication started with mesa isolation. The $Al_{0.45}Ga_{0.55}As/GaAs$ DGSL layer was etched by using $1H_2SO_4:1H_2O_2:8H_2O$ solution at

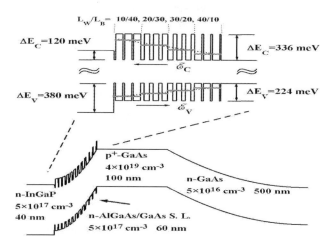

Figure 1. The band diagram of a new CEHBT with structural parameters a etching rate of 210 Å/s. Etching selectivity between GaAs (etched in

3NH$_4$OH:1H$_2$O$_2$:50H$_2$O) and InGaP (etched in 3NH$_4$OH:1H$_2$O$_2$:50H$_2$O) was employed throughout this work [4]. An HBT was formed to have a passivation layer composed of an InGaP layer and an Al$_{0.45}$Ga$_{0.55}$As/GaAs DGSL layer. The InGaP layer of the InGaP/DGSL-passivated CEHBT was then etched away to form a DGSL-passivated CEHBT. For comparison, an AlGaAs-passivated HBT is also fabricated.

3. Results and discussion

Figure 2 shows the common-emitter characteristics for the InGaP/DGSL- and DGSL-passivated CEHBT's. A small collect-emitter offset voltage of 55 mV and knee voltage of 0.3 V at a collector current of 1 mA were obtained for both devices. These phenomena reveal that the DGSL structure really eliminates the spike resulting from ΔE_C. Figure 3 shows the Gummel plots of InGaP/DGSL- and DGSL-passivated CEHBT's with V_{BC}=0 V.. We firstly refer to the DGSL- and the AlGaAs-passivated devices. An important merit of the DGSL-passivated HBT is its $V_{ON(B-E)}$ defined as V_{BE} at which the collector current exceeds 1 μA. The $V_{ON(B-E)}$ of DGSL-passivated CEHBT is 0.87 V, which is 0.4 V lower than the 1.27 V measured in an AlGaAs-passivated HBT over a wide range of current level.

We found the current gain of the DGSL-passivated CEHBT increases with the collector current and even reaches 385, which is higher than that of the AlGaAs-passivated HBT (250), as shown in Fig.4. We find that the existence of the InGaP does not change $V_{ON(B-E)}$ (i.e., 0.86 and 0.87 V). However, the InGaP/DGSL-passivated HBT exhibits a smaller current gain, indicating that optimized passivation-layer thickness is around 600 Å by selectively removing the InGaP layer, as suggested in the authors' previous work [5].

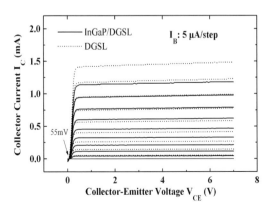

Figure 2. Common-emitter I-V curves for both CEHBT's.

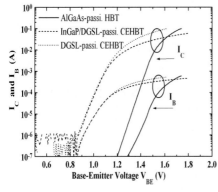
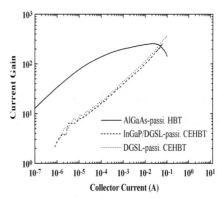

Figure 3. Gummel plots for all fabricated Devices

Figure 4. The calculated current gains deduced from Figure 3.

4. Conclusions

We have demonstrated and analyzed a new InGaP/DGSL composite emitter for use in heterojunction bipolar transistors. The studied devices exhibit superior direct-current characteristics to those of conventional GaAs-based HBT's. Moreover, due to simplified growth control for both MBE and MOCVD, the DGSL technology is a promising alternative to the conventional grading growth method for the AlGaAs/GaAs material system.

Acknowledgement
This work is partly supported by National Science Council under the contract No. NSC 90-2215-E-019-001.

References
[1] Pan N, Elliott J, Knowles J, Vu D P, Kishimoto K, Twynam J K, Sato H, M. T. Fresina M T and Stillman G E 1998 IEEE Electron Device Lett. 19 115-117.
[2] Yang K, Cowles J C, East J R and Haddad G I 1995 IEEE Trans. Electron Devices 42 1047-1057.
[3] Li N Y, Chang P C, Baca A G, Xie X M, Sharps P R and Hou H Q 2000 Electronics Lett. 36 81-83.
[4] Lour W S, Chang, W L, Shih Y M and Liu W C 1999 IEEE Electron Device Lett. 20 304-306.
[5] Lour W S and Hsieh J L 1998 Semicond. Sci. Technol. 13 847-851.

Properties of Electronic States at Free Surfaces and Schottky Barrier Interfaces of AlGaN/GaN Heterostructure

Hideki Hasegawa, Takanori Inagaki, Shinya Ootomo and Tamotsu Hashizume

Research Center for Integrated Quantum Electronics (RCIQE) and
Graduate School of Electronics and Information Engineering,
Hokkaido University, North 13, West 8, Sapporo, 060-8628, Japan

Abstract. Properties of electronic states at free surfaces and Schottky interfaces of AlGaN/GaN HFETs are investigated by gateless HFET, XPS and Schottky I-V-T measurements. A new model of near-surface electronic states including a U-shaped surface state continuum and N-vacancy related near-surface deep donors is proposed that can explain observed large gate leakage and current collapse in AlGaN/GaN HFETs.

1. Introduction

AlGaN/GaN heterostructure field effect transistors (HFETs) suffer from various surface-related problems, including I-V dispersion, drain current collapse, and large gate leakage currents whose mechanisms are not well understood. In the gated devices, it is difficult to decide whether problems arise from the gated or ungated region.

In this paper, properties of electronic states at free surfaces and those at Schottky barrier interfaces are investigated separately by gateless HFET, XPS and Schottky I-V-T measurements. A new model is proposed that can explain observed large gate leakage and current collapse in AlGaN/GaN HFETs in a unified way.

2. Experimental

Wafers had a structure of $Al_{0.28}Ga_{0.72}N$ (5nm)/n^+-$Al_{0.28}Ga_{0.72}N$ (20nm)/ $Al_{0.28}Ga_{0.72}N$ (5nm)/ undoped-GaN(1μm) grown on sapphire substrates by MOVPE with a Hall mobility of 900 cm^2/Vs and a sheet carrier density of $1.1 \times 10^{13} cm^{-2}$ at room temperature. The gateless HFET shown in **Fig.1** was fabricated by mesa isolation by UV light-assisted KOH wet etching and Ti/Al/Ti/Au (20/80/20/50 nm) ohmic metallization. It is extremely powerful in correlating various surface processing with the inner current transport [1].

Surface treatments on the air-exposed portion of the gateless HFET included H_2-plasma treatment, N_2-plasma treatment and formation of a SiO_2 passivation film. Both plasma treatments, which are typical for native oxide removal, were applied for 1 min at 200 °C under ECR plasma excited at a microwave frequency of 2.75GHz and a power of 50 - 100 W with gas flow rates of 5-10 sccm. Since current collapse takes in a pronounced way in SiO_2 passivated devices, a SiO_2 film formed by the standard plasma CVD process using SiH_4 and N_2O on the HF treated surface was chosen as the test passivation film.

Treated surfaces were analysed by *in situ* XPS measurements (Perkin Elmer PHI 1600C) without breaking UHV, using a monochromated Al Kα x-ray source (hυ=1486.6 eV). Circular Au/Ni Schottky structures were used for Schottky current transport study.

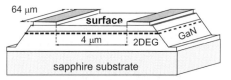

Figure 1. Gateless HFET.

3. Results and discussion

3.1 Electronic states on free surfaces

DC Ids-Vds curves of gateless HFETs before and after various treatments are summarized in **Fig. 2(a)**. Since the electric field was too small to cause velocity saturation, the observed highly non-linear DC I-V curves similar to a gated device can only be explained in terms of strong Fermi level pinning by surface states which behave like a "virtual gate". In fact, data could be reasonably well fitted to the theoretical curves based on the gradual channel approximation. This gave a surface Fermi level position of $E_{FS} = E_c - 1.4$ eV for the air exposed sample after taking account of the polarization effect. In the H_2-plasma sample and SiO_2 sample, the saturation current decreased, whereas it slightly increased in the N_2-plasma sample, indicating treatment-induced changes of the pinning position. Except the N_2-plasma sample, AC I_{ds}-V_{ds} curves showed frequency dependences, i.e., so-called I-V dispersion.

Results of current transient measurements are summarized in **Fig. 2(b)**. The quencient bias was kept at V_{DS} =0.5V in the linear region, and a positive voltage with a variable peak value of V_{DSp} was applied for a duration time of 50 s. Then, the current transients were measured. When the value of V_{DSp} was small, no change of I_{DS} was observed in all cases. When V_{DSp} went deep into saturation region, transients became visible especially in H_2-plasma and SiO_2 samples. The current showed a "fast" and dominant

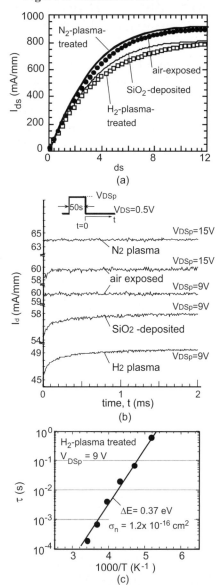

Figure 2. (a) Drain I-V characteristics, (b) pulse response and (c) time constant of the drain current.

exponential transient with a large amplitude followed by subsequent smaller, slow and highly non-exponential response. In the air-exposed sample, transient became visible at V_{DSp} = 15 V whereas no transient was seen even at V_{DSp} = 15 V in the N_2-plasma sample.

The time constant of the initial exponential transient is plotted in **Fig.2(c)** vs. inverse temperature for the H_2-plasma sample. This indicated that a discrete trap level with an activation energy of 0.37 eV and a capture cross section of 1.2×10^{-16} cm^2 plays a dominant role in the transient. The same trap was detected in SiO_2 and air-exposed samples. Subsequent small, slow and highly non-exponential response is a typical response of a surface state continuum which includes a wide range of time constants.

In order to get information on the origin of the discrete trap, *in situ* XPS measurements were made on the H_2-plasma and SiO_2 samples. Angle resolved photoemission analyses indicated that the V/III ratio of the AlGaN surface after treatments was far below unity near the surface, strongly indicating near-surface decrease of N atoms. On the other hand, no such decrease of N atoms was observed in the N_2-plasma sample.

These results indicate that current collapse in gated HFETs comes, in most part, from the ungated free surface region near the drain edge which contains N-vacancy related traps. As regards the energy position of the N-vacancy related traps, a theoretical calculation by Yamaguchi and Junnarkar [2] predicted that the N-vacancy defects can form s-like deep donor levels at around Ec-0.4 eV within the gap of GaN and AlGaN.

3.2 Current transport at Schottky interfaces

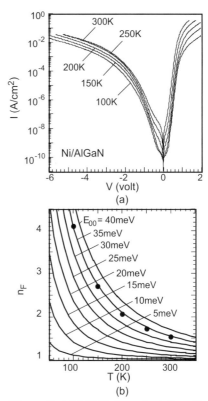

Measured forward and reverse I-V characteristics of Schottky diodes are shown in **Fig. 3 (a)** for various temperatures. Most surprisingly, temperature dependences of I-V curves were extremely small, deviating from the thermionic emission transport by many orders of magnitudes. Reverse currents were also anomalously large.

The measured ideality factor for forward currents, n_F, could be very well fitted into the following formula for the therminonic field emission [3], as shown in **Fig. 3(b)**.

$$n_F = \frac{E_{00}}{kT} \coth\left(\frac{E_{00}}{kT}\right)$$

with $E_{00} = (qh/\pi)(N_D/m^* \varepsilon_0 \varepsilon_s)^{1/2}$ (1)

However, the donor concentration N_D obtained by fitting is 8×10^{18} cm^{-3}, and this is much larger than the actual value of N_D = 3×10^{17} cm^{-3} of the sample used in the experiment.

Figure 3. (a) I-V-T characte-ristics of Schottky sample and (b) ideality factor vs. T.

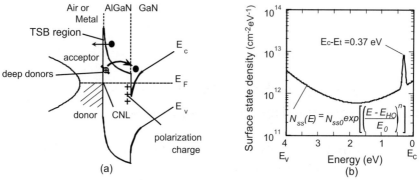

Figure 4. (a) A unified model for free surface and Schottky interface of AlGaN/GaN heterostructure and (b) state density distribution.

3.3 A new model for electronic states and a new mechanism for HFET current collapse

Our new model for near-surface electronic states is shown in **Fig.4 (a)** with a common effective surface state density distribution shown in **Fig.4 (b)**. Similarly to other III-V semiconductors [4], a U-shaped high density surface state continuum is formed which pins the surface Fermi level near the charge neutrality level (CNL). CNL lies at Ec - 1.6 eV for $Al_{0.28}Ga_{0.72}N$ according to the branch point energy data in ref.[5]. Depending on details of processings such as plasma treatments, passivation and metal deposition, high density deep donors related to N-vacancies may be created near surface. The value of N_D obtained by fitting Eq.(1) to n_F corresponds to the deep donor density. Due to Fermi level pinning near CNL both at free surfaces and at Schottky interface, these deep donors are ionized, supplying electrons to 2DEG together with those due to shallow donors and intrinsic and piezoelectric polarization [6]. This produces a thin surface barrier (TSB) region shown in **Fig. 4(a),** as we recently proposed for GaN Schottky diodes [7].

In the case of the Schottky interface, the TSB region gives rise to thermionic-field emission path for current transport, giving rise to large leakage currents in HFETs. In the case of HFET transport, injections of high energy electrons from the channel into the free surface region near the drain edge fill up near-surface deep donors, reducing 2DEG density and expanding the depletion width. This causes current collapse. The calculated pulse transients as well as forward and reverse Schottky I-V characteristics using the model shown in **Fig.4 (a)** and the state distribution shown in **Fig. 4(b)** reproduce the experimental behaviour remarkable well, as will be reported in detail elsewhere [8].

In conclusion, our new model explains observed large gate leakage and current collapse in AlGaN/GaN HFETs.

References

[1] Hasegawa H et al, J. Vac. Sci. Technol. B **5**, 1097 (1987).
[2] Yamaguchi E, and Junnarkar M R, J. Crystal Growth **189/190**, 570 (1998)
[3] Padovani F A and Stratton R, Solid State Electron 9, 695 (1966).
[4] Hasegawa H and Ohno H, J. Vac. Sci. Technol. B **4**, 1130 (1986).
[5] Mönch W, Appl. Sur. Sci. **117/118**, 380 (1997).
[6] Ibbetson J P et al, Appl. Phys. Lett. **77**, 250 (2000).
[7] Hasegawa H and Oyama S, J. Vac.Sci. Technol. B **20**, 1647 (2002).
[8] Inagaki T, Hashizume T and Hasegawa H, to be presented at ISCSI-02, Oct. 21-25, Karuizawa, Japan (2002)

Direct S-Parameter Extraction by Physical Two-Dimensional Device AC-Simulation

V Palankovski, S Wagner, T Grasser, R Schultheis*, and S Selberherr

Institute for Microelectronics, Technical University Vienna,
Gusshausstrasse 27-29, A-1040 Vienna, Austria
*Infineon Technologies AG, Wireless Products, Technology and Innovations,
Otto-Hahn-Ring 6, D-81730 Munich, Germany

Abstract. We present results from fully two-dimensional physical device simulation. Scattering parameters (S-parameters) are directly obtained from small-signal AC-analysis of real heterostructure devices. A comparison reveals very good agreement with measured data.

1. Introduction

Heterojunction Bipolar Transistors (HBTs) are among the most advanced semiconductor devices today. Two-dimensional device simulation proved to be valuable for understanding the underlying device physics [1] and for improving the device reliability [2]. Bias-dependent S-parameters hold the full small-signal RF-information about the device behavior and allow process control beyond the information about the DC-quantities.

There are several approaches to compute bias-dependent S-parameters, e.g. [3,4], applying quasi-static or equivalent-circuit parameter models. These approaches employ transformations in the time domain to extract S-parameters. All these methods are both more CPU-time consuming (steady-state has to be reached for each bias and frequency) and more inaccurate (only a limited number of time-steps in reasonable CPU-time, equivalent-circuit approximation, etc.) compared to AC-analysis [5].

We implemented a feature for direct extraction of either extrinsic or intrinsic (de-embedded) S-parameters from AC-simulation in the three-dimensional device simulator Minimos-NT [6]. Thus, we use a combination of rigorous III-V group and IV group semiconductor materials modeling and the ability to simulate in the frequency domain.

2. Physical models in Minimos-NT

Minimos-NT deals with different complex structures and materials, such as Si, Ge, GaAs, AlAs, InAs, GaP, InP, their alloys and non-ideal dielectrics. Various important physical effects, such as bandgap narrowing, surface recombination, transient trap recombination, self-heating, and hot electron effects, are taken into account. The models are based on experimental or Monte Carlo simulation data and cover the whole material composition range. The model parameters in Minimos-NT are checked against several independent HEMT and HBT technologies to obtain one concise set used in all simulations. Efficiency is proven by hydrodynamic DC-simulations with self-heating, e.g. see Fig. 1.

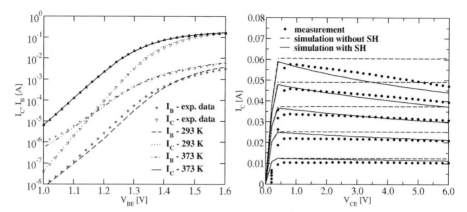

Figure 1. Forward Gummel plots at $V_{CB} = 0$ V for GaAs HBT (left): Comparison with measurement data at 293 K and 373 K. Output characteristics (right): Simulation with and without self-heating (SH) compared to measurement data at constant I_B stepped from 0.1 to 0.5 mA.

3. Simulation example

By means of two-dimensional device AC-simulation, we extracted the S-parameters for a one-finger InGaP/GaAs HBT with emitter area of 3 μm × 30 μm. Fig. 2 shows the simulated device structure and the pad parasitics (capacitances and inductances) in the two-port pad parasitic equivalent circuit, which is used to transform the intrinsic parameters to extrinsic ones. The parasitics result from measurements of open/short thru-test-structures [7]. Thus, the pad capacitances are $C_{pBE} = 150$ fF, $C_{pCE} = 75$ fF, and $C_{pBC} = 24$ fF, while the parasitic inductance values are $L_E = 1$ pH, $L_B = 75$ pH, and $L_C = 50$ pH. Any resistive parasitics are neglected, since we consider a rather small device and, therefore, only low currents.

The combined smith/polar charts in Fig. 3 show a comparison of simulated and measured S-parameters at $V_{CE} = 3$ V and $V_{CE} = 3.5$ V, with current densities $J_C = 2 \times 10^3$ A/cm^2, $J_C = 8 \times 10^3$ A/cm^2, and $J_C = 15 \times 10^3$ A/cm^2, respectively, for the frequency range between 50 MHz and 10 GHz.

4. Computational effort

The AC-simulation takes about 200 s CPU-time on a 2.4 GHz Linux Pentium machine for S-parameters computation with 20 frequency steps. For comparison, the conventional small-signal equivalent-circuit approach [3] takes about 590 s CPU-time at the same machine for 200 time steps at a given frequency. The time for post-processing of the transient simulation results to obtain the S-parameters at all frequencies is not included.

5. Conclusion

The good agreement with measured data and the speed-up achieved demonstrate the quality and the efficiency of our approach. At this instance, the shown approach enables further extensive optimization tasks with hundreds of runs in a reasonable time. We expect almost perfect match between simulated and measured S-parameters, as it has already been demonstrated for such devices by applying the standard small-signal equivalent-circuit modeling approach [7]. In addition, the two-dimensional physical simulation allows for a direct relation between the material properties and the high-frequency device behavior.

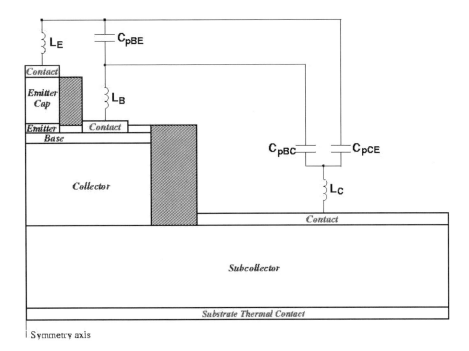

Figure 2. Simulated device structure together with pad parasitics used for S-parameter calculation.

Acknowledgment

The authors acknowledge inputs from R. Quay and K. Dragosits.

References

[1] Palankovski V, Selberherr S, Quay R, and Schultheis R, "Analysis of HBT Degradation After Electrothermal Stress", 2000 in Simulation of Semiconductor Processes and Devices (Seattle) p.245-248

[2] Palankovski V, Schultheis R, and Selberherr S, "Simulation of Power Heterojunction Bipolar Transistors on Gallium Arsenide", 2001 IEEE Trans. Electron Devices, vol.48, no.6, p.1264-1269

[3] Quay R, Reuter R, Palankovski V, and Selberherr S, "S-Parameter Simulation of RF-HEMTs", 1998 in EDMO (Manchester), p.13-18

[4] Anholt R, "HBT S-Parameter Computations Using G-PISCES-2B", 1999 GaAs Simulation and Analysis News, no.4, p.1-4

[5] Laux S, "Techniques for Small-Signal Analysis of Semiconductor Devices", 1986 IEEE Trans. Electron Devices, vol.ED-32, no.10, p.2028-2037

[6] Palankovski V, Quay R, and Selberherr S, "Industrial Application of Heterostructure Device Simulation", 2001 IEEE J. Solid-State Circuits, vol.36, no.9, p.1365-1370 (invited)

[7] Schultheis R, Bovolon N, Mueller J-E, and Zwicknagl P, "Modelling of Heterojunction Bipolar Transistors (HBTs) Based on Gallium Arsenide (GaAs)", 2000 Intl.J. of RF and Microwave Computer-Aided Engineering, vol.10, no.1, p.33-42

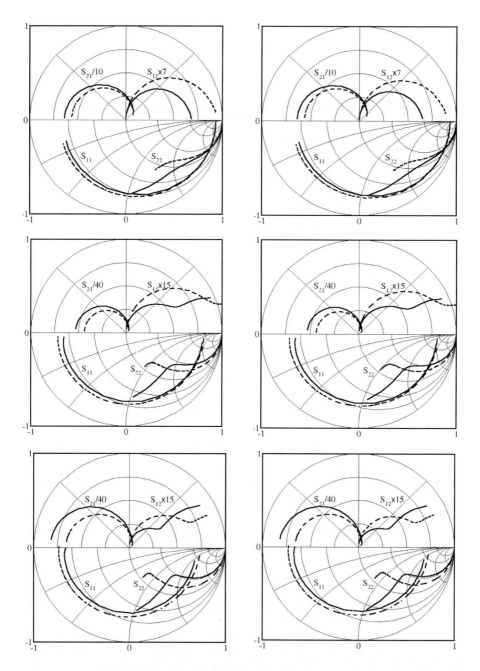

Figure 3. S-parameters in a combined Smith chart (S_{11} and S_{22}) and a polar graph (S_{21} and S_{12}) from 50 MHz to 10 GHz at V_{CE} = 3 V (left column) and V_{CE} = 3.5 V (right column), J_C = 2×10^3 A/cm^2 (row 1), J_C = 8×10^3 A/cm^2 (row 2), and J_C = 15×10^3 A/cm^2 (row 3): Simulation (solid lines) vs. experiment (dashed lines).

High-power Blue-Violet Lasers Grown On 3-inch Sapphire and GaN Substrate.

Shiro Uchida, Shinroh Ikeda, Takashi Mizuno, Shu.Goto, Tomomi Sasaki, Yoshio Ohfuji, Tsuyoshi Fujimoto, Osamu Matsumoto, Kenji Oikawa, Motonobu Takeya, Yoshifumi Yabuki, and Masao Ikeda

Development Center, Sony Shiroishi Semiconductor Inc., 3-53-2 Shiratori, Shiroishi, Miyagi, 989-0734 Japan

Abstract. This report presents a high-power blue-violet laser grown on a 3-inch sapphire substrate. The laser has a low threshold current of 27.5 mA, achieved by reducing the internal loss to 13.6 cm^{-1}, and an Al$_{0.18}$GaN electron-blocking layer inserted at an appropriate distance from the active layer reduces the absorption loss of the Mg-doped cladding layer while maintaining the internal quantum efficiency (0.94). The lifetime of these lasers is improved by employing an epitaxial lateral overgrown (ELO) substrate with low dislocation density of 3×10^6 cm^{-2}. The very low dislocation density was found to be essential to achieve stable operation for more than 5,000 h under 30 mW continuous wave operation at 60 °C. Lasers grown on GaN substrates are also discussed, and the lifetime of lasers with dislocation density of below 1×10^6 cm^{-2} is estimated to exceed 100,000 h.

1. Introduction

GaN-based 400-nm high-power lasers[1-3] have been developed as a light source for optical disk recording. Both the Blu-ray Disk system and DVD recording system require lasers with relative intensity noise (RIN) of below −125 dB/Hz, a beam profile with low aspect ratio, and reliability of greater than 5000 h. In addition, lasers for use in optical disks capable of capacities greater than 50 GB are required to operate at an output power exceeding 100 mW in order to write information to dual-layer disks. Therefore, lower operating current, a stable beam profile, and high reliability are required to satisfy these requirements. In the present report, the authors discuss and introduce some major techniques employed to achieve these laser characteristics.

2. Threshold current and RIN

The threshold current was reduced to 27.5 mA in a 600 μm cavity, which corresponds to 3.2 kA/cm². NEC reported[4] that an Mg-doped GaN layer has an absorption loss of 100 cm⁻¹. They also suggested that it was effective to reduce the confinement factor (Γ_{Mg}) of Mg-doped layer in the perpendicular optical field, in order to decrease this absorption loss. The present authors reduced the absorption loss in Mg-doped GaN/AlGaN super-lattice layers, leading to lower internal loss, by isolating these super-lattice layers from the active layer.

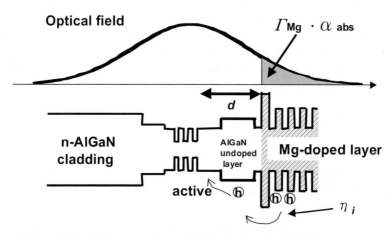

Figure 1. Band diagram and near-field profile of AlGaInN laser.

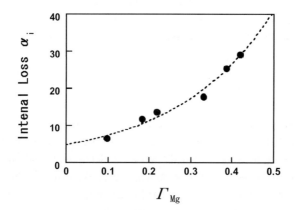

Figure 2. Relationship between α_i and Γ_{Mg}.

An GaInN interlayer and undoped $Al_{0.02}GaN$ layer were inserted between the multiple quantum wells (MQWs) and the $Al_{0.18}GaN$ electron-blocking layer, thereby effectively isolating the Mg-doped layers (including $Al_{0.18}GaN$ electron-blocking layer and super-lattice layers) from the MQWs so as not to degrade the efficiency of hole injection into the MQWs as shown in Fig.1.

In this structure, the confinement factor Γ_{Mg} was reduced from 0.386 to 0.219, affording a reduction in internal loss α_i from 25 cm^{-1} to 13.6 cm^{-1} (Fig. 2). This reduction in internal loss lead to an improvement in threshold current from 34.5 mA to 27.5 mA and slope efficiency from 1.1 W/A to 1.4 W/A while maintaining the internal quantum efficiency (0.94). The characteristic temperature T_0 was also improved from 145 K to 180 K, due mainly to the suppression of electron leakage currents. Laser diodes with this structure had lower operating current (110 mA) at 100 mW output power compared to previous devices, even at 70 °C ambient temperature. As a result of the lower threshold current, spontaneous emission is also reduced. As quantum noise is primarily caused by spontaneous emission, a lower RIN of –130 dB/Hz[5] at 2.5 mW output power is achieved without high-frequency modulation.

3. Far-Field Patterns

In order to achieve a stable laser beam profile with low aspect ratio, the authors have improved the laser structure[1-2] and developed methods for controlling the ridge stripe width (W) and remnant etching depth (d) (Fig. 3). These two factors dictate the beam divergence angle ($\theta_{//}$) in the direction parallel to the junction plane. In the present calculations, parameter d is strictly controlled to within ±20 nm so as to design $\theta_{//}$ in the range 9° to 11°. This controllability was achieved by setting appropriate condition of reactive ion etching (RIE). The stripe width must be precisely controlled to less than 1.4 μm in order to cut the first-order optical mode and suppress the kink in the L-I curve.[1]

Figure 4 shows the two-dimensional distribution of beam divergence angle for 500 laser diodes grown on 3-inch sapphire substrates. The standard deviations are still high compared to conventional DVD lasers, and the beam divergence angle in the perpendicular direction in particular has a large standard deviation, attributable to an interference fringe observed in the far-field pattern. The fringe is due to an optical leakage mode in the direction of the n-type GaN, which is transparent with respect to 400-410 nm laser light.

A thicker n-AlGaN cladding is expected to be effective for countering these negative affects. Figures 5(a) and (b) show the calculated far-field patterns for lasers with a 1.0 μm $Al_{0.04}GaN$ cladding layer or 2.0 μm $Al_{0.04}GaN$ cladding layer, and Fig. 5(c) shows the observed far-field pattern for a laser with 2.1 μm $Al_{0.045}GaN$ layer, which has a Gaussian-like profile.

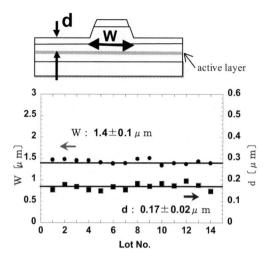

Figure 3. Reproducibility of ridge stripe process.

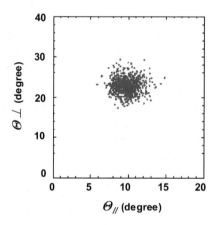

Figure 4. Distribution of the Far-field Patterns.

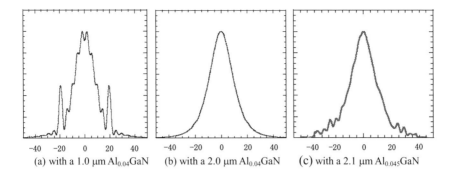

(a) with a 1.0 µm $Al_{0.04}GaN$ (b) with a 2.0 µm $Al_{0.04}GaN$ (c) with a 2.1 µm $Al_{0.045}GaN$

Figure 5. (a)(b) Calculated far-field patterns, and (c) experimental far-field pattern.

4. Reliability

The reliability of these lasers was improved by refining the epitaxial lateral overgrowth (ELO) [6-8] technique. The lifetime of GaN lasers has been found to be closely related to the dislocation density in the laser stripe region. To achieve lower dislocation density in this region, the window region has been extended, first by forming a 9 µm-wide wing in 2000, and more recently by fabricating a 18 µm-wide wing region with a minimum dislocation density of 2×10^6 cm^{-2}. The growth temperature on the 3-inch wafer was strictly controlled to 1070 ± 7 °C, and the distribution of thickness and Al content have been dramatically improved to 6.8 ± 0.3 µm and 6 ± 0.5%, respectively.

The advanced ELO technique employed successfully maintaining the total layer thickness of the laser structure to less than 7 µm and reduced wafer bending to as low as 60 µm. During photoresist processing, precise alignment is difficult on a wafer that exhibits significant bending. ELO with minimal bending therefore facilitates laser stripe alignment, allowing the laser to be constructed on the region of lowest dislocation density.

The relationship between the mean time to failure (MTTF) and the dark spot density (DSD), which corresponds to the dislocation density, was examined quantitatively, as shown in Fig. 6.

Dark spots on the surface were observed by photoluminescence of n-type GaN formed by RIE etching. This method is highly effective for measuring the dislocation density of laser devices directly, and allows the region on which the laser stripe is aligned to be visualized. As indicated in Fig. 6, the dislocation density should be less than 3×10^6 cm^{-2} to achieve a device lifetime of 5,000 h assuming 30 mW output power at 60 °C.

Lasers with internal loss of 14-16 cm^{-1} were tested under conditions of 50 mW output power at 70 °C for aging tests (Fig. 7). The estimated lifetime of these devices is 10,000 h, and the dislocation density is estimated to be $2-3 \times 10^6$ cm^{-2}.

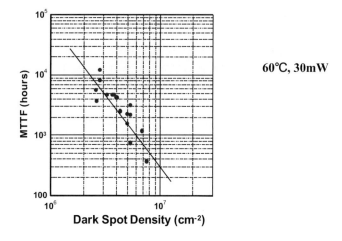

Figure 6. Dependence of lifetime on DSD.

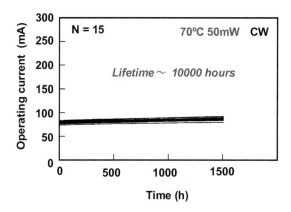

Figure 7. Lifetime test results of AlGaInN lasers grown on ELO substrate.

5. GaN substrate

As discussed above, the lifetime of GaN-based lasers is still governed by the density of dislocations. Therefore, GaN substrates with no dislocations are suitable for fabricating laser diodes. We have examined[9-10] lasers grown homoepitaxially on GaN substrates developed by Sumitomo Electric Industries, Ltd. The spatial distribution of dislocation density as measured by the growth pit technique revealed these devices to have dislocation densities of less than 1×10^6 cm^{-2} in 150 µm-wide periodic wing regions. The minimum dislocation density of 3×10^5 cm^{-2} is the lowest number that we have ever reported. Lasers fabricated with the conventional structure of lasers grown on ELO substrates, except for an n-type ohmic metal on the reverse side of the GaN substrate surface, exhibited a threshold current of 29 mA, a threshold voltage of 4.6 V, and a slope efficiency of 1.6 W/A. This slope efficiency is higher than that (1.4 W/A) of conventional lasers, due to the excellent facet cleavage of the GaN substrate.

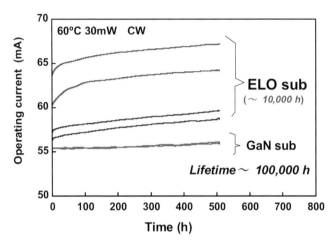

Figure 8. Lifetime test results of AlGaInN lasers grown on GaN substrate.

Figure 8 shows the results of a lifetime test under 30 mW continuous wave operation at 60 °C. These lasers exhibited extremely stable operations for more than 500 h without significant increase in operating current, and the lifetime of these two lasers is expected to exceed 100,000 hours. Considering the relationship between the dislocation density and lifetime as shown in Fig. 6, it is estimated these devices have less than 10 dislocations in the region of the laser stripes.

6. Conclusion

Threshold current of AlGaInN lasers was reduced to 27.5mA (3.2kA/cm^{-2}) by reducing the internal loss. The dominant factor of internal loss was found to be the absorption loss in Mg-doped super-lattice layers. The internal loss was decreased to 13.6cm^{-1} by isolating super-lattice layers from the active layer. Lasers with this structure have the improved characteristics such as slope efficiency (1.4 W/A), the internal quantum efficiency (0.94), and characteristic temperature T_0 (180 K). We also revealed the quantitative dependence of lifetime in AlGaInN lasers on the dislocation density. The lifetime of lasers employing an epitaxial lateral overgrown (ELO) substrate with low dislocation density of 3×10^6 cm^{-2} was estimated to 10,000h under conditions of 50 mW output power at 70 °C. Futhermore, the lifetime of lasers grown homoepitaxially on GaN substrate with low dislocation density of less than 1×10^6 cm^{-2} was estimated to 100,000h under conditions of 30 mW output power at 60 °C.

Acknowledgements

The authors would like to thank their colleagues at Sony Shiroishi Semiconductor Inc. and Sony Corporation for technical support and helpful discussion. Special thanks are extended to T. Asano, T. Tojyo, T. Hino, S.Kijima, K. Shibuya and S. Tomiya for their contribution. The authors also thank S. Tanemo and Dr. O. Kumagai for their encouragement during this work.

References

[1] Tojyo T, Uchida S, Mizuno T, Asano T, Takeya M, Hino T, Kijima S, Goto S, Yabuki Y, and Ikeda M 2002: Jpn. J. Appl. Phys. 41, 1829.
[2] Asano T, Takeya M, Tojyo T, Mizuno T, Ikeda S, Shibuya K, Hino T, Uchida S, and Ikeda M 2002: Appl. Phys. Lett. 80, 3497.
[3] Nagahama S, Iwasa N, Senoh M, Matsushita T, Sugimoto Y, Kiyoku H, Kozaki T, Sano M, Matsuura H., Umemoto H, Chocho K, and Mukai T2000: Jpn. J. Appl. Phys. 39, L647.
[4] Kuramoto M, Sasaoka C, Futagawa N, Nido M, and Yamaguchi A 2002: in: Proc. 4th Int. Symp. Blue Laser and Light Emitting Diodes, Cordoba, WeA5.
[5] Takeya M, Tojyo T, Asano T, Ikeda S, Mizuno T, Matsumoto O, Goto S, Yabuki Y, Uchida S, and Ikeda M 2002: in: Proc. 4th Int. Symp. Blue Laser and Light Emitting Diodes, Cordoba, WeB1.
[6] Nishinaga T, Nakano T, and Zhang S 1998: Jpn. J. Appl. Phys. 27, L964.
[7] H. Nam O, D. Bremser M, Zheleva T, and F. Davis R 1997, Appl. Phys. Lett. 71, 2638.
[8] Usui A, Sunakawa H, Sakai A, and Yamaguchi A 1997, Jpn. J. Appl. Phys. 36, L899.
[9] Takeya M, Yanashima K, Asano T, Hino T, Ikeda S, Shibuya K, Kijima S, Tojyo T, Ansai S, Uchida S, Yabuki Y, Aoki T, Asatsuma T, Ozawa M, Kobayashi T, Morita E, and Ikeda M 2000: J. Cryst. Growth, 221, 646.
[10] Goto S, Tamamura K, Matsumoto O, Tojyo T, Sasaki T, Yabuki Y, Naganuma K, Asatsuma T, Tomiya S, Uchida S and Ikeda M 2002: Late News Abst. 2002 Electronic Materials Conference, Santa Barbara, V9.
[11] Matsumoto O, Goto S, Sasaki T, Yabuki Y, Tojyo T, Tomiya S, Naganuma K, Asatsuma T, Tamamura K, Uchida S and Ikeda M 2002: Extended Abst. 2002 Int. Conf. on Solid State Devices and Materials, Nagoya, pp. 832.

III/V Nitride LEDs and lasers

B. Hahn, D. Eisert, J. Baur, M. Fehrer, S. Kaiser, H.-J. Lugauer, U. Strauss, A. Lell, V. Härle

Osram Opto Semiconductors, Regensburg, Germany

Abstract. On the recent LED designs three different main paths can be observed. First, low cost, shrinked devices operating at low voltage are focus of mobilecom application, second, high brightness devices driven at 20mA are targeted for mass markets such as the automotive market and third, high optical power devices for high flux outdoor light sources and general lighting are in the technological focus. All of these devices focus on high quantum and high extraction efficiencies leading to overall efficiencies as high as 25-30 lm/W and absolute light output as high as 30 lm of white light per single device. To reach such numbers new technologies on light generation and extraction have been developed. Another main path of research is focused on laser devices where applications as high optical data storage density, high resolution printing, spectroscopy & sensing, projection & display technology as well as general lighting are targeted. The laser research of today aims on especially long lifetime at elevated temperatures which is still the limiting factor to start mass market applications. Most likely these markets will develop in the next years, e.g. 2004 is addressed for "Blu-ray" DVD applications.

1. High brightness LEDs

Semiconductor light emitting devices have seen a rapid development over the last decade. Today high brightness LEDs are available for the whole visible spectral range. They compete with and even outperform conventional light sources like incandescent bulbs and halogen lamps. LEDs in increasing numbers are used in automotive, mobile and display applications where brightness, costs (size) and power consumption are key parameters. One important market for Osram OS is the automotive industry where interior lighting is already dominated by LEDs due to their high reliability. In the future LEDs will also be the choice for the exterior lights. White LEDs will strongly enter the market of general lighting. A low cost high performance solution of a white LED is realized by using blue Ga(In)N-chips combined with yellow phosphors. Key performance parameters of these patented device structures are luminous efficiency, which is already comparable to halogen lamps.

The success of LEDs in the lighting market was made possible by the emergence of new high brightness LEDs. The technological progress is due to the use of new material systems like InGaAlP and of course the nitride system giving access to the short wavelength range. The combination of advanced growth techniques like MOVPE and MBE on one hand and the design of high performance LED chips on the other hand lead to rather complex devices compared to conventional LEDs.

The SiC-based LEDs as they are produced by OSRAM have smaller size than sapphire-based chips, a true vertical current flow and an excellent stability against electrostatic discharge. These are the reasons why today the SiC-based chips dominate the automotive applications. However, the first SiC-based LEDs did not reach the brightness level of sapphire-based chips due to the worse light extraction and worse epitaxial quality. In the past large area displays and outdoor applications were dominated by sapphire-based LEDs. In 2000 OSRAM developed and patented the ATON® technology [2], which pushed the output power of blue LEDs on SiC above the limit of 5mW at 470 nm and 20mA [3][4][5]. The ATON® technology improves the light extraction efficiency from about 25% for a cubical chip design to about 55% for chips with a shaped substrate. The improved chip type has a strong emission at the chip side walls due to an undercut of the chip edges as shown in Fig. 1.

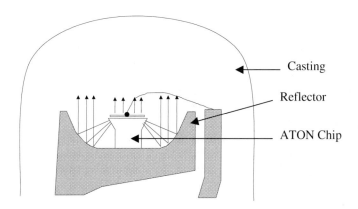

Fig. 1: Light emission of an ATON® chip in Radial housing. An efficient mirror is required for optimum operation of the chip.

By improving the quality of the epilayer and refining the chip design an output power of 9.5mW at 20mA has been demonstrated.

Applications in mobile electronics are pushing for miniaturisation of the LED packages. The reflectors which are necessary for efficient operation of the ATON® (Fig.1) can not be maintained in new miniature SMT package designs as the SMARTLED® [6]. In these packages the emitted light from the ATON® facets is mainly absorbed within the package or reflected back into the chip. For such devices the ATON® chips have to be improved. A straightforward way is a flipchip design, in this case the main light emission is toward the package surface and no reflector is needed. The light can leave the package on first incidence.

A new p-contact was developed for an optimum of light extraction with flip chip design. As p-GaN shows only neglible current spreading, a semitransparent metal p-contact has to be applied in the standard design. These current spreading layers consist of Ni, Au or Pt and are partially optically absorbing.

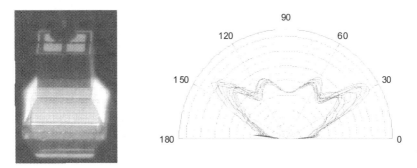

Fig. 2: Image of an ATON® flip chip. The main emission is directed upward. No external reflectors in the package are required.

Ray-Tracing analysis show that in standard chips >25% of the emitted light leaves the chip via the p-contact. Approx. 50% of this light is lost by absorption in the contact. These losses can be avoided by a flip chip design with highly reflective p-contacts. The mirror has to be carefully designed as non optimal mirror contacts can even increase these losses. By using Ag or Al based highly reflective contacts we could increase the device efficiency by 10-20%, yielding output powers of 10.5mW at 20mA in 5mm radial lamp.

However, the main progress is found in packages without reflector: here an increase of optical output power of the packaged device of 100% compared to ATON® or standard chips is observed. With the flip chip design light can leave the package on first incidence. The new design however requires also a more sophisticated mounting technology. As the pn-junction is separated less than 1μm from the mounting surface, standard gluing technologies can not be used anymore. Instead solder technologies have to be adopted and considered in the chip and package design.

2. High Power Applications

For general lighting high lumen values are required. This can hardly be achieved by accumulating appropriate numbers of "standard" chips. We need single chips which are able to emit tens of lumens. This can only be achieved by increasing the current in the single chip while maintaining the efficiency of a "standard" chip. For high current chips, the chip size must be drastically increased to avoid saturation effects in the active QW region and increased power loss by ohmic losses due to the increased current densities.

For the development of large chips the ATON® technology had to be adapted since the extraction probability for individual photons depends on the position where they are generated. With increasing distance from the shaped substrate edge, the light extraction probability is reduced. This is illustrated in Fig. 3: Area 1 marks the cone of angles, where light cannot be coupled into the substrate due to the difference in refractive indices between GaN (n=2,5) and 6H-SiC (n=2,7). This "dead"-cone covers angles up to 22,2° measured from the GaN/SiC-surface. Photons emitted into the cone of angles marked by area 2 are extracted from the chip, since this area fully overlaps with the light extraction cone of the shaped SiC surface for nearly all light generation positions on the chip. Photons emitted into the angles of area 3 are reflected on the vertical substrate side and cannot be extracted on first incidence. With increasing distance of light generation from the chip edge, the shaped substrate area is increasingly covered by the "dead"-angle area, where no photons are coupled into the substrate. Therefore, the benefits of the ATON® technology are strongly reduced for the inner area of large area chips. This is confirmed by ray-tracing calculations and experimental results, where the light extraction efficiency is reduced from 55% for a 290µm x 290µm chip to 28% for a 1000µm x 1000µm chip when the ATON® technology is applied to both chips. This means that by simply scaling the chip size, large area chips cannot compete in efficiency with standard ATON® chips.

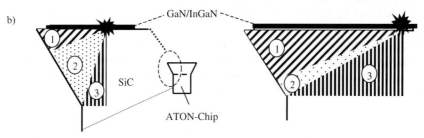

Fig. 3: Only photons emitted into the angle area (2) are extracted from the shaped substrate surface. This area is largest for photon generation close to the chip edge (a), while it decreases with increasing distance of light generation from the chip edge (b)

A technique to increase light extraction is to apply additional substrate shaping in the inner perimeter of the chip (Fig. 4a). However if the chip is mounted upside up, which is the standard technology for GaN on SiC chips, the light which is emitted in a inner groove can not be extracted. It is reflected at the leadframe and recoupled into the chip, where it is mostly absorbed within the structure. The problem is overcome when the chip is mounted upside down (Fig. 4b). In this case the light extraction efficiency is close to a "standard" chip and the chip area and therefore the lumen output is scalable.

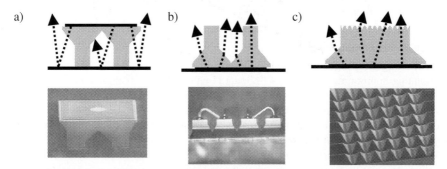

Fig. 4: a) Standard chip with additional substrate shaping; b) Flip Chip design; c) Final design with Fresnel structures

The chip design yields good light extraction, but in order to achive acceptable homogenous current densities in the active layer, each column has to be connected via an extra n-contact and an extra wire bond. This design is expensive and difficult to introduce in a mass production enviroment.

This disadvantage can be overcome when the substrate shaping is performed on a microscopic scale (similar to a Fresnel-lens; see Fig. 4c) or by roughening the surface in a suitable manner by a plasma process. Both techniques allow good current spreading either by metallic current spreading structures on the SiC surface (star-like or square-like contact geometry) or by using the anisotropic conductivity of the 6H-SiC substrate (e.g. $\rho\perp c \approx 3\times \rho\|c$) to achieve current spreading in the remaining substrate.

To complete the chip design well designed current spreading structures for high currents and highly reflective solderable ohmic p-contacts to reduce light absorbtion in the p-contacts on the leadframe side of the chips have to be applied.

In these chips the light extraction efficiency is only minimally affected by the chip size. The optimum size is defined by the applied current which determines the current density and therefore the internal quantum efficiency as well as the ohmic losses and the chip price which is kind of proportional to the chip area. To provide a proper thermal environment and prevent efficiency losses by thermal quenching a special moldable SMT package with a thermal series resistance well below 10 K/W was developed.

Viewing these parameters, we optimized a substrate shaped 1000µm×1000µm InGaN on SiC chip, which is designed for a forward current of 350mA. Soldered up-side down into the above mentioned SMT-package emission powers of 150mW of blue light (λ_{dom}=460nm) are achieved at a forward current of 350mA and a forward voltage of 3,9V. If the blue emission is converted to white light by adding a suitable amount of cerium doped yttrium aluminum garnet (YAG:Ce) into the applied resin, 33lm of white light are generated. These emission values are comparable to values achieved for large area chips bases sapphire substrates [9]

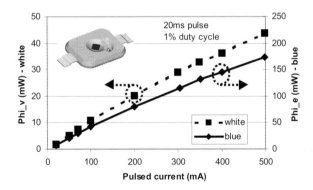

Fig. 5: Φ_e and Φ_v characteristics of a 1mm² InGaN on SiC chip mounted up side down into our special SMT power-package (see inset)

3. Laser

Research and development activities concerning gallium nitride based laser diodes are of high importance for the improvement of existing technologies as well as for new applications. Mainly the Blu-Ray Disc for data and video storage, but also laser printing, laser projection techniques like beamer or laser TV are regarded as key markets for GaN lasers.

In 1998 Osram Opto Semiconductors in Regensburg, the Fraunhofer Institute for Applied Solid State Physics in Freiburg, as well as the universities of Brunswick, Stuttgart and Ulm started the research project „Gallium nitride Laser Diodes" supported by the German BMBF. Following the first presentation of blue-violett laser diode [10] major steps towards a long-lived operation have been made [11][12].

Substrate and laser diode structure

SiC is an attractive choice until GaN substrates with sufficient quality and quantity are available. The conductive SiC offers in contrast to insulating sapphire the advantage of a true vertical chip structure with substrate n-electrode. This simplifies laser technology and packaging. InGaN technology on SiC also enables high quality cleaved laser mirrors (with facet roughness of less than 1nm). It does not need a thick lateral n-conductive layer and therefore avoids problems with a second waveguide which are critical in sapphire based technology. SiC (thermal conductivity of 4,9 W/cmK) is also an excellent heat spreader (see for comparison: sapphire: 0,46W/cmK and GaN 1,3W/cmK). [13]

A typical laser structure grown by metal organic vapour phase epitaxy consists of n/p-AlGaN cladding, n/p-GaN waveguide and three $In_{0.10}Ga_{0.9}N$ / GaN multiple quantum wells with an $Al_{0.2}Ga_{0.8}N$ electron blocking layer. Contacts are deposited on a p-GaN contact layer and on the n-SiC backside.

Ridge waveguide lasers are structured and laser facets with typical cavity length of 600μm are formed by cleaving. Finally laser facets are finished with high-reflectivity coatings (50% to 70% and 98%).

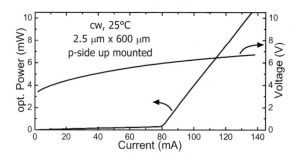

Fig 6. p-side up mounted ridge laser diode (2.5 μm x 600 μm) voltage-current (VI) and light-current (LI) characteristics under cw-operation at 25 °C. The cw threshold current (Ith) is 80 mA (threshold current density jth=5.3 kA/cm2), threshold voltage Uth = 5.9 V, slope efficiency ◻= 0.2 W/A and emission wavelength ◻= 407 nm

Laser results

In July 1999 the first pulsed laser operation [14] was observed for gain guided laser structures with 900µm cavity length and cleaved, highly reflecting laser facets. The laser structure was grown on a SiC substrate and exhibited pulsed laser emission at room temperature. The threshold current density was below 20 kA/cm^2 and the turn-on voltage was about 25V at an emission wavelength of 420nm. Following theoretical analysis the epitaxial structure could be optimised with respect to current injection and internal quantum efficiency. Together with an optimised index guiding structure a significant reduction of the laser threshold current from 1200mA to less than 100mA could be observed [15].

Thanks to the close co-operation of the BMBF partners, as well as the inhouse experience in GaN based LED technology and laser technology on different material systems, Osram Opto Semiconductors succeeded in cw-operation at room temperature in March 2001.

Actual results of InGaN laser diodes with a cavity lenght of 600µm and facet reflectivities of 70% and 98% are shown in Fig. 6. The lasers show cw threshold currents of 80mA (threshold current density j_{th} = 5,3kA/cm²) and a threshold voltage of 5,9V.

Up to now cw-operation @ 1mW optical output power could be extended from few minutes to more than 100 hours at room temperature [17] without sophisticated defect reducing structures (see Fig. 7). Further improvements will be expected by reducing the defect density.

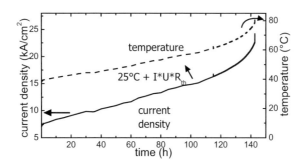

Fig 7 Current density and temperature during 143 h of cw lasing with 1 mW optical power and 25 °C ambient temperature. Ridge-Laser diode with 2.75 µm x 600 µm dimension mounted p-side up on heatsink.

Acknowledgement

This work was partially supported by the German Ministry of Education+Research (BMBF).

References

[1] S. Nagahama, N. Iwasa, M. Senoh, T. Matsushita, Y. Sugimoto, H. Kiyoku, T. Kozuki, M. Sano, H. Matsumura, H. Umemoto, K. Chocho, T. Mukai, Proceedings of the International Workshop of Nitride Semiconductors, ISBN 4-9000526-13-4, p. 899 (2000)

[2] Compound Semiconductors **7**, 7 (2001)

[3] D. Eisert, S. Bader, H.-J. Lugauer, M. Fehrer, B. Hahn, J. Baur, U. Zehnder, N. Stath and V. Härle, Proceedings of the International Workshop of Nitride Semiconductors, ISBN 4-9000526-13-4, p. 841 (2000)

[4] V. Härle, B. Hahn, H.-J. Lugauer, S. Bader, G. Brüderl, J, Baur, D. Eisert, U. Strauss, U. Zehnder, Lell, and N. Hiller, Phys. Stat. Solidi (a) **180**, 5 (2000)

[5] U. Zehnder, A. Weimar, U. Strauss, M. Fehrer, B. Hahn, H.-J. Lugauer, and V. Härle, J. Crystal Growth 230, 497 (2001)

[6] www.osram-os.com

[7] Compound Semiconductor 6, 11 (2000)

[8] T. Mukai, IEEE Journal of Selected Topics in Quantum Electronics 8, 264 (2002)

[9] D. Steigerwald, IEEE Journal of Selected Topics in Quantum Electronics 8, 310 (2002)

[10] S. Nakamura, M. Senoh, S. Nagahama, N. Iwasa, T. Yamada, T. Matsushita, H. Kiyoku, Y. Sugimoto, Jpn. J. Appl. Phys. **35**, L 74 (1996)

[11] S. Nagahama, N. Iwasa, M. Senoh, T. Matsushita, Y. Sugimoto, H. Kiyoku, T. Kozaki, M. Sano, H. Matsumura, H. Umemoto, K. Chocho, T. Mukai , Jpn. J. Appl. Phys. **35**, L 74 (1996)

[12] T. Asano, M. Takeya, T. Tojyo, T. Mizuno, S. Ikede, K. Shibuya, T. Hino, S. Uchida, M. Ikeda, Appl. Phys. **80**, 3497 (2002)

[13] V. Härle, B. Hahn, H.-J. Lugauer, S. Bader, G. Brüderl, J. Baur, D. Eisert, U. Strauss, A. Lell, N. Hiller, phys. Stat. Sol. (a) **180**, 5 (2000)

[14] S. Bader, B. Hahn, H.-J. Lugauer, A. Lell, A. Weimar, G. Brüderl, J. Baur, D. Eisert, M. Scheubeck, S. Heppel, A. Hangleiter, V. Härle, phys. stat. sol.(a) **180**, 177 (2000)

[15] V. Härle, B. Hahn, H.-J. Lugauer, G. Brüderl, D. Eisert, U. Strauss, A. Lell, N. Hiller, Compound Semiconductors 6(8), 1, Nov 2000

[16] A. Weimar, A. Lell, B. Brüderl, S. Bader, V. Härle, phys. stat. sol. (a) **183**, 169 (2000)

[17] V. Kümmler, G. Brüderl, S. Bader, S. Miller, A. Weimar, A.Lell, V. Härle, U. Schwarz, N. Gmeinwieser, W. Wegscheider (to be publised in phys. stat. sol.)

Recent advances in continuous wave quantum cascade lasers

D Hofstetter, M Beck, S Blaser, T Aellen, J Faist, U Oesterle, M Ilegems, E Gini, and H Melchior

University of Neuchâtel, 1 A.-L. Breguet, 2000 Neuchâtel, Switzerland

Abstract. Continuous wave (CW) operation of quantum cascade lasers is reported up to a temperature of 312 K. The junction down mounted devices were designed as buried heterostructure lasers with high-reflection coatings on both facets. This resulted in CW operation at an emission wavelength of 9.1 µm with an optical power ranging from 17 mW at 293 K to 3 mW at 312 K. A distributed feedback type device was fabricated and tested as well. It showed CW singlemode operation up to 260 K. These results demonstrate the potential of quantum cascade lasers as CW mid-infrared light sources for high-resolution spectroscopy and free space telecommunication systems.

1. Introduction

In the mid-infrared portion of the electromagnetic spectrum, it used to be difficult to produce convenient semiconductor lasers with sufficiently high output powers. So far, the highest operating temperature of a CW mid-IR semiconductor laser was 243 K, which is just high enough to be maintained by a Peltier cooler. Lead salt lasers [1] routinely achieve CW operation up to a temperature of 150 K, with a record value of 220 K. Using antimony-based III-V lasers, CW operation was obtained at room temperature for wavelengths as long as $\lambda = 2.6$ µm [2]. However, the performances of these devices degrade dramatically as the wavelength is extended beyond $\lambda = 3$ µm. Growing exponentially with temperature, Auger recombination is the most important limitation for high temperature operation of interband lasers.

Because quantum cascade (QC) lasers, in which photon emission is obtained by electrons undergoing optical transitions between confined intersubband levels [3], are limited by a completely different non-radiative mechanism, namely optical phonon emission, it quickly became clear that this new approach could lead to devices operating at higher temperatures [4]. High power pulsed operation of such devices was indeed achieved up to temperatures much higher than 400 K [5]. Continuous wave operation of these devices, however, was restricted to cryogenic temperatures with a maximum temperature of 243 K [6].

Although chemical sensing based on optical absorption has been successfully demonstrated using pulsed QC lasers [7], these systems are typically limited by the fairly wide emission linewidth of the light source (>500MHz); very high sensitivity can only be achieved with the narrow linewidth of a CW operated device which does not suffer from thermal chirping [8]. Quasi-CW lasers are also needed for mid-infrared free-space optical telecommunication systems [9], [10]. Such in-the-field applications will benefit greatly from non-cryogenic light sources and detectors.

In the early attempts, the lasers used to reach room temperature CW operation [11], [12], [13] had, at threshold and for 300 K, a typical power consumption of $P = j_{th} \times V = 5 \text{ kA/cm}^2 \times 10 \text{ V} = 50 \text{ kW/cm}^2$). Even assuming an idealized device geometry, this value was unacceptably high and resulted in an overheating of the device. For this reason, new approaches for the gain region such as devices based on double-phonon resonance [14] and the bound-to-continuum design [15] were developed. As a consequence of these novel gain region designs, the pulsed threshold current density at 300 K dropped to a value as low as 3 kA/cm^2 [5].

In the work presented here, QC lasers with a gain region based on a double-phonon resonance are grown in a narrow stripe, buried heterostructure geometry in which the multi-quantum well active region is vertically and laterally surrounded by InP [16]. The choice of a narrow stripe greatly improves the lateral heat transport by allowing heat extraction towards all sides of the active region. Moreover, it also decreases the total amount of strain that builds up in a material subjected to a very strong temperature gradient. In order to further reduce the produced heat in such lasers, the devices were cleaved to a relatively short cavity length but fitted with high-reflectivity (70%) coatings on both facets and mounted junction down on a diamond heatsink.

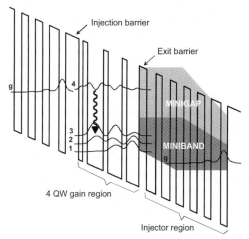

Figure 1. Schematic view of one period of the active region. Shown are the moduli squared of the relevant wavefunctions, a wavy line corresponding to the lasing transition, and the injector miniband/minigap. The layer sequence starting from the injection barrier is as follows: **40**/19/7/**58**/9/**57**/9/**50**/**22**/34/**14**/33/**13**/32/**15**/<u>31</u>/**<u>19</u>**/<u>30</u>/**<u>23</u>**/29/**25**/29 Å.
InGaAs wells are in roman, and InAlAs barriers in bold. Doped layers (Si, 2×10^{17} cm^{-3}) are underlined.

2. Experiment

Fabrication of the laser structure relied on molecular beam epitaxy (MBE) for the growth of the waveguide core (lower waveguide layer, active region and upper waveguide layer). Metalorganic vapor phase epitaxy (MOVPE) was used for the growth of the InP top cladding layer and for the re-growth of the buried heterostructure. The two ternary InGaAs and InAlAs semiconductor compounds were grown lattice matched to the low n-type doped InP substrate. As shown in Fig. 1, the material between the two waveguide layers consisted of 35 repetitions of alternating undoped double-phonon resonance gain regions with 4 quantum wells and partially n-doped injector regions. The injector ground state, g, is in resonance with the upper laser state, 4.

The gain region used a narrow QW/barrier pair just behind the injection barrier to increase the injection efficiency, a double-phonon resonance for the lower laser level (i.e. between levels 3->2 and 2->1) in order to shorten the lower state lifetime, and a vertical lasing transition at an energy of 135 meV (levels 4->3). Similar to the device presented in ref. [5], it was optimized for high temperature operation. After growth of the waveguide core, the QC structure was completed by epitaxial re-growth of the n-doped InP top cladding and contact layers. Laser samples were lithographically processed into 12 μm wide (Fabry-Pérot device) or 8 μm wide (distributed feedback device) and 5 μm deep mesa etched ridge waveguides using a SiO_2 mask. A planarized buried heterostructure was then formed by lateral overgrowth of 4 μm non-intentionally doped InP. After this last regrowth step, the SiO_2 film was removed and a 100 nm thin Si_3N_4 electrical isolation layer was deposited on top of the nominally undoped material in order to reduce parasitic currents outside the waveguide (the low 10^{15} cm^{-3} n-type background doping of the non-intentionally doped InP might have led to leakage currents when being in direct contact with a metal). Laser fabrication was completed by electron-beam evaporation of Ti/Ge/Au/Ag/Au contacts on the highly doped top contact layer and a standard Ge/Au/Ag/Au metallization on the back of the thinned substrate. The length of the optical cavity was defined by cleaving the samples into 750 μm long bars. Single lasers were then mounted junction down on diamond platelets, which were soldered on copper heatsinks. The diamond-on-copper soldering was done with PbSn while the lasers were soldered on the diamond using In. After soldering, both laser facets were high-reflection coated with one pair of ZnSe/PbTe layers, resulting in a facet reflectivity of R ~ 0.70. For our particular geometry, the threshold current decreased by about 30 % after the facet coatings. The maximal output power remained roughly at the initial value, however.

3. Measurement results

Despite the fact that the lasers were operated at or even above room temperature, a considerable amount of heat needed to be evacuated. For this reason, the lasers were mounted on the cold finger of an N_2-flow cryostat. The optical output power emitted from one facet was measured using a calibrated thermopile detector which was held directly in front of the laser facet. For the acquisition of emission spectra, the light was collected with f/0.8 optics and sent through a Fourier transform infrared spectrometer (FTIR). In this case, the detection was done with a liquid nitrogen cooled HgCdTe detector.

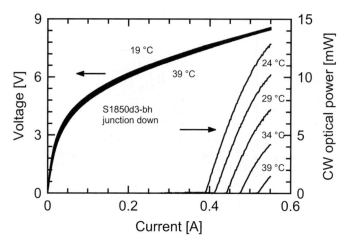

Figure 2. L-I-V-curves of a continuous wave FP QC laser at temperatures ranging from 19 °C to 39 °C. The intermediate curves are measured every 5 degrees. The uppermost I-V-curve corresponds to the lowest temperature.

Fig. 2 shows light intensity and bias voltage vs. current (L-I- and V-I-) curves for a 750 μm long and 12 μm wide laser under CW operation. At 292 K, the laser had a threshold current of 390 mA which corresponds to a threshold current density of j_{th} = 4.3 kA/cm^2. A threshold voltage of 7.65 V and a slope efficiency dP/dI of 101 mW/A were seen. At 550 mA, the optical power from one facet was 13 mW; this is equivalent to a wall plug efficiency of 0.55 % per facet. A laser with the same cavity length but a slightly larger stripe width of 15 μm emitted 17 mW per facet at a drive current of 600 mA. Its threshold current density was 3.7 kA/cm^2, and a slope efficiency of 118 mW/A was observed. The highest CW operating temperature of the narrower device was 39 °C; the threshold current at this higher temperature increased to 520 mA (j_{th} = 5.8 kA/cm^2), while still 2 mW output power were emitted. The device with the larger stripe width could be operated up to 38 °C with a maximum optical power of 3 mW and a threshold current of 540 mA (j_{th} = 4.8 kA/cm^2). From the temperature dependence of the threshold current density, we derived a T_0 value of 68 K.

Figure 3 shows, for the 12 μm wide laser, a series of emission spectra as a function of temperature. The series starts at 292 K and goes in steps of 2 K up to 312 K (except for 306 K to 307 K). When changing the injection current from 395 mA to 525 mA at a constant temperature of 292 K, we observed a linear tuning as function of the injected power and again singlemode operation. This particular device always lases in the same longitudinal mode; a fact that can be explained by a small defect within the cavity. We confirmed this hypothesis by high resolution sub-threshold spectra; they revealed a pronounced intensity modulation with a period of approximately twice the Fabry-Pérot mode spacing. Since the modal tuning is a direct measurement for the averaged active region temperature, the two tuning experiments allowed us the calculation of the thermal resistance of the laser.

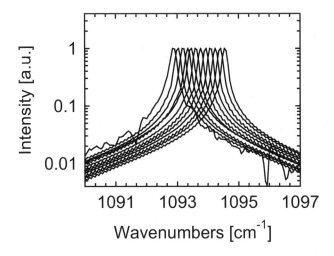

Figure 3. CW emission spectra of a 750 μm long and 12 μm wide Fabry-Perot laser at different temperatures and at 530 mA. The spectrum at the right is at a temperature of 292 K, the one at the far left at 313 K. Except for the spectra at 306 and 307 K, the temperature interval between subsequent spectra is always 2 K.

With $\beta(292\ K) = 1094.55\ cm^{-1}$ and $\beta(313\ K) = 1092.89\ cm^{-1}$, we found a temperature tuning value of $\Delta\beta/\Delta T = -0.078\ cm^{-1}/K$. Keeping in mind that the above temperature tuning is based on the averaged refractive index change across the entire laser structure, we can calculate a preliminary value for the thermal resistance using the formula

$$R_{th} = \frac{\Delta T}{\Delta P} = \frac{[\beta(I_1) - \beta(I_2)]/\left(\frac{\Delta\beta}{\Delta T}\right)}{I_1 V_1 - I_2 V_2}$$

where $I_1 = 395$ mA, $I_2 = 525$ mA, $V_1 = 7.65$ V, and $V_2 = 8.43$ V are corresponding current and voltage values on the V-I-curve at 19 °C. With these numbers, we find R_{th} = 19.4 K/W, or, using the device area of 750 μm x 12 μm, g_{th} = 574 W/Kcm². When using the sub-threshold spectra, we found slightly better, but nevertheless similar values of R_{th} = 16.9 K/W, and g_{th} = 658 W/Kcm². If we calculate now the expected temperature increase from zero injection based on the V-I-curves at the corresponding temperatures and the formula $\Delta T = R_{th} \times I \times V$, then we find ΔT = 58 K for 395 mA / 7.65 V / 19 °C and ΔT = 87 K for 525 mA / 8.43 V / 39 °C. With the same model, we can also calculate the highest possible temperature at which CW lasing could theoretically be achieved. This process yielded a value of 48 °C. During these spectral measurements, the laser was turned on for four hours without interruption. A control measurement at 395 mA at the end of the series showed that the laser did not suffer from any notable degradation.

By measuring a series of sub-threshold spectra and using a slightly modified Hakki-Paoli technique [17], we were able to determine an approximate value for the waveguide loss of the laser cavity. We came up with a value of about 19 cm⁻¹. This relatively high value can be explained by the slightly too thin InP upper cladding layer which results in an additional, top metal-induced absorption loss.

Far field distributions in the two directions parallel and perpendicular to the grown layers were measured on the 15 μm wide device under pulsed conditions and at four representative injection currents. In both directions, we observed Gaussian distributions which did not change at higher injection. In the direction parallel to the layers, we found a far field angle of 40 ° full width at half maximum (FWHM). In the other direction, the angle was on the order of 80 °. The Gaussian distributions prove that the device oscillates in its fundamental lateral and transverse mode, which makes this laser much more convenient for potential applications.

Using the same concepts of buried heterostructures, junction down mounting, facet coatings, and diamond heatsinking, we fabricated and tested also real singlemode distributed feedback lasers. Since the stripe width was reduced to a somewhat too small value, namely about 8-10 μm, the top contact resistance increased. This led to a slightly worse temperature behavior. CW operation was observed up to 260 K with several mW of singlemode output power. The facet coatings were this time based on two quarter wave layers of Al_2O_3 and Ge and had a reflectivity of about 70 %. The L-I-V-curves as well as an emission spectrum are shown in figure 4.

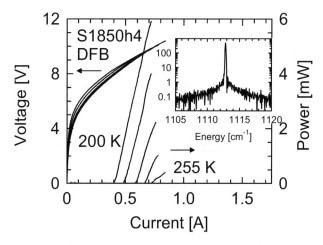

Figure 4. L-I-V-curves of a continuous wave DFB laser at temperatures ranging from 200 K to 255 K. The intermediate curves are measured at 220 K, 240 K, and at 250 K. The inset shows a singlemode spectrum of this device at 240 K. As usual, the uppermost I-V-curve corresponds to the lowest temperature

Why could the highest operating temperature suddenly be increased to such a high value? First and foremost, the material was of excellent quality and had a very low pulsed threshold current density. As mentioned in the introduction, this was mainly due to the new active region design based on a double-phonon resonance, but also due to efficient high reflection coatings on both facets. An additional point was, of course, that narrower stripes (now 12/15 μm instead of former 28 μm) greatly improved the heat transport in all directions and thus the thermal conductance of the laser. But there was another important point, which never received the attention it actually deserved: thermal stress.

By making narrower devices, one can significantly reduce the shear stress due to thermal expansion mismatch. For CW operation at the limiting operating conditions, where the active region is at a temperature of almost 150 °C, this point becomes a very crucial one. Using a finite element solver, we computed the shear stress of a 28 μm wide ridge waveguide laser and of a 12 μm wide buried heterostructure device. For the former, we found the highest stress value of 22 MPa at the top corners of the ridge, for the latter device, it was only 4 MPa. In agreement with this observation, the 28 μm wide device failed already at 4 kA/cm^2, the 12 μm laser presented in this paper supported nearly 6 kA/cm^2. Another indication for the thermal stress being the limiting factor is given by the observation that burnt lasers had always failed by short-circuiting across the facet. Within the cavity the thermally induced shear stress is distributed symmetrically. Since this symmetry is locally broken at the facet, it is no surprise that catastrophic damage occurs mostly there.

4. Conclusions

In conclusion, we have presented a QC laser with a novel type of active region which allowed CW operation above room temperature. The active region was based on a double-phonon-resonance which led to an extremely efficient extraction of the electrons in the lower lasing state. Junction down mounting, high reflection coatings, and narrow stripe buried heterostructure architecture helped to reduce thermal stress and threshold current density of this device. The maximal operating temperature was 39 °C, 62 K higher than what was published previously.

The authors gratefully acknowledge Martin Ebnöther for technical assistance with the lateral InP regrowth. This work was financially supported by the Swiss National Science Foundation and the Science Foundation of the European community under BRITE/EURAM project SUPERSMILE.

References

[1] M. Tacke, Infrared Physics. Technol., **36**, 447 (1995)

[2] D. Garbuzov, M. Maiorov, H. Lee, V. Khalfin, R. Martinelli, and J. Connolly, Appl. Phys. Lett., **74**, 2990 (1999)

[3] J. Faist, F. Capasso, D.L. Sivco, A.L. Hutchinson, and A.Y. Cho, Science **264**, 553 (1994)

[4] J. Faist, F. Capasso, D.L. Sivco, C. Sirtori, A.L. Hutchinson, and A.Y. Cho, Appl. Phys. Lett., **65**, 2901 (1994)

[5] D. Hofstetter, M. Beck, T. Aellen, J. Faist, U. Oesterle, M. Ilegems, E. Gini, and H. Melchior, Appl. Phys. Lett., **78**, 1964 (2001)

[6] C. Gmachl, A.M. Sergent, A. Tredicucci, F. Capasso, A.L. Hutchinson, D.L. Sivco, J.N. Baillargeon, S.-N.G. Chu, and A.Y. Cho, IEEE Photonics Technol. Lett., **11**, 1369 (1999)

[7] D. Hofstetter, M. Beck, J. Faist, M. Nägele, and M.W. Sigrist, Optics Lett., **26**, 887 (2001)

[8] B.A. Paldus, T.G. Spence, R.N. Zare, J. Oomens, F.J.M. Harren, D.H. Parker, C. Gmachl, F. Capasso, D.L. Sivco, J. N. Baillargeon, A.L. Hutchinson, and A.Y. Cho, Opt. Lett., **24**, 178 (1999)

[9] S. Blaser, D. Hofstetter, M. Beck, and J. Faist, Electron. Lett., **37**, 778 (2001)

[10] R. Martini, C. Gmachl, J. Falciglia, F.G. Curti, C.G. Bethea, F. Capasso, E.A. Whittaker, R. Paiella, A. Tredicucci, A.L. Hutchinson, D.L. Sivco, and A.Y. Cho, Electron. Lett., **37**, 111 (2001)

[11] C. Sirtori, J. Faist, F. Capasso, D.L. Sivco, A.L. Hutchinson, and A.Y. Cho, IEEE J. Quantum Electron., **33**, 89 (1997)
[12] C. Gmachl, F. Capasso, J. Faist, A.L. Hutchinson, A. Tredicucci, D.L. Sivco, J.N. Baillargeon, S.-N.G. Chu, and A.Y. Cho, Appl. Phys. Lett., **72**, 1430 (1998)
[13] C. Sirtori, J. Faist, F. Capasso, D.L. Sivco, A.L. Hutchinson, and A.Y. Cho, Appl. Phys. Lett., **66**, 3242 (1995)
[14] D. Hofstetter, M. Beck, T. Aellen, and J. Faist, Appl. Phys. Lett., **78**, 396 (2001).
[15] J. Faist, M. Beck, T. Aellen, and E. Gini, Appl. Phys. Lett., **78**, 147 (2001)
[16] M. Beck, J. Faist, U. Oesterle, M. Ilegems, E. Gini, and H. Melchior, IEEE Photonics Technol. Lett., **12**, 1450 (2000)
[17] D. Hofstetter and J. Faist, IEEE Photonics Tech. Lett., **11**, 1372 (1999)

Edge- and Surface-Emitting Photonic-Crystal Distributed-Feedback Lasers

I. Vurgaftman, W. W. Bewley, C. L. Canedy, J. R. Lindle, C. S. Kim, and J. R. Meyer

Code 5613, Naval Research Laboratory, Washington, DC 20375

S. J. Spector, D. M. Lennon, G. W. Turner, and M. J. Manfra

Lincoln Laboratory, Massachusetts Institute of Technology, Lexington, MA 02420

Abstract. 2nd-order and 1st-order optically pumped edge-emitting PCDFB lasers with "W" active regions are investigated. While the 2nd-order grating was not optimized due to the limitations of optical lithography, the 1st-order grating was fabricated using electron-beam lithography with a nearly optimum set of coupling coefficients. Unfortunately, detailed characterization revealed that the 1st-order grating resonance was detuned from the gain spectrum of the device's active region. In spite of the non-optimized coupling and substantial detuning, however, the beam quality for both grating types improved by a factor of 5 over angled-grating (α-DFB) lasers fabricated from similar material. Although the devices were operated in pulsed mode, a considerable improvement in the spectral purity was also realized, with linewidths as narrow as 7 nm for those temperatures where the gain spectrum was closest to the cavity resonance. We also discuss the surface-emitting PCDFB laser concept, and project the optimized parameter space for which a large fraction of the generated photons can be emitted into a near-diffraction-limited single mode.

1. Introduction

High-power semiconductor lasers that produce a near-diffraction-limited output beam are needed for a wide variety of commercial and military applications. Thus far, the goal of combining high output power with good beam quality and spectral purity in semiconductor lasers has been elusive. This is principally due to the filamentation that develops in the most straightforward implementation based on broad-area gain-guided devices[1,2] and the relative complexity of numerous other suggested approaches.[3,4] Recently, we theoretically investigated the performance of a novel class of coherent semiconductor lasers with wide stripes in the edge-emitting configuration: the photonic-crystal distributed-feedback (PCDFB) laser.[5] The conventional one-dimensional (1D) DFB laser and the angled-grating DFB (α-DFB) laser[6,7] are special cases of the PCDFB, which combines and enhances their best features.

Figure 1 schematically shows the top view of a rectangular grating for an edge-emitting PCDFB laser with an aspect ratio of $\tan\theta = \Lambda_2/\Lambda_1$ between the two periods. There are two equivalent propagation directions, $\mathbf{P}_1$ and $\mathbf{P}_2$, which form equal angles of θ with respect to the facet normal. The 2D pattern allows simultaneous diffraction in three directions (*e.g.*, from $\mathbf{P}_1$ to $\mathbf{P}_2$, $-\mathbf{P}_1$, and $-\mathbf{P}_2$), which leads to strong spectral filtering (similar to the 1D DFB action) and coupling of different points across the laser stripe (similar to the α-DFB laser).[8]

In Section 2, we briefly review the main principles of the edge-emitting PCDFB operation. This will be followed by our results on edge-emitting devices including both 2nd-order gratings fabricated by optical lithography and 1st-order gratings fabricated by electron-beam lithography in Section 3. In Section 4 we describe another novel PCDFB device capable of producing high-power output with good beam and spectral quality, the surface-emitting PCDFB laser. The main conclusions of this work will be presented in Section 5.

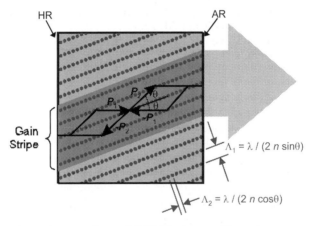

Figure 1. Schematic top view of PCDFB laser with various propagation directions described in detail in the text.

2. Edge-emitting PCDFB theory

Performing a transformation to the reciprocal space, the vectors corresponding to the four propagation directions of the rectangular lattice become: $\mathbf{P}_1 = \{l/2, m/2\}$, $-\mathbf{P}_1 = \{-l/2, -m/2\}$, $\mathbf{P}_2 = \{-l/2, m/2\}$, and $-\mathbf{P}_2 = \{l/2, -m/2\}$, where the notation $\{l, m\}$ defines the order of diffractive coupling in terms of the primitive reciprocal lattice vectors. That is, $\{l, m\} = l\mathbf{b}_1 + m\mathbf{b}_2$, $\mathbf{b}_1 = 2\beta(\sin\theta, 0)$ and $\mathbf{b}_2 = 2\beta(0, \cos\theta)$, where $\beta = \pi/(\Lambda_2\cos\theta)$ is the diagonal distance from the center to the corner of the 1st Brillouin zone (the P point). We use a shorthand of 1st-order and 2nd-order gratings for devices designed to operate in the $\{1,1\}$ and $\{2,2\}$ diffraction orders, respectively.

The coefficients κ_1', κ_2', κ_3' are introduced[8,9] to account for diffractive coupling of the $\mathbf{P}_1$ beam to $-\mathbf{P}_1$, to $\mathbf{P}_2$, and to $-\mathbf{P}_2$, respectively, as well as the other symmetry-equivalent couplings. In other words, κ_1' accounts for DFB-like distributed reflection (by 180°), κ_2' represents α-DFB-like diffraction by an angle 2θ from the "nearly-horizontal" Bragg planes, and κ_3' quantifies diffraction by an angle $(180°-2\theta)$ from the "nearly-vertical" Bragg planes. We conclude that: (a) *both* the 1D DFB and α-DFB configurations are special cases of the rectangular PCDFB lattice; and (b) a general PC grating is *not* "separable", *i.e.*, it is not equivalent to a superposition of two mutually-perpendicular line gratings whenever $\kappa_1 \neq 0$. In general, optimizing a grating of order higher than {1,1} requires a larger index perturbation Δn.

In order to calculate the near-field, far-field, and spectral characteristics of the PCDFB lasers, we employ the time-domain Fourier-transform (TDFT) approach developed specifically for that purpose.[8] By analyzing the propagation equations for the optical field, it is quite clear that the primary figure of merit governing the performance of PCDFB lasers is the product of the threshold gain and the linewidth enhancement factor (LEF). The threshold gain (scaled by the confinement factor Γ) comprises contributions from the internal loss of the lasing material, reflectivity losses from the uncoated facets, and any diffraction loss that is particularly important for tilted gain-guided laser stripes used in α-DFB and rectangular-lattice PCDFB lasers. Whereas a mid-IR PCDFB structure should perform relatively well even when this figure of merit is rather large, *for edge emitters it is always advantageous to minimize both the internal loss and the LEF as much as possible*. However, we will see below that this conclusion does not hold for surface-emitting PCDFB devices. Experimental results to date indicate that both the internal loss and LEF for current mid-IR antimonide "W" active regions may be relatively attractive at lower temperatures in the 77 K range, although they may become substantially larger at higher *T*.

3. Edge-emitting PCDFB Experimental Results

The fabrication and testing of a 2nd-order PCDFB laser with a "W" active region was reported previously.[10] The grating was oriented at $\theta = 20°$ with respect to the facet normal, and had an aspect ratio of tan(20°). The circular features, with diameter equal to 0.67Λ_2 ($\Lambda_2 = 1.49$ µm), were patterned by *i*-line projection lithography, yielding the following coupling coefficients: $\kappa_1' = -7$ cm^{-1}, $\kappa_2' = 40$ cm^{-1}, and $\kappa_3' = -4$ cm^{-1}. These non-optimal coupling coefficients (*e.g.*, κ_1' is too high) were constrained by the minimum resolution of the optical lithography.

The device with uncoated facets was cleaved to a cavity length of 2.5 mm and pumped by 100 ns pulses (1 Hz repetition) from a 2.1 µm Ho:YAG laser. While the PCDFB laser did not display single-mode output, the spectral linewidth was narrower than for any earlier antimonide mid-IR lasers pumped far above threshold, apart from an external-cavity study.[11] Figure 2 illustrates the spectra of PCDFB and Fabry-Perot (FP) devices fabricated from the same wafer at several temperatures. The FP linewidths ranged from 43 nm FWHM at 180 K to 69 nm at 240 K, and the earlier α-DFB devices showed very little narrowing, *e.g.*, to 39 nm at their resonance temperature of 150 K. Other mid-IR α-DFB lasers also failed to show significant spectral narrowing.[12] At lower temperatures such as $T = 135$ K, where the peak wavelength of the gain spectrum is much shorter than the grating resonance, the PCDFB spectrum contains a number of oscillations. While the device is able to sustain lasing in that region, the threshold is several times the Fabry-Perot value.

At a temperature where the gain spectrum is near the {2,2}-order resonance of the grating, the PCDFB emission collapsed to a single line (FWHM = 10 nm centered at $\lambda = 4.59$ µm). When the temperature was increased further, a second line emerged at $\lambda = 4.70$ µm. The presence of two separate resonances was attributed to an accidental near-degeneracy between the wavelengths corresponding to the {2,2} (nominally, $\lambda_c = 4.60$ µm) and {5,1} ($\lambda_c = 4.71$ µm) diffraction orders, owing to the particular choice of the angle $\theta = 20°$.[9]

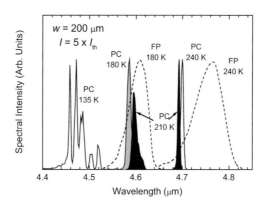

Figure 2. 2nd-order PCDFB laser emission spectra at several operating temperatures, for pulsed optical pumping with a 200-μm-wide Gaussian stripe at five times threshold. Spectra for Fabry-Perot (FP) lasers (no grating) from the same bar are also shown for comparison (dashed curves).

Designing the structure to operate in the {1,1} diffraction order eliminates the possibility of accidental degeneracies. The 1st-order grating results were obtained from a different sample with an integrated absorber (IA) active region grown on a GaSb substrate. The 1st-order rectangular PCDFB grating with a long-axis period of 1.66 μm, short-axis period of 0.63 μm, and 10% separation between the features along the short-period axis was patterned by electron-beam lithography, followed by reactive ion etching. Although nominally designed for $\theta = 20°$, the actual aspect ratio led to a slightly larger angle. The simulations indicate that this grating has a nearly optimized set of coupling coefficients: $\kappa_1' = 10$ cm^{-1}, $\kappa_2' = 118$ cm^{-1}, and $\kappa_3' = 13$ cm^{-1}. The devices were pumped using the tilted PCDFB cavity arrangement, as well as with the pump stripe normal to the facets. In the latter case, the high losses that would occur if the lasing beam passed through unpumped regions in its zigzag path prevented the PCDFB mode from dominating the spectral characteristics. The lasing was then apparently via "parasitic" Fabry-Perot modes that pass through the grating at off-resonance wavelengths.

The threshold pump intensity for the tilted-stripe configuration was substantially higher for temperatures lower than the maximum operating temperature of 200 K (probably limited by the valence band offset of the IA layers). This indicates that the gain spectrum does not come into resonance with the PCDFB mode for lower temperatures, and that the modal index estimate used in designing the structure was underestimated owing to the strong nonlinearity in the refractive index of GaInAsSb. In order to confirm this conclusion, we examine the spectra of the devices in the normal and tilted-stripe pumping configurations at $T = 180$ and 190 K in Fig. 3. The PCDFB device tends to lase $\lambda \approx 4.33$ μm, at the extreme long-wavelength end of the range where modal gain is available, as indicated by comparison with the normal-pump spectra. This demonstrates that even at $T = 180$-190 K, an optimum matching of the gain spectrum and the PCDFB resonance has not been reached. In spite of the spectral mismatch, the PCDFB lines are quite narrow (7 nm), as compared to our previous studies of α-DFB and Fabry-Perot lasers. The 2nd-order PCDFB lasers had yielded similar linewidths.

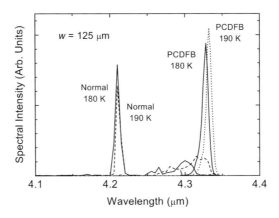

Figure 3. Lasing spectra for 1st-order-grating devices at temperatures of 180 and 190 K, pumped in both the tilted PCDFB configuration and normal to the facets. The stripe width was 125 μm.

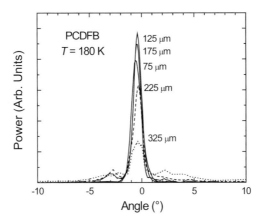

Figure 4. Far-field patterns at several stripe widths for the 1st-order PCDFB at $T = 180$ K in the tilted-stripe pumping scheme.

The spectral narrowing observed in Fig. 3 is also associated with greatly improved far-field characteristics. The far-field pattern for several stripe widths is shown in Fig. 4. For stripes with $W < 200$ μm, the output is primarily in a single, low-divergence lobe. By correlating the results of the far-field measurements with calculations of the near-field and far-field profiles, the etendue as a function of stripe width may be obtained, as described previously.[9,10,13] Results for the 1st-order PCDFB laser as well as for the 2nd-order PCDFB and α-DFB devices discussed above are shown in Fig. 5. The beam quality for the 1st-order PCDFB laser is essentially diffraction-limited at a stripe of 75 μm, and remains slightly better than the 2nd-order PCDFB at $W = 125$ μm. However, at stripes beyond 200 μm the 1st-order PCDFB is slightly less favorable, owing to the larger value of κ_2' in the 1st-order PCDFB that favors better beam quality in narrower stripes. We also note that the beam quality for both PCDFB devices is substantially better than that of α-DFB lasers at all the investigated stripe widths (by as much as a factor of 5 for stripes near 200 μm). Of course, it must be kept in mind that the excellent results for $W < 200$ μm for the 1st-order PCDFB laser were obtained when the modal gain at the resonance wavelength was apparently quite low, due to the unintended spectral mismatch.

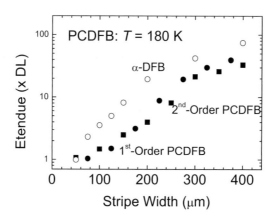

Figure 5. Etendue as a function of temperature for 1st-order PCDFB, 2nd-order PCDFB, and α-DFB lasers as a function of stripe width.

4. Surface-emitting PCDFB lasers

In spite of the great promise of edge-emitting PCDFB lasers, one drawback is the severe ellipticity of the output beam, which has a fast angular divergence along the growth axis (since the aperture size is on the order of the wavelength) and very slow divergence along the laser stripe (the aperture size is hundreds or thousands of wavelengths). The PCDFB fabrication must insure that the angular orientation of the grating with respect to the facets is precise to within a fraction of a degree, and that the laser stripe is aligned at the same angle. However, an alternative 2D grating configuration which can produce a circular output beam without external optics and also dispenses with the need for angular alignment with the facets (by eliminating the facets altogether!) is the surface-emitting (SE) PCDFB laser. Such devices have in fact been demonstrated, with two works employing semiconductor active regions.[14,15]

We have developed a novel time-domain Fourier-Galerkin (TDFG) formalism in order to describe the operation of SE PCDFB lasers. It has been applied to optically pumped devices, which are ideal in this case, since current injection inevitably introduces compromises associated with partial blocking of the emitting aperture by electrical contacts. We find that for active regions that have optical gain primarily in the TE polarization, a hexagonal lattice needs to be employed, whereas both hexagonal and square lattices may be used for devices that emit preferentially in the TM polarization. Focusing on the former case, we find a number of possible distinct near-field patterns, in which the device may operate. In all but one of them, strong destructive interference between the different plane-wave-like components of the optical field leads to a considerable reduction of the surface-emitting output power. By an empirical search of the parameter space we find that the following conditions must be satisfied in order to insure operation in the in-phase mode with the maximum output power:

(1) The mode must be allowed to self-pump regions somewhat beyond the nominal pump spot.

(2) The linewidth enhancement factor and the internal loss must both be sufficiently *large*, with the two minimum values being interdependent.

(3) The signs of the in-plane coupling coefficients for the TE polarization should be: $\kappa_1 > 0$, $\kappa_2 < 0$, and $\kappa_3 > 0$.

(4) The magnitudes of at least two of the coupling coefficients (taken to be κ_2 and κ_3) should range approximately from $1/D$ to $3/D$, while the third coefficient (taken to be κ_1) falls between $1/(5D)$ and $3/D$. Here D is the diameter of the pump spot.

(5) The surface-coupling coefficient κ_0 should not be much larger than 1 cm^{-1}.

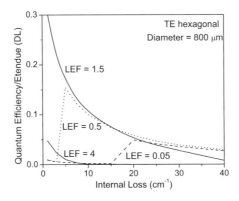

Figure 6. External differential quantum efficiency divided by the etendue as a function of internal loss for hexagonal-lattice TE-polarized SE PCDFBs. Results are given for LEF = 0.05 (dotted), LEF = 0.5 (dashed), LEF = 1.5 (solid), and LEF = 4 (dash-dot).

The somewhat counterintuitive condition (2) is illustrated further in Fig. 6, where the quantum efficiency divided by the etendue (normalized brightness) is shown as a function of internal loss for several values of the LEF. The explanation of this interesting behavior appears to be that the extra loss due to self-pumping favors the in-phase mode whenever the carrier-induced change in the refractive index (proportional to the loss-LEF product) is sufficiently large. We find that when the internal loss has a moderate value, the optimum active region for the SE PCDFB device should have a LEF ranging from 0.5 to 2. In that case, a diffraction-limited circular output beam may be produced with a quantum efficiency of up to 30% (60% if the radiation in the opposite direction is collected by some means, such as mirrors).

5. Conclusions

We have studied the spectral and far-field characteristics of 1st-order PCDFB lasers with a nearly optimal set of coupling coefficients and compared the results to non-optimized 2nd-order PCDFB devices and α-DFB lasers. It was determined that an underestimate of the modal index led to the resonance being shifted to wavelengths with weak gain at all of the available operating temperatures. Nonetheless, relatively narrow lines and low-divergence far-field profiles were observed. We have also explored the potential of surface-emitting PCDFB lasers, and confirmed that optimum results may be expected for an experimentally-accessible parameter space that would have been difficult to identify without detailed simulations.

References

[1] R. J. Lang, D. Mehuys, D. F. Welch, and L. Goldberg 1994 IEEE J. Quantum Electron. **30** 685.
[2] J. R. Marciante and G. P. Agrawal 1996 IEEE J. Quantum Electron. **32** 590.
[3] R. Parke, D. F. Welch, A. Hardy, R. Lang, D. Mehuys, S. O'Brien, K. Dzurko, and D. Scifres 1993 IEEE Photon. Technol. Lett. **5** 297.
[4] D. Mehuys, L. Goldberg, and D. F. Welch 1993 IEEE Photon. Technol. Lett. **5** 1179.
[5] I. Vurgaftman and J. R. Meyer 2001 Appl. Phys. Lett. **78** 1475.
[6] R. J. Lang, K. Dzurko, A. A. Hardy, S. DeMars, A. Schoenfelder, and D. F. Welch 1998 IEEE J. Quantum Electron. **34** 2196.
[7] A. M. Sarangan, M. W. Wright, J. R. Marciante, and D. J. Bossert 1999 IEEE J. Quantum Electron. **35** 1220.
[8] I. Vurgaftman and J. R. Meyer 2002 IEEE J. Quantum Electron. **38** 592.
[9] W. W. Bewley, C. L. Felix, I. Vurgaftman, R. E. Bartolo, J. R. Lindle, J. R. Meyer, H. Lee and R. U. Martinelli (in press) Solid State Electronics.
[10] W. W. Bewley, C. L. Felix, I. Vurgaftman, R. E. Bartolo, J. R. Lindle, J. R. Meyer, H. Lee, and R. U. Martinelli 2001 Appl. Phys. Lett. **79** 3221.
[11] H. Q. Le, G. W. Turner, J. R. Ochoa, M. J. Manfra, C. C. Cook, and Y.-H. Zhang, 1996 Appl. Phys. Lett. **69** 2804.
[12] R. E. Bartolo, W. W. Bewley, I. Vurgaftman, C. L. Felix, J. R. Meyer, and M. J. Yang, 2000 Appl. Phys. Lett. **76** 3164.
[13] I. Vurgaftman, W. W. Bewley, R. E. Bartolo, C. L. Felix, M. J. Jurkovic, J. R. Meyer, M. J. Yang, H. Lee, and R. U. Martinelli 2000 J. Appl. Phys. **88** 6997.
[14] M. Imada, S. Noda, A. Chutinan, T. Tokuda, M. Murata, and G. Sasaki 1999 Appl. Phys. Lett. **75** 316.
[15] S. Noda, M. Yokoyama, M. Imada, A. Chutinan, and M. Mochizuki, 2001 Science **293** 1123.

Quantum Well Infrared Photodetectors and Thermal Imaging Cameras

H Schneider, R Rehm, J Fleissner, M Walther, P Koidl, and G Weimann
Fraunhofer Institute for Applied Solid State Physics, Tullastrasse 72, D-79108 Freiburg, Germany

J Ziegler, R Breiter, and W Cabanski
AEG Infrarot-Module GmbH, Theresienstrasse 2, D-74072 Heilbronn, Germany

Abstract. We report on our QWIP focal plane array (FPA) developments for thermal-imaging applications in the 8 – 12 µm long-wavelength infrared (LWIR) and 3 – 5 µm mid-wavelength infrared (MWIR) regimes. Photovoltaic low-noise QWIP FPAs are best suited for long integration times (> 20 ms), resulting in the best noise-equivalent temperature difference (NETD) ever achieved with any detector technology. For short integration times, we use photoconductive QWIP arrays with higher carrier concentrations and quantum efficiency, with an NETD below 50 mK at only 2 ms integration time for FPAs with 24 µm pitch. For the MWIR, we have developed FPAs based on strained InGaAs/AlGaAs quantum wells. These 640 x 512 MWIR QWIP FPAs exhibit a high peak quantum efficiency above 10% and an excellent NETD of 14 mK at 88 K detector temperature. High-performance FPAs for the MWIR can thus be realized with GaAs technology.

1. Introduction

In the past twelve years, quantum well infrared photodetectors (QWIP) have covered the long way from the laboratory into commercial high-performance thermal imagers. A variety of camera systems with increasing performance and complexity has been demonstrated [1,2,3,4,5,6,7]. For commercial applications, QWIP-based infrared cameras have to compete with other sensor technologies, including HgCdTe, InSb, PtSi, and uncooled microbolometers. Each technology having its pros and cons, different detector materials are used according to specific system demands for each application. These requirements include thermal resolution, spatial resolution, wavelength band, integration time, and cost.

At present, QWIP-based thermal imagers have their main benefits if high thermal and spatial resolution is needed at integration times of about 10 ms or longer. Most QWIP FPAs operate in the 8 – 12 µm long-wavelength infrared (LWIR) regime, where we have obtained excellent thermal resolution as low as 5 mK and 7 mK (at 40 ms and 20 ms integration time, respectively) with a 256 x 256 low-noise QWIP FPA [5]. In addition to the excellent thermal and spatial resolution, further advantages include the maturity of GaAs-based technology, low fixed-pattern noise, low 1/f noise, and high pixel operability at moderate cost.

In this paper, we first give a general discussion of detector requirements in thermal imaging. We then describe a few specific QWIP structures and their detection properties. The third part focuses on focal plane technology and performance data achieved by our QWIP thermal imagers. We will also address some applications.

2. Detector requirements for thermal imaging

A state-of-the-art detector for thermal radiation needs background limited performance (BLIP), i. e., the signal-to-noise ratio S/N should not be deteriorated by the dark current noise. The detector signal S is proportional to the expression $\eta g \Phi \tau$, where η is the *internal* quantum efficiency, g the photoconductive gain (defined as the number of collected electrons per absorbed photon), τ the measurement time, and Φ the incident photon flux. The generation-recombination noise N of a standard photoconductor is proportional to $\sqrt{Sg} = g\sqrt{\eta \Phi \tau}$. The signal-to-noise ratio S/N therefore obeys the proportionality $S/N \propto \sqrt{\eta \Phi \tau}$, which implies that S/N is proportional to the square root of the number of detected photons. We point out that S/N does not depend on g.

Similarly, the gain does not enter into the noise-equivalent temperature difference (NETD), which is proportional to S/N. Nevertheless, g becomes relevant in the following situations: (i) If S is very small, then S/N might degrade due to other noise sources of the system. This problem, which becomes relevant at very small photon fluxes and/or integration time τ, can be avoided by working at high g. (ii) Due to the limited charge storage capacity of the readout integrated circuit (ROIC), it is often necessary to use an artificially short τ in order to reduce S to a tolerable value. In this case, a small g will allow us to increase τ and thus to improve the NETD.

Assuming a typical QWIP optimised to 8 – 9 µm wavelength and an f/2 objective, the readout capacitor goes into overload at $\tau = 10 - 20$ ms if the responsivity exceeds 0.2 A/W, corresponding to an *external* quantum efficiency ηg of only 3%. Gain reduction is highly desirable to avoid this readout limitation and to exploit higher η. In addition, a low gain increases the dynamic range of the sensor.

Different restrictions are valid in the 3 – 5 µm spectral regime, since the usable photon fluxes are about an order of magnitude smaller than in the LWIR. Therefore, readout limitation is not relevant for practical integration times. In the MWIR, the main goal is to obtain a high η for efficient collection of the incident photons.

3. QWIP device structures

3.1. Photoconductive QWIP

Figure 1 compares the responsivity, gain, and peak detectivity of a photoconductive QWIP as a function of bias voltage [8]. Here the gain is strictly proportional to the peak responsivity, which indicates that the internal quantum efficiency has a fixed value. The ratio R/g yields $\eta = 3\%$ for this test structure with a 45° facet geometry. The detectivity is fully developed already at 100 mV applied bias where the gain is only 10%.

Significantly higher values for R, η, and the detectivity D^* as compared to the 45° facet geometry used in Figure 1 are obtained for devices with two-dimensional grating couplers. This enhancement, giving rise to a quantum efficiency of $\eta > 10\%$, is due to the high coupling efficiency of the grating. At integration times of about 10 ms, a bias voltage of 150 – 300 mV is sufficient for operating the QWIP array in combination with

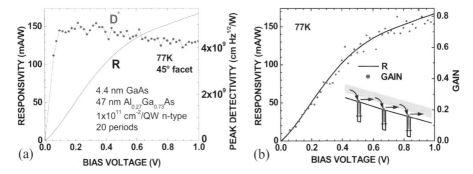

Figure 1: (a) Peak detectivity and responsivity, (b) gain and responsivity of a photoconductive QWIP vs. bias voltage. The inset of (a) and (b) indicate the growth parameters and the transport mechanism, respectively.

a ROIC. Excellent thermal resolution is thus obtained already for a relatively thin active region comprising only 20 QWIP periods, thus facilitating detector processing.

Aiming at applications with short integration times of 5 ms and below, necessary for fast moving objects and for cameras where sub-pixel spatial resolution is achieved by micro-scan methods, we have also developed QWIP FPAs with high quantum efficiency (HQE), where we use a higher carrier concentration of 4×10^{11} cm^{-2} electrons per quantum well (rather than 1×10^{11} cm^{-2} as used in the "standard" photoconductive QWIP). In these HQE QWIP FPAs, we achieve an increase of the internal quantum efficiency η to about $\eta = 30 - 40\%$, which accordingly improves the signal-to-noise ratio at short integration times. This advantage comes at the penalty that slightly lower (by a few Kelvin) detector temperatures are necessary in order to maintain background-limited operation.

3.2. Low-Noise QWIP for highest thermal resolution

In order to reduce the photoconductive gain even further and to enable longer integration times, we have developed a different detector structure, the *low-noise QWIP* [9]. The inset of Figure 2b shows the photoconduction mechanism of this device. Each period of the low-noise QWIP comprises four sections. In the excitation zone (E), carriers are optically excited and emitted into the quasi-continuum of the drift zone (D). These sections are in analogy to the quantum well and barrier of a photoconductive QWIP. The low-noise QWIP now contains two additional zones which control the relaxation of the photoexcited carriers, namely a capture zone (C) and a tunneling zone (T).

Figure 2 summarizes the detection properties of a typical low-noise QWIP [9, 10]. The peak responsivity is 11 mA/W at 0 V applied bias and about 22 mA/W in the range 2 – 3 V (see Figure 2a). We associate the latter bias region with the field regime where complete emission of the photoexcited electrons occurs, whereas the emitted carriers are still captured efficiently [9]. This behaviour results in a photocarrier mean free path of one period, i. e., in a photoconductive gain of 1/20 for this 20 period QWIP structure. This interpretation is further supported by the behaviour of the noise current. As indicated in Figure 2a, a gain of 5% is obtained from noise measurements [8].

From these data we calculate an internal peak quantum efficiency of $\eta = 7\%$ for a 45° facet geometry. For detector arrays with two-dimensional diffraction gratings, the quantum efficiency increases to about $\eta = 30 - 40\%$ due to the higher quantum efficiency of the gratings as compared to 45° facets. A responsivity of about 0.1 A/W is thus obtained for a low-noise QWIP FPA.

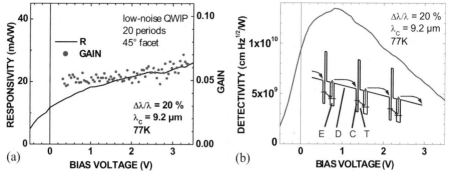

Figure 2: (a) Peak responsivity and gain, (b) peak detectivity of a low-noise QWIP vs. bias voltage. The inset of (b) indicates the conduction bandedge distribution and transport mechanism of a low-noise QWIP. Each period consists of an excitation zone (E), drift zone (D), capture zone (C), and tunneling zone (T).

The bias dependence of the detectivity obtained from the measured responsivity and noise is shown in Figure 2b. The detectivity has its maximum at around 0.8 V. Due to the inherent asymmetry of the structure, there is still an efficient photoconduction process without external bias voltage, such that the detectivity still reaches about 70% of its maximum value at 0 V. This behavior is in strong contrast to that of a conventional QWIP where the detectivity practically vanishes at 0 V (see Figure 1a).

The expression "low-noise" relates to the interesting property that this detector shows a shot-noise like behavior rather than generation-recombination noise. This phenomenon, which further reduces the noise current of the detector, arises from the complete determinacy of the capture process if the capture probability approaches unity. This determinacy suppresses the recombination noise, thus giving rise to an extra reduction of the noise current by a factor of $\sqrt{2}$ with respect to a photoconductor.

3.3. InGaAs/AlGaAs QWIPs for the MWIR

Beside the LWIR, QWIP FPAs are also interesting for the MWIR. An additional important motivation for our development of MWIR QWIP FPAs consists of the fact that high-performance MWIR QWIPs are critical building blocks for dual-band FPAs.

According to Planck's radiation law, the photon flux decreases exponentially with increasing photon energy, such that most of the photon flux in the MWIR occurs at long-wavelength part of the 3 – 5 µm window. In order to optimize these detectors with respect to camera applications in the MWIR, we have carefully optimized the well width and material compositions in order to hit the wavelengths between 4.5 µm and 5.0 µm. Figure 3a shows a collection of photocurrent spectra showing that essentially the whole 3 – 5 µm regime is accessible via InGaAs/AlGaAs QWIPs. The solid curve in Figure 3a originates from a 20 period QWIP with 2.6 nm $In_{0.3}Ga_{0.7}As$ quantum wells and 45 nm $Al_{0.32}Ga_{0.68}As$ barriers and exhibits the desired spectral properties.

The large intersubband energies of MWIR QWIPs result in extremely high dark resistances at usual detector temperatures, such that the carrier concentration can be increased drastically without obtaining too much dark current. Figure 3 shows the responsivity and the dark current at 2.5 V bias voltage and 77 K as a function of the carrier density n. At $n < 2.1 \times 10^{12}$ cm^{-2}, both the dark current and the responsivity are approximately proportional to n, while at $n > 2.1 \times 10^{12}$ cm^{-2}, the responsivity strongly saturates (and even slightly decreases) and the dark current increases exponential-like.

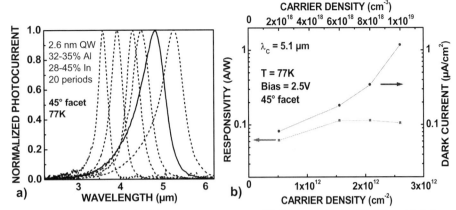

Figure 3: (a) Photocurrent spectra of InGaAs/AlGaAs QWIPs with 2.6 nm InGaAs quantum wells and different material compositions. (b) Peak responsivity and dark current of $In_{0.3}Ga_{0.7}As/Al_{0.32}Ga_{0.68}As$ MWIR QWIPs vs. carrier density. The latter growth parameters correspond to the solid curve in (a).

For our MWIR QWIP FPAs we therefore use $n = 2.1 \times 10^{12}$ cm^{-2} per quantum well. In a 45° facet geometry, the responsivity amounts to 114 mA/W at 2.5 V with a measured photoconductive gain of $g = 0.57$, which results in a quantum efficiency of $\eta = 4.8\%$. From this value we expect η to be well above 10% for a FPA pixel with grating.

4. QWIP-based thermal imagers

4.1. QWIP FPA fabrication

Our QWIP FPAs are manufactured in three different geometries, namely 256 x 256 pixels with 40 μm pitch, 640 x 512 pixels with 24 μm pitch, and 384 x 288 pixels with 24 μm pitch. Figure 4 displays micrographs of FPAs with 24 μm pitch. Dry etching allows us to realize small gaps of only 2 μm between adjacent mesas, resulting in a high filling factor of about 84%. The arrays are covered by a silicon nitride passivation layer with openings in the center of each mesa. An additional bond metallization provides electrical contacts for connection with the ROIC via indium solder bumps. Also visible are dry-etched two-dimensional diffraction gratings with a period of 2.95 μm and 1.65 μm, respectively, which have been covered by a mirror metallization before depositing the passivation layer. Details of QWIP FPA process technology can be found elsewhere [2].

Figure 4: Scanning electron microscope micrographs of 640 x 512 QWIP arrays with 24 μm pitch, comprising two-dimensional diffraction gratings for the LWIR (left) with 2.95 μm period, and for the MWIR (right) with 1.65 μm period.

The QWIP arrays are hybridized to Si-CMOS ROIC using an In re-flow technique. After hybridization, the GaAs substrate is removed by mechanical lapping and selective wet-chemical etching. Finally, system integration of the FPA sensors is performed. Our cameras use Stirling coolers to operate the LWIR and MWIR QWIP FPAs at 60 – 65 K and 88 K, respectively. The camera platform is equipped with a standard serial interface for easy connection with a personal computer and with a fast parallel port for real-time data transfer of the full 14 bit digitized images of the 256 x 256, 288 x 384, and 640 x 512 QWIP FPAs at frame rates up to 200 Hz, 120 Hz, and 60 Hz, respectively.

5. Camera performance

For thermal imaging applications, the extremely low noise levels of low-noise QWIP FPAs enable a higher dynamic range, longer integration time, and improved thermal resolution as compared to conventional photoconductive QWIPs. The experimentally determined NETD histogram of a 640 x 486 low-noise QWIP camera system with 24 μm pitch is depicted in Figure 5a. At 30 ms integration time, we observe a NETD as low as 9.6 mK, the best NETD ever obtained with a large-format thermal imager operating in the LWIR [11]. This FPA also exhibits an excellent linearity and correctability, and a very low fixed-pattern noise.

As a second example, Figure 5b depicts the NETD histogram of a typical 640 x 512 MWIR QWIP FPA at a detector temperature of 88 K, with an excellent NETD of only 14.3 mK. Also this FPA has turned out to have a very good correctability [12]. High-performance FPA can thus be realized also for the MWIR, thus providing an attractive alternative to HgCdTe, InSb, and PtSi.

Figure 6 shows sample images taken with both FPAs. In addition to the excellent thermal resolution and contrast, both thermal images show high detail which indicates a small optical crosstalk between adjacent pixels and a good modulation transfer function. A closer inspection of the imagery reveals a better sharpness of the LWIR image, which is not unexpected since the absorption quantum efficiency of the low-noise QWIP FPA is at least twice as high as compared to the MWIR QWIP FPA.

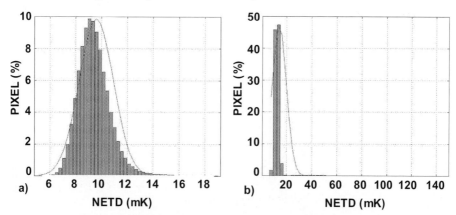

Figure 5: NETD-histogram of a 640 x 512 LWIR low-noise QWIP FPA at f/2 and 30 ms (a), and of a 640 x 512 MWIR QWIP FPA at f/1.5 and 20 ms.

LWIR　　　　　　　　　　　　　　　MWIR

Figure 6: 640x486 thermal images taken by a low-noise QWIP camera (left) and by a MWIR QWIP camera (right).

6. Applications

Besides the military market, low-noise QWIP FPAs are particularly promising for civil applications including medical imaging and environmental research. Most importantly, a market for Dynamic Area Telethermometry (DAT) [13] has recently developed, which is used, e. g., for breast cancer detection and for verifying the efficiency of cancer medication and therapy. The method is based on the detection of subtle temporal changes of the skin temperature, which exhibit characteristic modifications in the vicinity of a tumor. Exploiting both local and global signatures, the method also allows the detection of deeper lying cancerous lesions. To this end, thermal deviations with a modulation frequency of 0.1 – 2 Hz have to be detected. The spatial distribution of temperature modulation on the skin surface is obtained by quantitative numerical analysis of several thousands of subsequent thermal images. In contrast to the standard method of X-ray mammography, the new method of DAT enables early detection of cancerous lesions without mechanical stress or radiation exposure involved. The method can therefore be repeated arbitrarily often, thus allowing to monitor the efficacy of medication. Additional applications in medicine include brain surgery and pre-clinical testing.

The excellent thermal resolution of, in particular, low-noise QWIP cameras also makes them an interesting measurement tool in research. For instance, such a camera is used by C. Garbe et al. [14] in order to study convection processes on water surfaces and wind-water interactions, with the ultimate goal of understanding the exchange of CO_2 between the oceans and the atmosphere. This kind of investigations also aims towards a better understanding of the greenhouse effect induced by the combustion of fossil fuels.

QWIP being the only technology at present that commercially supplies large 640 x 480 staring arrays operating in the 8 – 12 μm spectral regime, further increasing potential exists also in other non-military markets, e. g., production monitoring, non-destructive testing, and fire fighting. A QWIP camera unit is presently being delivered for the inspection of turbine blades during production.

7. Conclusion

QWIPs represent by now a mature and versatile GaAs-based technology that allows us to realize thermal imagers with highest thermal and spatial resolution, both in the 8 – 12 μm and 3 – 5 μm regimes, excellent homogeneity, low fixed-pattern noise, and high pixel yield at tolerable production costs. Different QWIP structures are used in FPAs adapted to varying system requirements. Besides "standard" photoconductive GaAs/AlGaAs QWIPs, optimization of the quantum efficiency provides good NETD at short integration times of 4 ms and below. Photovoltaic low-noise QWIPs enable FPAs with record-high thermal resolution at 8 – 12 μm. InGaAs/AlGaAs QWIPs allow us to realize FPAs in the 640 x 512 format sensitive to the 3 – 5 μm regime with an excellent NETD of only 14.3 mK, thus indicating an interesting potential of QWIP technology also in the MWIR. In addition, high-performance MWIR QWIPs are critical building blocks for dual-band thermal imagers, like the 384 x 288 dual-band QWIP FPA detecting both LWIR and MWIR radiation simultaneously on each pixel that we will realize in the near future.

Acknowledgements - This work has been financed by the Federal Ministry of Defence.

References

[1] S. Gunapala and S. V. Bandara, *QWIP focal plane arrays*, in: *Intersubband Transitions in Quantum Wells*, edited by H. C. Liu and F. Capasso, Semiconductors and Semimetals Vol. **62** (Academic Press, 2000), p. 197.

[2] M. Walther, J. Fleissner, H. Schneider, C. Schönbein, W. Pletschen, E. Diwo, K. Schwarz, J. Braunstein, P. Koidl, G. Weimann, J. Ziegler, and W. Cabanski, Inst. Phys. Conf. Ser. No. 166, 427 (2000).

[3] B. Hirschauer. J. Alverbro, J. Andersson, J. Borglind, A. Bustamente, Z. Fakoor-Biniaz, U. Halldin, P. Helander, Y. Lindberg-Eriksson, H. Malm, H. Martijn, C. Nordahl, and O. Öberg, Infrared Physics and Technology 42, 329-332 (2001).

[4] W. Cabanski, R. Breiter, R. Koch, K.-H. Mauk, W. Rode, J. Ziegler, K. Eberhardt, R. Oelmaier, H. Schneider, and M. Walther, Proc. SPIE Vol. 4028, 113 (2000).

[5] H. Schneider, M. Walther, J. Fleissner, R. Rehm, E. Diwo, K. Schwarz, P. Koidl, G. Weimann, J. Ziegler, R. Breiter, and W. Cabanski, Proc. SPIE Vol. 4130, 353 (2000).

[6] S. D. Gunapala, S. V. Bandara, J. K. Liu, B. Rafol, J. M. Mumolo, F. M. Reininger, J. M. Fastenau, and A. K. Liu, Proc. SPIE Vol. 4369, 516 (2001).

[7] A. C. Goldberg, S. W. Kennerly, J. W. Little, H. K. Pollehn, T. A. Shafer, C. L. Mears, H. F. Schaake, M. L. Winn, M. F. Taylor, and P. N. Uppal, Proc. SPIE Vol. 4369, 532 (2001)

[8] R. Rehm, H. Schneider, C. Schönbein, and M. Walther, Physica E 7, 124 (2000).

[9] H. Schneider, C. Schönbein, M. Walther, K. Schwarz, J. Fleissner, and P. Koidl, Appl. Phys. Lett. 71, 246 (1997).

[10] H. Schneider, M. Walther, C. Schönbein, R. Rehm, J. Fleissner, W. Pletschen, J. Braunstein, P. Koidl, G. Weimann, J. Ziegler, and W. Cabanski, Physica E 7, 101 (2000).

[11] H. Schneider, M. Walther, J. Fleissner, R. Rehm, E. Diwo, K. Schwarz, P. Koidl, and G. Weimann, Inst. Phys. Conf. Ser. No. 170, 171 (2002).

[12] W. A. Cabanski, R. Breiter, R. Koch, K. Mauk, W. Rode, J. Ziegler, H. Schneider, and M. Walther, Proc. SPIE Vol. 4369, 547 (2001).

[13] W. A. Cabanski, R. Breiter, W. Rode, J. Ziegler, H. Schneider, M. Walther, and M. A. Fauci, Proc. SPIE Vol. 4721, 165 (2002).

[14] C. Garbe, U. Schimpf, and B. Jähne, Proc. SPIE Vol. 4710, 171 (2002).

Temperature sensitivity of high power GaSb based 2 µm diode lasers

M. Rattunde [a], **C. Mermelstein, J. Schmitz, R. Kiefer, W. Pletschen, M. Walther, and J. Wagner**

Fraunhofer Institut für Angewandte Festkörperphysik, Tullastrasse. 72, D-79108 Freiburg, Germany, a) e-mail: marcel.rattunde@iaf.fhg.de

Abstract. We have realized strained triple-quantum-well, large-optical-cavity GaInSb/AlGaAsSb/GaSb diode lasers emitting in the 2 µm wavelength range. In order to optimize these devices for power applications, we investigate samples at 2.2 µm wavelength with different Al-content in the barrier and seperate confinement layers, and thus different quantum-well barrier heights. Devices with 40 % Al revealed the highest value for the characteristic temperature T_0, which is attributed to a reduction in the heterobarrier leakage. On the other hand, the lasers with 20 % Al yielded the best power efficiency η_P with a maximum value of 30 %, reducing the thermal load generated in the active region and making this device structure well suited for power applications.

1. Introduction

There is an increasing need for room-temperature diode lasers emitting in the infrared wavelength region around 2 µm not only for trace gas analysis, but also for power applications such as welding of transparent plastic materials, photo-thermal therapy of skin diseases, laser surgery and pumping of solid state lasers. These applications require an optical output power exceeding 1 W per single emitter, which calls for a high power conversion efficiency as well as a good thermal stability of the laser performance. In this paper we present results on high power GaSb-based 2 µm lasers, which have already been used successfully as pump sources for solid state lasers [1].

2. Growth and processing

The laser structures under investigation were grown by solid-source molecular-beam epitaxy on n-type (100) GaSb:Te substrate. The active region consists of three compressively strained 10 nm thick $Ga_{1-x}In_xAs_ySb_{1-y}$ quantum wells (QWs), separated by 20 nm wide lattice matched $Al_xGa_{1-x}As_ySb_{1-y}$ barriers with a standard Al content of 30 %, and surrounded by 400 nm thick separate confinement (SC) layers out of the same quaternary material. This large optical cavity waveguide core is embedded between 2 µm thick lattice matched $Al_{0.84}Ga_{0.16}As_{0.06}Sb_{0.94}$ n- and p-doped cladding layers. For the set of samples described in section 4, the Al-content in the barrier and SC layers was varied between 20 % and 40 % while keeping these layers lattice matched; the other parts of the structure were identical. Edge emitting, index guided ridge-waveguide lasers with ridge-widths between 6 and 64 µm were fabricated. These devices were used to derive the internal parameters and characteristic temperatures of the laser. For higher output powers gain guided broad area (BA) lasers with a 150 µm wide active area were prepared. The laser facets were high-reflection/antireflection (HR/AR) coated with 95 % and 5 % reflectivity, respectively and the devices were mounted p-side down on copper heat sinks.

3. High power 2.0 µm diode lasers

Devices with an emission wavelength of 2.0 µm exhibit a high internal efficiency η_i of 77 %, internal losses α_i as low as 6 cm^{-1} and a threshold current density at infinite cavity length of 120 A/cm^2. Temperature dependent output power-versus-current (P-I) measurements revealed a characteristic temperature for the threshold current T_0 of 180 K and 81 K for the 250-290 K and 290-360 K temperature regime, respectively. The temperature dependence of the quantum efficiency η_d can be described by a characteristic temperature T_1 of 433 K, which drops to 204 K for temperatures above 320 K.

In order to reach high output powers, a high wall-plug efficiency, concomitant with a low thermal resistance of the mounted device are of prime importance. To model the laser performance including these thermal effects, we have derived an electro-optical-thermal model, described in [2]. In Fig.1 the measured P-I, η_P-I and V-I characteristics at 300 K are shown for a 1000x150 µm^2 BA lasers with HR/AR coated facets, mounted p-side down, together with the simulated dependencies. The laser reaches a maximum output power of 1.7 W, with a maximum power efficiency of 27 % and still 15 % at 1.7 W output power. The right bottom panel of Fig.1 shows the temperature of the active region extracted from the simulation of the device characteristics, indicating a temperature rise in the active region of about 100 K above heat sink temperature at thermal rollover. Any reduction of the thermal load or increase of the thermal stability of the laser will have a big effect on the power performance, shifting the thermal rollover to higher current values.

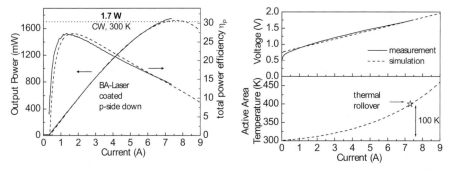

Fig.1: Comparison of measured (solid lines) and simulated (dashed lines) P-I, η_P-I and V-I characteristics for a 1000x150 µm^2 BA laser emitting at 2.0 µm with HR/AR coated facets, mounted p-side down. Data were taken in cw operation at a heatsink temperature of 300 K.

4. Diode lasers with different barrier heights

For the optimization of the barrier heights with respect to thermal stability and power efficiency, a set of three samples with identical quantum well composition, corresponding to a wavelength of 2.2 µm at 280 K, but different Al-content of 20 %, 30 % and 40 % in the barrier and SC layers were realized. With increasing Al-content, the conduction and valence band offsets between QW and barrier increased from 0.27 to

0.43 eV and from 0.18 to 0.30 eV, respectively. Fig.2a displays the temperature dependence of the threshold current for the three different Al-contents measured in pulsed operation. In the high temperature range above 310 K, the benefit of the higher band offset, and therefore reduced leakage of thermally activated carriers, leads to a high T_0 value of 82 K for the 40 % Al-content device, while the samples with 30 % and 20 % Al in the barrier and SC layers exhibit T_0 values of 66 K and 63 K, respectively. The absolute values for the threshold current are comparable for the 20 % and 30 % Al-content laser, whereas the 40 % Al device shows an increase in the threshold current which is not fully understood yet.

Due to the different refractive index of the SC layers with different Al-content, the confinement of the optical mode in the waveguide decreases with increasing Al-content, resulting in a confinement factor for the QWs Γ_{QW} of 4.3 % for the 20 % Al device which decreases to 4.1 % and 3.9 % for the higher Al-concentrations. The confinement factor of the cladding layers Γ_{Clad} shows the opposite trend, increasing from 12 % to 17 % with increasing Al-content. As the free carrier absorption in the highly doped cladding layers is proportional to Γ_{Clad}, as reported in [3], the internal losses α_i should increase with the Al-content. However, as the leakage out of the QWs is reduced with increasing Al-content, the carrier concentration in the SC layers should be lower, reducing the free carrier absorption in this part of the structure. Thus we explain the independence of α_i (6.2 - 6.8 cm^{-1}) from the SC layer composition by an interplay of these two mechanisms.

The internal quantum efficiency η_i reaches very high values between 83 % and 89 % for all three structures, again without systematical correlation to the Al-content. This is explained by the fact that the carrier density in the QWs, and therefore the amount of leakage current, clamps at threshold, keeping η_i constant, whose value is only affected by loss mechanisms which scale with the current above threshold. The temperature dependence of the external quantum efficiency η_d is shown in Fig.2b, together with its characteristic temperature T_1, which is almost identical for all three samples. As η_d is a function of α_i, η_i, and the mirror losses α_m, this is consistent with the above findings.

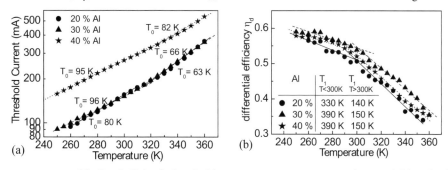

Fig.2: a) Pulsed threshold current vs. temperature data and b) pulsed differential quantum efficiency η_d vs. temperature for three 2.2 µm diode lasers with different Al-content in the barrier and SC layers.

An advantage of a low Al-content barrier and SC layer becomes evident when comparing the voltage-versus-current characteristics, shown in Fig.3a. The turn on voltage U_0 increases with Al-content which is a result of the band offsets: In Fig. 3b, the band edge profile of the three different samples is shown at zero bias. In order to achieve flat-band-condition in the active region, and thus high carrier injection in the QWs, higher voltages are required for the high Al-content devices. To first order, the operating voltage scales with the band gap E_G of the SC layers. The difference in E_G of 0.12 eV

(0.28 eV) between the 20 % and 30 % (20 % and 40 %) Al-layer, respectively, compare favorably with the observed differences in operating voltage, which amount to 0.14 V (0.29 V) at a current of 0.5 A.

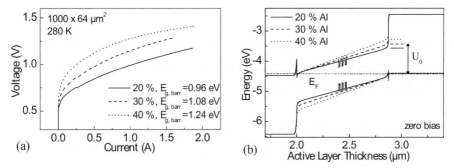

Fig.3: a) Voltage vs. current characteristic for the three diode lasers with different Al-concentration in the barrier and SC layers. b) band edge profile at zero bias for these devices.

In Fig.4 the 280 K cw P-I and η_P-I characteristics are shown, recorded from ridge waveguide lasers with uncoated facets and a 1000x64 μm^2 geometry, mounted p-side up: Due to the increase in operating voltage, the power efficiency drops significantly with increasing Al-content. This results in a thermal rollover of the P-I characteristic already at lower currents, thus severely limiting the maximum output power. The 20 % Al-content laser shows the highest power efficiency and therefore the smallest amount of heat dissipation in the active region. As the T_0 value is only slightly reduced for this sample, and the other characteristic parameters remain unchanged, this low barrier design seems to be most favorable for high power devices.

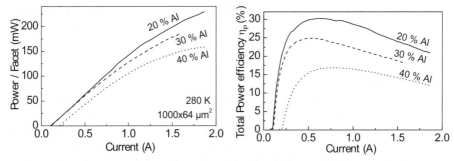

Fig.4: Output power and power efficiency vs. current characteristics of 1000x64 μm^2 ridge waveguide lasers with uncoated facets, mounted p-side up. The lasers were operated in cw mode at 280 K heatsink temperature.

References

[1] M. Mond et al., Opt. Lett. **27**, 1034 (2002)
[2] M. Rattunde, C. Mermelstein, J. Schmitz, R. Kiefer, W. Pletschen, M. Walther and J. Wagner, Appl. Phys. Lett. **80**, 4085 (2002)
[3] D. Garbuzov, R. Martinelli, H. Lee, P. York, R. Menna, J. Connolly, S. Narayan, Appl. Phys. Lett. **69**, 2006 (1996)

Long-wavelength, two-dimensional, WDM vertical-cavity surface-emitting laser arrays fabricated by nonplanar wafer bonding

J Geske, Y L Okuno, and J E Bowers

University of California, Electrical and Computer Engineering Department, Santa Barbara, CA 93106

D Leonard

Gore Photonics, Lompoc CA 93436

Abstract. We demonstrate the first long-wavelength, two-dimensional, wavelength division multiplexed vertical-cavity surface-emitting laser array. The eight-channel single-mode array covers the C-band from 1532 nm to 1565 nm. The devices are fabricated using two separate active regions laterally integrated using nonplanar wafer bonding. We achieved single-mode powers up to 0.8 mW, 2-dB output power uniformity across the array, and side-mode suppression ratios in excess of 43 dB. This fabrication technique can be used to maintain the gain-peak and cavity-mode alignment across wideband arrays and, with the use of non-traditional mirrors, can be extended to the fabrication of arrays covering the entire C, S, and L-bands as well as the 1310-nm transmission band.

I. Introduction

Vertical-cavity surface-emitting lasers (VCSELs) are of great interest due to their advantages in low-cost manufacturing and packaging. This is made possible by wafer-scale fabrication and testability, low-power dissipation, low-divergence circular-output beams, and the relative ease of fabricating one and two-dimensional VCSEL arrays. These qualities are compatible with the emerging market for coarse wavelength division multiplexing (CWDM) in low-cost, high-performance, optical networks. CWDM networks are being developed as a lower-cost alternative to dense wavelength division multiplexing, and cover the entire low-loss, low-dispersion window from 1470 nm to 1610 nm, typically in 20-nm increments [1]. Integrating all the multiple-wavelength sources required into a single package can achieve great cost savings for this application. Integrating the WDM sources themselves in a wafer-scale fabrication process could attain even further cost reductions. When considering this wide wavelength range for the integrated devices, it is important that the individual devices exhibit uniform device properties, such as threshold, differential efficiency, output power, and other laser emission properties. Any CWDM VCSEL array technology must address these requirements.

Several groups have been working on attaining multi-wavelength VCSEL arrays [2-5]. These techniques, which make use of crystal-growth techniques, have been successful in demonstrating multiple-wavelength VCSEL arrays operating up to 1.2 μm[5] and covering wavelength spans of up to 45 nm [4]. One-dimensional VCSEL arrays operating around 1.55 μm and 1.3 μm have also been demonstrated using superlattice thickness-adjustment etches in wafer-bonded VCSEL structures [6, 7]. All these techniques primarily adjust the cavity-mode wavelength across the surface of the wafer. Though this is sufficient to adjust the lasing wavelength of the devices in the array, the alignment between the gain peak and the cavity mode must also be maintained in order to control the device properties in wideband WDM VCSEL arrays. Thus, it is necessary to have equal control over both composition and thickness laterally across the surface of the wafer to achieve wideband WDM VCSEL arrays. We have previously reported a new technique called nonplanar wafer bonding which is capable of achieving this lateral-composition control [8].

In this paper we have applied nonplanar wafer bonding to demonstrate the first two-dimensional WDM VCSEL array covering the C-band. The array uses eight channels on a 4.5-nm pitch to extend from 1532 nm to 1565 nm. This technique should be extendable to the simultaneous integration of VCSEL active regions covering the C, S, and L bands as well as the 1310-nm transmission band.

2. Array fabrication

The fabrication of this device begins with two strained periodic-gain multi-quantum-well long-wavelength VCSEL active regions grown vertically on an InP wafer and separated by an etch-stop layer. The active regions have optical-cavity lengths of 2.5 wavelengths at 1565 nm and 1547 nm and have photoluminescence peaks at 1530 nm. In addition, the active regions have a two-period superlattice on one side that can be used for additional wavelength control in the second axis of the final two-dimensional array [6].

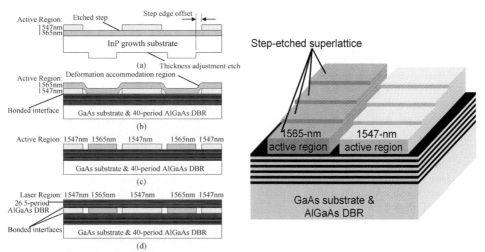

Figure 1. Process-flow cross-section schematics for the fabrication of the eight-channel VCSEL array

Figure 2. 3-D schematic of the device after the nonplanar bond and the superlattice etch.

The active-region wafer surface is etched with a step-shaped profile to reveal a different active region on each step level. The backside of the wafer is etched to have a profile complimentary to the step-etched active-region side of the wafer, as shown in Fig. 1(a). This thickness-adjustment etch is designed to yield an identical substrate plus epitaxial film thickness at each lateral point on the wafer. The lateral offset between the front-side and back-side step edges provides a region over which the substrate layers can accommodate the deformation. The nonplanar wafer is direct wafer bonded to a 40-period AlGaAs distributed Bragg reflector (DBR) grown on a GaAs substrate. The original InP growth substrate is removed, leaving the active regions attached to the AlGaAs DBR as depicted in Fig. 1(b). The excess active-region material and the deformation accommodation regions are removed, revealing a different active region at each lateral position along the first dimension of the AlGaAs mirror as represented by Fig. 1(c). At this point, the original superlattice that was grown on each active region is etched with a step-shaped profile to trim the cavity resonance of each of the two separate active regions in the second dimension of the wafer surface. Fig. 2 shows a schematic of a small section of the wafer surface after the superlattice etches. Fig. 3 is a photograph of the wafer surface in the same region shown in Fig. 2 and indicates the location of the eight channels in the final VCSEL structure. Each of the separate wavelength regions is about 500 μm wide. A 250-μm region where the deformation accommodation region was removed separates the two active regions from each other. A second, 27-period, AlGaAs DBR is bonded by traditional semiconductor-direct bonding [9] to create the structure shown in Fig. 1(d). One half period of the mirror is used as the etch-stop layer for the substrate removal, leaving a 26.5-period DBR. Fig. 1(d) shows a cross-section schematic of the eight-wavelength VCSEL array structure. Index guiding in the VCSEL structure is accomplished with 30-nm tall, circular post index guides that are etched into the surface of the top DBR prior to the second wafer bond. The index guides are 8 μm in diameter and are located at the bonded interface in the final device structure.

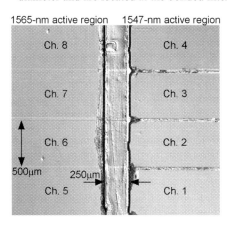

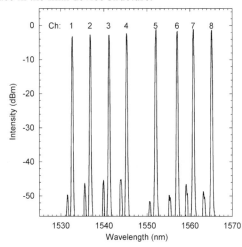

Figure 3. Photograph of the device area after the nonplanar bond and superlattice etch.

Figure 4. Superimposed lasing spectra of the eight-channel VCSEL chip

3. Results

The final structure is an eight-channel, two-dimensional, WDM VCSEL array. The array is optically pumped with a 980-nm pump laser. The periodic-gain active regions utilize about 680 nm of 1.35-μm InGaAsP barrier material and have a single-pass absorption of about 80%. Fig. 4 shows the room-temperature lasing spectra at a constant absorbed pump power of 10.5 mW for each device. There is a 2-dB variation in the output power at this constant pump power. The devices all exhibit single-mode operation with the worst-case side-mode suppression ratio of 43 dB occurring in the fourth channel. The excess wavelength separation between the fourth and the fifth channel is a result of a growth-rate error during the growth of the 1565-nm active region.

4. Conclusion

Nonplanar wafer bonding has been used to generate the first long-wavelength, two-dimensional, WDM VCSEL array. These devices exhibit high-quality single-mode emission over the range of 1532 nm to 1565 nm. By using two separate active regions laterally integrated on the surface of the wafer we have demonstrated that nonplanar wafer bonding can be used as a lateral-heterogeneous-integration technique in the fabrication of VCSEL arrays. This technique can be extended to combine VCSEL active regions with optimised gain-peak and cavity-mode alignment over a wide wavelength range, allowing for the future fabrication of uniform wideband WDM VCSEL arrays suitable for CWDM optical-network applications.

References

[1] B. E. Lemoff, OSA Optics and Photonics News, vol. 13, no. 3, pp. S8-S14, Mar. 2002.
[2] C. J. Chang-Hasnain, J. P. Harbinson, C. Zah, M. W. Maeda, L. T. Florez, N. G. Stoffel, and T. Lee, IEEE J. Quantum Electron., vol. 27, no. 6, pp. 1368-1376, June 1991.
[3] L. E. Eng, C. J. Chang-Hasnain, K. Bacher, M. Larson, G. Ding, and J. S. Harris, Proc. LEOS'94, 1994, pp. 261-262.
[4] F. Koyama, T. Mukaihara, Y. Hayashi, N. Ohnoki, N. Hatori, and K. Iga, Electron. Lett., vol. 30, no. 23, pp. 1947-1948, Nov. 1994.
[5] M. Arai, T. Kondo, M. Azuchi, T. Uchida, A. Matsutani, T. Miyamoto, and F. Koyama, 14th IPRM, 2002, pp. 303-306.
[6] A. Karim, P. Abraham, D. Lofgreen, Y. J. Chiu, J. Piprek, and J. E. Bowers, Electron. Lett., vol. 37, no. 7, pp. 431-432, Mar. 2001.
[7] V. Jayaraman, M. Soler, T. Goodwin, M. J. Culik, T. C. Goodnough, M. H. MacDougal, F. H. Peters, D. VanDeusen, D. Welch, Proc. 2000 CLEO, 2000, pp. 344.
[8] J. Geske, V. Jayaraman, Y. L. Okuno, and J. E. Bowers, Appl. Phys. Lett., vol. 79, no. 12, pp. 1760-1762, Sept. 2001.
[9] A. Black, A. R. Hawkins, N. M. Margalit, D. I. Babic, A. L. Holmes, Jr., Y. Chang, P. Abraham, J. E. Bowers, and E. L. Hu, IEEE J. Select. Topics Quantum Electron., vol. 3, no. 3, pp. 943-951, June 1997.

High Gain, Low Noise 4H-SiC UV Avalanche Photodiodes

B K Ng[1], J P R David[1], R C Tozer[1], G J Rees[1], F Yan[2], C Qin[2], and J H Zhao[2]

[1] Department of Electronic and Electrical Engineering, University of Sheffield, Mappin Street, Sheffield, S1 3JD, UK

[2] Department of Electrical and Computer Engineering, Rutgers University, 94 Brett Road, Piscataway, NJ08540, USA

Abstract. High gain, low noise visible-blind avalanche photodiodes are demonstrated using sub-micron 4H-SiC as the avalanche medium. The multiplication characteristics of the devices from UV light of various wavelengths showed unambiguously that the ionization coefficient of holes (β) is significantly larger than that of electron (α). Very low excess noise corresponding to a value of k_{eff} (= α/β for hole multiplication) of 0.1 in the local noise model is achieved.

1. Introduction

Silicon carbide (SiC) is an excellent candidate for UV detectors owing to its wide band gap. SiC photodetectors are expected to exhibit very low dark current, even at high temperature, and good visible-blind performance. 4H-SiC has been reported to have a large ratio of hole to electron ionization coefficients (β/α) [1] and consequently should be a good material in which to make UV avalanche photodiodes (APDs). In this work, we report the performance of sub-micron 4H-SiC APDs with nominal avalanche widths, w, of 0.1 μm and 0.2 μm. Thin avalanche widths are chosen to help reduce the operating voltage, increase the device speed and possibly further reduce the excess noise by increasing the effect of dead space.

2. Measurement technique

Two 4H-SiC APD structures were used in this study. The first comprises a 2 μm n layer, a 0.1 μm i-region, a 0.2 μm p layer and a thin 0.1 μm p^+ cap, grown on an n^+ substrate. The second structure is made up of a 2 μm n^- reach-through layer, a 0.11 μm n layer, a 0.2 μm i-region, a 0.25 μm p layer and a 0.2 μm p^+ cap grown on an n^+ substrate. Mesa diodes with optical access were fabricated using inductive coupled plasma etching and a 2° positive bevel edge termination technology [2] for the first APD structure while a multistep junction extension termination [3] was used to fabricate the reach-through structure. All diodes were passivated with either a thin layer of SiO_2/Si_3N_4 or SiO_2.

The photo-responses of the APDs in the wavelength range 230 – 375 nm were measured using a mercury-xenon lamp, a 0.22 m monochromator and a lock-in amplifier. The spectral responsivity of the diodes at unity gain was determined by calibrating the optical system with a commercial UV-enhanced Si photodiode. Photomultiplication characteristics of the APDs for different carrier injection conditions were measured using UV light from the mercury-xenon lamp. The excess avalanche noise was measured at a center frequency of 10 MHz and a noise effective bandwidth of 4.2 MHz using the phase sensitive technique, as described by Li *et al.* [4].

3. Results and discussions

Modeling the capacitance-voltage (CV) measurements of the diodes by solving Poisson's equation gave *i*-region widths of $w = 0.105$ μm and $w = 0.285$ μm for the bevel edge and multistep junction extension terminated APDs respectively. Secondary ion mass spectroscopy performed for the bevel edge terminated APD structure corroborates the parameters extracted from CV measurement. The CV measurement on the $w = 0.285$ μm structure indicates that the reach-through layer is not depleted up to device breakdown. This was confirmed by the lack of a characteristic "step" in the photocurrent characteristic of the reach-through structure when illuminated with weakly absorbed UV light.

Figure 1 shows the unity gain responsivity curves of both APD structures. Peak responsivities of 144 mA/W and 130 mA/W at the wavelength of 265 nm are achieved for the APDs with $w = 0.105$ μm and $w = 0.285$ μm respectively, corresponding to external quantum efficiencies of 67% and 61%. Neither structure exhibits any photo-response for wavelengths > ~380 nm, in good agreement with the minimum band gap of 4H-SiC (3.25 eV).

The dark current and photo-response characteristics of the APDs are depicted in Fig. 2. Both structures exhibit low dark currents prior to avalanche breakdown and have dark current densities < 8 μA/cm^2 at 95% of the breakdown voltage. Consequently the measured photocurrents are at least an order of magnitude larger than the dark current, as seen in Fig. 2. The breakdown voltages estimated from the dark current characteristics of the APDs with $w = 0.105$ μm and $w = 0.285$ μm are 58.5 V and 124.0 V respectively.

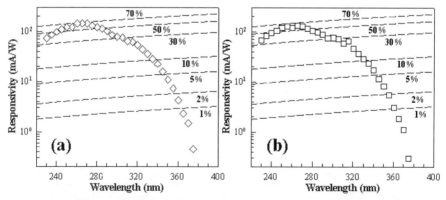

Figure 1. Measured spectral responsivity curves (symbols) of the 4H-SiC APDs with a) $w = 0.105$ μm and b) $w = 0.285$ μm at unity gain. The responsivity curves (dashed lines) corresponding to external quantum efficiencies of 1 – 70% are also shown for reference.

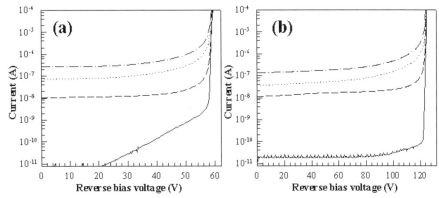

Figure 2. Typical photo-response characteristics of the 4H-SiC APDs with a) $w = 0.105$ μm and b) $w = 0.285$ μm illuminated with 230 nm (dashed lines), 297 nm (dot-dashed lines) and 365 nm (dotted lines) light. The dark currents (solid lines) of the diodes are also shown for comparison.

Figure 3 shows the normalised multiplication characteristics of the APDs illuminated with 230 nm, 265 nm and 365 nm light. It should be noted that the 365 nm light creates carriers almost uniformly in the APD structures while it is estimated that about 95% of the 230 nm light is absorbed in the p/p^+ cladding layers. The multiplication characteristics measured using longer wavelength light are always higher than those from light of shorter wavelength. Since the injection condition changes from that of almost pure electrons to mixed carriers as the illumination wavelength varies from 230 nm to 365 nm, the results show unambiguously that $\beta > \alpha$ in 4H-SiC. A conservative estimate of the β/α ratio in 4H-SiC can be made from the multiplication characteristics by assuming that the 365 nm light results in pure hole injection. This gives β/α values in the range 2.5 – 20 and 5 – 40 for the $w = 0.105$ μm and $w = 0.285$ μm structures respectively.

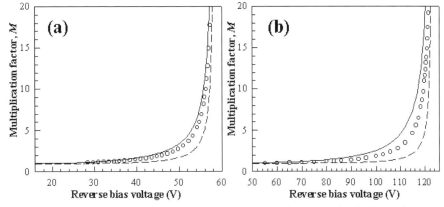

Figure 3. Multiplication characteristics of the 4H-SiC APDs with a) $w = 0.105$ μm and b) $w = 0.285$ μm measured using 230 nm (dashed lines), 265 nm (O) and 365 nm (solid lines) light.

The excess noise characteristics of the APDs illuminated with 365 nm and 230 nm light are depicted in Fig. 4. The corresponding k_{eff} values in McIntyre's local prediction, given by

$$F = k_{eff}M + (2 - 1/M)(1 - k_{eff}), \qquad (1)$$

where $k_{eff} = \beta/\alpha$ (α/β) for electron (hole) injection, are also shown for reference. The excess noise of both APD structures corresponds to a very low k_{eff} value of ~0.1 when illuminated by the weakly absorbed 365 nm light. When mainly electrons are injected using 230 nm light to initiate the avalanche process, the excess noise becomes much higher with $k_{eff} = 0.8$ and $k_{eff} = 2.7$ for the structures with $w = 0.105$ μm and $w = 0.285$ μm respectively. This behaviour corroborates the deduction of $\beta > \alpha$ from the multiplication characteristics of both structures and indicates that injecting holes to initiate avalanche multiplication is essential to achieve low excess noise in 4H-SiC, even when the avalanche region is thin.

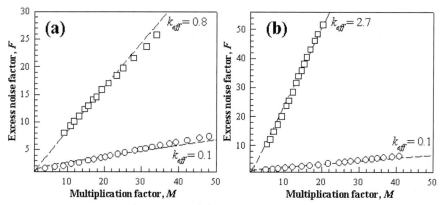

Figure 4. Excess noise characteristics of the 4H-SiC APDs with a) $w = 0.105$ μm and b) $w = 0.285$ μm measured using 365 nm (○) and 230 nm (□) light. Dashed lines are McIntyre's local predictions.

4. Conclusions

High gain, low noise UV APDs were demonstrated using sub-micron 4H-SiC as the avalanche medium. The photomultiplication results show unambiguously that $\beta > \alpha$ in 4H-SiC and that the β/α ratio remains large even in thin avalanche regions. Consequently very low excess noise corresponding to $k_{eff} = 0.1$ is achieved when the multiplication process is initiated by mixed carrier injection from using 365 nm light. The excess noise is expected to be lower when only holes are injected.

References

[1] Konstantinov A O, Wahab Q, Nordell N and Lindefelt U 1997 App. Phys. Lett. 71 90–92
[2] Yan F, Qin C, Zhao J H, Alexandrov P and Weiner M 2001 Tech. Dig. Int. Conf. on SiC and Related Materials 652–653
[3] Yan F, Luo Y, Zhao J H, Bush M, Olsen G H and Weiner M 2001 Electron. Lett. 37 1080–1081
[4] Li K F, Ong D S, David J P R, Rees G J, Tozer R C, Robson P N and Grey R 1998 IEEE Trans. Electron Devices 45 2102–2107

Novel Microcavity Light Emitting Diodes

R. P. Stanley[1], P. Royo[2], U. Oesterle[3], R. Joray[3], M. Ilegems[3]

[1] Swiss Center for Electronics and Microtechnology (CSEM),
   CH-2007 Neuchâtel, Switzerland
[2] Avalon Photonics Ltd, CH-8048 Zürich, Switzerland
[3] Institute of Quantum Electronics and Photonics, Swiss Federal Institute of
   Technology Lausanne (EPFL), CH-1015 Lausanne, Switzerland

Abstract. This paper presents a novel microcavity structure which uses two /8 phase shift layers which form a short microcavity. Using this principle surface emitting devices were made where the top mirror is just the air/semiconductor interface. Record external efficiencies were measured.

1. Introduction

The efficiency of planar light–emitting diodes (LEDs) is usually low due to total internal reflection at the semiconductor–air (or epoxy for encapsulated devices) interface. The current generation of high efficiency LEDs are all non-planar using some variation on a cube where the light is extracted from all sides, or non-parallel faces such as tapers or trapezoids where the totally reflected light is redirected towards the normal direction by one or more internal reflections [1, 2]. Certain quasi-planar techniques using textured surfaces are also very promising but rely on substrate removal [3]. While these non-planar solutions can give extremely high external quantum efficiencies >50%, a completely planar technology would be preferable.

Microcavities are a planar solution to the problem of light extraction. A Fabry-Pérot cavity around the emitting region uses interference to enhance the emission in the normal direction and to inhibit it in other directions. As well as increasing the efficiency, microcavities also improve brightness (power per area per solid angle) and can reduce both linewidth and spectral variation with temperature [4].

For microcavity light emitting diodes (MCLEDs), the main limitations to the extraction efficiency are: (a) The order of the cavity including the penetration of the light field into the mirrors. (b) The enhancement of the guided mode. (c) Leakage of light into the substrate at angles greater than the angular bandwidth of the distributed Bragg reflector (DBR) mirrors. In this paper, we describe a novel microcavity structure that improves on the first two issues.

2. General Case

The standard cavity has a high refractive index (low bandgap) layer of λ thickness sandwiched between two DBR mirrors. A $\lambda/2$ low index cavity would be better optically having both a shorter cavity and reduced coupling to guided modes. This is not practical because efficient electrical injection requires a low band gap active layer (high refractive index). In the GaAs/AlAs system, defects associated with AlAs layers would lower the internal quantum efficiency. Instead we propose to use two $\lambda/8$ low index layers surrounding a $\lambda/4$ high index layer. This gives most of the advantages of the $\lambda/2$ cavity while remaining compatible with the needs of electrical injection and growth. We denote this new cavity a phase-shift cavity.

Figure 1 shows three generic cavities: (1) the standard λ high index cavity, (2) the $\lambda/2$ low index cavity and (3) the proposed $\lambda/8$ phase-shift cavity. The indices of the layers are 3.5 and 2.9 for the GaAs and AlAs, respectively. At normal incidence, all three cavities show a Fabry-Pérot (FP) mode in transmission. It is somewhat surprising to find a FP mode for the phase–shift structure as there is no apparent cavity. If the low index layers had zero thickness, the cavity would be $3\lambda/4$ in length, as the layer thickness increase, we eventually reach $\lambda/4$ in thickness for each layer. In both cases the structure is just one long DBR with no resonance. For intermediate lengths close to $\lambda/8$ a virtual $\lambda/2$ cavity forms. Such phase-shifting layers are commonly used in distributed feedback (DFB) lasers to create a single mode in the center of the DFB stopband. However, for DFB lasers the phase-shift layers are placed far apart in order to distribute the optical field throughout the laser structure, while in these microcavities the phase shift layers are acting to concentrate the optical field.

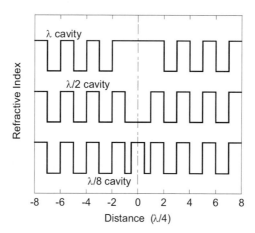

Figure 1. Three cavity structures with (1) λ cavity, (2) $\lambda/2$ cavity and (3) $\lambda/8$ phase-shift cavity. The dashed line denotes the source position.

Figure 2 (a) shows the optical field of the FP mode as a function of position. The field has a maximum at the center of the cavities and the penetration of the optical field into the DBR mirrors is apparent. The field falls off most rapidly for the $\lambda/2$ low index cavity, and the $\lambda/8$ phase-shift cavity better than the λ cavity. A full numerical calculation shows that fraction of emission into the FP mode is 20% for the λ cavity, 36% for the $\lambda/2$ cavity and 28% for the phase-shift cavity. The difference between the $\lambda/2$ cavity and the phase-shift cavity is due to the low refractive index emitter in the former case.

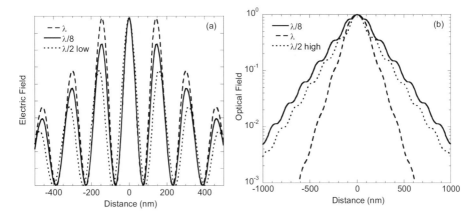

Figure 2. The optical field of (a) Fabry-Pérot mode & (b) guided mode as a function of position for a λ cavity (dashed), a λ/2 cavity (dotted) and the λ/8 phase-shift cavity (solid line). λ/2 cavity has (a) low, (b) high index.

The optical fields of the guided modes are compared in Fig. 2(b). The λ/2 low index cavity has no guided mode so we show a λ/2 high index cavity for comparison. The λ cavity has the most confined guided mode while the λ/8 phase-shift cavity is less confined than the λ/2 high index cavity. The confinement factor is significantly lower for the λ/8 phase-shift cavity making it a poor candidate as an edge-emitting laser. The guided mode competes with the Fabry-Pérot mode in MCLEDs so the reduced confinement factor gives a second advantage to the λ/8 PS cavity. The numerical calculation shows that the percentage of light going into the guided mode reduces from 36% for the λ cavity, to 28% in the λ/8 phase-shift cavity.

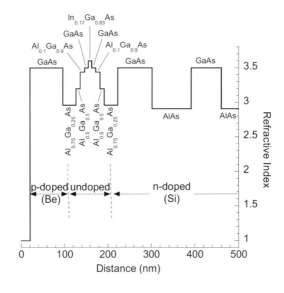

Figure 3. Schematic MCLED structure

3. Surface emitters with air interface

The phase-shifted principle can be adapted to any existing microcavity structure with the advantage that the effective order decreases by one. We have examined surface emitting structures where the top mirror is simply the GaAs/air interface (see Figure 3). The low reflectivity (30%) is still sufficient to give a microcavity effect when combined with a highly reflective bottom mirror. There is no penetration depth for the GaAs/air interface.

An additional property of this structure is that when gold is deposited on it, the phase change is sufficient to inhibit emission in the normal direction. Instead the light is redirected into leaky and guided modes. The result is that the light has a tendency to move sideways out from underneath the metal contact. If the light is reabsorbed, then photon assisted current spreading occurs.

Devices were fabricated following the above design and their performance is shown in Figure 4 for a square mesa with a 400 μm aperture. The maximum extraction efficiency is 19% and exceeds that from any other similar sized planar LED. The maximum efficiency occurs at very low current densities, of the order of 2 A/cm^2. The efficiency rolls over above 2 A/cm^2 due to a combination of current crowding and device heating. The calculated extraction efficiency is 25.5% showing that the internal quantum efficiency is high and that the effect of contact shadowing is small. Using the same principle devices using AlO$_x$ mirrors give 28% external quantum efficiency [5].

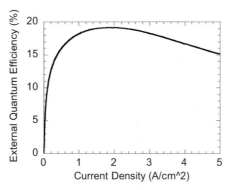

Figure 4. Measured external quantum efficiency as a function of drive current for a phase-shift MCLED.

4. Conclusion

In conclusion, we present a novel microcavity structure which uses a phase-shift cavity to form a microcavity which has advantages over standard MCLEDs. Additionally, the air/semiconductor interface is used as the top mirror. High efficiencies (19%) were measured.

References

[1] Krames M R, et al. 1999 Appl. Phys. Lett. 75 2365-2367
[2] Schmid W, et al. 2002 IEEE J. Select. Topics Quantum Electron. 8 256-263
[3] Windisch R, et al. 2002 IEEE J. Select. Topics Quantum Electron. 8 248-255
[4] Royo P, Stanley R P and Ilegems M 2002 IEEE J. Select. Topics Quantum Electron. 8 207-218
[5] Rattier M, et al. 2002 IEEE J. Select. Topics Quantum Electron. 8 238-247

High Extraction Efficiency AlGaInP Microcavity Light Emitting Diodes at 650 nm with AlGaAs-AlO$_x$ DBR

R. Joray[1], J. Dorsaz[1], R. P. Stanley[2], M. Ilegems[1], M. Rattier[3], C. Karnutsch[4], and K. Streubel[4]

[1] Institute of Quantum Electronics and Photonics, Swiss Federal Institute of Technology Lausanne (EPFL), CH-1015 Lausanne, Switzerland

[2] Swiss Center for Electronics and Microtechnology (CSEM), CH-2007 Neuchâtel, Switzerland

[3] Laboratoire de Physique de la Matière Condensée, Ecole Polytechnique, F-91128 Palaiseau Cedex, France

[4] OSRAM Opto Semiconductors, D-93049 Regensburg, Germany

Abstract. Microcavity light emitting diodes emitting at 650 nm, which make use of an $Al_{0.5}Ga_{0.5}As$-AlO_x bottom DBR formed by lateral wet oxidation of $Al_{0.98}Ga_{0.02}As$, are fabricated and characterized. The active region consists of a compressively strained GaInP quantum well in the center of a 2λ-width AlGaAs/AlGaInP cavity. The maximum measured external quantum efficiency is 12.5% into air. The measurements are compared with simulations.

1. Introduction

Red light emitting diodes (LEDs) are mainly used for lighting and plastic optical fiber (POF) communication. Some of the basic requirements are high efficiency, high brightness and minimal power consumption, as well as low manufacturing costs. Good candidates to fulfil these needs are microcavity LEDs (MCLEDs). A standard MCLED consists of an active region placed inside a Fabry-Pérot cavity made of two distributed Bragg reflectors (DBRs) separated by a distance on the order of the optical wavelength [1]. It shows high efficiency, high brightness, control of the angular emission pattern and at the same time low manufacturing costs due to the fact that no additional fabrication steps such as substrate removal, epitaxial lift-off or shaping are necessary. The highest reported external quantum efficiencies for Microcavity LEDs (MCLEDs) with GaAs-AlAs distributed Bragg reflectors (DBRs) are 23% at 980 nm [2] and 9.6% at 640 nm [3], both for emission into air.

The use of high contrast GaAs-AlO$_x$ DBRs brings several advantages. The low refractive index of the aluminum oxide (n≈1.6) leads to a high refractive index difference, which means a drastically reduced number of periods necessary for a given reflectivity and a much wider stopband. Contrary to dielectric DBRs, devices with oxide DBRs can be fabricated in one single epitaxial growth step. Due to the decreased penetration depth, the effective cavity length can be significantly reduced by using an oxide DBR, that in turn means a higher optical field at the emitter and a higher overall efficiency. On the other hand, contacting the bottom n-doped layer and heat dissipation are more critical than with nonoxidized DBRs. An external quantum efficiency of 28% was reported by Rattier et al. for MCLEDs at 980 nm with a GaAs-AlO$_x$ DBR [4]. We present initial results on the realization of AlGaInP MCLEDs emitting at 650 nm with an Al$_{0.5}$Ga$_{0.5}$As-AlO$_x$ DBR.

2. Growth and fabrication

The structure consists of a 2λ cavity with an Al$_{0.5}$Ga$_{0.5}$As-AlO$_x$ bottom DBR whereas the interface GaAs-air acts as the top mirror. It was grown by OSRAM Opto Semiconductors by metalorganic vapor phase epitaxy (MOVPE) on 4" GaAs substrates tilted by 6° toward <111>$_A$. The undoped bottom distributed Bragg reflector (DBR) is made of four periods of Al$_{0.5}$Ga$_{0.5}$As-Al$_{0.98}$Ga$_{0.02}$As. Subsequently the n-doped Al$_{0.7}$Ga$_{0.3}$As n-contact layer is deposited. The undoped active region consists of a single compressively strained GaInP quantum well with (Al$_{0.5}$Ga$_{0.5}$)$_{0.51}$In$_{0.49}$P cladding and (Al$_{0.7}$Ga$_{0.3}$)$_{0.51}$In$_{0.49}$P confinement layers. The structure is completed by a p-doped Al$_{0.7}$Ga$_{0.3}$As current spreading layer and a thin highly p-doped GaAs p-contact layer.

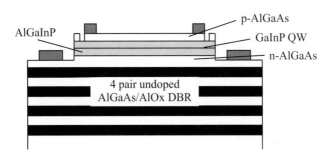

Figure 1. Schematic device structure

The high aluminum content layers in the bottom DBR are oxidized by lateral wet oxidation [5] at 450 °C during two hours after deep trenches have been etched in order to expose the layers to the oxidizing atmosphere. During the oxidation process the surface is protected by a silicon nitride layer deposited by plasma enhanced chemical vapor deposition (PECVD). Circular mesas are wet etched to expose the n-contact layer and the top p-doped Al$_y$Ga$_{1-y}$As layers are etched away on a small ring at the mesa exterior in order to suppress current spreading under the p-contact and non-radiative surface recombination at the mesa sidewalls. A Ni/GeAu/Ni/Au n-contact is deposited and alloyed at 380 °C; polyimide is applied as a dielectric coating between n- and p-contact and a nonalloyed Ti/Au p-contact ring is deposited. Some of the devices include a p-contact current injection grid.

The critical issue of this structure is the aluminum content of the n-contact layer. The aluminum oxide is electrically isolating, which means that the electrons need to be injected laterally. To avoid optical absorption in the n-contact layer, $Al_xGa_{1-x}As$ with $x \geq 0.5$ has to be used. The resistivity and the vertical oxidation rate of this layer are increasing with aluminum content. As a compromise the aluminum content of this layer was fixed to 70%. The lateral oxidation rate of $Al_xGa_{1-x}As$ layers increases exponentially with aluminum content and therefore high aluminum content layers can be selectively oxidized [6]. In this structure the $x = 0.7$ n-contact layer is on top of the last $x = 0.98$ layer, which means that, as the oxidation of the $x = 0.98$ layer proceeds, the $x = 0.7$ layer oxidizes vertically from the interface. Since this layer is quite thin, it will be completely oxidized within a short time, even though its oxidation rate is about two orders of magnitude lower than for the $x = 0.98$ mirror layers. Therefore the oxidation has to be stopped before the layer is completely oxidized, otherwise no current could be injected. As a result the $x = 0.98$ layers in the DBR were not completely oxidized in the present experiment. In the center of the diodes a square of about 15-20% of the total surface remains therefore nonoxidized.

3. Results and discussion

The devices show an emission peak at 653 nm and a FWHM of 23 nm at room temperature. By comparing the reflectivity and far field measurements with simulations we estimate that about 20 nm of the n-intracavity contact layer are transformed into AlO_x. The emission properties of the structure were simulated using a method based on the plane wave expansion of an electrical dipole inside a multilayer structure [7]. The simulations predict a theoretical external quantum efficiency of 18% for emission into air and 24% for emission into epoxy. If we take into account the vertical oxidation the theoretical efficiency into epoxy drops to 21%, whereas the efficiency into air stays at 18%.

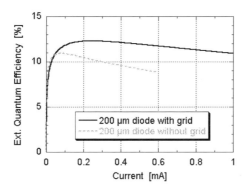

Figure 2. Measured external quantum efficiencies as a function of diode current for a 200 µm diode with p-contact current injection grid (solid line) and a 200 µm diode without grid (dashed line)

The L-I curves and the measured external quantum efficiencies as a function of diode current show that the addition of a p-contact current injection grid allows to achieve higher optical powers and efficiencies thanks to the improved current injection although this leads to an increased contact shadowing. The maximum measured efficiencies are 10.9% for the 200 μm devices without a p-contact grid and 12.5% for the 200 μm devices with a grid. To our knowledge these are the highest reported effiencies for red MCLEDs. Note that the devices start to emit light at very low current densities, which shows that grown material is of extremely good quality. The efficiency starts to roll over already at very low current injection levels. The roll over is attributed to current crowding due to the reduction in thickness of the n-current injection layer by the vertical oxidation

4. Conclusions

The advantages and the feasibility of red AlGaInP MCLEDs with an $Al_{0.5}Ga_{0.5}As/AlO_x$ DBR were demonstrated. The measured external quantum efficiencies exceed 12%, which is to our knowledge the highest reported efficiency for red MCLEDs. The simulations show that external quantum efficiencies up to 18% for emission into air and 24% for emission into epoxy are possible. The critical issue is the aluminum content of the n-intracavity layer. It needs to be high enough to avoid optical absorption in the cavity but should not be too high since otherwise the layer is oxidizing vertically during the mirror oxidation and the current injection and spreading are insufficient. The structure will be further optimised to achieve full mirror oxidation and improve the current injection.

References

[1] Schubert E F, Wang Y-H, Cho A Y, Tu L-W and Zydzik G J 1992 Appl. Phys. Lett. 60 921-923
[2] De Neve H, Blondelle J, Van Daele P, Demeester P, Baets R and Borghs G 1997 Appl. Phys. Lett. 70 799-801
[3] Wirth R, Karnutsch C, Kugler S and Streubel K 2001 IEEE Photon. Technol. Lett. 13 421-423
[4] Rattier M, Benisty H, Stanley R P, Carlin J-F, Houdré R, Oesterle U, Smith C J M, Weisbuch C and Krauss T F 2002 IEEE J. Select. Topics Quantum Electron. 8 238-247
[5] Dallesasse J M, Holonyak Jr. N, Sugg A R, Richard T A and El-Zein N 1990 Appl. Phys. Lett. 57 2844-2846
[6] Choquette K D, Geib K M, Ashby C I H, Twesten R D, Blum O, Hou H Q, Follstaedt D M, Hammons B E, Mathes D and Hull R 1997 IEEE J. Select. Topics Quantum Electron. 3 916-926
[7] Benisty H, Stanley R and Mayer M 1998 J. Opt. Soc. Am. A 15 1192-1201

Orientation-mismatched wafer bonding for polarization control of 1.3 μm-wavelength vertical cavity surface emitting lasers (VCSEL)

Yae L. Okuno, Jon Geske, Yi-Jen Chiu, Steven P. DenBaars, and John E. Bowers

Department of Electrical and Computer Engineering, University of California, Santa Barbara, CA 93106

Abstract. We propose and demonstrate a new type of long-wavelength (LW) VCSEL which consists of a (311)B InP-based active region and (100) GaAs-based distributed Bragg reflectors (DBRs), with an aim to control the in-plane polarization of output power. Crystal growth on (311)B InP substrates was performed under low-migration conditions to achieve good crystalline quality. The VCSEL was fabricated by wafer bonding, which enables us to combine different materials regardless of their lattice- and orientation-mismatch without degrading their quality. The VCSEL showed polarization dependent performance with a maximum power extinction ratio of 37 dB.

1. Introduction

LW VCSELs have been extensively studied as low-cost, high-performance light sources for telecommunications. However, compared to edge-emitting lasers, VCSELs have the disadvantage of not having fundamental selection rules for the polarization axis of output power due to its crystal symmetry when fabricated on the conventional (100) plane. To make polarization fixed to one axis of the VCSEL, there has to be some asymmetry introduced in its structure. For example, it has been shown theoretically [1] and experimentally [2] that multi-quantum wells (MQWs) grown on an asymmetric crystal plane have asymmetric in-plane gain, which leads to a fixed in-plane polarization axis. Such polarization control research has been mainly done for short-wavelength VCSELs, but relatively little has been done on LW VCSELs.

To fabricate LW VCSELs, various techniques have been investigated [3-8]. Among them, wafer bonding is a technique that has enabled LW VCSELs to operate continuously up to 115°C [6-8]. With this technique we can combine two dissimilar materials without degrading their quality. Therefore, it is possible for a wafer-bonded VCSEL to have an InP-based active region and GaAs-based DBRs. Also, with wafer bonding, we have the freedom to choose the crystal planes of active region and DBRs independently [9]. Here we present the fabrication of a wafer-bonded LW VCSEL which utilizes (311)B InP-based active region and (100) GaAs-based DBRs and investigate its polarization properties.

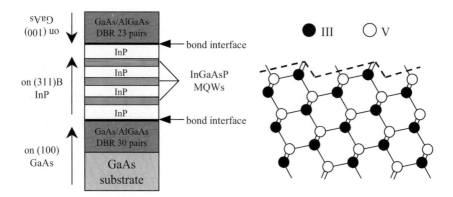

Figure 1. Structure of VCSEL

Figure 2. Schematic of atomic structure of (311)B

2. VCSEL structure and its fabrication procedure

Figure 1 is a schematic of our VCSEL structure. By changing the crystallographic orientation of the active region to (311), we can expect to have not only an asymmetric in-plane gain, but also an asymmetric stress provided from two bonded interfaces of (311)B InP and (100) GaAs. That is, at the lattice- and orientation-mismatched interface, effective lattice mismatch would be very different between two orthogonal in-plane axes [9]. We expect this asymmetric stress to have an effect on controlling the polarization axis. To focus on this effect, the MQWs were designed to have a small lattice mismatch so that the asymmetry of in-plane gain prior to bonding would be small.

The 2.5 λ-cavity active region consisted of three identical sets of InGaAsP MQWs and each set consisted of five 50Å thick wells and six 100Å thick barriers. It was grown by metalorganic chemical vapor deposition (MOCVD) using trimethylindium, trimethylgallium, tertiarybutylarsine, and tertiarybutylphosphine as the source materials. DBRs consisted of GaAs and $Al_{0.9}Ga_{0.1}As$, and were grown by conventional solid-source molecular beam epitaxy. The VCSEL was intended to be optically pumped, hence, all structures were grown undoped.

After the material growth, wafer bonding was performed to fabricate the VCSEL. First, the surfaces of the active region and one DBR were chemically cleaned and then they were placed face-to-face and annealed at 650°C with an applied pressure of about 3MPa. The InP substrate was selectively etched by a mixture of HCl and H_2O from its back side, i.e., (311)A plane. Then another DBR was wafer-bonded in the same way. A more detailed procedure about wafer bonding can be found elsewhere [8].

3. Growth on (311)B InP substrate

Figure 2 is a schematic drawing of the (311)B atomic structure, showing the step-like features of this surface. Phenomena such as In atom accumulation at steps during growth have been reported on this surface [2]. Therefore, it is important to perform growth under low-migration conditions such as low temperature and high V/III ratio. Figure 3 compares photoluminescence (PL) spectra from MQWs grown under two different conditions. As can be seen, when grown under high-migration conditions, PL intensity degraded to half of that from MQWs grown on (100) substrate at the same time.

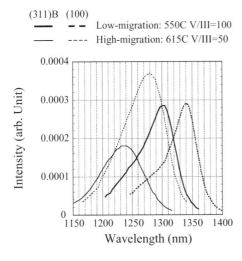
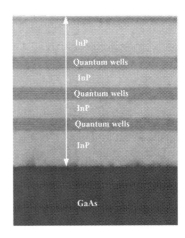

Figure 3. PL spectra from MQWs grown under different conditions.

Figure 4. Cross-section TEM of VCSEL active region bonded to GaAs substrate.

The PL intensity and shape is comparable to that of (100) when grown under low-migration condition, but a notable fact is a blue-shift of the emission peak by about 40nm. By comparing bulk InGaAsP materials grown on (311)B and (100), it was found that this blue-shift is due to a large reduction in In incorporation and a small reduction in As incorporation on the (311)B surface. It is believed that In tends to desorb from step corners on the (311) surface [10]. Since the strain in the MQWs was small, little piezoelectric effect was expected. A low temperature PL experiment showed no peak shift by changing excitation power, confirming this expectation.

4. Wafer bonding of (311)B InP and (100) GaAs

Figure 4 shows a cross-section image of the active region bonded onto a plain GaAs substrate taken by transmission electron microscope (TEM). No dislocations can be observed in the structure. A high-resolution TEM observation was also performed, and atomic bonding between these two materials was confirmed.

Figure 5 compares PL spectra from the active region grown on (311)B InP. After the bonding, the PL intensity decreased to less than half of that of the as-grown active region, and the emission peak blue-shifted by about 20 nm. Since these changes are also visible on the annealed active region, they are due to annealing during the bonding at a temperature 100°C higher than the growth temperature. It is assumed that there was interface mixing occurring at well/barrier interfaces during the annealing process [11]. These PL changes can be suppressed by changing barrier material to InP.

5. VCSEL performance

Figure 6 shows the polarization-dependent lasing characteristics of the VCSEL measured by an optical spectrum analyser. The VCSEL was optically pumped by a 980nm edge-emitting laser via free-space optics. Since the pump laser was TE polarized, the VCSEL polarization behavior was examined by aligning the pump laser polarization axis to one of the two orthogonal crystal axes of the VCSEL.

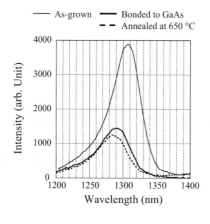
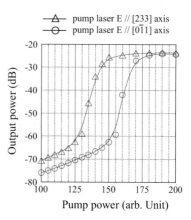

Figure 5. PL spectra from active region   Figure 6. Output power of VCSEL

The power extinction ratio between the two axes was 37 dB at the largest point. The VCSEL had no index guiding structure, and showed a single mode spectrum at 1273 nm with side-mode suppression ratio of 41 dB.

6. Conclusion

We have fabricated a LW VCSEL which utilized a (311)B InP-based active region instead of a conventional (100) InP-based active region, in order to achieve polarization control. MOCVD growth on (311)B InP showed distinctive properties due to the step-like feature of this surface. TEM observations showed no signs of defects in the active region wafer-bonded to (100) GaAs, while PL showed its intensity degradation due to annealing during the bonding process. The active region was double-bonded to (100) GaAs-based DBRs to form a VCSEL. A power extinction ratio between the two orthogonal axes of the VCSEL was 37 dB at its maximum point, showing polarization dependent behavior.

Acknowledgement

The author would like to thank Mr. Kohl Gill for his help on low-temperature PL measurement. This research was supported by National Science Foundation (NSF) and Walsin Lihwa Corporation.

References

[1]   Ohtoshi T et al. 1994 Appl. Phys. Lett. 65 1886-7
[2]   Nishiyama N et al. 2001 IEEE J. Select. Topics Quantum Electron. 7 242-8
[3]   Nakagawa S et al. 2001 IEEE J. Select. Topics Quantum Electron. 7 224-30
[4]   Chang-Hasnain C J 2000 IEEE J. Select. Topics Quantum Electron. 6 978-987
[5]   Fischer M et al. 2000 IEEE Photon. Tech. Lett. 12 1313-5
[6]   Jayaraman V et al. 2000 IEEE Photon. Tech. Lett. 12 1595-7
[7]   Karim A et al. 2001 Appl. Phys. Lett. 78 2632-3
[8]   Black A et al. 1997 IEEE J. Select. Topics Quantum Electron. 3 943-51
[9]   Okuno Y et al. 1997 IEEE J. Quantum Electron. 33 959-69
[10] Takahashi M et al. 1996 Jpn. J. Appl. Phys. 35 6102-7
[11] Teng J H et al. 2001 Mat. Sci. Semiconductor Processing 4 621-4

Population inversion enhancement by resonant magnetic confinement in THz quantum cascade lasers

G. Scalari, S. Blaser, L. Ajili, M. Rochat, H. Willenberg, D. Hofstetter, J. Faist

Institute of Physics, University of Neuchâtel, CH-2000 Neuchâtel, Switzerland

H. Beere, G. Davies, E. Linfield, D. Ritchie

Cavendish Laboratory, University of Cambridge, Madingley Road, Cambridge CB3 0HE, UK

Abstract. Effects of a strong perpendicular magnetic field on a terahertz quantum cascade (QC) laser are presented. A reduction of the threshold current density to 50% of the zero-field value together with an increase of the slope efficiency by a factor of 2.7 was observed. Experimental results suggest a strong influence of the intersubband Landau resonances in the extraction process of the laser structure. A new THz QC laser design that relies on this effect is presented.

1. Introduction

Optical phonon scattering is a key ingredient in the design of a quantum cascade laser [1]. Resonant optical phonon emission is used in many mid-infrared (MIR) QC laser designs to reduce the lower state lifetime. Even in designs in which it is not explicitly used to engineer the upper and lower states lifetimes [2,3], optical phonon emission is the dominant electron cooling mechanism. In the far-infrared (FIR), intersubband electroluminescence experiments clearly showed that electron-electron scattering is the main scattering mechanism in clean structures at low temperature [4]. As electron-electron scattering does not by itself lower the energy of an ensemble of electrons but tends to equilibrate the population between subbands, the achievement of population inversion is even more challenging in the FIR than in the MIR.

2. THz QC laser in magnetic field: magnetotransport measurements

A possible approach to quench the carrier-carrier scattering and realize a simplified way to study the effect of in-plane confinement in QC structures is by applying a strong perpendicular magnetic field. Luminescence intensity enhancement and a substantial emission linewidth narrowing together with oscillations in emitted radiation were observed in Thz QC emitters [5,6]. Those issues led us to perform measurements of THz QC laser in strong perpendicular magnetic field (parallel to growth axis).
The design used is a three quantum well chirped-superlattice active region [7,8]. The computed band diagram and squared wavefunctions of sample A2672 are displayed in Fig. 1 . Details on this device can be found in Ref [9]. Magnetotransport measurements

were performed on sample A2672 to determine energy level spacing [10] and are shown in Fig. 2 (Left Panel). Bias on the sample has been fixed and current flow has been measured as a function of magnetic field. When the resonance condition is satisfied between the cyclotron energy of the host material and Landau Levels attached to each of the states of the structure, intersubband Landau resonances occurs (ISLR) and a peak in the current is observed due to elastic scattering between these levels. This allows us to identify three states almost equally spaced in the lower miniband of the structure just computing the transition energy from field values at current peaks and then comparing it with the calculated energies of electronic states. The measured energy spacing are E_{43}=19.2 meV, E_{32}=6.3 meV, E_{21}=6.2 meV.

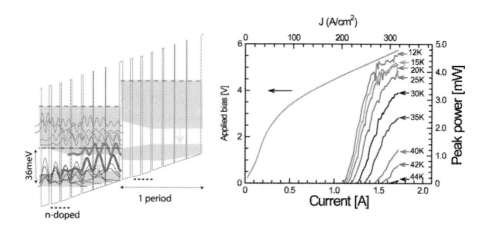

Figure 1: Left panel: Computed conduction band profile of one stage of the structure under an average applied electric field of 3.3kV/cm. Right panel: LI-IV curves for THz QC laser in pulsed mode. Light versus current measurements have been performed at various temperatures. Very low duty cycles were used, in order to avoid saturation on our He-cooled Si bolometer. Peak output powers as large as 5 mW have been obtained at low temperatures, with still 1mW measured at 40 K.

2. THz QC laser in magnetic field: emission measurements

The experimental setup used for magnetotransport and electroluminescence experiment comprises as fundamental parts a liquid He cryostat equipped with a superconducting magnet (max field of 14 T) together with a cyclotron resonance tunable InSb detector and all measurements are performed at a temperature of 4.2 K or below [6,11]. A scheme of the Landau levels (Landau fan) originating from the electronic states of the structure determined by magnetotransport measurements is shown in Fig.2 together with laser emission in function of the magnetic field. The magnetic field has a strong influence on the laser performances as the signal undergoes several oscillations and the emission is enhanced also at low fields, with a maximum at 3.4 T. An increasing of light intensity is visible in proximity of resonances between the three levels of the lower miniband (grey shaded areas on Fig. 2). On the contrary, where ISLR occur between the levels attached to upper miniband edge and lower miniband Landau states (hatched areas on Fig. 2), a decrease in light intensity is observed.

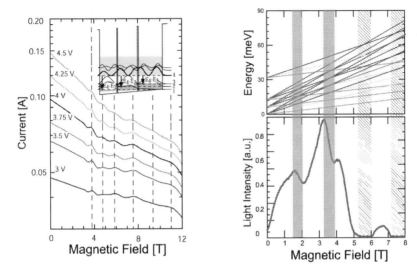

Figure 2: Left Panel: Magnetotransport measurements on sample A2672 at different biases: dotted lines indicate current maxima corresponding to ISLR. Inset: calculated bandstructure of three quantum-well active region showing the three levels in lower miniband and one level in upper miniband extracted from magnetotransport measurements. Right Panel: Landau fans of the active region and laser emission vs magnetic field.

3. Threshold current and slope efficiency analysis

Together with increased emission there is a drastic reduction in threshold current. A plot of threshold currents measured at relevant magnetic field values is shown in Fig.3. Threshold current values oscillate with opposite phase in respect to light intensity and their values are always under the value at zero field ($J_{th}^{B=0}$). In detail, $J_{th}^{B=1} \cong 3/4\, J_{th}^{B=0}$ and for B = 4.1 T it reaches its minimum value, nearly 50% of $J_{th}^{B=0}$. Minimum threshold current density for this device (2.7 mm-long, 200 μm-wide with HR backfacet coating) is 110 A/cm^2. An analysis of slope efficiency dP/dI has also been carried out. Slope efficiency is directly related to population inversion in a laser structure, and the measured value at zero magnetic field is 13 mW/A, much lower than the value ($\cong$1W/A) extrapolated from an expression in which lower state population is neglected [1,9]. This suggests that the laser action relies on a marginal population inversion probably limited by population of the lower state due to poor extraction. The increasing of the slope efficiency measured in function of applied magnetic field (Fig. 3), with a maximum increment of 2.7 times the zero field value, points towards an enhancement of the extraction of carriers from lower state due to resonant magnetic confinement. A new structure for THz QC laser (see Fig 3 for bandstructure) has been designed exploiting this effect: we use the resonance between the Landau states attached to the ground state and first excited of the well to help the extraction of carriers after the radiating transition between levels 3 and 2 in the same well. At the same time the transition relative to levels 3 and 2 is antiresonant with the applied magnetic field, quenching the non radiative channels The designed energy spacing of the structure is $E_{21}=10.6$ meV, $E_{32}=14.8$ meV, requiring a field of 6.1 T to reach resonance condition for the extraction states. This design should allow us to observe a huge increasing of laser emission and a strong threshold current diminution in correspondence of the resonant magnetic field 6.1 T, thus proving the strong influence of resonant magnetic confinement on the extraction process.

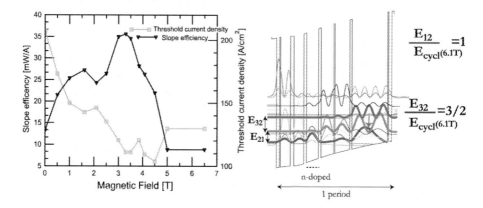

Figure 3: Left: threshold current densities for applied magnetic fields and computed slope efficiencies. Right: Design of a THz QC laser exploiting the resonant extraction by mangnetic confinement. Energy spacing between extraction levels 1 and 2 is designed to be resonant at 6.1 T, where the radiative transition 3-2 is completely antiresonant

References

[1] J. Faist, F. Capasso, C. Sirtori, D. Sivco, and A. Cho, Quantum cascade lasers, in *Intersubband transitions in quantum wells: Physics and device applications II*, edited by H. Liu and F. Capasso, volume 66, chapter 1, pages 1-83, Academic Press, 2000.
[2] G. Scamarcio et al., Science **276**, 773 (1997).
[3] J. Faist et al., Nature **387**, 777 (1997).
[4] M. Rochat, J. Faist, M. Beck, U. Oesterle, and M. Ilegems, Appl. Phys. Lett. **73**, 3724 (1998).
[5] J. Ulrich, R. Zobl, K. Unterrainer, G. Strasser, and E. Gornik, Appl. Phys. Lett. **76**, 19 (2000).
[6] S. Blaser, M. Rochat, M. Beck, D. Hofstetter, and J. Faist, Appl. Phys. Lett. **81**, 67 (2002).
[7] R. Köhler et al.,Nature **417**, 156 (2002).
[8] A. Tredicucci et al., Appl. Phys. Lett. **72**, 2388 (1998).
[9] M. Rochat et al., Appl. Phys. Lett. **81**, 1381 (2002)
[10] S. Blaser et al., Physica E **13**, 854 (2002).
[11] E. Gornik Landau emission, in *Landau level spectroscopy*, edited by G. Landwehr and E. I. Rashba, chapter 16, pages 911-996, Elsevier Science, New-York, 1991.

Interferometric temperature mapping of GaAs-based quantum cascade laser

C. Pflügl, M. Litzenberger, W. Schrenk, S. Anders, D. Pogany, E. Gornik, G. Strasser

Institut für Festkörperelektronik, TU Wien, Floragasse 7, A-1040 Vienna, Austria

Abstract. We investigated the thermal dynamics of GaAs-based quantum cascade lasers by a scanning interferometric thermal mapping technique. The current-induced heating of the working devices causes a change of the semiconductor refractive index, which is probed by an infrared laser beam. The interferometrically detected phase shift provides a quantitative information on the thermal characteristics in the devices with a nanosecond time resolution and a spatial resolution in the microns range. Analyzing these experimental results with a two-dimensional model of heat diffusion reveals the heat distribution in the lasers as well as the anisotropic heat conductivity in the multilayered active region.

1. Introduction

Quantum cascade lasers (QCLs) based on intersubband/interminiband transitions in GaAs/AlGaAs are unipolar semiconductor lasers [1], [2]. Compared to InP-based QCLs [3, and references within], GaAs-based QCLs offer the advantage of higher flexibility in the engineering of the electronic states. In spite of the improvements concerning output power, threshold current, single mode operation and maximum operating temperature [4], [5], [6] their performance is still limited by strong heating of the active region due to poor heat dissipation. Whereas continuous wave operation was early achieved for InP-based QCLs [7] and is already demonstrated at room temperature [3], there is still only one report about a GaAs-based QCL operating in continuous wave up to 30 K [8]. Improving the lasers with respect to better thermal characteristics requires knowledge about the thermal dynamics in the laser under operation. Microluminescence measurements of the thermal behavior of QCLs have been reported by Spagnolo et al [9], [10]. This method enables to obtain a spatially resolved temperature distribution in the active area and thus allows the investigation of the thermal resistance in the active region and the efficiency of different mounting methods. Nevertheless the time evolution of the heat distribution cannot be conceived directly.

We want to report on a technique, namely the interferometric thermal laser thermal laser mapping technique, which we used to investigate GaAs-based QCLs under operation. Comparing the experimental results with a thermal model enables to determine the heat distribution of the working devices and reveals the heat conductivity of the multilayered active region.

2. Experimental technique

An infrared laser probe beam (wavelength of 1.3μm, well below the GaAs bandgap), is directed to the sample from the device backside, passes through the substrate and laser active area and is back reflected on the surface metalization (see Fig. 2b). The current-induced heating causes a temperature increase in the active region, which induces changes in the semiconductor refractive index. The resulting phase shift, which is detected interferometrically, provides a quantitative information on the thermal dynamics [11], [12].

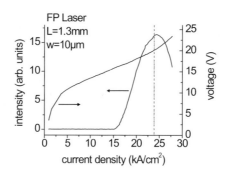

Figure 1. L-I and U-I characteristics at room temperature of the investigated QCL structure. The current density applied during the measurements is indicated.

The investigated QCL has an active region consisting of 50 periods of a chirped $Al_{0.45}Ga_{0.55}As$ /GaAs superlattice. This active region is embedded into a double plasmon enhanced waveguide. Ridge waveguides with a width of 10 μm are fabricated by etching 10 μm trenches. The extended TiAu contacts are insulated with SiN. The length of the laser is 1.3 mm. The L-I and U-I characteristics of one of these lasers operated at room temperature are shown in Fig. 1.

3. Results

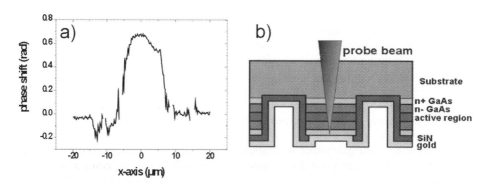

Figure 2. a) Measured phase shift across the laser ridge at the end of a 100 ns electrical pulse. Outside the trenches the phase shift is zero. b) Schematics of a cross sectional view of a QCL. The position of the reference beam is indicated.

We measured the transient phase shift during pulsed mode operation at room temperature, with a typical pulse length for GaAs based QCLs of 100 ns. Figure 2 shows a scan across the laser ridge at the end of the pulse where the phase shift has a maximum. The asymmetric heat distribution across the ridge is due to an asymmetric top isolation.

The time evolution of the phase shift in the middle of the laser ridge (see Fig. 3a) can be used to determine the anisotropic heat conductivity of the multilayered active region, $k_{ar}=(k_{\|}, k_{\perp})$, where $k_{\|}$ is the in-plane heat conductivity and $k_{\perp}$ the cross-plane component perpendicular to the layers. For all parameter in this model standard literature data are used except the anisotropic heat conductivity of the active region, which was fitted. A best fit was obtained with an anisotropic heat conductivity $k_{\|}=0.25$ W/Kcm and $k_{\perp}\approx 0.015$ W/Kcm. After the heating during the pulse (t<100 ns) the first strong cooling is determined by the in-plane heat conductivity. After the in-plane heat fluxes are mostly saturated the further cooling depends on the low cross-plane heat conductivity. This model also reveals the temperature distribution of the working devices. Figure 3b shows the calculated temperature increase in the middle of the active region. The maximum temperature increase in the active region at the end of the pulse, we found to be 77 K at an applied current of 3.1 A corresponding to a dissipated power in the active region of 63 W (1.24 W/cm^3) during the pulse.

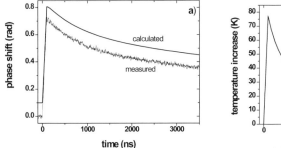
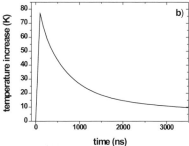

Figure 3. a) Calculated and measured phase shift in the middle of the laser ridge. The calculated graph is shifted by 0.1 rad for clarity. b) From the phase shift extracted temperature increase in the middle of the active region.

4. Discussion

The reduced in-plane heat conductivity compared with the weighted average of its constituents ($k=0.31$ W/Kcm [13]) can be explained by partly diffusive scattering of the phonons at the interfaces in the multilayered active region. The cross-plane conductivity was found to be much smaller than the heat conductivity of an Al$_{0.17}$Ga$_{0.83}$As alloy ($k=0.18$ W/Kcm [13], the average Al-content of the investigated active region is 17 %). In our superlattice structure the width of the single layers is in the range of a few nm and thus is much smaller than the mean free path of thermal phonons in GaAs at 293 K ($\Lambda\approx 50$ nm [14]). In this case the heat conductivity is no longer determined by the properties of the involved materials but rather it must be considered that the effect of the superlattice is to modify the phonon-dispersion relation [15]. This crucial difference of the two components shows that the best way to improve the lasers with respect to a better thermal behavior is to support the in-plane fluxes. This can be done e.g. by fabricating the lasers as buried heterostructures [3] or thicker gold layers in the trenches.

5. Conclusion

In conclusion, we have shown that the presented technique is a valuable tool to investigate the thermal dynamics in GaAs-based QCLs. Comparing the experiment with a thermal model enables us to extract the anisotropic heat conductivity of the multilayered active region as well as the temperature distribution in the working devices. The ratio of the two components of the heat conductivity $k_{\parallel}/k_{\perp}$ was found to be in the range of 15-20. The maximum temperature increase in the investigated active region is up to 80 K depending on the applied current.

6. Acknowledgements

This work was supported by the European Community-IST project SUPERSMILE, by the Austrian Microelectronics society and the Austrian FWF (ADLIS).

References

[1] C. Sirtori, P. Kruck, S. Barbieri, P. Collot, J. Nagle, M. Beck, J. Faist, and U. Oesterle, Appl. Phys. Lett. **73**, 3486 (1998)
[2] G. Strasser, S. Gianordoli, L. Hvozdara, W. Schrenk, K. Unterrainer, and E. Gornik, Appl. Phys. Lett. **75**, 1345 (1999)
[3] M.Beck, D. Hofstetter, T. Aellen, J, Faist, U. Oesterle, M. Ilegems, E. Gini, and H. Melchior, Science **295**, 301 (2002)
[4] H. Page, C. Becker, A. Robertson, G. Glastre, V. Ortiz, and C. Sirtori, Appl. Phys. Lett. **78**, 3529 (2001)
[5] S. Anders, W. Schrenk, E. Gornik, and G. Strasser, Appl. Phys. Lett. **80**, 1864 (2002)
[6] C. Pflügl, W. Schrenk, S. Anders, C. Becker, C. Sirtori, G. Strasser "High temperature performance of GaAs-based chirped superlattice quantum cascade lasers"; to be published
[7] J. Faist, F. Capasso, C. Sirtori, D. L. Sivco, A. L. Hutchinson, and A. Y. Cho, Appl. Phys. Lett. **67**, 3057 (1995)
[8] W. Schrenk, N. Finger, S. Gianordoli, E. Gornik, and G. Strasser, Appl. Phys. Lett.**77**, 3328 (2000)
[9] V. Spagnalo, M. Troccoli, and G. Scamarcio, C. Becker, G. Glastre, and C. Sirtori, Appl. Phys. Lett. **78**, 1177 (2001)
[10] V. Spagnalo, M. Troccoli, and G. Scamarcio, C. Gmachl, F. Capasso, A. Tredicucci, A. M. Sergent, A. L. Hutchinson, D. L. Sivco, and A. Y. Cho, Appl. Phys. Lett. **78**, 2095 (2001)
[11] C. Fürböck, D. Pogany, M. Litzenberger, E. Gornik, N. Seliger, H. Goßner, T. Müller-Lynch, M. Stecher and W. Werner, J.Electrostatics **49** 195 (2000)
[12] C. Fürböck, K. Esmark, M. Litzenberger, D. Pogany, G. Groos, R. Zelsacher, M. Stecher, E. Gornik, Microel. Reliab. **40** (8-10), 1365 (2000)
[13] S. Adachi, *GaAs and related materials*, World Scientific (1994)
[14] J. S. Blakemore, J. Appl. Phys. **53** (10), 123 (1982)
[15] G. Chen, Phys. Rev. B **57**, 14958 (1998)

Sensitivity of intersubband absorption linewidth and transport mobility to interface roughness scattering in GaAs quantum wells

Takeya Unuma[1,*], **Masahiro Yoshita**[1], **Takeshi Noda**[2], **Hiroyuki Sakaki**[2], **Motoyoshi Baba**[1], **and Hidefumi Akiyama**[1]

[1]Institute for Solid State Physics, University of Tokyo, 5-1-5 Kashiwanoha, Kashiwa, Chiba 277-8581, Japan. *E-mail: unuma@issp.u-tokyo.ac.jp

[2]Institute of Industrial Science, University of Tokyo, 4-6-1 Komaba, Meguro-ku, Tokyo 153-8505, Japan

Abstract. We study experimentally and theoretically the intersubband absorption linewidth due to scattering by interface roughness (IFR), phonons, and alloy disorder in modulation-doped GaAs quantum wells, and compare it with the transport mobility. Microscopic calculations agree quantitatively with experimental data, and show that high sensitivity of absorption linewidth to IFR scattering is the key to understanding apparent lack of correlation between linewidth and mobility in temperature and alloy composition dependences.

1. Introduction

The intersubband absorption linewidth in semiconductor quantum wells (QWs) closely relates fundamental problems in the physics of optical transition, such as relaxation [1], many-body effects [2,3], and disorder [4,5]. Furthermore, it is a key factor in improving the performance of quantum cascade lasers [6] and QW infrared photodetectors [7].

The absorption linewidth is said to have little correlation with electron mobility. Campman *et al.* reported that linewidth is sensitive to interface roughness (IFR) scattering, while it is insensitive to alloy disorder scattering, to which mobility is sensitive [8]. Moreover, the collective nature of intersubband transition and its effect on linewidth were demonstrated by Warburton *et al.* in InAs/AlSb QWs [3].

In this work, the effects of scattering by IFR, phonons, and alloy disorder on absorption linewidth were investigated. We measured linewidth in comparison with mobility in a modulation-doped GaAs/AlAs QW at temperatures $T = 4.5 - 300$ K, which were indeed different in both absolute values and temperature dependence. Calculations of microscopic scattering rates make it clear that linewidth is one order of magnitude more sensitive to IFR scattering than mobility is; this is the key to understanding the apparent lack of correlation between linewidth and mobility.

2. Theory of intersubband absorption linewidth

A general theory of intersubband absorption linewidth due to elastic scatterers in two-dimensional (2D) systems was formulated by Ando [1]. According to Ando's theory, the dynamical conductivity, describing absorption lineshape of single-particle excitation, can be expressed as $\text{Re}\,\sigma_{zz}(\omega) \propto \int dE\, f(E)\Gamma_{op}(E)/[(\hbar\omega - E_{10})^2 + \Gamma_{op}(E)^2]$ with $\Gamma_{op}(E) = \frac{1}{2}[\Gamma_{intra}(E) + \Gamma_{inter}(E)]$. Here, ω is the photon frequency, E is the in-plane kinetic energy, E_{10} is the intersubband energy separation, $f(E)$ is the Fermi distribution function, $\Gamma_{intra}(E)$ is the width due to the difference in intrasubband scattering matrix elements for the two subbands, and $\Gamma_{inter}(E)$ is the width due to the intersubband scattering [1]. The dynamic screening effect can be included by calculating $\tilde{\sigma}_{zz}(\omega) = \sigma_{zz}(\omega)/\varepsilon_{zz}(\omega)$ with $\varepsilon_{zz}(\omega) = 1 + i\sigma_{zz}(\omega)/\varepsilon_0 \kappa_0 d_{eff}$, and $\text{Re}\,\tilde{\sigma}_{zz}(\omega)$ describes absorption lineshape of collective excitation [9]. Here, ε_0 is the vacuum permittivity, κ_0 is the static dielectric constant of 2D material, and d_{eff} is the effective thickness of the 2D electron gas. At low electron concentrations, $\tilde{\sigma}_{zz}(\omega)$ is approximately equal to $\sigma_{zz}(\omega)$. We denote the full width at half maximum (FWHM) of the absorption spectrum as $2\Gamma_{op}$.

Using a roughness model that characterises the GaAs-on-AlGaAs interface by the Gaussian correlation function with a mean height of Δ and a correlation length of Λ, we have [10]

$$\Gamma_{intra}(E) = m^* \Delta^2 \Lambda^2 \hbar^{-2} (F_{00} - F_{11})^2 \int_0^\pi d\theta\, e^{-q^2\Lambda^2/4}, \tag{1}$$

$$\Gamma_{inter}(E) = m^* \Delta^2 \Lambda^2 \hbar^{-2} F_{01}^2 \int_0^\pi d\theta\, e^{-\tilde{q}^2\Lambda^2/4}, \tag{2}$$

with $F_{mn} = \sqrt{(\partial E_m/\partial L)(\partial E_n/\partial L)}$. Here, m^* is the electron effective mass, L is the well width, E_n is the quantisation energy of the n-th subband, and q and $\tilde{q}$ are the absolute values of the 2D intrasubband and intersubband scattering vectors, respectively [10]. $\tilde{q}$ is much larger than q, which makes $\Gamma_{inter}(E)$ much smaller than $\Gamma_{intra}(E)$. In Eq. (1), the static screening factor that appeared in the original theory [1] is omitted because its contribution is negligible in the following analysis. On the other hand, the transport relaxation rate is given by [11]

$$2\hbar/\tau_{tr}(E) = 2m^* \Delta^2 \Lambda^2 \hbar^{-2} F_{00}^2 \int_0^\pi d\theta\, e^{-q^2\Lambda^2/4} (1-\cos\theta)/\varepsilon(q,T)^2, \tag{3}$$

where $\varepsilon(q,T)$ is the static dielectric function [1].

Note that F_{11} is much larger than F_{00} because E_1 is more sensitive to L than E_0 is. (In the infinite-barrier approximation, F_{11} is four times larger than F_{00}.) This makes $\Gamma_{intra}(E)$ much larger than $2\hbar/\tau_{tr}(E)$. Calculations for phonon and alloy disorder scattering are shown in Ref. 12.

3. Temperature dependence

The sample we used to investigate temperature dependence of intersubband absorption linewidth was a modulation-doped GaAs/AlAs single QW with $L = 80$ Å grown by molecular beam epitaxy. The mobility μ and sheet electron concentration N_S, determined by a Hall measurement, were 2.9×10^4 cm^2/V s and 9.8×10^{11} cm^{-2} at 4.2 K, respectively. Absorption spectra were measured at temperatures ranging from 4.5 to 300 K using a Fourier transform infrared spectrometer with microscope (μ-FTIR) via a modulation technique.

Figure 1 (a) shows the observed absorption spectrum at 4.5 K [10], and its FWHM was $2\Gamma_{op} = 11.1$ meV. To compare this value with mobility, we define transport energy broadening as $2\Gamma_{tr} = 2\hbar/\tau_{tr} = 2\hbar e/m^*\mu$, where τ_{tr} is an average relaxation time. From the low-temperature mobility of 2.9×10^4 cm^2/Vs, we get $2\Gamma_{tr} = 1.2$ meV. Note that the linewidth $2\Gamma_{op}$ was about an order of magnitude larger than transport broadening $2\Gamma_{tr}$.

The temperature dependences of absorption linewidth and transport broadening (or mobility) are shown in Fig. 1 (b) by solid and open circles, respectively. It is found that

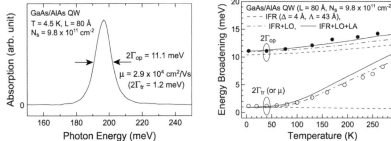

Figure 1. (a: left) Intersubband absorption spectrum of a modulation-doped GaAs/AlAs single QW observed at 4.5 K.

(b: right) Temperature dependences of the absorption linewidth $2\Gamma_{op}$ and transport broadening $2\Gamma_{tr}$ (or mobility μ) in a modulation-doped GaAs/AlAs single QW. Measured values are plotted by circles, and calculated ones are shown by lines, for which IFR, LO phonon, and LA phonon scattering are considered.

linewidth slowly increases as temperature increases up to 300 K, while transport broadening rapidly increases.

To explain the experimental results, we calculate the contributions of scattering by IFR ($\Delta = 4$ Å and $\Lambda = 43$ Å), polar optical (LO) phonons, and acoustic (LA) phonons via deformation potential coupling [10]. The calculated results for linewidth versus temperature are also shown in Fig. 1 by dashed (IFR), dash-dotted (IFR+LO), and solid (IFR+LO+LA) curves, in comparison with transport broadening.

At low temperatures, IFR scattering dominates both linewidth and transport broadening, and the $2\Gamma_{op}$ of 10.4 meV is about an order of magnitude larger than the $2\Gamma_{tr}$ of 0.73 meV. This is because the contribution from intrasubband scattering in the first excited subband is much larger than that in the ground subband (i.e. $F_{11} \gg F_{00}$). LO phonon scattering contributes 0.7 meV to linewidth via intersubband spontaneous emission. Even at room temperature, LO phonon scattering contributes little to $2\Gamma_{op}$ because the difference in intrasubband scattering matrix elements for the two subbands is small in LO phonon scattering [12]. In contrast, LO phonon scattering rapidly lowers electron mobility with increasing temperature above 80 K. LA phonon scattering makes comparable contributions to linewidth and mobility.

As a result of such characteristics of IFR and LO phonon scattering, linewidth and mobility have very different dependences on temperature. The calculations are in quantitative agreement with experimental results.

4. Alloy composition dependence

The alloy composition dependences of absorption linewidth and mobility were measured by Campman *et al.* in modulation-doped In$_x$Ga$_{1-x}$As/Al$_{0.3}$Ga$_{0.7}$As QWs with $L = 100$ Å and $N_S \cong 8 \times 10^{11}$ cm^{-2} at low temperatures [8]. The result for linewidth is plotted in Fig. 2 (a) by solid circles and shows that linewidth is hardly affected by alloy composition x. This is in contrast with remarkable drop in mobility, plotted in Fig. 2 (b) by open circles.

To understand the apparent lack of correlation between linewidth and mobility, we calculate the contributions of scattering by IFR, LO phonons, and alloy disorder (AD). The numerical results are also shown in Fig. 2 by dashed (IFR), dash-dotted (IFR+LO), and solid (IFR+LO+AD). IFR scattering contributes about 1.5 meV to $2\Gamma_{op}$ and 0.1 meV

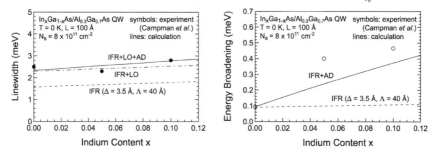

Figure 2. Alloy composition dependences of linewidth (a: left) and transport broadening (b: right) in modulation-doped In$_x$Ga$_{1-x}$As /Al$_{0.3}$Ga$_{0.7}$As single QWs. Calculated values are plotted by lines, for which IFR, LO phonon, and alloy disorder (AD) scattering are considered. Circles show experimental values reported by Campman et al [8].

to $2\Gamma_{tr}$. LO phonon scattering only contributes approximately 1 meV to linewidth. Alloy disorder scattering makes comparable contributions of about 0.3 meV to $2\Gamma_{op}$ and $2\Gamma_{tr}$ at $x = 0.1$, which is negligible for $2\Gamma_{op}$ but large enough for $2\Gamma_{tr}$ compared with the contributions of IFR scattering [12]. This explains very different experimental results for linewidth and mobility. A small discrepancy in mobility may be due to clustering in alloy layers.

5. Summary

We calculated the energy-dependent relaxation rate $2\Gamma_{op}(E)$ and lineshape of intersubband absorption for microscopic scattering by IFR, phonons, and alloy disorder in GaAs-based QWs. Significant agreement between numerical and experimental results was obtained for both absorption linewidth and mobility. It was shown that linewidth is an order of magnitude more sensitive to IFR scattering than mobility is, because the contribution from intrasubband scattering in the first excited subband is much larger than that in the ground subband. This is the key to understanding the apparent lack of correlation between measured linewidth and mobility in temperature and alloy composition dependences. Phonon scattering processes contribute little to linewidth even at room temperature, while mobility is limited by LO phonon scattering above 80 K. The contribution of alloy disorder scattering is negligible for linewidth but predominant for mobility compared with that of IFR scattering.

References

[1] Ando T 1985 J. Phys. Soc. Jpn. 54 2671
[2] Nikonov D E et al. 1997 Phys. Rev. Lett. 79 4633
[3] Warburton R J et al. 1998 Phys. Rev. Lett. 80 2185
[4] Riemann R N et al. 2002 Phys. Rev. B 65 115304
[5] Luin S et al. 2001 Phys. Rev. B 64 041306

[6] Faist J *et al.* 1994 Science 264 553
[7] Levine B F 1993 J. Appl. Phys. 74 R1
[8] Campman K L *et al.* 1996 Appl. Phys. Lett. 69 2554
[9] Ando T 1977 Z. Phys. B 26 263
[10] Unuma T *et al.* 2001 Appl. Phys. Lett. 78 3448
[11] Sakaki H *et al.* 1987 Appl. Phys. Lett. 54 2671
[12] Unuma T *et al.* submitted to J. Appl. Phys.

Improved temperature performance of GaAs/AlGaAs quantum cascade lasers

W. Schrenk, S. Anders, C. Pflügl, E. Gornik, G. Strasser
Solid state electronics, Vienna University of Technology, Austria

C. Becker, C. Sirtori
Thales Research and Technology, Orsay, France

Abstract. The temperature performance of quantum cascade lasers was improved by using a bound to continuum design. This result in pulsed operation close to 100°C for GaAs/AlGaAs quantum cascade lasers.

1. Introduction

Quantum cascade lasers (QCLs) [1-3] are powerful light emitters in the mid infrared. The strong light emission in this spectral region is interesting for chemical sensing [4]. In this work, we intent to present our latest results on the improvement of the highest working temperature of GaAs/AlGaAs QCLs.

2. Laser design

The design of the bandstructure of quantum cascade lasers is a challenging task. The population of the individual levels is determined by different scattering processes like electron-electron scattering, LO-phonon emission or absorption, and photon emission or absorption. The population of the energy levels together with the position of the doping atoms modify the band edge energy. Almost all laser structures are designed so far by very simple one band calculations, based on an effective mass approximation. The bandstructures are calculated neglecting the electron population and assuming a constant electric field across each cascade. The non-parabolicity of the energy dispersion is taken into account by an energy dependent effective mass [5]. Only for a few designs more sophisticated calculations were done afterwards, because they are too time consuming to be used as design tool [6,7].

To achieve a high working temperature it is important to prevent carrier escape into the continuum and reduce thermal activated backfilling. Calculations showed that the electron temperature in QCLs can be in the range of $T_e = 600$ K [8], therefore the barriers should be at least kT_e higher than the upper laser level. The band offset in the AlGaAs/GaAs material system increases linearly with the Al-content, but at the same time the X-point energy is going down in AlGaAs. For an Al-content of 45% the Γ-point and X-point energy are both 390 meV above the GaAs Γ-point. The use of an Al-content of 45 % allowed the first room temperature operation of GaAs/AlGaAs QCLs based on a three well design [9] and on a chirped superlattice design [10].

The active material of our structure consists of 50 periods of an AlGaAs/GaAs bound to continuum design (Fig. 1). A bound to continuum design combines the efficient extraction from the lower laser level, typical for superlattice structures with the single state in the upper laser level, typical for three well structures. The emission wavelength is designed for 11 μm. Thermal activated backfilling is reduced by an increased energy spacing between the lower laser level and the upper laser level of the following period. Further, population of the level above the upper laser level is reduced by an increased energy spacing in comparison to a similar chirped superlattice structure [10].

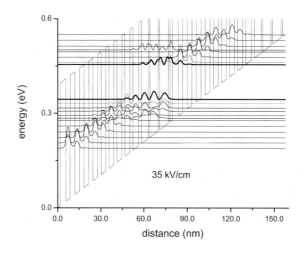

Figure 1. Calculated conduction-band structure of the bound to continuum structure at an applied field of 35 kV/cm.

A double plasmon enhanced waveguide is used for vertical optical confinement. The laser material is grown by solid source molecular beam epitaxy. We fabricated deep etched ridge lasers by reactive ion etching, resulting in a rectangular shaped cross section. This results in strong lateral electrical and optical confinement. SiN is used for insulation and Ti/Au is sputtered for both contacts.

3. Results

The highest working temperature of a 1.38 mm long and 40 μm wide FP laser in pulsed mode operation was found to be 95°C. This is significantly higher than the reported values for FP lasers (36°C) based on a three well design [9], and microcylinder lasers (50°C) based on a chirped superlattice [10]. Short FP lasers work also at room temperature with reduced currents (Fig. 2). We observed lasing for cavity lengths as short as 0.25 mm. The light is guided by internal total reflection for short cavities, similar to the light guiding in micro disk lasers. Long FP laser operate on the lowest lateral mode, because the mirror losses are indirectly proportional to the cavity length and the losses due to the SiN insulation layer are smaller for lower order lateral modes.

The crossover between these two kinds of laser operation occurs for cavity lengths around 1 mm.

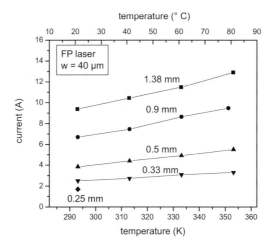

Figure 2. Threshold current of 40 μm wide FP lasers in pulsed operation as a function of the heat sink temperature for different cavity lengths. The cleaved facets are left uncoated.

We have also fabricated narrow (w = 10, 20 μm) FP and DFB lasers. The light and voltage vs. current curves of 10 μm wide and 20 μm wide FP lasers are compared in Fig. 3. The lasers are 1.3 mm long and the cleaved facets are left uncoated. The threshold current density is only slightly higher for 10 μm wide ridges than for 20 μm wide ones.

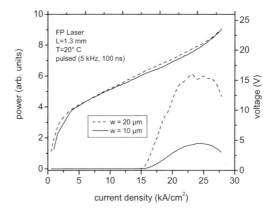

Figure 3. Light vs. current and voltage vs. current curves of 10 and 20 μm wide FP lasers (length 1.3 mm) at 20°C.

4. Conclusions

We have demonstrated high temperature operation of GaAs/AlGaAs quantum cascade lasers close to 100°C using a bound to continuum design.

5. Acknowledgement

This work was supported by the European Community-IST project SUPERSMILE, by the Austrian Microelectronics society and the Austrian FWF (ADLIS).

References

[1]   J. Faist, F. Capasso, D. L. Sivco, C. Sirtori, A. L. Hutchinson, and A. Y. Cho, Science. 264, 553-556 (1994).
[2]   J. Faist, C. Gmachl, F. Capasso, C. Sirtori, D. L. Sivco, J. N. Baillargeon, and A. Y. Cho, Appl. Phys. Lett., 70, 3486-3488 (1998).
[3]   C. Sirtori, P. Kruck, S. Barbieri, P. Collot, J. Nagle, M. Beck, J. Faist, and U. Oesterle, Appl. Phys. Lett., 73, 3486-3488 (1998).
[4]   A. Edelmann, C. Ruzicka, J. Frank, B. Lendl, W. Schrenk, E. Gornik, and G. Strasser, J. of Chromatography A. 934, 123-128 (2001).
[5]   C. Sirtori, F. Capasso, J. Faist, S. Scandolo, Phys. Rev. B, 8663-8674 (1994)
[6]   R. Iotti, and F. Rossi, Phys. Rev. Lett., 87, 146603/1-4 (2001)
[7]   A. Wacker, S. C. Lee, Physica B, 314 (1-4), 327-331 (2002)
[8]   R. Iotti, and F. Rossi, Appl. Phys. Lett., 78, 2902-2904 (2001)
[9]   H. Page, C. Becker, A. Robertson, G. Glastre, V. Ortiz, and C. Sirtori, Appl. Phys. Lett. 78, 3529-3531 (2001).
[10] S. Anders, W. Schrenk, E. Gornik and G. Strasser, Appl. Phys. Lett. 80, 4094-4096 (2002).

Electron-Photon Strong Coupling in Intersubband Resonators

G. Biasiol[1], L. Sorba[1,2], D. Dini[3], R. Köhler[3], A. Tredicucci[3], and F. Beltram[3]

[1] Laboratorio Nazionale TASC-INFM, AREA Science Park, SS 14, Km 163.5 Basovizza, 34012 Trieste (Italy); e-mail: biasiol@tasc.infm.it

[2] Università di Modena e Reggio Emilia, Via Campi 213/A, 41100 Modena (Italy)

[3] NEST-INFM and Scuola Normale Superiore, Piazza dei Cavalieri 7, 56126 Pisa (Italy)

Abstract. We have designed and successfully fabricated an AlAs/AlGaAs microcavity resonator tuned to the intersubband electronic transition of an embedded modulation-doped GaAs/AlGaAs multiple quantum well system. Optimisation of the structure design and of the growth protocol yielded a perfectly tuned system, which shows electron-photon strong coupling, manifested by a Rabi splitting of 15meV at 10K.

Intersubband transitions are attracting increasingly larger interest both for the implementation of new photonic devices (infrared cameras, quantum cascade lasers) and for fundamental physics studies, such as collective phenomena and many-body interactions in the excitations of the two-dimensional electron gas (2DEG). For these investigations and applications, it is of crucial importance to understand and control the interactions of such transitions with the electromagnetic radiation in photon confinement structures. However, the development of microcavity structures is by far less developed than in the case of interband transitions, despite the interest related to this new spectral range, and to the peculiar potential advantages of this system (such as ultrafast relaxation times, tailorable oscillator strength, 2D coupling, etc.).

We have developed a modulation-doped GaAs/$Al_{0.33}Ga_{0.67}As$ multiple quantum well (MQW) structure embedded in a microcavity resonator tuned to the intersubband electronic transition of the QW, with the aim of investigating modifications of the intersubband resonance due to the coupling with the cavity mode. Since the effective oscillator strength of the MQW structure must be high enough to observe substantial coupling effects [1], we had to incorporate a large number of QWs in the structure (17 in our case). For the same reason, we had to make use of a relatively high carrier density in the individual wells (around $4 \times 10^{11} cm^{-2}$). For photon confinement, a monolithic wavelength-size high-Q resonator was designed in the form of an oblique prism-shaped microcavity exploiting total internal reflection at the semiconductor-air interface on one side (surface), and at an $Al_{0.33}Ga_{0.67}As/AlAs$ interface on the other. The cavity resonance was tuned on the intersubband transition energy (chosen to be about 140 meV) for an

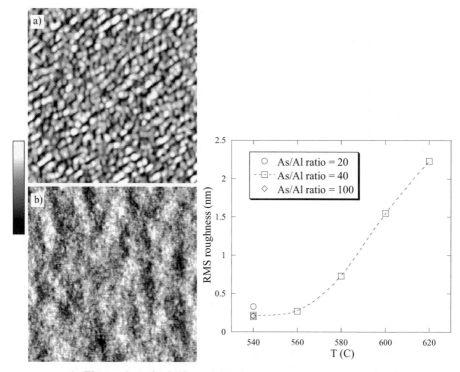

Figure 1. Left: 3X3μm AFM images of 800nm-thick AlAs layers (with 15nm GaAs cap on top) grown a) at 620C, b) at 540C, with an As/Al flux ratio of 40. The greyscale bar corresponds to 10nm height for image a) and 1nm for image b). Right: RMS roughness of different AlAs layers as a function of growth temperature and As/Al ratio.

incidence angle around 60°, also with the help of accurate many-body calculations of the intersubband excitation energy [2]. More details on the resonator structure will be given in Ref. [3].

Epitaxial growth was performed with solid source MBE in a Veeco Applied EPI GenII reactor, on semi-insulating GaAs (001) substrates. Optimisation of the sample structure was performed in two stages. First, we explored the growth conditions in order to obtain an AlAs surface as smooth as possible. The quality of the inverted GaAs/AlAs interface is known to depend critically on the growth conditions already for AlAs layers much thinner than ours (~100nm), and non-optimised parameters may result in a very rough interface, thus degrading the interface reflectivity [4,5]. Optimisation was performed on a series of test samples consisting of 800nm-thick AlAs layers, capped with 15nm GaAs to prevent surface oxidation after air exposure, grown at different growth temperatures and As/Al flux ratios. We studied the surface morphology of the different samples with Atomic Force Microscopy in air in non-contact mode. For the highest temperature studied (620C, as determined by a thermocouple calibrated with the GaAs de-oxidation temperature), the surface resulted to be very rough, and characterised by 5-10nm high hillocks (see Figure 1, image a). As we decreased the growth temperature, the surface flattened out, until we could obtain a monolayer-smooth surface for T=540C (see Figure 1, image b). At this temperature, we found that an As/Al flux

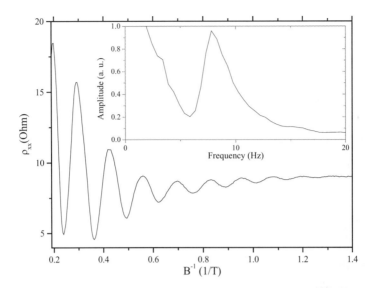

Figure 2. SdH oscillations of a 17 MQW structure with average carrier density of $4.17 \times 10^{11} cm^{-2}$, as a function of 1/B. Inset: FFT of the oscillations.

ratio between 40 and 100 yielded the smoothest surfaces, with a typical RMS roughness of about 0.2nm (see Figure 1, right). The same roughness was measured at the surface of the final structure, consisting of 3900nm AlAs, followed by the MQW structure.

We optimised as well the QW doping, with the aim of obtaining a carrier density as uniform as possible from well to well. Achieving a good uniformity is crucial, since the QW intersubband resonance energy is expected to depend on their sheet electron density [1]. The MQW structure consisted of 17 modulation-doped, 7.2nm-thick GaAs/Al$_{0.33}$Ga$_{0.67}$As QWs, and was grown at 640°C, with an As/Ga ratio of 20, i.e., the growth conditions yielding the highest electron mobility in modulation-doped heterostructures grown in our system. The QWs were alternated with Si δ-doping layers ($n_{Si}=6 \times 10^{11} at.cm^{-2}$) with 65nm- and 42nm-thick spacers below and above the wells, respectively. The asymmetry between the lower and upper spacers was introduced to compensate for the tendency of Si atoms to migrate towards the surface in the AlGaAs layers at these growth temperatures [6], and ensured a symmetric charge distribution within the wells. We checked the uniformity of the carrier density among the 17 QWs by measuring the Shubnikov-de Haas (SdH) oscillations on Van der Pauw structures at 1.5K and magnetic fields up to 5T. The SdH oscillations are plotted in Figure 2 as a function of 1/B and show clearly a single oscillation frequency. This is seen also in the Fourier transform of the oscillations shown in the inset, exhibiting a single peak. From the period of the oscillations we inferred a carrier density of $4.17 \times 10^{11} cm^{-2}$. We have also checked, by intentionally varying the Si amount in some of the δ-doping layers, that our measures are sensible to doping variations as small as 10% from well to well.

Optical characterization was performed via infrared Fourier transform spectroscopy using a broadband glowbar source and a liquid-nitrogen-cooled MCT detector. Reflectivity spectra were recorded in single-pass waveguide geometry with a resolution of 2 cm^{-1}. Results at a temperature of 10K are reported in Figure 3 for different angles of

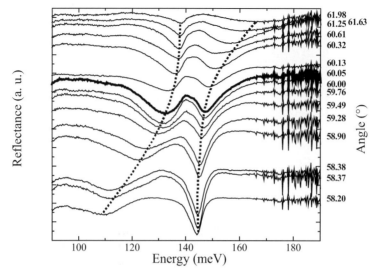

Figure 3. Reflectance spectra at 10 K of the microcavity- intersubband resonance for various angles of incidence. The anti-crossing of the strongly coupled intersubband polaritons is clearly apparent. Dotted lines are added as a guide to the eye.

incidence within the GaAs waveguide, around a value of 60°. The dispersion of the reflectivity dips shows clearly an anti-crossing behaviour, due to the strong coupling of the photonic cavity mode with the intersubband plasmon, thanks to the perfect tuning of the intersubband resonator. At resonance (about 60.05°, thick spectrum) the reflectance shows two anti-crossed modes of mixed photon-plasmon character with comparable linewidths. The modes are separated by a Rabi splitting of 15 meV that, to our knowledge, was never observed before for intersubband transitions, although theoretically predicted. This strong coupling regime is observable also at room temperature where the splitting drops to around 10 meV.

The result might open the way to the development of intersubband cavity electrodynamics. Further developments in the near future include the possibility of controlling externally the splitting of the transition by varying the 2DEG density in the wells, e.g., with the application of a gate voltage to the MQW structure.

This work was partially supported by INFM through the PAIS project "ICONS".

References

[1] Liu A S 1997 Phys Rev B 55 7101-7109
[2] Luin S, Pellegrini V, Beltram F, Marcadet X, and Sirtori C 2001 Phys Rev B 6404 art. no.-041306
[3] Dini D, Koehler R, Tredicucci A, Biasiol G, Sorba L, and Beltram F Phys Rev Lett (submitted)
[4] Asom M T, Geva M, Leibenguth R E, and Chu S N G 1991 Appl Phys Lett 59 976-978
[5] Wohlert D E, Chang K L, Lin H C, Hsieh K C, and Cheng K Y 2000 J Vac Sci Technol B 18 1590-1593
[6] Pfeiffer L, Schubert E F, West K W, and Magee C W 1991 Appl Phys Lett 58 2258-2260

Demonstration of 640x512 Pixel Four-Band Quantum Well Infrared Photodetector (QWIP) Focal Plane Array

S. D. Gunapala, S. V. Bandara, J. K. Liu, S. B. Rafol, M. Jhabvala[*] and K. K. Choi[+]

Jet Propulsion Laboratory, California Institute of Technology
4800 Oak Grove Drive, Pasadena, CA 91109, USA
[*] NASA Goddard Space Flight Center, Greenbelt, MD 20771, USA
[+] Army research Laboratory, Adelphi, MD 20783, USA

Voice: (818) 354-1880, Fax: (818) 393-4540,
Email: sarath.d.gunapala@jpl.nasa.gov

ABSTRACT. We have demonstrated the first monolithic spatially separated four-band 640x512 pixel QWIP focal plane array. The four spectral bands cover 4-5 µm, 8.5-10 µm, 10-12 µm, and 13-15.5 µm spectral regions with 640x128 pixels in each band. In this paper, we discuss the development of this very sensitive four-band focal plane array based on a GaAs/ $Al_xGa_{1-x}As/In_yGa_{1-y}As$ QWIPs and its performance in quantum efficiency, noise equivalent differential temperature, uniformity, and operability.

1. Introduction

The absorption lines of many gas molecules, such as ozone, water, carbon monoxide, carbon dioxide, and nitrous oxide occur in the wavelength region from 3 to 16 µm. Thus, hyper-spectral and multi-spectral measurements in 3 to 16 µm region will enable remote sounding of numerous geophysical quantities such as cloud (height and fraction, emissivity, particle size and phase), surface (soil and vegetation type, temperature, emissivity, pollutants), and atmospheric (temperature and composition) parameters. The need for smaller, lighter and lower cost imaging radiometers is now apparent, particularly in missions that combine different types of remote sensing instruments. Instruments requiring different detectors for different channels impose a severe cost and complexity burden on the focal plane, and the specific detector can further drive the optical design to larger and more costly elements to obtain sufficient signal. The key to developing smaller, lighter and less costly imagers is the development of a more integrated, multi-band FPAs.

A quantum well designed to detect infrared (IR) light is called a quantum well infrared photodetector (QWIP) [1]. When the quantum well is sufficiently deep and narrow, its energy states are quantized (discrete). The potential depth and width of the well can be adjusted so that it holds only two energy states: a ground state near the well bottom, and a first excited state near the well top. A photon striking the well will excite an electron in the ground state to the first excited state, then an externally-applied voltage sweeps it out producing a photocurrent. Only photons having energies corresponding to the energy separation between the two states are absorbed, resulting in a detector with a narrow responsivity spectrum. Thus, QWIP is an ideal detector for fabrication of multi-band detectors.

2. 640x512 Pixel Four-Color Spatially Separated Focal Plane Array

One unique feature of this spatially separated four-band focal plane array is that the four infrared bands are independently readable on a single imaging array. This feature leads to a reduction in instrument size, weight, mechanical complexity, optical complexity and power requirements since no moving parts are needed. Furthermore, a single optical train can be employed, and the focal plane can operate at a single temperature.

This four-band device structure was achieved by the growth of multi stack QWIP structures separated by heavily doped n^+ contact layers, on a GaAs substrate. Device parameters of each QWIP stack were designed to respond in different wavelength bands. Figure 1 shows the schematic device structure of a four color QWIP focal plane array. A typical QWIP stack consists of a MQW structure of GaAs quantum wells separated by Al_xGa_{1-x} As barriers. The actual device structure consists of a 15 period stack of 4-5 μm QWIP structure, a 25 period stack of 8.5-10 μm QWIP structure, a 25 period stack of 10-12 μm QWIP structure and a 30 period stack of 13-15.5 μm QWIP structure. Each photosensitive MQW stack was separated by a heavily doped n^+ (thickness 0.2 to 0.8 μm) intermediate GaAs contact layer (see Figure 1). Since the dark current of this device structure is dominated by the longest wavelength portion of the device structure, the VLWIR QWIP structure has been designed to have a bound-to-quasibound intersubband absorption peak at 14.0 μm. Other QWIP device structures have been designed to have a bound-to-continuum intersubband absorption process, because the photo current and dark current of these devices are relatively small compared to the VLWIR device. This whole four-band QWIP device structure was then sandwiched between 0.5 μm GaAs top and bottom contact layers doped with n = 5 x 10^{17} cm^{-3}, and was grown on a semi-insulating GaAs substrate by MBE.

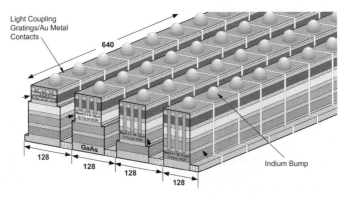

Figure 1. Layer diagram of four-band QWIP device structure and the deep groove 2-D periodic grating structure. Each pixel represent a 640x128 pixel area of the four-band focal plane array.

The individual pixels were defined by photolithographic processing techniques. Four separate detector bands were defined by a deep trench etch process and the unwanted spectral bands were eliminated by a detector short-circuiting process. The unwanted top detectors were electrically shorted by a gold coated reflective 2-D etched gratings as shown in the Fig. 1. In addition to shorting, these gratings serve as light couplers for active QWIP stack in each detector pixel. Design and optimization of these 2-D gratings to maximize QWIP light coupling were extensively discussed in reference 2. The unwanted bottom QWIP stacks were electrically shorted at the end of each detector pixel row.

Typically, quarter wavelength deep (h = $\lambda_p/4n_{GaAs}$) grating grooves are used for efficient light coupling in single-band QWIP FPAs. However, in the four-band FPA, the thickness of the quarter wavelength deep grating grooves are not deep enough to short circuit the top three MQW QWIP stacks (e.g.: three top QWIP stacks on 14-15.5 μm QWIP in Fig. 1). Thus, three-quarter wavelength groove depth 2-D gratings (h = $3\lambda_p/4n_{GaAs}$) were used to short the top unwanted detectors over the 10-12 and 14-15.5 microns bands. This technique optimized the light coupling to each QWIP stack at corresponding bands while keeping the pixel (or mesa) height at the same level which is essential for indium bump-bonding process used for detector array and readout multiplexer hybridization. Figure 2 shows the normalized spectral responsivities of all four spectral bands of this four-band FPAs.

A few QWIP FPAs were chosen and hybridized to a 640x512 CMOS multiplexer (ISC 9803) and biased at V_B = -1.1 V. At temperatures below 83 K, the signal to noise ratio of the 4-5 μm spectral band is limited by array non-uniformity, multiplexer readout noise, and photo current (photon flux) noise. At temperatures above 45 K, temporal noise due to the 14-15.5 μm QWIP's higher dark current becomes the limitation. The 8-10 and 10-12 μm spectral bands have shown BLIP performance at temperatures between 45 and 83 K. The FPAs were back-illuminated through the flat thinned substrate membrane (thickness ≈ 1300 Å). This initial array gave excellent images with 99.9% of the pixels working (number of dead pixels ≈ 250), demonstrating the high yield of GaAs technology.

A 640x512 pixel four-band QWIP FPA hybrid was mounted onto a 84-pin lead-less chip carrier and installed into a laboratory dewar which is cooled by liquid helium to demonstrate a 4-band simultaneous imaging camera. The FPA was cooled to 45 K and the temperature was stabilized by regulating the pressure of gaseous helium. The other element of the camera is a 100 mm focal length anti-reflection coated germanium lens, which gives a 9.2°x6.9° field of view. It is designed to be transparent in the 8-12 μm wavelength range. The digital data acquisition resolution of the camera is 14-bits, which determines the instantaneous dynamic range of the camera (i.e., 16,384), however, the dynamic range of QWIP is 85 Decibels. Video images were taken at a frame rate of 30 Hz at temperatures as high as T = 45 K, using a ROC capacitor having a charge capacity of 11×10^6 electrons (the maximum number of photoelectrons and dark electrons that can be counted in the time taken to read each detector pixel).

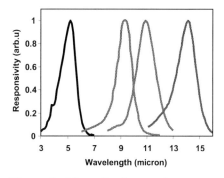

Figure 2. Normalized spectral response of the four-band QWIP focal plane array.

Figure 3 shows one frame of a video image taken with four-band 640x512 pixel QWIP camera.

The peak detectivities of all four bands at 300K background with f/5 optics are shown in Fig. 4. Based on this single element test detector data, the 4-5, 8-12, 10-12, and 13.5-15.5 µm spectral bands show BLIP at temperatures 40, 50, 60 and 120 K respectively for a 300 K background with f/5 cold stop. As expected (due to BLIP), the estimated and experimentally obtained NEDT values of all spectral-bands do not change significantly below their BLIP temperatures. The experimentally measured NEDT of 4-5, 8-12, 10-12, and 13.5-15.5 µm detectors at 40 K are 21.4, 45.2, 13.5, and 44.6 mK, respectively. These experimentally measured NEDT agree reasonably well with the estimated NEDT values based on the single element test detector data.

Figure 3. One frame of video image taken with the 4-15.5 microns cutoff four-band 640x512 pixel QWIP camera. The image is barely visible in the 14-15.5 microns spectral band due to the poor optical transmission of the anti-reflection layer coated germanium lens.

3. Acknowledgements

The research described in this paper was performed by the Jet Propulsion Laboratory, California Institute of Technology, and was sponsored by the NASA Code R Micro & Nano Technology Program, and the NASA Code Y Advance Technology Initiative Program.

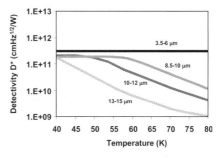

Figure 4. Detectivities of each spectral-band of the four-band QWIP FPA as a function of temperature.

References

[1] S. D. Gunapala, J. K. Liu, J. S. Park, T. L. Lin, and M. Sundaram "Infrared Radiation Detecting Device", US Patent No. 6,211,529.

[2] S. D. Gunapala and S. V. Bandara, "Quantum Well Infrared Photodetector (QWIP) Focal Plane Arrays," Semiconductors and Semimetals, Vol. 62, 197-282, Academic Press, 1999.

Giant polarized photoluminescence and photoconductivity in type-II GaAs/GaAsSb multiple quantum wells induced by interface chemical bonds

Y.F.Chen, Y.S.Chiu, M.H.Ya, and T.T.Chen

Department of Physics, National Taiwan University, Taipei, Taiwan

Abstract. Anisotropic property of type-II GaAs/GaAsSb heterostructures was studied by photoluminescence (PL) and photoconductivity (PC). It was found that the PL and PC spectra exhibit a strong in-plane polarization with respect to <011> axis with polarization degrees up to 40%. We showed that the polarization does not depend on the excitation intensity as well as temperature, which excludes any extrinsic mechanisms related to the in-plane anisotropy. The observed polarized optical properties of GaAsSb/GaAs multiple quantum wells was attributed to the intrinsic property of the orientation of chemical bonds at heterointerfaces.

1. Introduction

The GaAsSb/GaAs heterostructure system has been revived with the development of advanced technology recently. Particularly, its type-II band alignment provides an excellent opportunity to improve the performance of both heterojunction bipolar transistors and optoelectronic devices.[1,2] Besides, its technological potential, it serves as a model system for investigating the atomic ordering and compositional modulation expected in III-V-V alloys. In spite of its importance, the optical properties of this system have not been clearly understood due to the difficulties in growing high quality samples as well as the detection of type-II luminescence. In this paper a peculiar optical anisotropy of GaAsSb/GaAs multiple quantum wells (MQWs) has been discovered. Based on the studies of the dependents of excitation intensity and temperature, we suggest that our observed anisotropy is an inherent property of the type-II band alignment in GaAsSb/GaAs MQWs.

2. Experiment

The type-II $GaAs_{0.7}Sb_{0.3}$/GaAs MQWs were grown on semi-insulating GaAs(100) substrates using a VG V-80MKII molecular beam epitaxy (MBE). Besides the Ga beam and As_4 beams, the Sb source was supplied using an EPI model 175 standard cracker K-cell. The cracker zone temperature was $1050°C$, while the bulk zone temperature was about $430°C$. The As source was supplied from an 150 c.c. K-Cell. The structure of MQWs contains a 500 nm GaAs buffer layer, 5 periods of $GaAs_{0.7}Sb_{0.3}$ 50Å/GaAs 300Å quantum well, and a 1000Å GaAs cap layer. The growth temperature of the buffer layer and MQWs were $600°C$ and $500°C$, respectively. The composition of antimony in the $GaAs_{0.7}Sb_{0.3}$ layer was determined from double crystal x-ray diffraction (DXRD) measurement. For the PL measurement, the spectra were dispersed by a Spectra Pro 300i monochromator, and detected by an InGaAs detector. An Ar-ion laser was used as the excitation source. For the PC measurement, ohmic contacts were formed by depositing indium drops to the four corners of the samples, and annealing the sample at

250 °C for 10 min. A tungsten lamp dispersed by triple-grating monochromator was used as the light source. A constant current was supplied to the sample by a Keithley 236 source measure unit. The conductivity signal was detected as a change in the voltage drop across the sample using a lock-in amplifier. A detailed description of the experimental setup has been given elsewhere.[3]

3. Results and discussion

In contrast to type-I semiconductor structures, in type-II structures the energy minima for electrons and holes lie in different layers. Spatially separated electrons and holes are easily realized in such a system, in which electrons are confined in the GaAs layer and holes are localized in the GaAsSb layer. For the spatially indirect transition, it is caused by the wave function overlap between electron and hole across interface. Because some Sb atoms may replace the As atoms in zinc-blende structure, the interfaces in this quaternary GaAsSb/GaAs heterostructure consist of Sb-Ga and As-Ga bonds or As-Ga and As-Ga bonds. It is obvious that the lower and the upper interfaces of the quantum well are not equivalent with respect to the bond directions. Their contributions to the anisotropy can not compensate each other[4-6]. Therefore, the in-plane anisotropy inherently exists in the GaAsSb/GaAs MQWs studied here.

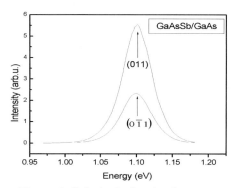

Figure 1. Polarized photoluminescence spectra of GaAsSb/GaAs MQWs under 74 mW excitation.

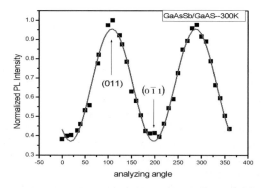

Figure 2. Photoluminescence intensity vs analyzer angle for GaAsSb/GaAs multiple quantum wells. The solid curve is a $\cos^2\theta$ fit.

Anisotropic property of the PL spectra of the emitted radiation along (011) and $(0\bar{1}1)$ direction is shown in Fig.1. The peak at 1.05 eV corresponds to the spatially indirect transition. As we can see, the PL intensity is very sensitive to the polarization. Figure 2 shows the PL intensity versus the angle of the analyzer. Solid dots are experimental data, and the curve is fit to $\cos^2\theta$. The polarization degree defined as

$$P_l = (I_{011} - I_{0\bar{1}1}) / (I_{011} + I_{0\bar{1}1}) \quad (1)$$

can be as large as 40%. This value is too large to be explained by strain or electric field effects[7]. In order to clarify that the observed polarized PL spectra are induced by the anisotropic nature of the interface chemical bonds, we have performed the pumping power and temperature dependences of the polarization. It is found that the degree of polarization is not sensitive to the change of pumping power in the range of 45 mW to 168 mW. The degree of polarization is also very stable with respect to the change of temperature from 20 K to 300 K. These results can be used to rule out any extrinsic mechanisms related to the in-plane anisotropy. For example, the built-in electric fields caused by unintentional doping will be screened under light irradiation. We can also exclude a significant role of localized states and nonradiative channels in the formation of the in-plane anisotropy, since they will be gradually saturated by the pumping source. We thus conclude that the polarization of the spatially indirect PL in GaAsSb/GaAs is an inherent nature of the interface chemical bonds. Polarized PL measurements therefore provide a simple tool to probe interface anisotropy in quaternary heterostructures. We believe that this result is very important for the application of GaAsSb/GaAs in optical devices.

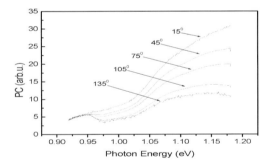

Figure 3. Polarized photoconductivity spectra of GaAsSb/GaAs MQWs.

The PC spectra for the incident radiation at different polarization angles are shown in Fig.3. We can also see that the PC spectra display a strong polarization dependence. Since the energy that corresponds to the spatially indirect transition in the GaAsSb/GaAs multiple quantum wells is about 1 eV, we therefore choose the photoresponse signal at around 1 eV to analyze the polarization behavior. Using the same definition of the polarization degrees defined for the PL spectra, we obtain that the value of P can also be as high as 40%. In addition, we can see that the polarization dependence of the PC spectra covers a wide range of wavelengths as shown in Fig.3.

Figure 4. shows the polarization dependence of the PC response at around 1 eV. The data can also be described by the $\cos^2\theta$ rule. The maximum and minimum signal correspond to the polarization along (011) and $(0\bar{1}1)$, respectively, which is similar to the behavior of the PL spectra. It indicates that the anisotropic PC response arises from the interface chemical bonds. However, the wide range response of the PC spectra as shown in Fig.3 does not solely come from the type-I or type-II band to band transitions.

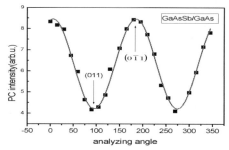

Figure 4. Polarized photoconductivity signal around 1 eV vs analyzer angle for GaAsSb/GaAs MQWs. The solid curve is a $\cos^2\theta$ fit.

The transition of the continuous energy states can be attributed to interface imperfections. For example, there may exist wrong bonds at interfaces (e.g., Sb-Ga and As-Ga bonds or As-Ga and As-Ga in GaAsSb/GaAs heterostructures), which are different from the bonds in constituent layers. These wrong bonds can be considered as a kind of localized distortion, i.e., a certain "defectlike" impurity, with respect to the host structure. Thus, polarized PC spectra provide a very good opportunity to search the effects of microscopic interface defects in semiconductor heterstructures.

4. Conclusion

In conclusion, a strong polarization dependence has been observed in type-II GaAsSb/GaAs MQWs by photoluminescence and photoconductivity measurements, and the polarization degrees can be as large as 40%. This finding should be very important for the application of GaAsSb/GaAs MQWs in optoelectronic devices. We have shown that the effects of anisotropy arise from the inherent nature of the interface chemical bonds in a type-II heterostructure. In addition, we have pointed out that polarized PC measurement is a very sensitive tool to probe the effects of interface imperfections.

This research was supported by the National Science Council and Ministry of Education of the Republic of China.

References

[1] O. M. Khries, K. P. Homewood, and W. P. Gillin, J. Appl. Phys 84, 4017 (1998).
[2] T. C. Mclinn, T. N. Krabac, and M. V. Klein, Phys. Rev. B 33, 8396 (1985).
[3] S. M. Tseng, Y. F. Chen, Y. T. Cheng, C. W. Hsu, Y. S. Huang, and D. Y. Lin, Phys. Rev. B 64, 195311 (2001).
[4] H. J. Chang and Y. F. Chen, H. P. Lin, and C. Y. Mou, Appl. Phys. Lett. 78, 3791 (2001).
[5] A. V. Platonov, V. P. Kochereshko, E. L. Ivchenko, and G. V. Mikhailov, R. Yakovlev, M. Keim, W. Ossau, A. Waag, and G. Landwehr, Phys. Rev. Lett. 83, 3546 (1999).
[6] M. Schmidt, M. Grun, S. Petillon, E. Kurtz, and C. Klingshirn, Appl. Phys. Lett. 77, 85 (2000).
[7] B. V. Shanabrook and B. R. Bennet, Phys. Rev. B 50, 1695 (1994).

Efficient nitride-based short-wavelength emitters with enhanced hole injection

J. M. Zavada[1], S. M. Komirenko[2], K. W. Kim[2], and V. A. Kochelap[3]

[1]U.S. ARO, Research Triangle Park, NC 27709, USA; [2]North Carolina State University, Raleigh, NC 27695-7911, USA; [3]Institute of Semiconductor Physics, Kiev, 252650, Ukraine

Abstract. We propose two novel designs of efficient emitters in deep UV range. Improvement in the emitter efficiency can be achieved by enhancement of hole injection in the nitride-based heterostructures.

1. Introduction

Further progress in III-N based UV emitters requires solution of an important problem – obtaining high-density hole currents. The difficulties in achieving high hole concentrations mainly originate from high values of activation energy of known acceptors (about 250 meV for Mg in GaN). Use of Al-contained compounds leads to further increase in acceptor activation energy. To overcome *the low acceptor activation problem*, it was suggested [1] that the average hole concentration could be enhanced in p-doped ternary compounds with a spatially modulated chemical composition [e.g., a superlattice (SL)]. Calculations [1] and measurements [2]-[4] show improved acceptor efficiency in Mg-doped SLs: the average hole concentration can be increased up to one order of magnitude. Nevertheless, most of the holes ionized from the acceptors are localized inside the quantum wells (QWs) with the potential barriers as high as 100-400 meV. These barriers hinder participation of the holes in vertical transport required in traditional light-emitting devices. In this report we propose two novel solutions to enhance hole injection in wide gap semiconductors.

2. Superlattice hole injector

To increase the overbarrier hole concentration and the vertical hole current, we propose to modify the traditional design of LEDs by introducing a two-terminal hole injector schematically illustrated in Fig. 1(a). The injector consists of a *p*-doped SL base and two contacts S and D. The injector is separated from the rest of the device by an *i*-region. A bias voltage applied between the S and D contacts provides *lateral* hole acceleration and increases the effective temperature of the holes T_h. An increase in T_h results in significant enhancement of overbarrier hot-hole concentration. This is known as the real-space transfer effect [5]. The proposed device can be thought of as a *three terminal device, or light-emitting triode (LET)* schematically shown in Fig. 1(b), with a hole-injector region, an intrinsic *i*-layer, and an *n*-doped region (contact C). If a *p*-doped SL is used as a hot-hole injector, the device can operate as a charge injection transistor [6]. With the D contact as ground and a positive voltage V_S applied to the S contact, the

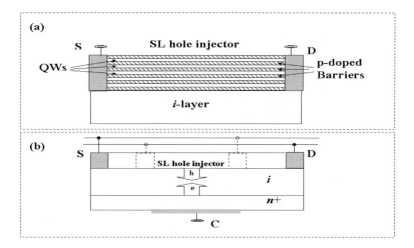

Figure 1: Schematic illustration for (a) the proposed two-terminal SL hole injector and (b) a light-emitting triode with the SL hole injector.

lateral hole current heats the holes and, consequently, increases the overbarrier hole concentration in the injector. If a negative bias V_C is applied to the cathode C, both the hot holes from the SL and the electrons from the n-region are injected into the *i*-layer, as illustrated in Fig. 1(b). Emission of light occurs in the *i*-layer as a result of electron-hole recombination. The proposed AlGaInN-based LET can significantly increase the intensity of emission comparing to the traditional LED structure, especially in deep UV range. In AlGaN structures, a modest lateral electric field of about 3-5 kV/cm may lead to more than an order of magnitude increase in overbarrier hole concentration at room temperatures (and more than two orders of magnitude increase at higher barriers or lower temperatures).

3. Lateral current pumped emitters (LACE)

To achieve high-density electron-hole plasma (EHP) and interband population inversion in group-III nitrides, we propose a planar two- dimensional (2D) *p-i-n* structure, shown in Fig. 2. It can be created in selectively-doped SLs and QWs. The QW layers are confined by selectively-doped barriers. Each barrier is doped laterally: a region doped with acceptors is followed by an undoped i-region and, finally, by the region doped with donors. Thermal activation of the dopants in the barrier supplies carriers into the QW layers. The QW layers accumulate both types of free carriers. This leads to the formation of a *lateral p-i-n* structure, as sketched in Fig. 2. It can be fabricated by using the re-growth techniques, position-dependent implantation methods, etc.

When a forward bias is applied to the lateral *p-i-n* structure, the planar double injection gives rise to non-equilibrium 2D EHP in the *i*-region. Radiative recombination of the plasma in the active *i*-region results in light emission.

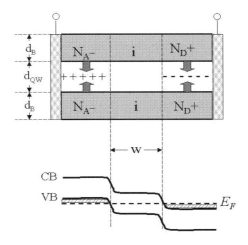

Figure 2. Schematic illustration of the proposed multi-layered lateral *p-i-n* structure and energy band diagram in the lateral direction with no bias. CB and VB indicate the lowest populated energy levels in conduction and valence bands, respectively.

A SL arranged from QWs considered above will form an efficient source of radiation − LAteral Current Pumped Emitter.

We developed the theory of electron-hole double injection in the lateral *p-i-n* structure illustrated in Fig. 2. The results of calculations are given in Table 1. We use the following designations: p_0, n_0 are the electron and hole area densities in the QW layers at *p*-side and *n*-side of the structure, respectively; p_m and p_M are minimum and maximum injected carrier concentrations in the *i*-region; J is the electric current per a single QW; Δ_p, Δ_n the energy barriers at *p*- and *n*-sides, respectively; U is the voltage drop across the *i*-region.

Table 1
Parameters characterizing the lateral double injection in the nitride-based *p-i-n* structures; (a) $p_0=n_0/2=5\times10^{12}$ cm^{-2}; (b) $p_0=n_0=10^{13}$ cm^{-2}; (c) $p_0=n_0=2\times10^{13}$ cm^{-2}. The barrier materials are assumed to be AlInN or AlGaN.

QW Material	T K	J mA/mm	Δ_p, Δ_n, U meV	p_m, p_M 10^{12} cm^{-2}
InN				
(a)	80	37	12, 153, 74	1, 2.9
(b)	300	110	27, 73, 285	2.4, 6.6
GaN				
(a)	80	37	11, 92, 66	0.9, 2.3

Extension of the *i*-region is assumed to be w=3 μm. As an example, we calculated the double injection for the lateral structure with 5 nm QWs and 2D-carrier concentrations in *p*- and *n*-regions equal to 5×10^{12} cm^{-2}. The temperature was assumed to be 80 K.

For InN QWs we found that at the current density 37 mA/mm (per a single QW) the EHP concentration in the i-region reaches magnitudes of $(1-3)\times 10^{12}$ cm^{-2}. The population inversion for photo-transitions centered at the wavelength $\lambda=587$ nm occurs across the entire i-region. For GaN QWs at the currents density above 16 mA/mm and the total voltage drop across the structure 4.22 V, we found the plasma concentrations to be $(1-2)\times 10^{12}$ cm^{-2} and the population inversion for $\lambda=344$ nm. These examples show that in the Lateral Current-pumped Emitters the population inversion can be reached at quite modest currents and biases. For GaN-LACE with ten QWs, a strip of area of 100×3.6 μm^2 can be inverted in the currents less than 16 mA at 80 K. If the *p*- and *n*-regions are doped above 10^{13} cm^2, the population inversion can be reached at the room temperature.

4. Summary

In conclusion, contemporary group-III nitride technology allows fabrication of novel structures for light emitting devices. We propose two novel designs of efficient emitters in deep UV range. First, improvement in the emitter efficiency is expected to be achieved by implementing the nitride-based heterostructures with enhanced hole injection. Second, in laterally, selective doped QWs and SLs, composing a planar *p-i-n* regions, double injection can provide for high densities of the electron-hole plasma and interband population inversion. We suggest that this planar double injection is an efficient method for electrical pumping of short-wavelength nitride-based light emitting diodes and lasers.

The work was supported by US Army Research Office and the ERO of US Army under Contract N62558-02-M-6381.

References

[1] Schubert E F, Grieshabert W, and Goepfert I D 1996 Appl. Phys. Lett. **69** 3737
[2] Kozodoy P, Hansen M, Denbaars S P, and Mishra U 1999 Appl. Phys. Lett. **74** 3681
[3] Sheu J K, Chi G C, and Jou M J 2001 Electron. Lett. **22** 160
[4] Waldron E L, Graff J W, and Schubert E F 2001 Appl. Phys. Lett. **70** 2737
[5] Gribnikov Z S, Hess K, and Kosinsky G A 1995 J. Appl. Phys. **77** 1337
[6] Luryi S 1991 Appl. Phys. Lett. **58** 1727

Study of polarisation switch in a three-contacts vertical-cavity surface-emitting laser

V. Badilita[1], J.-F. Carlin[1], M. Ilegems[1], M. Brunner[2], G. Verschaffelt[3], K. Panajotov[3]

[1] Swiss Federal Institute of Technology, Institute for Quantum Electronics and Photonics, CH-1015, Lausanne, Switzerland, vlad.badilita@epfl.ch;

[2] Avalon PhotonicsLtd., Badenerstrasse 569, CH-8048, Zürich, Switzerland;

[3] Vrije Universiteit Brussel, Department of Applied Physics and Photonics (TW-TONA), Pleinlaan 2, B-1050 Brussels, Belgium.

Abstract. We report a novel three-contacts vertical-cavity surface-emitting laser (VCSEL) for use in polarization-sensitive optical applications. The device consists in two vertical optically active cavities, coupled by a common mirror. We demonstrate that one can independently choose both the power - through the current in the first cavity - and the polarization state of the output beam - through the bias applied to the second cavity. The control of the polarization state is performed with a control of voltage required in the range -2V to 0V. Within this interval the structure exhibits bistable behaviour.

1. Introduction

Many VCSEL applications require a well-defined polarization state of the emitted light, therefore considerable efforts have been made in order to achieve polarization control of the light output in vertical-cavity lasers. We report new experimental results on control of polarisation state in a three-contacts vertical-cavity surface-emitting laser by means of an external applied bias.

2. Device and experiment

The structure and the working principles of the coupled-cavities VCSEL (BiVCSEL) containing two slightly asymmetric cavities that share a common coupling mirror were previously presented [1,2]. The measurements presented in this letter were performed on oxidised devices with 10 µm oxide-apertures. A particularity of this device is that both cavities are optically active, containing 4 (top cavity) and 2 (bottom cavity) InGaAs strained quantum wells. This structure exhibits two longitudinal coupled non-degenerate optical modes localized at 927nm and 953nm. The device has a double-threshold point for I_{TOP}=0.5mA, I_{BOT}=0.5mA - the values for the currents in the two cavities for which both wavelengths begin to lase simultaneously [3].

The bias of the two cavities may be independently modified therefore achieving different domains of operation [3]. In this paper we present a representative sample studied under the following conditions: constant current injected in the top cavity and bias of the bottom cavity modified between –5V and +1V, always operating the bottom cavity in reversed bias regime. Therefore only the 927nm mode lases with a higher threshold current around 1mA. We performed pulsed measurements for both cavities with 1kHz frequency and 1% duty cycle, which corresponds to a period of 1ms and a pulse-length of 10µs. The maximum output power under these experimental conditions was about 15µW.

Modifying the bias of the bottom cavity in the range -5V to + 1V while maintaining a constant current I_{TOP}=3mA in the top cavity, a polarisation-switch (PS) - Figure 1 - between two

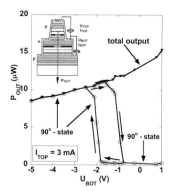

Figure 1. Polarization resolved optical output power vs. reverse bias in the bottom cavity.

orthogonal linear polarization states (0^0- and 90^0-state) of the radiation emitted by the top cavity is observed. Figure 1 presents the total light output and the output for one of the two polarization states (90^0- state) as well as bistable operation with respect to the bottom cavity bias. The curve in Figure 1 is representative for around 30 out of 60-70 tested devices from the same manufacturing run, all of them showing dual lasing. The switch is very abrupt, the transition taking place within a 0.2V interval. The device exhibits excellent polarization selectivity, the contrast ratio between the output powers of the two polarizations being around 1000 or higher. For $U_{BOT} = 0V$ the light output has the 0^0-state as the dominant polarization state. This means that the epitaxial wafer has specific polarization selectivity. The inset of Figure 1 shows a schematic of the device.

3. Experimental data

We present the very first experimental evidence of non-thermal, electrically induced polarisation switching in these special design VCSELs – BiVCSELs. We demonstrate the possibility to control the polarization state within the interval -2V to 0V of the applied voltage, as well as bistable behaviour of the light output. As shown in Figure 1, the bistable region extends for 1V and the output power is about 10µW. In order to be able to control the polarization state of the device, it is important to know the evolution of the bistable region with respect to the external applied parameters. In Figure 2 we show that the width of the bistable region increases with increasing value of the top current (I_{TOP}) and is centred on $U_{BOT} \cong -0.9V$. If the device is operated for this value of U_{BOT} close to threshold – the bistable region is very narrow. It becomes larger for higher injection currents so that at 4 times the threshold value it extends between –0.2 and –2V.

In Figure 3 the I-V curve of the bottom cavity is plotted for a constant $I_{TOP} = 3$ mA. An interesting aspect is the fact that the bottom cavity acts as a built-in detector. Therefore, the current in the bottom cavity is the photocurrent induced by the radiation emitted in the top cavity. One can notice that across the bistable region the current in the bottom cavity is constant (I_{BOT}=-0.8 mA) with respect to the applied voltage. So we can conclude that the polarization switch is influenced neither by the current in the bottom cavity, nor by the current in the top cavity (which is maintained constant).

This method of changing the polarization state using a third electrical contact offers an additional degree of freedom compared to the classic current-driven polarization switch since the output power and the polarization state can be independently modified. In the L-I curves presented in Figure 4 we demonstrate that the output power for each of the two states increases monotonically with the I_{TOP} current with nearly equal efficiencies.

Depending on the modulation on the bottom-applied bias, we can operate the device in different regimes from the point of view of polarization. The device can work in the triggered-mode (Figure 5), or in the bistable mode (Figure 6).

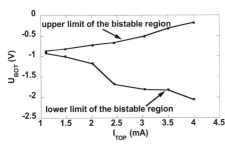

Figure 2. Width of the hysteresis region vs. bias current in the top cavity.

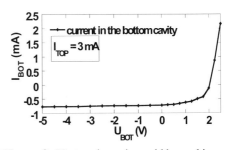

Figure 3. Hysteresis region width vs. bias current in the top cavity.

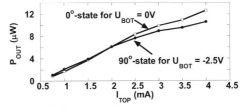

Figure 4. Polarization resolved L-I curves.

In both cases the top cavity is direct biased with a 10μs pulse for a 1 ms – long cycle. In the triggered-mode, both the baseline and the pulse of the inverse bias of the bottom cavity are outside the hysteresis-working region of the device. The pulse in the bottom cavity is sent within the pulse in the top cavity so that the polarization state changes only during the bottom-cavity pulse, returning to its original state after the pulse.

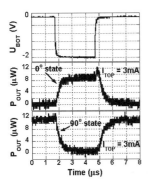

Figure 5. Triggered-mode: Resolved output power in response to a step variation of U_{BOT}.

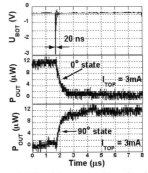

Figure 6. Bistable-mode: Resolved output power in response to a 20 ns pulse of U_{BOT}.

In the bistable mode, the baseline of the inverse bias of the bottom cavity is placed inside the hysteresis region and the amplitude of the pulse crosses this region. The polarization state of the device changes after a pulse as short as 20 ns. In the triggered-mode, the device changes its polarization state only during the pulse, returning to its original state after the pulse. Both pulse and baseline correspond to stable polarization states. In the bistable mode a pulse of 20 ns is sufficient to change the polarization state of the device after the pulse, while in the triggered regime we obtained a change in about 300 ns limited by the detector time response. It indicates that the switching is intrinsically fast taking into account the fact that neither the experimental set-up, nor the processing of the device was adapted for high-speed operation.

4. Interpretation

One possible physical mechanism to explain the observed polarization switch is the (linear) electro-optic birefringence induced by the reverse bias [4,5]. In this case the frequency splitting induced by the electric field along [001] changes its sign and, consequently, leads to a change of the gain-preferred mode and henceforth to polarization switching. To test this hypothesis we have measured the frequency shift of both modes as a function of the reverse bias voltage with a high-resolution spectrum analyser (Figure 7). The presented measurement was performed on another device for a value for the current in the top cavity very close to threshold. Therefore we are in the region (Figure 2) where the bistable region is extremely small. It can be seen that both modes red shift with the increasing magnitude of the reverse bias voltage, but their frequency difference essentially remains constant (~13GHz). As the measured electro-optic birefringence does not change with voltage, this mechanism cannot explain the occurrence of polarization switching in our VCSELs. Another hypothesis is the difference in absorption of the two frequency-split modes that changes with reverse bias, through the Quantum Confined Stark Effect in the InGaAs QWs of the bottom cavity. More experimental and theoretical work to test this conjecture is in progress.

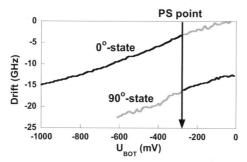

Figure 7. Frequency split between the polarization modes vs. reverse-bias in the bottom cavity

5. Conclusions

The polarization state of the coupled-cavities VCSEL was successfully controlled by the voltage applied to the bottom cavity while maintaining a constant current in the top cavity. A polarization switch occurs with a high contrast ratio between the output power for each intensity. We report bistable behaviour of the device that could be used for different polarization-sensitive applications: a signal applied within the bistable region succeeds to change the polarization state of the device. We show that an appropriate pulse applied within the stable working region is followed by the change in polarization limited by the time response of the detector used in this experiment. Therefore, by varying the bias of the bottom cavity we can select the desired polarization mode while by adjusting the current in the top cavity we can modify the intensity of this polarization mode. These results suggest that the device may be interesting for various system applications since controlling the polarization state of the emitted light in a simple, reliable and fast way gives new perspectives to different applications.

6. References

[1] J.-F. Carlin et al. - Appl. Phys. Lett. **75**, 908 (1999).
[2] M. Brunner et al. - IEEE Photonics Technol. Lett. **12**, 1316 (2000).
[3] V. Badilita et al., - Proc. of SPIE, vol. **4649**, 87 (2002).
[4] Min Soo Park et al., Appl. Phys. Lett. **76**, 813 (2000).
[5] B. Ryvkin et al. - Optics in Computing'98, Brugge, Belgium, 17-20 June 1998.

Analysis of dynamics and intensity noise of semiconductor lasers under strong optical feedback

S. Abdulrhmann, †M. Ahmed, T. Okamoto, W. Ishimori and M. Yamada

Dept. of Electrical and Electronic Engineering, Kanazawa University, Japan.
†Department of Physics, Faculty of Science, Minia University, Egypt.

Abstract. We report on a new theoretical approach to investigate dynamics and intensity noise of semiconductor lasers subjected to strong optical feedback. The influence of intrinsic fluctuations in the intensity and optical phase of the lasing field on the laser dynamics were taken into account. The laser operates in pulsation under strong feedback, and the associated intensity noise attains levels as low as the quantum noise at high injection levels.

1. Introduction

Recently InGaAs lasers emitting in a wavelength of about 1 μm have received much interest as pumping sources for fibre amplifiers in fibre communication systems. In such systems, an external cavity is formed between the laser front facet and a fibre grating. The pumping lasers are designed with a low front-facet reflectivity compared with that of the gratings and are then subjected to strong optical feedback (OFB) [1], which changes the laser operation dramatically [2]. Moreover, the laser operation is influenced by intrinsic fluctuations in the intensity and optical phase of the lasing field due to transition of electrons between the conduction and valence bands [3], which may modulate the influence of the OFB on the laser dynamics. Most previous models of OFB were applicable under weak or moderate OFB and were based on a small-signal approximation [4,5], which overlooks time variations of the lasing parameters. In this work, we present a new theoretical approach to analyze operation of lasers under an arbitrary amount of OFB. The laser dynamics under strong OFB are simulated by numerical integration of the modified rate equations which are superposed by Langevin noise sources to include the influence of the intrinsic fluctuations.

2. Theoretical Model

Our model of analysis of OFB is shown in Fig. 1. A_1 and B_1 are the forward and backward amplitudes of the electric component of the optical field in the laser cavity, respectively, while A_2 and B_2 are the corresponding components in the external cavity formed between the front facet of reflectivity R_f and an external mirror of reflectivity R_{ex}. R_b is the reflectivity of the back facet, t_{12} and t_{21} are coefficients of transmission. The emitted light from the front facet travels a round trip between the facet and the external mirror with a trip time of $\tau = 2n_{r2}\ell/c$, with n_{r2} being the refractive index of the external cavity and ℓ as its length, and is then re-injected into the laser cavity. The injected light at each instant t is the light emitted at $t-\tau$. The oscillation conditions of lasers are modified to

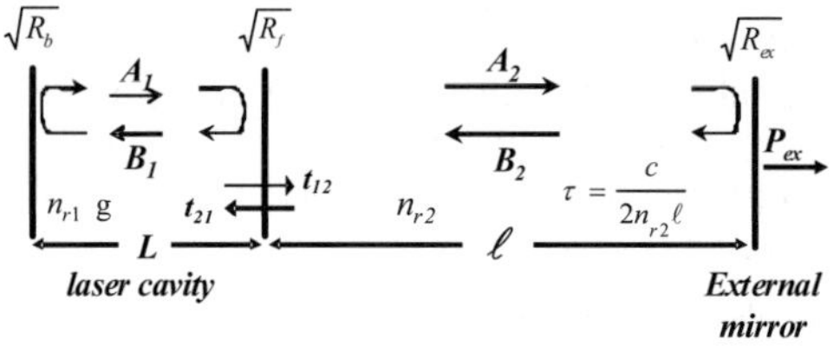

Figure 1. Scheme of our model of lasers under optical feedback.

$$G_{th} = G_{th0} - \frac{c}{n_{r1}L} \ln|T|, \qquad (1)$$

$$2\beta_1 L + \phi_b + \phi_1 + \phi = 2s\pi, \qquad (2)$$

where G_{th} and G_{th0} are the threshold gain levels with and without OFB, respectively. n_{r1}, L and β_1 are the refractive index, length and the propagation constant in the laser cavity, respectively. ϕ_1 and ϕ_b are phase changes due to reflection on the front and back facets, respectively. s is an integer. The complex coefficient T describes the influence of OFB on the threshold conditions and is given by [2]

$$T = 1 - (1-R_f)\sqrt{\frac{R_{ex}}{R_f}} e^{-j\psi} \sqrt{\frac{S(t-\tau)}{S(t)}} \frac{e^{j\theta(t-\tau)}}{e^{j\theta(t)}} = |T|e^{-j\phi}. \qquad (3)$$

ψ is the phase difference between delayed light in the external cavity and the reflected field at the front facet of the laser cavity. When the OFB is very weak, its effect is approximated by the following equation, which has been used in many papers.

$$\ln T \approx -(1-R_f)\sqrt{\frac{R_{ex}}{R_f}} e^{-j\psi} \sqrt{\frac{S(t-\tau)}{S(t)}} \frac{e^{j\theta(t-\tau)}}{e^{j\theta(t)}}, \qquad \text{for } R_f \ll R_{ex}$$

In our calculations we used Eq. (3) to include the case of the strong OFB.
The dynamics of lasers are described by the rate equations of the photon number $S(t)$, the carrier number $N(t)$ and the phase $\theta(t)$, which are modified to include the OFB to:

$$\frac{dS}{dt} = \left\{ \frac{a\xi}{V}(N-N_g) - BS - G_{th0} + \frac{c}{n_{r1}L}\ln|T| \right\} S + \frac{a\xi}{V}N + F_S(t), \qquad (4)$$

$$\frac{d\theta}{dt} = \frac{\alpha a\xi}{2V}(N-\bar{N}) - \frac{c}{2n_rL}(\phi-\bar{\phi}) + F_\theta(t), \qquad (5)$$

$$\frac{dN}{dt} = -\frac{a\xi}{V}(N-N_g)S - \frac{N}{\tau_s} + \frac{I}{e} + F_N(t), \qquad (6)$$

where $a\xi(N-N_g)/V$ is the linear gain coefficient with a and N_g as material constants and ξ as the confinement factor of the optical field into the active region of volume V. α is the linewidth enhancement factor, and I is the injection current. The coefficient B describes the nonlinear suppression of gain, and is given by [6],

$$B = 9\hbar\omega_o \big/ (4\varepsilon_o n_{r1}^2)(\tau_{in}/\hbar)^2 (\xi/V)^2 a|R_{cv}|^2 (N-N_s),\qquad(7)$$

where R_{cv} is the dipole moment, τ_{in} is the intraband relaxation time and N_s is an injection level characterizing the nonlinear gain. The terms $F_s(t)$, $F_\theta(t)$ and $F_N(t)$ are Langevin noise sources and are added to the rate equations to account for the intrinsic fluctuations in S, θ and N. The procedures of generating the noise sources are described in [7].

3. Numerical Simulation and Discussion

We numerically integrated the time-delay rate equations (4)-(6) by the fourth-order Runge-Kutta method using a time interval of 5 ps over a long period of 10 μs. The relative intensity noise (RIN) was calculated as the power density function of the fluctuations in the output power $\delta P(t) = P(t) - \overline{P}$ divided by $\overline{P}^2$ [7], where $\overline{P}$ is the dc-value of $P(t)$. The calculations of RIN were carried out when the operation becomes stable ($t \approx$ 8~10 μs). The applied numerical values of InGaAs lasers are: $\xi=0.1$, $a=2.21\times10^{-12}$ m^3/s, $L=800$ μm, $V=400$ μm^3, $n_{r1}=3.5$, $|R_{cv}|^2=2.8\times10^{-57}$ C^{2}m^2, $\tau_{in}=0.1$ ps, $N_g=4.08\times10^8$, $N_s=1.53\times10^8$, $\tau_s=2.79$ ns, $R_b=0.98$, $R_f=0.02$, and $\alpha=2$. ℓ and n_{r2} were set as 1.0 m and 1.5, respectively.

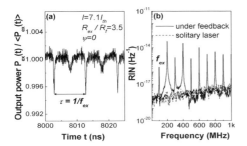
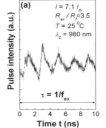
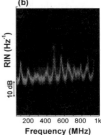

Figure 2. Simulated characteristics of the laser output: (a) time variation, and (b) RIN frequency spectrum.

Figure 3. Experimental observations: (a) output waveform, and (b) RIN frequency spectrum.

3.1. Pulsing operation under strong optical feedback

The simulation results of the power $P_{ex}(t)$ emitted from the external mirror under strong OFB corresponding to $R_{ex}/R_f=3.5$ and at high injection current of $I=7.1I_{th}$ are shown in Fig. 2. When there is no OFB, i.e., operating as a solitary laser, the laser shows conventional DC output ($P_{ex}(t)/<P_{ex}(t)>=1$). However, the laser shows pulsing operation as shown in Fig. 2(a) under strong OFB. Figure 2(b) is the corresponding frequency spectrum of the RIN. The pulsing oscillation is confirmed in the RIN spectrum, which exhibits peaks separated by the external frequency $f_{ex}=1/\tau=100$ MHz. Such pulsing peaks are not represented in the RIN spectrum of the solitary laser, however the levels of the low-frequency noise in both spectra almost coincide.

We confirmed the above simulated results in experiments. The observed results of an InGaAs laser sample emitting in 980 nm at the room temperature are given in Fig. 3. The waveform of the output power and the frequency spectrum of RIN are plotted in Fig. 3(a) and (b), respectively. The observed results confirm the pulsing operation at the external frequency f_{ex} under strong OFB and at high injection levels. Figures 2 and 3 indicate good correspondence of our simulated results with the experimental observations.

3.2. Chaos operation under strong optical feedback

Chaos operation is an attractive subject not only from the aspect of the physical interest in the laser instability problem but also from the aspect of practical necessity to stabilize the laser operation under OFB [8]. Figure 4 shows an example of the simulated characteristics of chaotic operation under strong optical feedback but at a lower current of $I=1.5I_{th}$. Figure 4(a) indicates a typical waveform of the chaotic operation, which is shown as random variation of the output power $P(t)$ with the variation of time. The corresponding RIN spectrum is shown in Fig. 4(b) in which one may identify peaks at both the external frequency f_{ex} and the relaxation frequency f_r as well as at their higher harmonics but with random noise levels. This means that the laser operates unstably under two threshold conditions; one corresponds to the laser cavity and the other corresponds to the external cavity, which enhances the time variations of both the optical phase and the electron number. In this case, the low-frequency level of RIN is enhanced above the quantum noise level.

3.3. Dependence on the injection current

In this section, we characterize the RIN induced under strong OFB over a wide range of injection current: $I=I_{th}\sim10I_{th}$. Figure 5 plots the bifurcation diagram of the power emitted from the external mirror $P_{ex}(t)$ under strong external reflectivity of $R_{ex}/R_f=3.5$. The figure plots also on the right axis the corresponding RIN values averaged over $f<10$ MHz for the solitary laser. The shown results were calculated under a phase difference of $\psi=0$ and were found to be independent of ψ. As shown in the figure, the laser operation is almost chaotic near the threshold current I_{th} and is characterized with RIN values higher than the RIN values of the solitary laser. The laser operates in pulsation when it is injected well above the threshold level. The enhanced noise levels in the range $I=2I_{th}\sim7I_{th}$ are attributed to induced low-frequency fluctuations. An attractive feature of the pulsing operation is that the feedback noise is suppressed to the quantum noise level of the solitary laser when the injection current increases beyond $7I_{th}$.

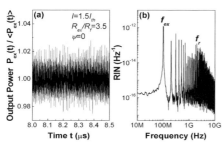
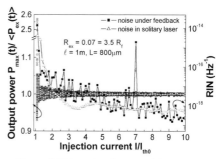

Figure 4. Simulated results of power P_{ex} under chaos operation: (a) time variation, and (b) spectrum of RIN.

Figure 5. Bifurcation diagram of the output power and the RIN levels averaged over $f<10$MHz.

4. Conclusions

We theoretically analyzed operation of semiconductor lasers under optical feedback and the induced feedback noise basing on a newly proposed model. The model is versatile and is applicable under an arbitrary amount of feedback. We showed that the laser exhibits pulsing operation under strong optical feedback and confirmed the results in experiments. The pulsing operation is characterized with low relative intensity noise in the limit of the quantum noise when the injection current exceeds seven times the threshold.

References

[1] Mukai T and Ohtsuka K 1985 Phys. Rev. Lett. 55, 1711-1714
[2] Abdulrhmann S G, Ahmed M, Okamoto T and Yamada M 2002 18th IEEE International Semiconductor Laser conference, Germany
[3] Ahmed M and Yamada M 2002 IEEE J. Quantum Electron. 38, 682-693
[4] Roar D, Rolf A and Beausoleil R G 1991IEEE J. Quantum Electron 27 352-372
[5] Lang R and Kobayashi K 1980 IEEE J. Quantum Electron. 16 347-355
[6] Yamada M and Suematsu Y 1981 J. Appl. Phys. 52 2653-2664
[7] Ahmed M, Yamada M and Saito M 2001IEEE J. Quantum Electron. 37 1600-1610
[8] Kikuchi N, Liu Y, and Ohtsubo J 1997 IEEE J. Quantum Electron. 33 56-65

High performance optically pumped 1.55 µm VCSELs for novel telecom applications

Alexei Syrbu[1], Grigore Suruceanu[1], Vladimir Iakovlev[1], Alok Rudra[1,2] Alexandru Mereuta[1], Claude-Albert Berseth[1], Andrei Mircea[1], Cristian Bungarzeanu[2] and Eli Kapon [1,2]

1 BeamExpress SA, Lausanne, CH-1015, Switzerland, e-mail: alexei.sirbu@beamexpress.com

2 Swiss Federal Institute of Technology Lausanne (EPFL), CH-1015, Switzerland

Abstract. High performance single wavelength and tuneable optically pumped 1.55µm VCSELs were fabricated by localized wafer fusion. For single wavelength devices high temperature operation (up to 93°C) and 2.5 Gb/s transmission were demonstrated using micro-assembled VCSELs with 980 nm pump laser chips. Optically pumped VCSELs with 38 nm tuneability at tuning voltage values below 4 V and 1 mW single mode output power with 40 dB side mode suppression ratio were demonstrated as well.

1. Introduction

Single wavelength vertical cavity surface emitting lasers (VCSELs) emitting in the 1.5-1.6 µm waveband are considered as a low cost alternative to 1.5 µm DFB lasers for applications in metro networks and coarse wavelength division multiplexing (CWDM) communication systems. Direct high bit-rate modulation and operation in a wide temperature range are important requirements for lasers used in these novel systems. Both electrically and optically pumped 1.55 µm VCSELs have been previously used for high bit rate transmission experiments over standard single mode fibers [1-3]. The problem with 1.55µm electrically pumped VCSELs is that, because of resistive heating and related optical losses, these devices have not demonstrated so far power values above 0.4 mW at elevated temperatures close to 80°C. Optically pumped 1.55 µm VCSELs have better chances of reaching high temperature operation as the self-heating is considerably reduced by pumping directly into the active region.

Tunable 1.5-1.6 µm lasers have great potential in dense wavelength division multiplexing (DWDM) systems allowing a considerable reduction of inventory costs by replacing single wavelength lasers and offering new possibilities in reconfigurable all optical networks[4]. In order to realize this potential these lasers should have a wide tuning range to cover the C or L bands with enough output power and at cost comparable to that of standard communication lasers. Existing solutions for edge emitting tunable lasers are very complex and the cost is high. VCSELs combined with micro-electro-mechanical (MEMs) technology offer an ideal solution for low cost, mode-hop free and wide wavelength tuning. So far there are reports on both electrically pumped VCSELs with

16 nm tuning range and tuning voltage of 35V[5] and optically pumped VCSELs with 50 nm tuning range at 40V[6]. The advantage of optically pumped VCSELs is that emission power comparable to that of DFB communication lasers can be obtained. High tuning voltage prevents using standard low-voltage driving electronics and increases cost. In this paper we present data on fabrication and operation of optically pumped 1,55μm single wavelength and tuneable VCSELs that make use of double wafer fusion technology.

2. Fabrication

In our fabrication approach we combine AlGaAs-based DBRs and InP-based active cavity material in a single device using the wafer fusion technique. This allows a monolithic integration of high performance InAlGaAs/InP multiquantum well (MQW) active cavity material and AlGaAs/GaAs based DBRs that have very good thermal conductivity, that allow efficient dissipation of the heat generated due to the energy difference between 0.98 μm pump and 1.5 μm emission photons and nonradiative recombination. Another advantage of AlGaAs/GaAs DBRs is the wide stop-band of about 100 nm making possible cavity mode tuning in intervals as wide as 60-70 nm. This is quite enough taking into consideration the restriction imposed by the spectral dependence of the gain.

Both tuneable and single wavelength device structures we realized include InGaAlAs/InP MQW active cavity material sandwiched between bottom and top AlGaAs/GaAs n-type DBRs. Lateral refractive index variation for mode confinement is obtained in-situ during the localized fusion process[7]. In the tuneable device structure (see Fig.1) an air gap cavity is obtained by selective etching of a spacer layer in the top DBR, which also serves for blocking the current when applying the voltage to the structure. The top DBR is suspended by one or several narrow beams (one beam suspension is shown in Fig.1 schematics) formed in the same DBR using reactive ion etching. Adjusting the thickness, width and length of suspending beams, the flexibility of the MEM structure can be controlled allowing low voltage deflection of the top DBR.

In our device structures we use InAlGaAs quantum wells with a conduction band offset of $0.72E_g$. Such structures have superior high temperature performance compared with standard InGaAsP wells with a band offset of only $0.4E_g$. Both active material and DBRs were grown by MOCVD.

3. Static characteristics of single-wavelength VCSELs

The single VCSEL structures are optically pumped using Nortel G4 pump lasers. The 980 nm pump light is focused on the VCSEL active region and the 1.55 μm output light is collected through the GaAs substrate into a single mode optical fiber.

We obtain 3 mW emission power at 24°C and about 1.2 mW at 73°C (see Fig.2) which, to our knowledge, is the best result obtained so far for any 1.5 μm VCSEL.

As shown on Fig.3, narrow line-width single mode emission with high side mode suppression ratio is obtained at all pump power levels. As expected, the emission wavelength increases linearly with temperature (Fig. 4)

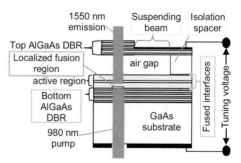

Fig. 1.Tuneable VCSEL structure

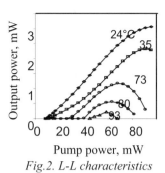
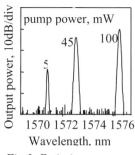
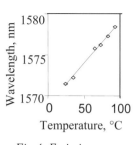

Fig.2. L-L characteristics   *Fig.3. Emission spectra*   *Fig.4. Emission wavelength at threshold*

4. Modulation and transmission experiment with single-wavelength VCSELs

We have developed a simple packaging technique for micro-assembling VCSEL structures and pump lasers into standard TO-56 packages. Pigtailed modules containing such VCSELs were mounted on driver PCB. Transmission experiments were performed at average optical power in the fiber of 0.6 mW and the dynamic extinction ratio was 5,5 dB. Data stream was generated by direct current modulation of pump laser at a rate of 2.5 Gb/s with 2^{31}-1 word length pseudorandom bit stream (PRBS NRZ) from a pattern generator. As shown in Fig.5, a clear eye-diagram was obtained at 2.5 Gb/s transmission data rate over 50 km of standard single mode SMF-28 optical fiber.

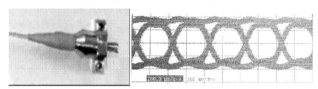

Fig.5. Packaged VCSEL and eye diagram

5. Tuneable VCSELs characteristics

We have tested tuneable VCSELs with 5 μm apertures for lateral optical confinement and top DBR suspended on 4 beams (see insert in Fig.6). The 980 nm pumping was performed through the GaAs substrate and the VCSEL emission was collected through the top-DBR side.

Fig. 6 shows the tuneable VCSEL emission wavelength versus the applied voltage at 40mW pump power. A tuning range of 38 nm was obtained for a voltage change between 1.5 and 3.4 V. The tuning curve is continuous, repeatable and well fitted with a quadratic function as expected from theory.

Emission spectra of the device at different tuning voltage values under a constant pump power of 40 mW (Fig.7) show that intensity variation in the tuning range is quite small. Maximum output power of 1 mW is obtained for every wavelength in the tuning range by slightly varying the incident pumping power. Light-light characteristic of the

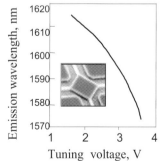

Fig. 6. Tuning characteristic

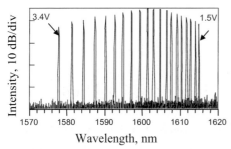
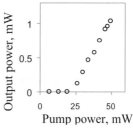

Fig. 7. Emission spectra at different tuning voltages

Fig. 8. L-L characteristic of the device at 2.9V tuning voltage (1593.1 nm).

device at 2.9V tuning voltage is shown on Fig.8. The reason for lower output power compared with single wavelength test devices based on the same active material is the large offset between the emission wavelength of the tuneable VCSELs and the peak of the gain spectral distribution, which is at 1540 nm. This may be easily corrected in future designs to obtain output power around 3 mW. As in the case of single wavelength devices, the emission of tuneable VCSELs is single mode with SMSR in excess of 40 dB.

Conclusions

We presented high performance single wavelength and tuneable optically pumped 1.55µm VCSELs fabricated by localized wafer fusion. CW operation up to 93°C and wide-opened eye-diagram at 2.5 Gb/s transmission over 50 km of standard single mode optical fiber are obtained using micro-assembled VCSELs with 980 nm pump laser chips. Optically pumped VCSELs with 38 nm tuneability at tuning voltage values below 4 V and 1 mW single mode output power with 40 dB side mode suppression ratio were demonstrated as well. All fabrication steps of these devices are simple and reliable, allowing to obtain high performance comparable to that of standard DFB lasers but at much lower cost.

References

[1] A.Keating, A. Black, A. Karim, Y.-J. Chiu, P. Abraham, C. Harder, E. Hu, J. Bowers, IEEE Photon. Technol. Lett., **12**, pp. 116-118, 2000

[2] S.Z. Zhang, N.M. Margalit, T.E. Reynolds, J. Bowers, IEEE Photon. Technol. Lett., **9**, pp. 374-376, 1997.

[3] R.J. Stone, R.F. Nabiev, J. Boucart, W. Yuen, P. Kner, G.S. Li, R.Carico, L. Scheffel, M. Jansen, D.P. Worland, C.J. Chang-Hasnain, Electron. Lett., **36,** pp. 1790-1791, 2000.

[4] E. Zouganelly, WDM solutions, June 2000,pp. 26-28.

[5] G.S. Li, R.F. Nabiev, W.Yuen, M. Jancen, D. Davis, C.J. Chang-Hasnain, Proc. 27th European Conference on Optical Communication, Vol. 2 (2001), pp. 220-221.

[6] D. Vakhshoori, P. Tayebati, Chih-Cheng Lu, M. Azimi, P. Wang, Jiang-Huai Zhou, E. Canoglu, Electron. Lett., 35 (1999), N 11. 900-901.

[7] A.V.Syrbu, V.P. Iakovlev, C.-A. Berseth, O. Dehaese, A. Rudra, E. Kapon, J. Jacquet, J. Boucart, C. Stark, F. Gaborit, I. Sagnes, J.C. Harmand, R. Raj, Electron. Lett., **34**, p.1744. (1998).

GaN-based Single Mirror Light Emitting Diodes

Ch. Zellweger, J. Dorsaz, J. F. Carlin, H. J. Bühlmann, and M. Ilegems
Institute for Quantum Electronics and Photonics, Swiss Federal Institute of Technology, CH-1015 Lausanne EPFL, Switzerland.

R. P. Stanley
Swiss Center for Electronics and Microtechnology, CSEM, CH-2007 Neuchâtel, Switzerland.

Abstract. We present a study on the dependence of the external Quantum Efficiency (QE) of Light Emitting Diodes (LEDs) on the device properties and p-contact metallisation. The external QE could be doubled by changing the p-type contact from oxidised Ni/Au to non-annealed Ag. The best value for the external QE of an unpackaged device is 13.5% and is obtained when the reflection from the Ag mirror contact is in phase with the QW emission.

1. Introduction

One important issue for the optimisation of Light Emitting Diodes is the extraction of the light. Similar to conventional III-V semiconductors different designs like Microcavity LEDs (MCLEDs) have been proposed for GaN [1]. The approach is to place the active InGaN quantum well (QW) emitter in a λ-cavity between two mirrors. The reflected and the emitted light are in phase, leading to an increased directional emission through one mirror. This mirror should be partially transparent and not absorbing. The only real option is a Bragg reflector, preferably grown monolithically, which is difficult in the case of GaN. The low refractive index change between GaN and AlGaN requires a large number of Bragg reflector pairs in order to achieve a sufficiently high reflectivity. Additionally, the crystallographic and electrical properties of AlGaN have a detrimental effect on the LED performance due to the increasing thermal and electrical resistances when additional AlGaN layers are introduced.

As an alternative we propose a layout using just one metallic mirror being placed at a distance such that the reflection from the mirror is in phase with the QW emission [2]. This mirror serves additionally as p-type contact. This Single Mirror LED (SMLED) has the advantage of being less demanding on growth in terms of complexity and precision as the overlap of the spontaneous emission spectrum and the cavity resonance is less important.

2. Experimental

Three different types of substrate emitting devices were fabricated from the same wafer. Type I is a standard LED with an oxidised Ni/Au ohmic p-contact; type II is the same structure with a non-annealed Ag mirror top contact; for type III, the Mg-doped top GaN layer was etched down prior to the Ag mirror contact deposition, such that the reflection from the Ag mirror adds in phase with the QW emission.

The structures were grown on sapphire in a 2'' AIXTRON MOVPE reactor. The active area consists of five 3.1 nm thick $In_{0.13}Ga_{0.87}N$ QWs separated by 12.4 nm Si-doped GaN barriers and a 14.5 nm thick $Al_{0.29}Ga_{0.71}N$ electron blocking layer.

After growth the Mg-doping was activated in a two-step anneal under nitrogen atmosphere. The processing was done using standard photolithography. First, 520 nm high mesas were etched in a RIE system using Cl_2. As etching mask, PECVD-deposited SiO_2 was used.

The contacts were deposited by lift-off technique. Prior to the metal evaporation the surface was deoxidised in a 20 s $HCl:H_2O$ 1:1 dip. As n-type contact Ti/Al/Ni/Au 15/220/40/50 nm was evaporated by E-beam.

Different metallisation schemes were used for the p-type contacts. As a standard contact, E-beam-evaporated Ni/Au 20/100 nm, oxidised for 5 min at 600 °C in O_2 [3], was used. In order to have a high reflective top mirror, non-annealed Ag 200 nm was deposited as an alternative p-type contact. The third scheme consisting of a grid of the standard Ni/Au metallisation covered by a large surface Ag mirror is designed in order to combine good current injection with a high reflective mirror.

A thick Si_3N_4 layer was deposited by PECVD on top, contact windows etched by BHF and finally large area Ti/Au 20/300 nm contact pads were evaporated.

The specific contact resistance of the standard and Ag contacts on the LED layers are around $1.7\ 10^{-4}\ \Omega cm^2$ and $2.1\ 10^{-3}\ \Omega cm^2$ respectively, where the Ag contact show a strong rectifying behaviour.

The measurements were performed with unpackaged devices of different size under uncooled operation. The light emission was collected through the sapphire substrate by a calibrated large area Si-photodiode.

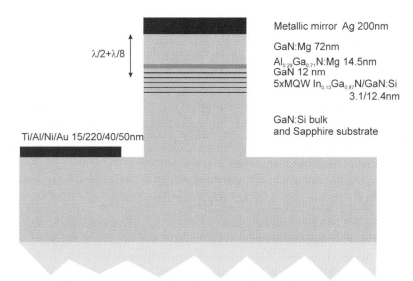

Figure 1. Layout of a Single Mirror Light Emitting Diode.

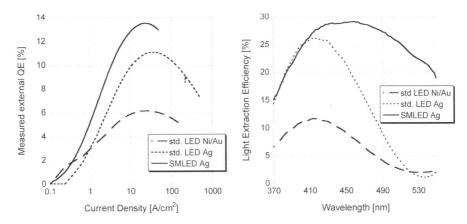

Figure 2. Measured external QE and simulated LEE of a standard LED with an oxidized Ni/Au or Ag p-type contact respectively, and a SMLED designed in a matter that the reflected light from the top Ag contact is in phase with the emitted light from the QE (p.m.).

3. Results and discussion

The devices on the standard LED show an emission maximum around 450 nm. The light output of the standard LED with the oxidized Ni/Au contact is ~3 mW at 20 mA through the substrate under uncooled cw operation. The maximum external QE (η_{ext}) for this structure was 6.2% at 25 A/cm².

If the Ni/Au contact is replaced by the high reflective Ag contact, an improvement by a factor of ~2 is observed. The best values of η_{ext} for the standard LED structure with Ag contacts are 11% as it can be seen in figure 2 a).

The combination of the Ni/Au grid and Ag as mirror results in a QE equal to that of the diode with the Ni/Au contact. It is attributed to the small current spreading from the grid. As a consequence the additional Ag mirror does not contribute to the reflection and thus the performance is not improved.

A improvement of the standard LED with Ag mirror can be obtained, if the design is changed to a SMLED design, i.e. the p-type GaN on top of the QWs corresponds to $\lambda/2 + \lambda/8$. On the standard LED layer this was achieved by dry etching of the surplus p-GaN before processing. The specific contact resistance is slightly increased, due to defects created by RIE, but the ext. QE improved to 13.5% with an Ag mirror at the same current density.

Figure 2b) shows the calculated Light Extraction Efficiency (LEE) of the LED as a function of the emission wavelength of the QWs. The LEE for the standard LED is 9% at the wavelength of 450 nm. With the Ag contact, the simulated LEE is 25%. The highest value is obtained by the SMLED with a maximum ext. QE of 30%. Thus our experimental results suggest an internal QE in the range of 50% in these samples.

4. Conclusion

We have shown that the external quantum efficiencies of standard GaN-LEDs can be significantly improved by introducing a high reflecting Ag mirror contact. By placing the mirror at a half-wavelength distance from the quantum well emitter (SMLED design) an increase in light extraction efficiency by more than a factor of 2 could be obtained.

References

[1] Diagne M, He Y, Zhou H, Makarona E, Nurmikko AV, Han J, et al. Vertical cavity violet light emitting diode incorporating an aluminum gallium nitride distributed Bragg mirror and a tunnel junction. Appl. Phys. Lett. 2001;79,3720-3722.

[2] Hunt NEJ, Schubert EF, Sivco DL, Cho AY, Zydyik GJ. Power and efficiency limits in Single-Mirror Light Emitting Diodes with enhanced intensity. Electron. Lett. 1992;28,2169 - 2171.

[3] Chen L-C, Chen F-R, Kai J-J, Chang L, Ho J-K, Jong C-S, et al. Microstructural investigation of oxidized Ni/Au ohmic contact to p-type GaN. J. Appl. Phys. 1999;86,3826 - 3832.

Nonlinear Semiconductor Materials for A Fully Passive Low-Loss Optical Combiner

G. Zhao and J. J. G. M. van der Tol

Opto-Electronic Devices Group, COBRA, Faculty of Electrical Engineering, Eindhoven University of Technology, Netherlands; Email: G.Z.Zhao@tue.nl

Abstract. A fully passive low-loss nonlinear optical combiner on semiconductor materials is shown based on an integrated Mach-Zehnder interferometer. For this device the nonlinear optical properties of the materials need to be optimized. Intensity dependent absorption and nonlinear refractive index change are calculated.

1. Introduction

With the rapid development of optical fibre communication, photonic integrated circuits (PIC) are becoming an important piece of optical hardware to be used in fibre networks. The optical waveguide in the PIC confines the beam of light into such small area that the intensity of these beams is large enough to cause optical nonlinearities in semiconductors [1,2]. Such optical nonlinearity is very significant for applications such as optical bistability [3,4] and all-optical switching [5,6]. An essential function in optical fibre networks is the combining of optical signals. Usually the combiner is a passive function, obtained with fused fibre couplers or planar Y-junctions. This combiner gives an unwanted 3-dB loss. This is avoided with optical switches, but these need control functions to synchronize with the optical signals. A recently proposed device [7] provides the combiner function without control signals, by using a nonlinear Mach-Zehnder interferometer (MZI). This combiner was realized with fibre components, with semiconductor optical amplifiers (SOA) acting as the nonlinear phase shifting elements [8]. An integrated Mach-Zehnder combiner with SOAs is under the investigation [9]. In this paper, a fully passive low-loss nonlinear optical combiner is evaluated based on optimisation of semiconductor materials.

2. The concept of device

The combiner mentioned above is a nonlinear Mach-Zehnder interferometer in a 2-to-1 port configuration, operating as a splitter in one direction and as a self-routing combiner in another, seen as Fig. 1. It has been shown [7] that the absorption of the nonlinear sections have to be small and the phase shift between two branches should reach at least a half of π in order to get a low-loss combiner function. For this

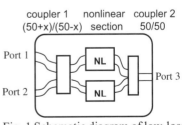

Fig. 1 Schematic diagram of low-loss combiner

purpose, we have calculated the absorption coefficient and refractive index change of the nonlinear sections. A simple, two-band model has been proved extremely useful for a first prediction of the intensity and the frequency dependence of the nonlinear refractive index of potential materials.

3. Model for Optical Nonlinearities in Semiconductor

In the semiconductor, due to the momentum conservation, each interband optical transition relates to two states in the conduction band and the valence band with the same wave-vector, which can be regarded as a two-band system. The macroscopic polarization of semiconductors can be calculated by averaging statistically over the microscopic dipole matrix element. Furthermore, the intraband interactions can be described as the relaxation. Similar to the derivation in our previous works [10], the absorption coefficient and the change of refractive index can be calculated by

$$\alpha(\omega, I) = \frac{2e^2}{\hbar c \varepsilon_0 V} \sum_{\vec{k}} \frac{\omega \tau_{in} |\mu_{cv}|^2 (1 - f_h - f_e)}{1 + [\omega_{cv}(\vec{k}) - \omega]^2 \tau_{in}^2 + I/I_s}$$

$$\Delta n(\omega, I) = \frac{e^2 c \tau_{in}}{n_0 \hbar c \varepsilon_0 V} \sum_{\vec{k}} \frac{[\omega_{cv}(\vec{k}) - \omega] \tau_{in} |\mu_{cv}|^2 (1 - f_h - f_e)}{1 + [\omega_{cv}(\vec{k}) - \omega]^2 \tau_{in}^2 + I/I_s} \quad (1)$$

here τ_0 indicates the depopulation time, τ_{in} the dephasing time, and the intensity I, n_0 the linear refraction index, c the light speed, ε_0 the vacuum dielectric constant, $\hbar$ Planck's constant. μ_{cv} the dipole transition moment between the conduction and valence band, f_e and f_h are, respectively, the population probability of electrons and holes, V the volume of semiconductor, and we have taken account of the spin by indicating a factor of 2. The characteristic intensity I_s is defined as

$$I_s \equiv \frac{\hbar^2 c \varepsilon_0 n_0}{2e^2 |\mu_{cv}|^2 \tau_{in} \tau_0}. \quad (2)$$

In fact, the optically induced change of the refractive index should be defined by $\Delta n_{NL}(\omega, I) = \Delta n(\omega, I) - \Delta n(\omega, I=0)$.

4. Results and Discussions

From Eq. (1), it is seen that the absorption coefficient and index change depend not only on the frequency of the optical field, but also on its intensity, that is, the nonlinear effect is expressed explicitly. On the other hand, it can be seen that the intensity dependent optical nonlinearities are measured by a characteristic intensity I_s. If the intensity I is much less than I_s, the optical nonlinearities can be neglected. Only if I is comparable or larger than I_s, the nonlinear absorption and the change of refractive index become notable. Eq.(2) indicates that the characteristic intensity is determined by the dipole matrix element and the relaxation time. The dipole matrix element can be calculated from the momentum matrix element [11]

$$\mu_{cv} = \frac{\hbar}{im_0 E_g} M_{cv}, \quad (3)$$

here m_0 is the free electron mass, E_g is the bandgap energy, whereas the momentum matrix element of bulk semiconductor can be expressed as [2]

$$|M_{cv}|^2 = \frac{m_0}{6}\left(\frac{m_0}{m^*} - 1\right)\frac{E_g(E_g + \Delta_0)}{E_g + \frac{2}{3}\Delta_0}, \qquad (4)$$

m^* is the effective mass of the electron, Δ_0 the spin-orbit split-off energy of the valence band. At the same time, the relaxation mechanism is important to the nonlinear optical properties in semiconductors. Due to the complexity of relaxation processes, however, the relaxation times cannot be discussed in a general way. Theoretical simulation can be fitted to the experimental measurement.

Table 1 show the bandgap E_g, the dipole matrix element μ_{cv}, the relaxation time τ_0, τ_{in}, and the characteristic intensity I_s for a number of plausible materials. The relaxation times τ_0 and τ_{in} are estimated as the carrier lifetime and the intraband relaxation time [10,12]. It can be seen that the nonlinear effect in semiconductors becomes more important when the intensity is much large than the characteristic intensity.

In our design of a fully passive combiner, the size of the waveguide is about 2 μm². If the power of light is 1mW, a large intensity 50 kW / cm² can be obtained at the input port of the waveguide. Due to the unequal intensities in the two arms of MZI, moreover, an intensity dependent nonlinear refractive index difference can be reached in our design of optical combiner.

Table 1. Some characteristic parameters related with the optical nonlinearities of bulk semiconductors. "Q" indicates the quaternary compound InGaAsP, with the number referring to the band edge wavelength.

materials	E_g (eV)	$\|\mu_{cv}\|^2$ (10^{-18}m²)	τ_0 (ns)	τ_{in} (ps)	I_s (kW/cm²)
GaAs	1.42	0.13	2.0	1.0	0.65
InP	1.35	0.11	10	0.1	1.66
Q(1.30)	0.95	0.24	10	0.1	0.77
Q(1.50)	0.83	0.34	10	0.1	0.56
Q(1.55)	0.80	0.37	10	0.1	0.51

Using Eq. (1) we have calculated the absorption coefficient and change of refractive index of Q(1.54) as a function of photon energy (wavelength) at different intensities, see Fig. 2.

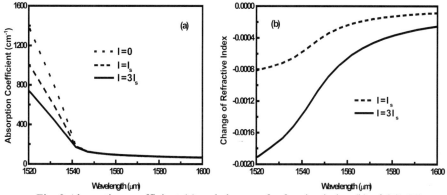

Fig. 2 Absorption coefficient (a) and change of refractive index (b) of Q(1.54) as a function of photon energy (wavelength)

It can be seen, from Fig. 2, the absorption edge locates at the bandgap. The absorption coefficient is small in the transparent region away from the absorption edge, such as 1.55 µm. A low-loss material is possible in this case. On the other hand, the change of refractive index is large near the absorption edge. So one can get a small absorption and large index change at certain optical frequency (wavelength). It is possible to get an optimisation of the absorption edge by designing the semiconductors bandgap so that the required operating wavelength (1.55 µm) falls in the transparency region.

On the other hand, the change of refractive index as a function of intensity is plotted at three different photon energies (wavelengths) in Fig. 3. It shows that the change of refractive index increase with intensity in an almost linear way in the transparency region. Under a suitable intensity, such as 1.6 kW/cm^2, the phase shift of half π can be reached with a change of refractive index of 7.75×10^{-3} at the optical communication wavelength 1.55 µm and length L = 500 µm of the sample.

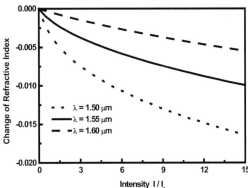

Fig. 3 Intensity dependent change of refractive index for Q(1.54) at different operating wavelength.

5. Conclusions

In conclusion, the possibility of a fully passive nonlinear optical combiner with low-loss is examined by optimizing materials. Based on the two-band model, a simplified modelling can be used to calculate the absorption coefficient and the change of refractive index as a function of frequency and intensity of incident optical field. The optical nonlinearities of bulk semiconductors are simulated. An optimization strategy is proposed for the design of low absorption and large index change of the nonlinear optical device.

References

[1] Haug H, Optical Nonlinearities and Instabilities in Semiconductors, 1988, AcademicPress, London.
[2] Chow W W, et al, Semiconductor Laser Physics, 1994, Springer-Verlag, Berlin.
[3] Gibbs H, Optical Bistability: Controling Light with Light, 1985, Academic press, USA
[4] Jeannes F, et al, Opt. Lett. 134 (1997), 607.
[5] Aitchison J S, et al, Opt. Lett. 18 (1993), 1153.
[6] Blair S, et al, J. Opt. Soc. Am. B 13 (1996), 2141.
[7] Tol J, in Proceedings of IEEE/LEOS Symposium Benelux Chapter, 1999, pp. 69-72.
[8] Tol J, et al, *IEEE Phot. Tech. Lett.* 13 (2001), 1197.
[9] Tol J, in Proceedings of SAFE 2000, 2000, pp. 43.
[10] Zhao G, et al, Sol. Stat. Comm. 99 (1996), 595
[11] Burt M G, J. Phys. C 5(1993), 4051.
[12] Alfano R R, Semiconductors Probed by Ultrafast Laser Spectroscopy, 1984, Academic Press, Inc.

Optically Pumped Vertical External Cavity Semiconductor Thin-Disk Laser with CW Operation at 660 nm

Moritz I. Müller, Christian Karnutsch, Johann Luft, Wolfgang Schmid, Klaus Streubel, Norbert Linder
OSRAM Opto Semiconductors, Wernerwerkstr. 2, 93049 Regensburg, Germany

Svent-Simon Beyertt, Uwe Brauch, Adolf Giesen
Institut für Strahlwerkzeuge, Univ. Stuttgart, Pfaffenwaldring 43, 70569 Stuttgart, Germany

Gottfried H. Döhler
Institut für Technische Physik, Univ. Erlangen, Erwin-Rommel-Str. 1, 91058 Erlangen, Germany

Abstract. We report on continuous wave operation of an optically pumped vertical external cavity semiconductor thin-disk laser emitting at 660 nm. A maximum output power of 55 mW could be achieved in TEM_{00} mode. The active region is a multi quantum well structure. The laser device is pumped by a dye laser emitting at 630 nm, corresponding to a quantum defect of only 90 meV between pump and laser photon energies. The internal efficiency of the laser material has been analyzed and found to be higher than 70%.

1. Introduction

High power lasers for medical, display or communication applications require good beam quality. Solid-state and gas lasers offer both, but are costly, large and power-consuming. Semiconductor lasers, on the other hand, usually suffer from either poor beam profiles or low output power. A way to get around these limitations is to combine an external cavity design with the advantages of optical pumping. By using an appropriate pump spot profile and a spherical output mirror the laser is forced into the fundamental mode, while laser output power is scalable in a wide range by varying the pump spot size and the pump power. This concept has initially been developed for diode-pumped solid state lasers [1] and more recently been applied to semiconductor systems in the infrared wavelength regime [2,3].

In this paper we demonstrate the first implementation of this concept for visible lasers. CW operation at λ = 660 nm has been achieved in a thin-disk laser structure based on MOVPE-grown AlInGaP material.

2. Laser Setup

A standard laser setup, consisting of the semiconductor gain element, an external mirror and a pump laser source, was used for the experiments (Fig. 1A). Pump light was focussed under an angle of 45° onto the semiconductor chip, creating a spot of size 54μm x 40μm ($1/e^2$ diameter) on the chip. The chip was clamped onto a copper heatsink, which is temperature controlled by a peltier stack. Typical heatsink temperature was -30°C. A semi-confocal resonator was formed using an external spherical mirror with R = 50 mm.

The gain element is an AlInGaP multiple quantum well (MQW) structure grown on top of an AlGaAs DBR (Fig. 1B). We used a variable number of strained InGaP quantum wells for the active layers, which are arranged in groups each containing 4 QWs. The groups form a resonant periodic gain structure with each group positioned at an antinode of the laser mode with an optical distance of λ/2 between the groups. This arrangement provides maximum overlap between the electrical field of the laser radiation and the periodic gain structure.

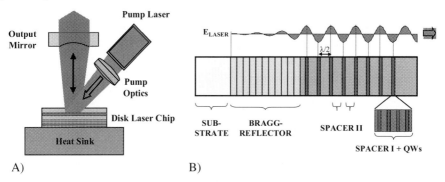

Figure 1. A) Schematic diagram of the laser device. The resonator is build by the DBR on the semiconductor chip and the concave mirror. The semiconductor chip is mounted to a heatsink.

B) Schematic layer structure of the semiconductor chip and the electric field of the longitudinal laser mode.

The structure was pumped with a dye laser at 630 nm and had an emission wavelength of 660 nm. In this pump configuration the defect energy between pump and laser photons is only 90 meV, minimizing the heat dissipation by carrier relaxation. Moreover, the DBR is reflective for both the emitted laser light and the pump light, resulting in double pass transmission for the pump beam and a total absorption of more than 80% of the incident pump light. An additional advantage of this pump configuration is a more homogeneous pumping of the active layers as compared to single pass absorption, where the quantum wells near the DBR receive less pump power than the quantum wells near the surface.

3. CW laser characteristics

The laser characteristics for various temperatures are shown in Fig. 2A. A maximum output power of approximately 55 mW in continuous wave operation was achieved at a temperature of -35°C, with an absorbed pump power of 400 mW at 629 nm. The pump threshold is 140 mW and the differential efficiency approximately 25%. Limiting the optical output power here is the thermal roll over. Pulsed experiments have shown that output powers of several 100 mW are possible if the thermal resistance of the structure is low enough [5].

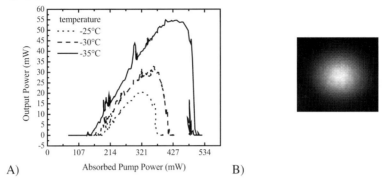

Figure 2: A) Output power versus absorbed pump power for various heatsink temperatures. The pump wavelength was 632 nm, the pump spot size 54µm x 40µm. Transmission of the outcoupling mirror was 0.46%. B) Intensity plot of the beam at 50 mW output power. M^2 was measured to be 1.15.

Fig. 2B shows the intensity plot of the laser mode at P_{out} = 50 mW. The resonator length is slightly below the critical length and has been adjusted such that fundamental mode operation was achieved. The beam parameter value M^2 was measured to be 1.15.

4. Analysis of Laser Characteristics

To reduce the influence of thermal effects when determining the inherent chip properties, pulsed measurements were used. As the QW barriers in AlInGaP lasers are low, knowledge about the internal quantum efficiency is important to determine the physical limits of device performance. The differential efficiency is given by [5]

$$\eta_{diff} = \eta_{pump} \cdot \frac{\lambda_{laser}}{\lambda_{Pump}} \cdot \frac{\ln(R_2)}{\ln(R_1 R_2 T_{RT})}$$

where λ_{pump} is the wavelength of the pump light, λ_{laser} the emission wavelength, R_1 the reflectivity of the DBR, T_{RT} the round trip transmission factor, and R_2 the reflectivity of the outcoupling mirror. The pump efficiency η_{pump} is given by

$$\eta_{pump} = (1-R) \cdot \eta_{abs} \cdot \eta_{int} \cdot \eta_{geo}$$

with η_{abs} the absorption efficiency, η_{int} the internal conversion efficiency, η_{geo} the geometrical overlap factor between pump spot and laser spot and R the reflection loss of the pump light at the semiconductor surface. The differential efficiency η_{diff} was measured versus R_2 and then fitted with the equations above (Fig. 3). The reflection loss is directly measured and the absorption efficiency is determined by transmission experiments to be η_{abs} = 85% at 514 nm. To obtain η_{geo} we use a two dimensional model which reproduces the output characteristics with η_{pump} and T_{RT} as fit parameters. The analysis was carried out for operation at –40°C, which was the optimum temperature of operation. We obtain values of η_{int} = 74% for the internal conversion efficiency, T_{RT} = 97.8% for the round-trip transmission factor, and η_{geo} = 54% for the overlap factor. These data indicate that losses due to carrier leakage out of the quantum wells are small. A room temperature analysis could not be performed yet for experimental reasons.

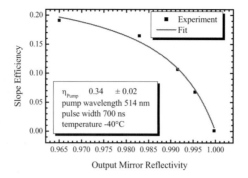

Figure 3: Slope efficiency versus output mirror reflectivity. The fit of the data leads to a pump efficiency of $\eta_{pump} = 0.34$.

5. Conclusions

We demonstrate the first optically pumped semiconductor thin disk laser with 55 mW output power operating at 660 nm in the TEM$_{00}$ mode. The beam propagation ratio was measured to be M^2 = 1.15 at 50 mW. A pump wavelength of 630 nm was used, leading to a defect energy of only 90 meV between pump and laser photons. This configuration leads to more efficient carrier capture inside the QWs, less energy dissipation by carrier relaxation and a more homogeneous vertical pump profile, because the pump light is reflected by the DBR and passes the active layers twice. Analysis of the laser slope efficiency shows that efficient laser operation is possible.

References

[1] A. Giesen, H. Hügel, A. Voss, K. Wittig, U. Brauch, and H. Opower, Appl. Phys. B **58**, 363 (1994)
[2] J. V. Sandusky and S. R. J. Brueck, IEEE Photon. Technol. Lett. **8**, 313 (1996)
[3] M. Kuznetsov, F. Hakimi, R. Sprague, and A. Mooradian, IEEE Photon. Technol. Lett. **9**, 1063 (1997)
[4] M. I. Mueller, N. Linder, C. Karnutsch, et al, Optically pumped semiconductor thin-disk laser with external cavity operating at 660 nm, Proc. of SPIE, **4649**, San Jose, 265-271, (2002)
[5] M. Kuznetsov, F. Hakimi, R. Sprague, and A. Mooradian, IEEE J. Sel. Topics Quantum Electron. **5**, 561 (1999)

Polarization-sensitive photo-detectors based on strained M-plane GaN

Sandip Ghosh*, Pranob Misra, O Brandt, and H T Grahn
Paul-Drude-Institut für Festkörperelektronik, Hausvogteiplatz 5-7,
10117 Berlin, Germany

Abstract. The energy gap of strained M-plane GaN films changes with the angle of in-plane linear polarization. We calculate the oscillator strengths of the three band-edge transitions and the resulting polarization anisotropy of the absorption coefficient for an M-plane ($[1\bar{1}00]$ oriented) GaN film as a function of an arbitrary in-plane strain. We show that for a particular range of M-plane strain both the wavelength range, over which the polarization anisotropy in the absorption occurs, and the magnitude of the polarization anisotropy are enhanced. Consequently, strained M-plane GaN represents a promising candidate for realizing polarization-sensitive photo-detectors.

1. Introduction

Several applications require the detection of the state of polarization of an incident light beam. Generally, one uses a combination of a detector and an external polarizing element, such as sheet polarizers and prisms. This is inconvenient for large-scale integration including magneto-optical readout and optical computation based on polarization-coded logic, both from the point of view of economics and mechanical aspects. A better solution would be a detector, which is intrinsically sensitive to optical polarization. The usual group IV and III-V semiconductor detectors do not show any significant in-plane anisotropy in their optical absorption coefficient (α) due to their cubic crystal structure, which has high symmetry. There are several other approaches to create an artificial polarization anisotropy such as chemically etched surface corrugations for the infrared spectral region [1] and ordered $In_{0.5}Ga_{0.5}P$ alloys for the long-wavelength regime of the visible spectral region [2]. Using these approaches, polarization-sensitive photo-detectors (PSPD's) have been realized. However, higher-density magneto-optical storage and readout require shorter operating wavelengths.

Compound semiconductors, which intrinsically have a low-symmetry crystal structure such as the wurtzite (WZ) structure, are another possible solution. However, currently all devices are based on C-plane ([0001] oriented) GaN films (cf. figures 1), which do not show any significant in-plane optical polarization anisotropy. In contrast, M-plane ($[1\bar{1}00]$ oriented) films, which have lower symmetry, have recently been of great interest in the context of obtaining quantum-well-based light emitters with higher quantum efficiency [3]. These M-plane GaN films may have greater potential for PSPD applications. Due to the combination of a large mismatch in lattice constants and thermal

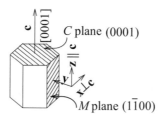

Figure 1. Schematic diagram of the C and M planes of a WZ-GaN unit cell and the choice of coordinates.

expansion coefficients between GaN and commonly used substrates, even thick GaN films are usually strained. Since strain affects the electronic band structure (EBS), it further modifies the polarization selection rules for optical absorption.

2. Polarization dependence of the transmittance

We have measured the transmittance of a strained 1.22 μm thick M-plane GaN film on a γ-LiAlO$_2$(100) substrate, which was grown by plasma-assisted molecular-beam epitaxy, for different in-plane polarization angles φ at 295 K. The light propagation direction is parallel to the [1$\bar{1}$00] direction, which is also the growth direction. The polarization angle is varied from φ = 0° (**E** ⊥ **c**) to φ = 90° (**E** ∥ **c**). The experimental results shown in figure 2 demonstrate that the energy gap increases by about 40 meV, when the polarization is changed from **E** ⊥ **c** to **E** ∥ **c**. At low temperatures, the increase of the energy gap is even larger [4]. Calculations of the strain dependence of the interband transition energies using the **k**•**p** perturbation approach reveal that the energy gap T_1 for **E** ⊥ **c** is determined by the uppermost valence band with wave functions of $|X\rangle$ symmetry, while for **E** ∥ **c** the next lower valence band with wave functions of $|Z\rangle$ symmetry defines the energy gap T_2 [5]. We have also calculated the oscillator strengths for these two polarization directions and confirmed the observed dependence for φ = 0° (**E** ⊥ **c**) and 90° (**E** ∥ **c**) with an energy shift of about 40 meV. For the other polarization angles, the transmittance follows from a superposition of the absorption coefficients of the transitions T_1 and T_2. The detailed dependence of the transmittance on the in-plane polarization angle φ is currently under investigation, and the corresponding results will be published elsewhere. The in-plane strain was determined to be $\varepsilon_{xx} = -0.56\%$ and $\varepsilon_{zz} = -0.31\%$ by comparing the calculated transition energies with the values obtained from polarized photoreflectance spectra [5] corresponding to an out-of-plane dilatation of $\varepsilon_{yy} = -0.29\%$ in agreement with x-ray diffraction measurements.

3. Calculated oscillator strengths for different polarizations

The two important figures of merit for PSPD applications are $\Delta f = f_x - f_z$, where f_x and f_z denote the oscillator strength for x- and z-polarized transitions, and $\Delta \lambda$, which defines the wavelength range, over which Δf is non-zero [6]. Taking GaN as an example, we will discuss the strain dependence of Δf and $\Delta \lambda$. The transition energies for the onset of absorption between the three valence bands and the conduction band are labelled T_i (i=1, 2, 3) in order of decreasing transition wavelength λ_i. For $\lambda < \lambda_3$, all three transitions

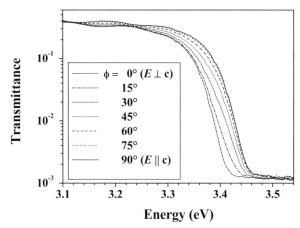

Figure 2. Measured transmittance for an *M*-plane GaN film on LiAlO$_2$ at 295 K for different polarization angles ϕ as indicated.

contribute to $\alpha(\lambda)$. However, due to the sum rule for the oscillator strengths $\Sigma^3_{i=1} f_{i\beta}=1$ with $\beta=x$, y, and z, we obtain $\Delta f = \Sigma^3_{i=1}[f_{ix} - f_{iz}] = 0$. Therefore, a significant polarization anisotropy in α can exist only for $\lambda > \lambda_3$, i.e., between λ_1 and λ_3. We identify three regimes in this wavelength range. Regime I is the range between λ_1 and λ_2. Here, only T_1 contributes to α so that $\Delta f = \Delta f^{12} = |f_{1x} - f_{1z}|$. The strain values, for which Δf^{12} is large, are depicted by the darker regions of the grey-scale plot in figure 3(a). In the regions marked *x* (*z*), α will be large for polarization $\| \boldsymbol{x}$ ($\| \boldsymbol{z}$). For a strain value with a large Δf^{12}, one can determine from the adjacent contour plots the central operating wavelength $\lambda_o = \lambda_o^{12} = (\lambda_1 + \lambda_2)/2$ and the operating range $\Delta\lambda = \Delta\lambda^{12} = \lambda_1 - \lambda_2$ for PSPD applications. Regime II corresponds to wavelengths between λ_2 and λ_3. Here in general, both T_1 and T_2 contribute to α so that $\Delta f = \Delta f^{23} = |f_{1x} + f_{2x} - f_{1z} - f_{2z}|$. The strain values, for which Δf^{23} is large, are shown by the darker regions of the grey-scale plot in figure 3(b). The adjacent contour plots show $\lambda_o = \lambda_o^{23} = (\lambda_2 + \lambda_3)/2$ and $\Delta\lambda = \Delta\lambda^{23} = \lambda_2 - \lambda_3$.

Note that there are strain values, for which both Δf^{12} and Δf^{23} are large. Therefore, a large and uniform polarization anisotropy in α can exist for these strain values from λ_1 up to λ_3. We call this regime III, in which we take Δf to be the smaller one of the two values Δf^{12} and Δf^{23}. This represents the minimum anisotropy attainable over the whole range between λ_1 and λ_3. The strain dependence of Δf in regime III is shown by the grey-scale plot in figure 3(c). The adjacent contour plots show $\lambda_o = \lambda_o^{13} = (\lambda_1 + \lambda_3)/2$ and $\Delta\lambda = \Delta\lambda^{13} = \lambda_1 - \lambda_3$. The reason that one can achieve $\Delta f \approx 1$ between λ_1 and λ_2 as well as between λ_2 and λ_3 is because for these strain values $f_{2y} \approx 1$. Then according to the sum rule $f_{2x} + f_{2y} + f_{2z} = 1$, $f_{2x} \approx 0$ and $f_{2z} \approx 0$ so that even between λ_2 and λ_3 it is essentially Δf^{12} that determines Δf, i.e., T_2 does not contribute to α of the *M*-plane film. Because of the difference in lattice constants and thermal expansion coefficients along $\boldsymbol{x}$ and z, *M*-plane films invariably experience asymmetric strain, and it may even be possible to achieve anti-symmetric in-plane strain as required for regime III with an appropriate choice of the substrate. Unlike for ordered $In_{0.5}Ga_{0.5}P$-alloy-based PSPD's, λ_o can be varied for *M*-plane GaN over a much larger wavelength range simply by alloying it with either In or Al to change the energy gap. At the same time, $\Delta\alpha$ for *M*-plane GaN is increased by one order of magnitude over the value for $In_{0.5}Ga_{0.5}P$-alloy-based PSPD's.

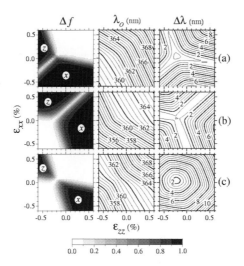

Figure 3. Δf (left), λ_o (center), and $\Delta\lambda$ (right) as a function of M-plane strain ε_{xx} and ε_{zz} for the three regimes for PSPD application of strained M-plane GaN. (a) regime I between λ_1 and λ_2; (b) regime II between λ_2 and λ_3; (c) regime III between λ_1 and λ_3, where Δf is taken to be the smaller value of the two amongst Δf^{12} and Δf^{23}. In the dark regions marked x (z), the absorption is large for polarization $\| x$ ($\| z$).

4. Summary

We have experimentally and theoretically investigated the polarization properties of the fundamental band-edge transitions of a strained M-plane GaN film. The calculations reveal that the in-plane polarization anisotropy in α is enhanced with strain, even for small values of the in-plane strain. This makes strained M-plane GaN useful for realizing polarization-sensitive photo-detectors. We have also determined the strain dependence of the operating wavelength characteristics for such an application.

Acknowledgement

We would like to thank P. Waltereit for sample growth.

References

*Present address: Tata Institute of Fundamental Research, Mumbai 400005, India

[1] Chen C J, Choi K K, Rokhinson L, Chang W H, and Tsui D C 1999 Appl. Phys. Lett. 74 862-864
[2] Greger E, Riel P, Moser M, Kippenberg T, Kiesel P, and Döhler G H 1997 Appl. Phys. Lett. 71 3245-3247
[3] Waltereit P, Brandt O, Trampert A, Grahn H T, Menniger J, Ramsteiner M, Reiche M, and Ploog K H 2000 Nature 406 865-868
[4] Ghosh S, Waltereit P, Brandt O, Grahn H T, and Ploog K H 2002 Appl. Phys. Lett. 80 413-415
[5] Ghosh S, Waltereit P, Brandt O, Grahn H T, and Ploog K H 2002 Phys. Rev. B 65 075202 (7 pages)
[6] Ghosh S, Brandt O, Grahn H T, and Ploog K H 2002 Appl. Phys. Lett. 81 3380-3382

Investigation of the modulation efficiency of depleted InGaAsP/InP ridge waveguide phase modulators at 1.55μm

H S Park, J C Yi, and Y T Byun[1]
Hong Ik University, [1]KIST, Seoul, Korea.

Abstract. Single mode depleted p-n InGaAsP/InP ridge waveguide phase modulators have been fabricated and investigated at 1.55μm. The phase modulation efficiency of the device was characterized by using the Fabry-Perot resonance method with a tunable laser. The measured phase modulation efficiency for a 2 mm long device was determined as high as 34 deg/V·mm for TE mode. This value corresponds to the highest experimental electrooptic modulation efficiency reported so far for InGaAsP/InP p-n phase modulators at the said wavelength region.

1. Introduction

III-V semiconductor optical waveguide devices have been heavily investigated for applications in photonic integrated circuits, and one important application area is the electrooptic modulators (EOM)[1]. Although the III-V semiconductor EOMs have demonstrated successful operation with high modulation bandwidth up to 80GHz, one of their drawbacks, however, turned out to be the high operating voltage. This is mainly due to the small electrooptic coefficients of the III-V semiconductor materials, which are usually ten times smaller than other conventional electrooptic materials. The highest electrooptic modulation efficiency of III-V semiconductors reported so far is 96°/V·mm and 48.9°/V·mm at the wavelength of 1.06 μm and 1.31 μm, respectively [2]. This value dramatically drops to 12 °/V·mm as the operation wavelength increases to 1.55 μm [3]. Although some InGaAs/InP MQW ridge phase modulators have reported this value around 39 °/V·mm [4] at 1.55μm, such devices mainly utilized the QCSE, which is usually not linear and not immune to chirping.

In this paper, we propose a P-p-n-N InGaAsP/InP DH ridge waveguide phase modulator where a normally doped p-n guiding layer is sandwiched by a heavily doped P- and N- cladding layers. Usually, the waveguide core consists of undoped layers to reduce the optical propagation loss. In such structures, the electric field intensity for a given applied voltage is inversely proportional to the core thickness whose nominal value is usually larger than 0.5 μm. On the other hand, the proposed epilayer structure is designed as such that a half of the core is p doped and the other half n doped. By applying a reverse bias to this device, one can obtain a depletion layer at the center of the core. Thus much higher electric field intensity will be available for a given bias over a very thin depletion layer that is usually smaller than 0.5 μm. Consequently, the electrooptic effects, which are linearly or quadratically proportional to the electric field intensity, will be enhanced.

The overall electrooptic phase modulation efficiency has been characterized using the Fabry-Perot (FP) resonance method. For various reversely biased voltages, the FP pattern was measured by using a very widely tunable laser and a precision optical spectrum analyser. From the measured FP pattern, the effective refractive index was estimated and the electrooptic phase modulation efficiency was evaluated.

2. Waveguide Design and Fabrication

Fig. 1 shows the P-p-n-N InGaAsP/InP DH waveguide phase modulator structure realized by inserting p- and n-InGaAsP guiding layers between P- and N-InP cladding layers.

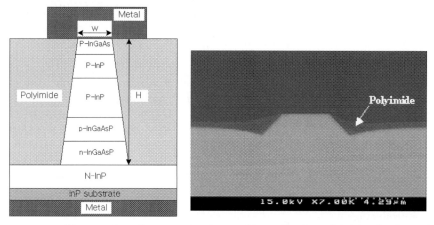

Figure 1. Cross-sectional profile of the fabricated P-p-n-N InP/InGaAsP/InP ridge waveguide (left) and its SEM picture (right).

The precise control of the doping concentration of the guiding layer is still a challenging task, thus we investigated only one epilayer structure where the p-n guiding layer is doped at $1\times10^{17}\,cm^{-3}$. The epitaxial layers were grown by metalorganic chemical vapor deposition (MOCVD) on (100) InP substrate [5]. It consists of a 0.25μm thick N-InP ($3\times10^{17}\,cm^{-3}$) lower cladding on a n$^+$-InP ($2\times10^{18}\,cm^{-3}$) substrate, a 0.25 μm thick n-InGaAsP ($1\times10^{17}\,cm^{-3}$) guiding layer, a 0.25 μm thick p-InGaAsP ($1\times10^{17}\,cm^{-3}$) guiding, a 0.75 μm thick P-InP ($5\times10^{17}\,cm^{-3}$) upper cladding, a 0.25 μm thick P-InP ($3\times10^{18}\,cm^{-3}$) upper cladding, a 0.2 μm thick p$^+$-InGaAs ($1\times10^{18}\,cm^{-3}$) cap layer. The composition of the guiding layers is controlled to be lattice matched layers with λ_g=1.25μm. The ridge waveguide was fabricated along $[01\bar{1}]$ direction using the conventional photolithography and wet chemical etching. To guarantee a single mode condition for the ridge waveguide, the ridge width was chosen to be 3 μm and the etching depth was determined as 1.7 μm. In this structure, the etching depth is deeper than the p-n junction interface. In this way, one can reduce the junction capacitance as well as enhance the optical mode confinement. Finally, the wafer was cleaved with a length of 2 mm, and the electrodes were soldered with silver paste to a submount for external electrical connection.

Fig. 2 shows the optical mode (contour lines) and electric field intensity (arrows) profiles calculated by a self-consistent finite-element method [2]. One can see that the electric field intensity becomes the maximum at the center of the optical mode profile.

Fig. 2 also shows the near-field pattern of the guided mode at 1.55 μm measured by using a ×40 objective lens and an IR CCD camera.

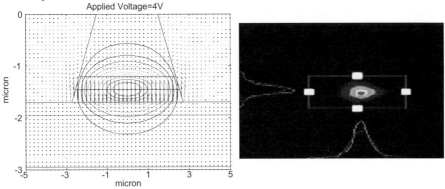

Figure 2. Calculated (left) and measured (right) optical mode profiles.

3. Characteristics of the Fabry-Perot resonance in the phase modulator

Since the cleaved facets of the fabricated phase modulator have finite reflectivity, they naturally form an FP cavity. Fig 3 shows a part of the measured FP pattern between 1550nm and 1550.5 nm.

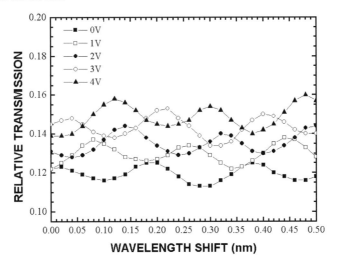

Figure 3. The measured Fabry-Perot pattern of the fabricated phase modulator for various applied voltages at 1550~1550.5nm.

The transmitted optical power at the FP waveguide resonator is given as following:

$$T = \frac{P_{out}}{P_{in}} = \frac{(1-R)^2 e^{-\alpha l}}{1 - 2R'\cos(4\pi n l/\lambda) + R'^2} \quad , \quad R' = R \cdot e^{-\alpha l} \tag{1}$$

Here, R is the reflectivity of the cleaved facets, l, the optical pass length of the cavity, λ, the wavelength, , the optical cavity loss, and n, the effective refractive index of the waveguides. From the equation (1), one can see that the output power of FP waveguide resonator will change as the wavelength of the tunable laser shifts. By measuring the FP fringe spacing ($\Delta\lambda_{FP}$) from two adjacent FP peak wavelengths, λ_1 and λ_2, one can evaluate the effective refractive index as follows:

$$n(V) = \frac{\lambda_2 \cdot \lambda_1}{2(\lambda_2 - \lambda_1)l} = \frac{\lambda_2 \cdot \lambda_1}{2 \cdot \Delta\lambda_{FP} \cdot l} \qquad (2)$$

The measured fringe spacing ($\Delta\lambda_{FP}$) was 0.185nm for the zero bias; hence the refractive index was 3.24. The optical loss was determined as 4.5 dB/mm.

4. Phase modulation efficiency

By varying the applied voltage, the overall phase shift can also be evaluated as:

$$\Delta\phi(V) = 2 \cdot \frac{2\pi}{\lambda}\{n(V) - n(0)\} \cdot l = \pi \cdot \frac{\Delta\lambda(V)}{\Delta\lambda_{FP}} \qquad (3)$$

Here, $\Delta\lambda$ is the peak wavelength variation for a given voltage from the zero bias position. From Fig. 3, one can see that the fringe peak position shifts by 0.30nm when the applied voltage is increased from 0 to 4V. Thus the overall phase shift is 0.30/0.185=1.6π. Fig. 4 shows the overall phase modulation efficiency of the device vs. the bias voltage. From this result, the phase modulation efficiency is estimated as 34.6°/ V·mm for the 2 mm long device and the switching voltage is only 2.6V. This value is the highest one among the InGaAsP/InP electrooptic phase modulators reported so far. This is approximately 3 times higher than a similar epitaxial structure described in reference [3] that uses an undoped guiding layer, and almost the same efficiency with the MQW devices [4].

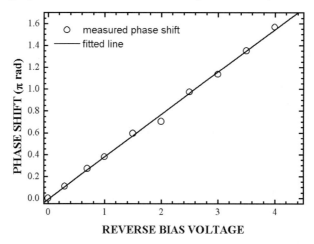

Figure 4. The measured phase shift of the device vs. the applied voltage.

References

[1] Dagli N, 1999 IEEE Tran. Microwave Theory Tech. 47 1151
[2] Byun Y T Park K H Kim S H Choi S S, Yi J C and Lim T K 1998 Appl. Opt. 37 497
[3] Koren U Koch T L Presting H and Miller B I 1987 Appl. Phys. Lett., 50, 368
[4] Tsang H K Soole J B D LeBlanc H P Bhat R Koza M A and White I H 1990 Appl. Phys. Lett., 57, 2285
[5] Saitoh M Takenaka M.Byongjin M and Nakano Y 2001 IEICE Trans. Electron. E84-C 1975

Continuous wave operation of far-infrared quantum cascade lasers

L.Ajili, G. Scalari, H. Willenberg, D. Hofstetter, M. Beck, J. Faist

Institute of Physics, University of Neuchâtel, CH-2000 Neuchâtel, Switzerland

H. Beere, G. Davies, E. Linfield, D. Ritchie

Cavendish Laboratory, University of Cambridge, Madingley Road, Cambridge CB3 0HE, UK

Abstract. Continuous Wave operation of Quantum Cascade Lasers (QCL) in the far-infrared emission is demonstrated for the first time. The devices designed using a three-quantum-well chirped-superlattice active region in a waveguide based on a single interface plasmon and a buried contact, with high reflectivity backfacet coating, results in continuous wave operation with optical output power of 250 mW at 10 Kelvin, at an emission wavelength of 66 µm. The lasers work continuously up to 30 Kelvin.

1. Introduction

Quantum cascade lasers (QCL)[1], are unipolar semiconductor laser nowadays showing very high performances in the mid-infrared range, such as self modelocking, continuous wave operation at room-temperature [2]. Powerful light sources in the far-infrared (FIR) are also highly desirable because of their potential use in chemical spectroscopy or astronomy and the lack of convenient emitters in this wavelength range. Intersubband electroluminescence at 88 µm has been shown with QC structures in 1998, but it is only recently that lasing action working in pulsed mode was demonstrated [3][4]. This milestone could be achieved by a careful re-design of the optical cavity. Indeed previously used double plasmon waveguides had very large overlap factor value (0.95), but also for the optical losses (about 50 cm^{-1}). In the recent approach, the light is reflected on the top side by a metallic layer, and only slightly confined by a heavily doped buried contact and leaks out in the semi-insulating GaAs substrate, on which the whole structure is grown. The overlap factor is consequently strongly decreased (0.22), but since a large portion of the light travels trough an undoped semiconductor, the optical losses are tremendously reduced, down to an estimated value of 2.7 cm^{-1}.

In this letter, we report on continuous wave operation of QCL in the therahertz frequency range (4.6THz or 66 µm).

2. Chirped superlattice structure and sample processing.

The structure consist of 120 periods (Fig1a) grown by molecular beam epitaxy (MBE) on a semi-insulating GaAs substrate. Each period is composed by three optically active GaAs/Al$_{0.15}$Ga$_{0.85}$As quantum wells followed by a four quantum well relaxation/injection region as already described in the references [4]. The waveguide is based on a single interface plasmon.

For the device fabrication, as schematised in the Fig1b the first step is the etching of the active region, down to the buried contact layer, creating 200 µm wide ridges. Two bottom contacts (Ge/Au/Ag/Au 12/27/50/400 nm alloyed at 400 °C during 1 minute) are then evaporated on both sides of the stripes. The top contact consists of two 10 µm wide contacts (Ge/Au 6/13 nm alloyed at 320°C during 1 minute) deposited along the edges of the unetched semiconductor. A Ti/Au layer is furthermore evaporated and completely covers the top of the ridge. Thinning (250 µm), and backside metallization (Ti/Au) end the processing of the devices. They are then cleaved and soldered on a copper mount with indium. One of the facet is coated with 110/60 nm ZnSe/Au to decrease the mirror losses.

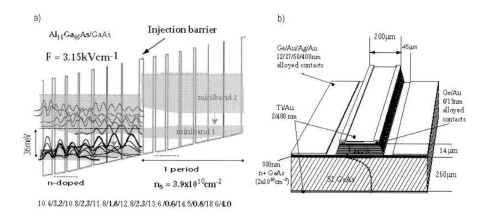

Figure 1: a) Self consistent computation of the energy conduction band of two periods of the chirped supperlattice (the tickness of different layers is indicated in nanometers). b) Schematic of the sample processing. The dashed line is the optical mode intensity in the transverse direction.

3. Low threshold current density.

A 2.8 mm long and 200 µm wide device was processed in the way described above and could be operated up to 60 K in pulsed mode with a maximum optical power of 15 mW and a threshold current density $J_{th} = 137$ A/cm^2 at 10 K temperature (Fig2a). This sample could not be operated in continuous wave although the threshold current density was very low. The large device area required a relatively high threshold current (767mA). For this reason we mounted a 1.85mm long and 200µm wide device.

The light versus current (L-I) curves in the pulsed mode from this device are shown in Fig 2b for temperature 10K. Knowing the duty cycle of (0.005%), the peak power was computed from the measurement of the average power which we determined with a liquid He-cooled Si-bolometer. The slope efficiency is $dP/dI = 45$mW/A at 10 K, while the threshold current density is 180 A/cm^2 with a maximum power of about 20 mW at this temperature. The pulsed threshold current of 670 mA at 10 K seemed to be low enough to try continuous mode operation.

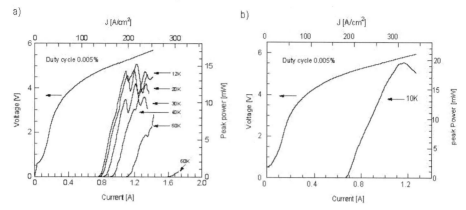

Figure 2: Peak optical output power versus injected current in pulsed mode at a duty cycle of 0.005 % at various temperatures, as indicated. The sample is processed into: a) a 2.8-mm-long and 200μm-wide waveguide b) a 1.85-mm-long and 200μm-wide waveguide.

4. Continuous wave operation.

The CW optical output power emitted from the front facet with a 22 % collection efficiency was measured with a calibrated thermopile detector mounted directly in front of the cryostat window and is shown in Fig3a. At 10K temperature, the laser exhibited a threshold current of 724 mA (corresponding to a threshold current density J_{th} = 195 A/cm^2 at a bias voltage V=5.95V) and a slope efficiency dP/dI = 877mW/A (where P is optical power and I is current). A maximum optical power of 250 mW at 10K was detected at a driving current of 1A. CW operation was observed up to 30K. The electrical transport characteristics of the device are shown in the same figure. The CW spectral properties were analyzed collecting the light by an off-axis parabolic mirror and sending it through a FTIR spectrometer operated in rapid scan mode. The emission spectra of 1.5 mm length and 200μm wide wave guide (Fig 3b) was measured at a constant heat sink temperature (10K) and at various currents (between 560 and 700mA).

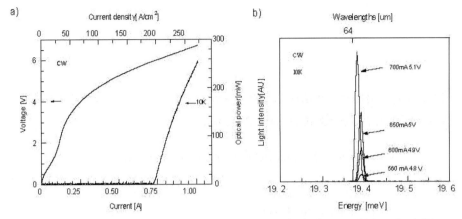

Figure 3: a) CW optical power from a single facet and bias of a 1.85 mm long and 200 μm wide laser stripe as function of drive current at 10K heat sink temperature. b) High-resolution (0.1 cm^{-1}) CW spectra as a function of injection current. The emission spectra were measured at a constant temperature of 10 K for various drive currents ranging from 560 mA up to 700 mA.

References

[1] R.Kazarinov and al, Sov. Phys.Semicond. 5, 707, (1971)
[2] M. Beck and al, Science, Continuous-wave operation of a mid-infrared semiconductor laser at room-temperature, volume 295 pages 301-305, {2002}
[3] R.Kohler et al., Nature, Terahertz semiconductor-heterostructure laser, volume 417 page 156 (2002).
[4] M. Rochat et al., Appl. Phys. Lett. **81**, 1381 (2002)

Resonant phonon-assisted depopulation in type-I and type-II intersubband laser heterostructures

M. V. Kisin[1], M. A. Stroscio[2], G. Belenky[1], and S. Luryi[1]

[1]Department of Electrical & Computer Engineering,
SUNY at Stony Brook, NY
[2]Departments of Electrical & Computer Engineering and Bioengineering,
University of Illinois at Chicago, IL

Abstract. We show that LO phonon-assisted interband tunneling in type-II intersubband laser heterostructures is more efficient for the fast depopulation of the lower lasing states than the corresponding intersubband process in type-I double quantum wells (DQW). The main peak of the electron-phonon resonance in type-II DQW corresponds to electron transitions from the lowest electron-like subband to the top of the highest heavy-hole subband, which is strongly spin-split and displaced from the center of the Brillouin zone due to the heterostructure asymmetry. Phonon-assisted depopulation can be conveniently employed even when the lower lasing level is designed near the upper edge of the heterostructure leaky window, where direct interband tunneling depopulation becomes inefficient. This design is beneficial for the laser performance providing the highest value of the matrix element for intrawell optical lasing transition and simultaneously preventing thermal backfilling of the lower lasing states.

The InAs/GaSb/AlSb material system is very promising for the implementation of high-temperature mid-infrared intersubband lasers covering the 3-5 and 8-12 μm atmospheric windows. Comparing with InGaAs/InAlAs type-I heterostructures, the higher conduction band offset at InAs/AlSb interface allows an extension of the laser operation to shorter wavelengths, while the cross-gap alignment between InAs and GaSb contributes to the better blocking of the injected electrons in the upper lasing states. The lower lasing state depopulation, which is crucial for achieving the inverse population in intersubband lasers, in type-II heterostructures can be favourably accomplished by two efficient processes: the direct interband tunneling through the InAs/GaSb "leaky window" ($0<\varepsilon<\delta$) and LO-phonon assisted interband electron transition from the lower electron-like lasing subband $c1$ into the highest hole-like subband $hh1$; see Figure 1. In type-II lasers, the direct interband tunneling has always been considered as a basic depopulation mechanism, while the interband LO-phonon assisted process is habitually treated as an inefficient one due to a symmetry difference between the initial (electron-like) and final (hole-like) states involved in the transition. In our previous work [1] we have shown that the efficiency of the direct interband tunneling depopulation, being

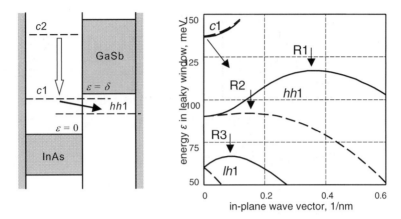

Figure 1. Left: schematic band diagram of an asymmetric InAs/GaSb DQW modeling an active region of intersubband type-II cascade laser. Black arrow depicts the interband LO-phonon assisted depopulation process. Right: subband splitting in the upper part of the leaky window δ. Short vertical arrows indicate the positions of the Van Hove singularities.

determined by a ratio $\varepsilon(\delta-\varepsilon)/E_{gInAs}E_{gGaSb}$, significantly decreases when the depopulated electron-like level $c1$ is located near the upper edge of the heterostructure leaky window $\varepsilon \sim \delta$. The latter configuration, however, is the most desirable from the standpoint of ensuring the highest oscillator strength for the lasing transition and preventing thermal backfilling of the lower lasing states.

In this paper, we analyze the process of the lower lasing state depopulation in type-II cascade laser heterostructures and show that the band mixing and the subband nonparabolicity effects inside the heterostructure leaky window essentially remove the symmetry constraint for interband phonon-assisted transition and can provide for a sufficiently strong overlap between electron and phonon states participating in the transition. Inherent asymmetry of type-II InAs/GaSb double quantum well (DQW) lifts the degeneracy of the hole-like subbands in the heterostructure leaky window and results in strong Van Hove singularities in the joint density of states. The peak value of the density of states at the top of the upper hole-like subband can be favourably combined with the maximum electron-phonon overlap, thus providing for the fast electron relaxation from the $c1$ into the $hh1$ subband. Subsequent tunneling from $hh1$ states into the InAs reservoir does not limit the depopulation rate since there is only one interface to penetrate in this tunneling process. The phonon-assisted depopulation can also be designed as a two-step process with the secondary resonant *hole* transition from deeper lying $lh1$ subband states or from an adjacent secondary GaSb QW, which is often used as an additional blocking layer for the electrons in the upper lasing subband.

Figure 1 illustrates the splitting of the hole-like subbands in the leaky window of a model isolated InAs/GaSb DQW with $d_A = 9$ nm and $d_B = 5$ nm. This "Rashba-type" splitting is inherent to any asymmetric DQW heterostructures and is especially strong in type-II heterostructures based on the narrow-gap InAs/GaSb material system [2]. Subband splitting displays the subband extremities from the Brillouin zone center and results in strong Van Hove singularities in the density of states, which can be used to engineer the phonon-assisted transitions of high efficiency. Electron states inside the

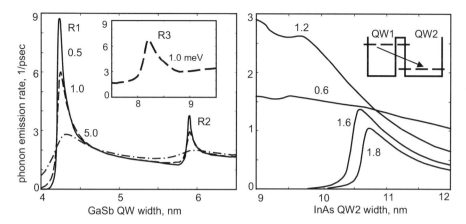

Figure 2. Comparison of the LO-phonon assisted depopulation rates in type-II InAs/GaSb DQW (left, curves are labeled with the level broadening in meV) and type-I InAs/AlSb/InAs DQW (right, curves are labeled with the width of AlSb barrier in nm, level broadening is 1 meV).

leaky window are quasibound with the width Γ determined by the interband tunneling, $\Gamma \propto \Gamma_{tun}$, which is about 1 meV in the upper part of the window [1]. Such level broadening can be accounted for in the Γ_{ph} rate calculations by including a Lorentzian lineshape function for the phonon emission transition. The phonon-assisted depopulation rate, Γ_{ph}, was calculated by assuming that final states for electron transitions are unoccupied, which is a rather good approximation for the uppermost states in the upper hole-like subband $hh1$. Since the LO-phonon energies are similar in both constituent materials, $\hbar\omega_{LO} \approx 30\text{meV}$, we can neglect the polar mode confinement and calculate the phonon emission rate within the 3D phonon approximation, that is assuming that the confined 2D electrons interact with dispersionless 3D bulk LO phonons. For a general discussion of phonon-assisted processes in nanostructures, see Ref. 3.

In Figure 2 we compare the LO-phonon spontaneous emission rates Γ_{ph} in type-II InAs/GaSb (left) and type-I InAs/AlSb/InAs (right) DQWs. For type-II DQW Γ_{ph} is shown as a function of the GaSb QW width d_B, with the InAs QW width kept constant at $d_A = 9$ nm. This value of d_A allows for the $c1$ states to be located near the upper edge of the leaky window δ. The increase of d_B in the range from 4 nm to 10 nm, while keeping the energy position of the initial electron-like subband $c1$ practically unchanged, makes it possible to scan the final states in the hole-like subbands E_f, here - the heavy-hole subband $hh1$ and the light-hole subband $lh1$, which thus move toward the upper part of the heterostructure leaky window. Type-II DQW clearly demonstrates three peaks in the LO-phonon emission rate, R1-R3, which are related to three consecutive resonances indicated in Figure 1. The first, most important resonance, R1, corresponds to the onset of spatially indirect (interwell) $c1 \rightarrow hh1$ phonon-assisted transition. R1 transition is indirect also in the momentum space since the top of the subband $hh1$ is noticeably displaced from the Brillouin zone center due to the subband spin splitting enhanced by the heavy-hole/light-hole mixing. Final momentum transfer in this resonance is important for the high phonon emission rate, because the interband transition can be engineered so that the peak of the density of states at the top of $hh1$ subband is complemented by the maximum value of the electron-phonon overlap integral $I(q)$; see Figure 3. This design explains the higher value of Γ_{ph} in R1 resonance comparing to the next resonance R2,

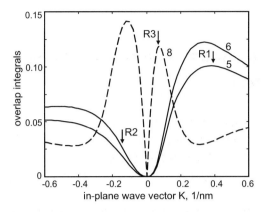

Figure 3. Electron-phonon overlap integrals $I(q)$ for $c1 \to hh1$ (solid lines) and $c1 \to lh1$ (dashed line) transitions vs. transferred (phonon) wave vector. Each curve is labeled with GaSb QW width d_B in nm and with a vertical arrow marking the position of the Van Hove singularity.

which is less pronounced due to the smaller electron-phonon overlap. With d_B increasing, R1 resonance becomes less efficient, firstly, because of the corresponding decrease of the final density of states for phonon-assisted transitions away from the top of the $hh1$ subband, and secondly, due to the suppression of the electron-phonon overlap integral $I(q)$ both at small and large momentum transfers. Finally, the increase of the phonon emission rate in the region R3 (see inset) corresponds to the onset of interwell electron transitions to the top states of $lh1$ subband, which are also displaced from the Brillouin zone center; see Figure 1. For type-I DQW (see Figure 2, right) the depopulation rate is shown as a function of the second InAs QW width d_{A2}, with the first InAs QW width d_{A1} = 9 nm. The decrease of d_{A2} from 12 nm to 9 nm reduces the separation between the initial and the final states for the phonon-assisted electron transitions and increases the rate of the spontaneous phonon emission. The cutoff of the phonon emission, which is typical for type-I DQW with sufficiently wide barrier layer d_b, corresponds to the subband separation below the optical phonon energy. For narrow barrier, d_b = 1.2 nm, the anticrossing gap dominates the separation and makes the emission possible in the whole range of d_{A2} values. Note, however, the overall remarkable decrease of the emission rate for the smaller barrier width, d_b = 0.6 nm, which is due to the increased subband separation under the anticrossing. It is readily seen also that the overall scale of the depopulation rate is in favor of the interband vs. intersubband depopulation.

In conclusion, we show that in type-II intersubband laser heterostructures the interband LO-phonon-assisted scattering can be used as an efficient complementary process for the fast depopulation of the lower lasing states.

This work was supported by ARO grant DAAD190010423 and AFOSR MURI grant F496200010331.

References

[1] Kisin M V, Stroscio M A, Luryi S and Belenky G 2001 Physica E **10** 576-586.
[2] Kisin M V, Dutta M and Stroscio M A 2002 in Advanced Semiconductor Heterostructures (World Scientific, Singapore, in press).
[3] Stroscio M A and Dutta M 2001 Phonons in Nanostructures (Cambridge University Press, Cambridge).

Mid-infrared quantum cascade lasers operation above room temperature

Q. K. Yang, C. Mann, F. Fuchs, R. Kiefer, K. Köhler, and H. Schneider

Fraunhofer Institute for Applied Solid State Physics (IAF), Tullastrasse 72,
D-79108 Freiburg, Germany
Phone: 0049-761-5159464, Fax: 0049-761-5159400, Email: yang@iaf.fhg.de

Abstract. Above room-temperature operation of $\lambda \sim 5$ µm quantum cascade lasers with three different active regions is reported. Comparison between the three different designs, namely the conventional design, the adoption of blocking barriers into the active region, and the adoption of strain-compensated active regions, is presented.

1. Introduction

Quantum cascade (QC) lasers have made tremendous progress in the past few years since their first demonstration in 1994 [1]. The wide wavelength coverage (3.4 –24 µm) of QC lasers has made this novel intersubband-transition based semiconductor laser in particular suitable for chemical sensing as well as for free-space data transmission. In this contribution we report on the above room-temperature operation of $\lambda \sim 5$ µm quantum cascade lasers.

2. Experiments

The conduction-band profiles (Γ valley) of two active regions connected by injectors are shown in Fig. 1 for three different designs. The laser transitions are indicated by the wavy arrows. The manifold of wave functions (shaded area) in the injector (Bragg reflector) constitute the "miniband" which allows electrons to tunnel through the injector into the next active region. In the conventional design (Fig. 1 (a), as published in Ref. [2]), the active regions are composed of lattice-matched $Al_{0.48}In_{0.52}As/Ga_{0.47}In_{0.53}As$ layers, and electrons are confined in the upper state of the laser transition by the injection regions serving as Bragg reflectors. The electrical confinement can be enhanced by the adoption of AlAs-blocking barriers [3], as shown in Fig. 1(b). The third structure (Fig. 1(c)) employs strain-compensated active regions using $Al_{0.6}In_{0.4}As$ as barrier and $Ga_{0.38}In_{0.62}As$ as quantum wells [4]. In this structure, the higher conduction-band offset ($\Delta E_c \sim 700$ meV) allows better electrical confinement compared to the lattice-matched active region ($\Delta E_c \sim 510$ meV). A one period layer sequence of the active region plus injector, in nanometers, from left to right starting from the injection barrier (left-most layer) is: 5.0, 0.9, 1.5, 4.7, 2.2, 4.0, 3.0, 2.3, 2.3, 2.2, 2.2, 2.0, 2.0, **2.0, 2.3, 1.9**, 2.8, and 1.9 for structure (a); 5.0, 1.0, 1.5, 2.0, 0.7 (InAs), 2.0, 2.2, 4.1, 0.9, 0.7 (AlAs), 0.9, 2.5, 2.3, 2.3, 2.2, 2.0, 2.0 **2.0, 2.3, 1.9**, 2.8, and 1.9 for structure (b); and 4.4, 0.9, 0.9, 4.8, 1.7, 4.4, 2.8, 3.1, 1.2, 2.9, 1.4, 2.8, 1.6, 2.7, 2.0, **2.5, 2.3, 2.3, 2.7**, 2.1, 3.0, and 1.9, for structure (c). The layers in bold face are Si doped to $n=2\times10^{17}$ cm^{-3} for structures (a) and (b), and $n=3\times10^{17}$ cm^{-3} for structure (c).

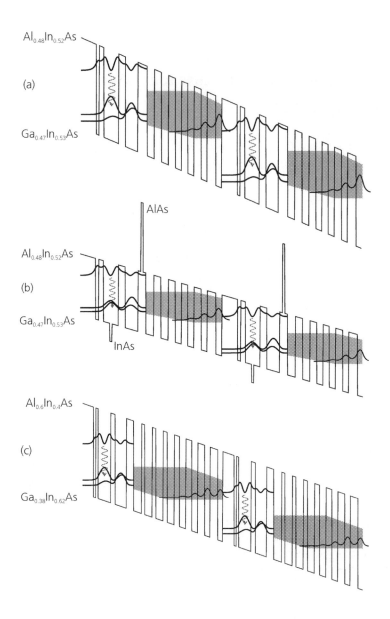

Fig. 1. Schematic conduction-band profiles (Γ valley) of the InP-based $\lambda \sim$ 5 μm quantum cascade lasers. (a) Conventional design with active regions composed of lattice-matched $Ga_{0.47}In_{0.53}As/Al_{0.48}In_{0.52}As$. (b) With AlAs blocking barriers in the lattice-matched $Ga_{0.47}In_{0.53}As/Al_{0.48}In_{0.52}As$ active regions. (c) With strain-compensated $Ga_{0.38}In_{0.62}As/Al_{0.6}In_{0.4}As$ active regions.

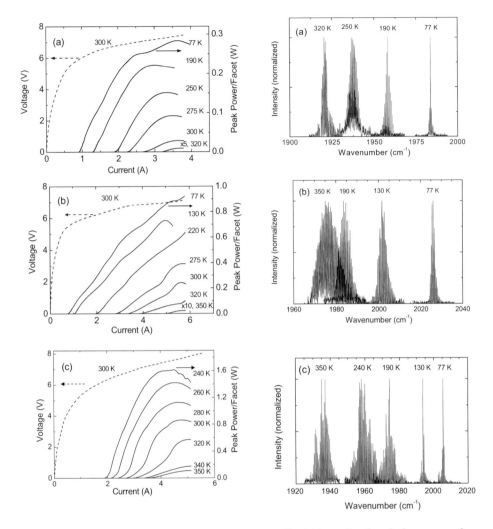

Fig. 2 L-I curves at various heat sink temperatures and V-I dependence at 300 K. (a), (b), and (c) have the same meaning as that in Fig. 1.

Fig. 3 Normalized emission spectra for the QC lasers with different designs. (a), (b), and (c) have the same meaning as that in Fig. 1.

Twenty-five periods of alternating active regions and injectors, surrounded by GaInAs separate confinement layers, were grown by molecular beam epitaxy (MBE) on S-doped $n=2\times10^{17}$ cm^{-3} (001)-oriented InP substrates. The wafers were then transferred into a metalorganic chemical vapor deposition (MOCVD) system to grow Si-doped InP layers serving as the upper waveguide cladding and contact layers. After growth the wafers were processed into 16 μm wide ridge waveguide structures by etching down to the active region. Then a 350 nm thick Si_3N_4 passivation layer was deposited, and windows were opened for top contact formation. Finally the lasers were cleaved into 3 mm long bars leaving the facets uncoated. After mounting the samples substrate-side down onto copper heatsinks and wire bonding, the lasers were driven by current pulses of 100 ns pulse width at a repetition rate of 5 kHz. The optical power emitted into a solid angle of about $\pi/10$ was detected with an InSb detector. The emission spectra were analyzed by a Fourier transform spectrometer.

3. Results and Discussion

The typical light output versus injection current (L-I) dependence at various heat-sink temperatures as well as the voltage-current (V-I) dependence at 300 K are shown in Fig. 2 (a)-(c) for the three different designs presented in Fig. 1. The QC lasers based on the design shown in Fig. 1(a) operate in pulsed mode up to 320 K, with the maximum peak power per facet of 0.29 W at 77 K, and 2 mW at 320 K. Both the maximum operation temperature and the output power are improved by the adoption of the blocking barriers (Fig. 1(b)). These devices operate up to 350 K in pulsed mode. The maximum peak power per facet is 0.9 W at 77 K, 0.1 W at 320 K, and 2 mW at 350 K. With the adoption of the strain-compensated active region as shown in Fig. 1(c), the performance of the QC lasers is further improved. The output power is 0.86 W at 300 K. Even at 350 K, which is the maximum temperature achievable with our cryostat, we achieve an output power of 0.11 W for this laser, and higher operation temperature is therefore expected.

The normalized emission spectra for these QC lasers are shown in Fig. 3 (a)-(c). The peak wavelengths at 77 K are 5.05 µm, 4.94 µm, and 4.98 µm, respectively. The shift of the wavelength between (a) and (b) is attributed to the deeper well caused by the introduction of the InAs layer into the active region.

4. Summary

In summary, we reported above-room-temperature operation of $\lambda \sim 5$ µm quantum cascade lasers based on three different designs.

The authors are grateful to J. Schaub and N. Rollbühler for material growth; J. Schleife, K. Schwarz, and R. Moritz for technical support; J. Wagner and G. Weimann for encouragement and continuous support.

References:

[1] J. Faist, F. Capasso, D. L. Sivco, C. Sirtori, A. L. Hutchinson, and A. Y. Cho, Science **264**, 553 (1994).
[2] J. Faist, F. Capasso, C. Sirtori, D. L. Sivco, J. N. Baillargeon, A. L. Hutchinson, and A. Y. Cho, Appl. Phys. Lett. **68**, 3680 (1996).
[3] Q. K. Yang, C. Mann, F. Fuchs, R. Kiefer, K. Köhler, N. Rollbühler, H. Schneider, and J. Wagner, Appl. Phys. Lett. **80**, 2048 (2002).
[4] R. Köhler, C. Gmachl, A. Tredicucci, F. Capasso, D. L. Sivco, S. N. G. Chu, and A. Y. Cho, Appl. Phys. Lett. **76**, 1092 (2000).

Scattering transport and electron temperature evaluation in a Terahertz GaAs/AlGaAs quantum cascade laser

D Indjin‡, P Harrsion, R W Kelsall and Z Ikonić

Institute of Microwaves and Photonics, School of Electronic and Electrical Engineering, University of Leeds, Leeds LS2 9JT, United Kingdom

Abstract. In this work, electron transport in a terahertz GaAs/AlGaAs quantum cascade laser is calculated self–consistently using an intersubband scattering model. Subband populations and carrier transition rates are calculated and all relevant electron–electron and electron-LO phonon scatterings between the injector/collector and active region levels (13–levels model) are included. Employing an energy balance equation which includes the influence of both electron-LO phonon and electron-electron scattering, the method also enables evaluation of the average electron temperature of the non–equilibrium carrier distributions in the device. The output characteristics, in particular the threshold currents and electric field–current density characteristics, are in good qualitative and quantitative agreement with experiment [Köhler et al, Nature, **417**, 156 (2002)].

1. Introduction

The quantum cascade laser (QCL) [1], has demonstrated an impressive extension of the infrared frequency range [2–4] and until recently could be operated at wavelengths as long as 24μm [5, 6]. This has stimulated a number of experimental [7, 8] and theoretical [9] studies of QCL structures designed for emission at THz frequencies well below the forbidden phonon Reststrahlenband. Finally, considerable research effort has resulted in theoretical prediction of population inversion by Monte–Carlo studies [10], followed by luminescence measurements [11] and the laser action [12] at $\lambda \sim 69\mu$m (4.4 THz) in a GaAs/AlGaAs QCL. Most recently, lasing at $\lambda \sim 66\mu$m has been demonstrated in similar structures with very low waveguide losses [13].

The charge transport in QCLs is due mainly to incoherent–scattering mechanisms [14], and in any realistic model of carrier transport, all principal electron-LO phonon and electron-electron scattering mechanisms have to be included [15]. The analysis of the operating characteristics of Terahertz QCLs is more complex, i.e. computationally demanding, than for mid–infrared devices. This is because the energy separation between most of subbands is smaller than the LO phonon energy and electron-electron scattering becomes the dominant scattering mechanism [7], hence necessitating a large number of possibly relevant scattering processes to be accounted for. Additionally, the electron temperature is expected to exceed significantly the lattice temperature.

2. Method

The electronic structure of a Terahertz QCL [12] is illustrated in Fig. 1a. Radiative transitions

‡ E-mail: d.indjin@ee.leeds.ac.uk

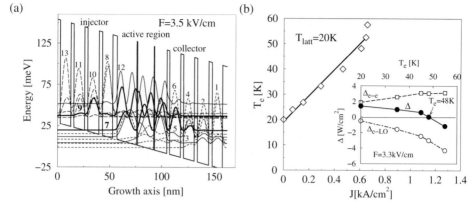

Figure 1. (a) A schematic diagram of quasi–bound energy levels and associated wave functions squared for $1\frac{1}{2}$ periods of a GaAs/Al$_{0.15}$Ga$_{0.85}$As Terahertz QCL: injector (levels 8, 10, 11 and 13)–active region (3, 5, 7, 9 and 12)–collector (1, 2, 4 and 6). The layer sequence of one period of the structure, in nanometers, from left to right starting from the injection barrier is **4.3**, 18.8, **0.8**, 15.8, **0.6**, 11.7, **2.5**, 10.3, **2.9**, 10.2, **3.0**, 10.8, **3.3**, 9.9, (Ref. [12]). The normal script denotes the wells, bold script the barriers and underscore the doped region, with a nominal donor sheet density $N_s = 4.08 \times 10^{10}$ cm^{-2} per period. (b) Calculated electron temperature vs current density in the terahertz QCL at lattice temperature $T_{\text{latt}} = 20$ K. The symbols are the calculated results and the line represents the least square fit used to derive the values of coupling constant β. An example of electron temperature evaluation for specific value of electric field ($F = 3.3$ kV/cm) is shown in inset (see Eq. (2)).

occur between the fourth and third state in the active region, denoted as 9 and 7 respectively. Simulating electron flow in a cascade requires one to have scattering rates for all intra– and inter–period processes, and an assumption is made that each period of the cascade behaves in the same manner (periodic boundary conditions). Calculation of all these scattering rates formally necessitates for energy states/wavefunctions in two full periods, but in view of negligible active–to–active region coupling, $1\frac{1}{2}$ period (Fig. 1a) suffices. The rate equations were written so that carriers were cycled around the 13–level system, i.e. electrons scattering outside the considered period reappear in equivalent states inside. All relevant electron-electron processes (∼ 400) are included: the accuracy of this approximation has been confirmed in an one-off calculation in which every electron-electron process (∼ 8000) was considered. The steady state subband populations of the i-th active region level, and ground collector level 1 (similarly for other collector/injector states) are written as

$$n_i = \left[\sum_{j=1, j\neq i}^{13} \frac{n_j}{\tau_{ji}}\right]\left[\sum_{j=1, j\neq i}^{13} \frac{1}{\tau_{ij}}\right]^{-1}; \quad n_1 = \left[\sum_k \frac{n_k}{\tau_{k1}} + \sum_s \frac{n_s}{\tau_{s8}}\right]\left[\sum_k \frac{1}{\tau_{1k}} + \sum_s \frac{1}{\tau_{8s}}\right]^{-1} \quad (1)$$

where $i = 3, 5, 7, 9$, and 12; $k = 2, \ldots, 7, 9, 12$ and $s = 3, 5, 7, 9, 12$. The scattering time τ_{if} is a function of both n_i and n_f, the initial and final subband populations. Hence, this set of equations has to be solved self–consistently using an iterative procedure. [9, 16]

At equilibrium, the rate at which the electron distributions gain kinetic energy (relative to the particular subband minimum) through scattering, will balance with the rate at which they lose kinetic energy to the lattice. Despite the fact that electron–electron scattering is elastic as far as total energy is concerned, intersubband electron–electron transitions do convert potential energy into kinetic energy (or vice-versa). From the viewpoint of this work this would lead to an increase (decrease) in the total kinetic energy of a subband population, because the potential energy as defined here includes the quantised component of the kinetic energy. Hence, the kinetic energy balance condition can be written as [17]:

$$\Delta = \sum_{\text{em.,abs.,e-e}} \sum_f \sum_i \frac{n_i}{\tau_{if}}(E_i - E_f + \delta E) = \Delta_{\text{e-LO}} + \Delta_{\text{e-e}} = 0 \qquad (2)$$

where $E_i - E_f$ is the subband separation, and the change in energy δE is equal to $-E_{\text{LO}}$ for phonon emission (em.), $+E_{\text{LO}}$ for phonon absorption (abs.) and zero for electron–electron (e-e) scattering. Hence, $\Delta_{\text{e-LO}}$ is net electron-LO phonon, and $\Delta_{\text{e-e}}$ is net electron–electron contribution in the balance equation. The next step of the procedure, is to vary the electron temperature (assumed to be the same for all subbands) until the kinetic energy balance equation Eq. (2) is satisfied self-consistently. From the self–consistent solution, the population inversion $\Delta n = n_9 - n_7$ in the steady–state condition is obtained and the modal gain can be calculated [16]. To extract the output characteristics of QCLs, one has to change the electric field F (i.e. the applied voltage) and calculate the modal gain G_M and the total current density J for each value of the field. The threshold current density J_{th} is found according to $G_M(J_{th}) = \alpha_M + \alpha_W$, where α_M and α_W are the mirror and waveguide losses, respectively.

3. Results and discussion

Calculated values of the average electron temperature are shown in Fig. 1b. T_e increases from the lattice temperature, under very low injection, up to ~ 60 K when the negative differential resistance (NDR) occurs. The electron temperatures are relatively more sensitive to the current density than in mid–infrared devices [17]. This is due to the fact that electron–electron scattering is by far the fastest scattering mechanism in Terahertz QCLs, with their relatively small spacing between states. Therefore, the inelastic electron–LO phonon scatterings, which couples electron distributions to the lattice heat bath, does not manage to cool the electrons as efficiently as in mid–infrared devices. The calculations also show that, up to the NDR feature, T_e can be approximated as linear function $T_e \approx T_{\text{latt}} + \beta J$, where $\beta \sim 52$ K/(kAcm^{-2}).

Figure 2a (right axis) shows the calculated electric field–current density characteristics at lattice temperature of 20K for the Terahertz QCL [12]. The current is predicted to saturate at ~ 680 A/cm^2 in reasonable agreement with that measured at ~ 820 A/cm^2 [11, 12]. The discrepancy of about 20 % is probably related to the fact that we have restricted consideration to a 13–level system (in order to reduce the number of scattering processes which could be extremely large if additional levels were included), i.e. the parasitic leakage currents from the injector to higher states are neglected [18]. In Fig. 2a (left axis) the modal gain vs. current density dependence at lattice temperatures of 5 K, 20 K and 40 K is calculated with $\lambda = 70 \mu$m, $\underline{n} = 3.28$, $\Gamma = 0.47$, $L_p = 104.9$ nm, and $2\gamma_{97} = 2 - 2.2$ meV [11]. In accordance with the experimentally obtained losses [12] from the intersection points of the total loss line $\alpha_M + \alpha_W \approx 20$ cm^{-1} and the $G_M(J)$ lines, we obtain threshold currents $J_{th} = 250 - 300$ A/cm^2, in good agreement with experiment [12]. The optical power start to decrease quite abruptly at about ~ 650 A/cm^2, also in accordance with experimental findings, i.e. close to the saturation of the injected (non–parasitic) current. The model predicted that the laser is able to run up to ~ 50 K which is close to the experimental maximum of ~ 55 K [12]. The population inversion $\Delta n = n_9 - n_7$, and subband populations of active laser levels as a function of lattice temperature are shown in Fig. 2b. The inversion decreases with temperature not only due to the thermal filling of the lower state n_7, but also as a result of increased nonradiative electron-LO scattering from the upper into the lower laser state, hence decreasing of n_9.

The authors would like to thank A. Tredicucci and R. Köhler, from NEST-INFM and Scuola Normale Superiore in Pisa, Italy, for useful discussions and communicating their

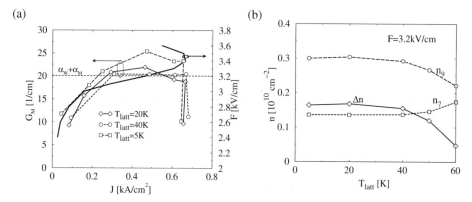

Figure 2. (a) Calculated modal gain vs. current density dependence at cryogenic lattice temperatures. The horizontal dashed line denotes the total losses ($\alpha_M + \alpha_W \approx 20$ cm^{-1}) [12](left axis). Electric field vs. current density characteristics at lattice temperature $T_{\text{latt}} = 20$ K (right axis). (b) The population inversion $\Delta n = n_9 - n_7$, and the subband populations of the upper (n_9) and the lower (n_7) laser levels as function of lattice temperature.

results before publication. This work is supported by EPSRC grant No. GR/R04485.

References

[1] J. Faist, F. Capasso, D. L. Sivco, C. Sirtori, A. L.Hutchison, and A. Y. Cho, *Science*, **264**, 553-556 (1994).
[2] C. Gmachl, D. L. Sivco, R. Colombelli, F. Capasso, and A. Y. Cho, *Nature*, **415**, 883-887 (2002).
[3] C. Sirtori, H. Page, C. Becker, and V. Ortiz, *IEEE. J. Quantum Electron.*, **38**, 547-558 (2002).
[4] J. Faist, D. Hofstetter, M. Beck, T. Aellen, M. Rochat, and S. Blaser, *IEEE. J. Quantum Electron.*, **38**, 533-546 (2002),
[5] J. Ulrich, J. Kreuter, W. Schrenk, G. Strasser, and K. Unterrainer, *Appl. Phys. Lett*, **80**, 3691-3693 (2002). *Appl. Phys. Lett.*, **80**, 3060-3062 (2002).
[6] R. Colombelli, F. Capasso, C. Gmachl, A. L. Hutchinson, D. L. Sivco, A. Tredicucci, M. C. Wanke, A. M. Sergent, and A. Y. Cho, *Appl. Phys. Lett.*, **78**, 2120-2622 (2001).
[7] M. Rochat, J. Faist, M. Beck, U. Oesterle, and M. Ilegems, *Appl. Phys. Lett.*, **73**, 3724-3726 (1998).
[8] R. Colombelli, A. Straub, F. Capasso, C. Gmachl, M. I. Blakey, A. M. Sergent, S. N. G. Chu, K. W. West, and L. N. Pfeiffer, *J. Appl. Phys.*, **91**, 3526-3529 (2002).
[9] P. Harrison, R. A. Soref, *IEEE J Quantum Electron.*, **37**, 153-158 (2001).
[10] R. Köhler, R. C. Iotti, A. Tredicucci, F. Rossi, *Appl. Phys. Lett*, **79**, 3920-3922 (2001).
[11] R. Köhler, A. Tredicucci, F. Beltram, H. E. Beere, E. H. Linfield, A. G. Davies, and D. A. Ritchie, *Appl. Phys. Lett*, **80**, 1867-1869 (2002).
[12] R. Köhler, A. Tredicucci, F. Beltram, H. E. Beere, E. H. Linfield, A. G. Davies, D. A. Ritchie, R. C. Iotti, and F. Rossi, *Nature*, **417**, 156-159 (2002).
[13] M. Rochat, L. Ajili, H. Willenberg, J. Faist, H. E. Beere, A. G. Davies, E. H. Linfield, and D. A. Ritchie, *Appl. Phys. Lett.*, **81**, 1381-1383, (2002).
[14] R. C. Iotti and F. Rossi, *Phys. Rev. Lett.*, **87**, Art. No. 146603-1–146603-4 (2001).
[15] P. Harrison, *Quantum Wells, Wires and Dots: Theoretical and Computational Physics*, (Wiley, Chichester, 1999).
[16] D. Indjin, P. Harrison, R. W. Kelsall, and Z. Ikonić, *J. Appl. Phys.*, **91**, 9019-9026 (2002).
[17] P. Harrison, D. Indjin, and R. W. Kelsall, *J. Appl. Phys.*, in press.
[18] D. Indjin, P. Harrison, R. W. Kelsall, and Z. Ikonić, *Appl. Phys. Lett.*, **81**, 400-402 (2002).

Graded interface 9.3 μm quantum cascade lasers

Thierry Aellen, Mattias Beck, Daniel Hofstetter, Jérôme Faist
Institute of Physics, University of Neuchâtel, 2000 Neuchâtel, Switzerland

Ursula Oesterle, Marc Ilegems
IMO, Physics Department, EPFL, Ecublens, 1015 Lausanne, Switzerland

Emilio Gini, Hans Melchior
Institute of Quantum Electronics, ETHZ, 8093 Zürich, Switzerland

Abstract. Pulsed operation of two different designs with abrupt and graded interfaces in the active region are compared. Junction up mounted devices are tested in pulsed mode at room temperature for both designs. Lasers based on the graded interface design exhibited a lower threshold current density but also a lower optical output power due to a higher operating voltage compared to the abrupt interface design. Junction down mounted graded interface devices which were fabricated as buried heterostructure lasers with high-reflection coatings on both facets could be operated in continuous wave up to a temperature of 268°K with an optical power of 0.5 mW.

1. Introduction

Mid–infrared quantum cascade (QC) lasers have undergone a remarkable development during the past 8 years and have recently demonstrated continuous wave (CW) operation at room temperature [1]. In the work presented here, QC lasers based on digital graded interfaces are compared to a structure with the identical 4 quantum wells (QW) active region but abrupt interfaces. The goal of the digital alloying is to reduce the interface roughness in the active region.

2. Experiment

The devices were grown by molecular beam epitaxy on InP substrates using lattice-matched InAlAs/InGaAs active layers and an InP top cladding. Except for the interfaces, both types of devices are identical and based on a 4 QW active region designed for a vertical lasing transition at 9.3 μm with a double phonon resonance for efficient electron extraction from the lower laser level [3] as shown in the Fig. 1. The graded interfaces were obtained by digital alloying of 1 and 2 Å thick InAlAs and InGaAs layers.

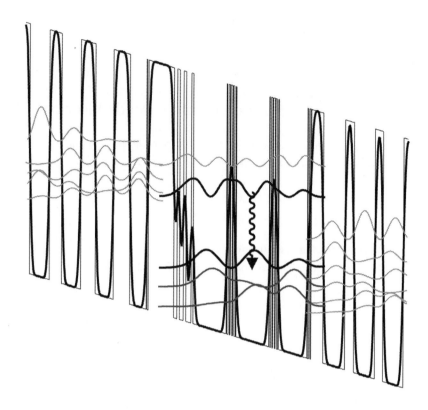

Figure 1. Schematic view of the conduction band of one period of the active region with the relevant, moduli squared wavefunctions. The wavy arrow corresponds to the vertical lasing transition. The layer sequence starting from the first quantum well from the left is as follows: 31/**19**/30/**23**/29/**25**/26/**40**/5/**4**/6/**4**/8/**3**/52/**3**/51/**3**/44/**19**/34/**14**/33/**13**/32/**15** Å InGaAs wells are in roman, and InAlAs barriers in bold. Doped layers (Si, 3×10^{17} cm^{-3}) are underlined. The barrier of the active zone are digitally alloyed with 1 and 2 Å thick layers over a 6 Å width. The potential calculated by taking into account of an interface diffusion length of 3 Å is plotted in bold.

3. Measurement results

Lasers based on the graded interface design exhibit lower threshold current densities over the whole temperature range between -30°C and 60°C, with for example 3 kA/cm^2 instead of 3.5 kA/cm^2 at room temperature for a 3 mm long and 28 µm wide laser. On the other hand, the room temperature optical peak output power was typically around 200 mW for the graded design and more than 1W for the abrupt interface device due to a better operating voltage (Fig. 2).

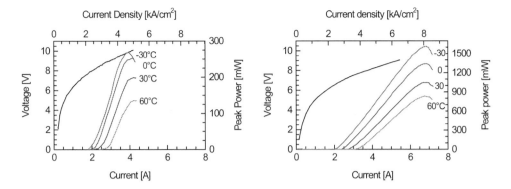

Figure 2. Applied bias and measured peak output power from a single facet as a function of injected current for various heat sink temperatures at 1.5% duty cycle. The devices were junction-up mounted, 3 mm-long, 28 μm-wide and based on graded (left) and abrupt (right) interfaces in the active region.

Narrower luminescence spectra at both cryogenic (12 meV FWHM instead of 16 meV at 4 K) and room temperature (19 meV FWHM instead of 23 meV at 300 K) were observed for the graded interface device as well (Fig. 3). This effect is due to the changing interface roughness [4] but depends also on the injection barrier thickness wich is larger for the graded interface design [5].

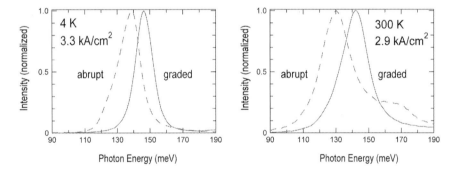

Figure 3. Comparison of the intersubband electroluminescence spectra from samples based on structures with graded and abrupt interfaces at 4 K (left) and 300 K (right). The figures exhibit a narrower peak for the digital alloyed interfaces due to the changing of the interface roughness but also to a different injection barrier design.

A buried heterostructure geometry in which the multi-quantum well active region is vertically and laterally surrounded by InP [2] was applied to the structure based on the graded interface design. Lasers were fabricated into 12 μm wide buried stripes by wet etching and selective metalorganic vapor-phase epitaxy regrowth (MOVPE) of InP. In order to further reduce the produced heat in such lasers, the devices were cleaved into 0.75 mm long lasers, soldered junction down onto a diamond platelet, and finally facet-coated by a ZnSe/PbTe high reflectivity (R=0.7) layer pair on both facets. The CW output power emitted from one facet was collected by f/0.8 optics and measured with a calibrated thermopile detector (Fig. 4). At –30°C, the laser exhibited a threshold current of 365 mA (corresponding to a current density of 4 kA/cm^2) and a slope efficiency of 25 mW/A. This device emitted 3.4 mW of optical power from one facet and CW operation was observed up to –5°C. At this temperature, the threshold current increased to 495 mA (5.5 kA/cm^2). Due to the high operating voltage, these lasers couldn't be operated at temperatures above –5°C in CW mode.

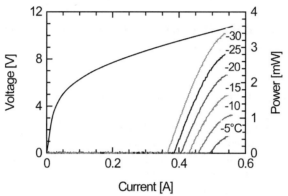

Figure 4. CW optical power from a single facet of a buried stripe graded interface QCL as a function of drive current for various heat sink temperatures. The device was junction-down mounted, 0.75 mm-long and 12 μm-wide.

4. Conclusion

We have presented a novel type of active region for QC lasers based on digitally alloyed interfaces which exhibit a reduced threshold current density and narrower luminescence linewidth in pulsed mode than devices based on abrupt interfaces. Lasers based on this novel type of active region could be operated up to –5°C in CW mode limited by the operating voltage.

The authors gratefully acknowledge Martin Ebnöther for technical assistance with the lateral InP regrowth. This work was financially supported by the Swiss National Science Foundation.

References

[1] M. Beck, D. Hofstetter, T. Aellen, J. Faist, U. Oesterle, M. Ilegems, E. Gini, H. Melchior, Science **295**, 301 (2002)
[2] M. Tacke, Infrared Physics. Technol., **36**, 447 (1995)
[3] D. Hofstetter, M. Beck, T. Aellen and J. Faist, Appl. Phys. Lett., **78**, 396 (2001)
[4] K. L. Campman, H. Schmidt, A. Imamoglu, and A. C. Gossard, Appl.Phys. Lett., **69**, 2554 (1996)
[5] S. Barbieri, C. Sirtori, H. Page, M. Stellmacher, J. Nagle, Appl.Phys. Lett., **78**, 282 (2001)

Lasing properties of GaAs/(Al,Ga)As quantum cascade lasers as a function of injector doping density

M Giehler, R Hey, H Kostial, T Ohtsuka, L Schrottke, and H T Grahn

Paul-Drude-Institut für Festkörperelektronik, Hausvogteiplatz 5–7, 10117 Berlin, Germany

Abstract. The lasing properties of GaAs/Al$_{0.33}$Ga$_{0.67}$As quantum cascade lasers (QCL's) are investigated as a function of the injector doping concentration n_s between 2×10^{11} and 1×10^{12} cm^{-2} per period. A minimal threshold current density j_{th} and maximum operating temperature T_{max} are observed for an intermediate doping level of about 6×10^{11} cm^{-2}, while QCL's with lower and higher doping levels exhibit a much higher value of j_{th} and lower value of T_{max}. The lasing energy E_0 of the QCL with the best value of j_{th} and T_{max} is blue-shifted to about 135 meV with respect to E_0 for the QCL's with the smallest and largest values of n_s, which emit at 115 and 120 meV, respectively.

1. Introduction

The development of quantum cascade lasers (QCL's) as light emitters for the mid-infrared region has very recently achieved a number of major breakthroughs such as cw-operation at room temperature [1], ultra-broadband emission [2], and lasing in the THz region [3]. While early studies of QCL's focused on the design of the optically active region in order to maximize the gain, later investigations varied the design of the injector region in order to optimise the electron transfer from the injector reservoir into the upper laser level as well as the electron transfer out of the active region into the injector.

We perform a detailed study of the influence of the doping density n_s in the injector region of GaAs/Al$_{0.33}$Ga$_{0.67}$As QCL's on their lasing properties such as the energy of the laser line E_0, the threshold current density j_{th} at low temperatures, and its temperature dependence $j_{th}(T)$. Furthermore, we determine the maximum operating temperature T_{max} and the temperature parameter T^* derived by fitting the measured quantity $j_{th}(T)$.

2. Samples and experimental

For the investigation of the influence of n_s on the lasing parameters, we used the well-established GaAs/Al$_{0.33}$Ga$_{0.67}$As QCL design, introduced by Sirtori et al. [4]. The number of periods was set to 30 in each QCL. The samples were grown by molecular-beam epitaxy at a substrate temperature of 600 °C. The Si-doping sheet concentration n_s in the four centre layers of the injector region was varied from 2.3×10^{11} to 1.0×10^{12} cm^{-2} per period. All QCL's were characterized by double-crystal x-ray diffraction to demonstrate that the actual layer thicknesses as well as the Al-content agree within 2% with the nominal values. Capacitance-voltage measurements on step-etched pieces were carried out in order to verify the nominal values of n_s. After growth, laser stripes with typical dimensions of 19×2400 μm² were prepared by plasma etching. The side walls of the stripes were formed by 7 μm deep and 2.5 μm broad trenches refilled with a photo-resist.

The laser emission was studied by Fourier-transform spectroscopy using pulse-mode operation (100 ns at 5 kHz). All spectra were recorded at 8 K with the sample placed in a He cryostat. The lasing energy E_0 was determined by taking the energy, for which the integral over the laser modes with $E \leq E_0$ is one half of the integral over all laser modes. The maximum operating temperature T_{max} is defined by the temperature at which the QCL is still operating for a threshold current of 10 A, while the temperature parameter T^* was obtained by fitting the measured $j_{th}(T)$ curves.

3. Experimental results and discussion

For the QCL with $n_s = 2.3\times10^{11}$ cm^{-2}, lasing was not observed. Figure 1 shows from top to bottom the lasing spectra of a weakly, an intermediately, and a strongly doped QCL. With increasing n_s up to the intermediate range, E_0 exhibits at first a blue-shift from 115.3 to 134.5 meV, followed by a red-shift back to 120.3 meV for the highest doping level. We note that this variation of E_0 by up to 20 meV is much larger than the fluctuations of E_0 across the wafer ($\Delta E_0 \leq 1$ meV) as well as due to different current levels for each QCL ($\Delta E_0 \leq 2.5$ meV). This observation implies that the observed dependence of E_0 on n_s is neither caused by a variation of the layer thicknesses nor by sample heating. The measured $j_{th}(T)$ data of the different QCL's are plotted in figure 2. The QCL with $n_s = 5.3\times10^{11}$ cm^{-2} experiences the lowest threshold current density at low temperatures and the largest value of T_{max}. In addition, figure 2 shows the fits of $j_{th}(T) = j_0 + A \exp(T/T^*)$ to the measured data, where j_0, A, and T^* denote fit parameters. This expression gives excellent fits for all samples over the whole operating temperature range with $j_0 \approx j_{th}(T = 8$ K$)$ in contrast to the usually applied formula $j_{th}(T) = j_0 \exp(T/T_0)$, which works well only for $T > 150$ K. Fitting previously published data [5] with our expression,

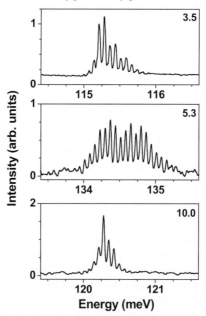

Figure 1. Emission spectra of GaAs/Al$_{0.67}$Ga$_{0.33}$As QCL's with $n_s = 3.5$, 5.3, and 10.0 in units of 10^{11} cm^{-2} for $T = 8$ K and $j = 1.1\, j_{th}$.

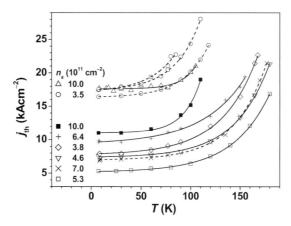

Figure 2. $j_{th}(T)$ of the QCL's with different values of n_s as indicated. The symbols indicate measured data, the lines are fits to these data. The solid squares display the values for a modified QCL as described in the text.

we obtain $T^* \approx T_0/4$. For comparison, a modified QCL with $n_s = 10.0 \times 10^{11}$ cm^{-2} was studied, in which in the first six periods the thicknesses of the (Al,Ga)As layers were increased by a factor of 1.13 and the Al content was increased to 0.41. This QCL displays a significant decrease of j_{th} at low temperatures (cf. figure 2) and an increase of E_0 up to 126.9 meV, while the temperature performance is about the same as the one for the original QCL with the same doping level. The better $j_{th}(T = 8$ K$)$ value for the modified QCL can neither be explained by the decreasing optical confinement due to a smaller number of periods nor by a remarkable reduction of the optical losses due to free carrier absorption in contrast to the discussions in Refs. [6] and [7].

In figures 3(a), 3(b), 3(c), and 3(d), we show E_0, j_{th}, T^*, and T_{max} versus n_s, respectively. These plots demonstrate that in the intermediate doping range for an n_s value of about 6×10^{11} cm^{-2} E_0 exhibits a maximum, j_{th} a minimum, and T^* as well as T_{max} a maximum. The optimal values for all parameters are in good agreement with data published by Sirtori *et al.* [4]. The minimum for j_{th} as well as the maximum for T^* and T_{max} at intermediate doping levels can be explained as follows. On the one hand, n_s and the resulting current j, which is proportional to n_s, have to be sufficiently large so that the gain becomes larger than the losses. Therefore, a QCL does not operate below a certain value of n_s. On the other hand, n_s should be as low as possible, because a larger n_s increases both the scattering rate of electrons from the injector back into the active region and the optical losses due to free-carrier absorption. Both effects will reduce the gain. Consequently, we expect a certain range of doping levels, in which the lasing parameters are optimal, i. e., the value of j_{th} becomes minimal and those of T^* and T_{max} maximal.

The non-monotonic dependence of E_0 on n_s cannot be explained by an increasing inter-diffusion of Al into GaAs at the interfaces with increasing n_s, since this would result in a monotonic dependence $E_0(n_s)$. In addition, self-consistent calculations based on the Schrödinger and Poisson equations show that the variation of E_0 is less than 3 meV for doping levels between 1.0 and 7.0×10^{11} cm^{-2} and electric field strengths between 46 and 58 kVcm^{-1}. Therefore, the observed dependence of E_0 on the doping level n_s cannot be explained by the influence of the electric field on the laser transition. At the same time, such calculations show that additional levels 10 to 25 meV above the upper laser level may become more and more occupied with increasing n_s and may cause the observed

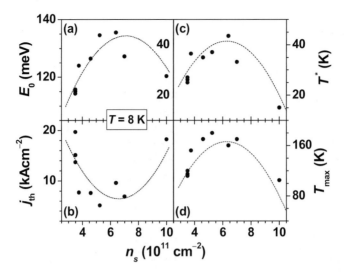

Figure 3. (a) Lasing energy E_0 and (b) threshold current density j_{th}, at $T = 8$ K as well as the temperature parameters (c) T^* and (d) T_{max} vs. n_s. The dashed lines are guides to the eye fitting a quadratic polynomial.

blue-shift. Furthermore, the non-parabolicity of the band structure as well as electron-electron interactions, such as the depolarisation shift and the intersubband exchange interaction, result in a carrier-density-dependent line position. However, for the laser transition, which occurs under non-equilibrium conditions, these effects cannot be estimated as in the case of intersubband absorption, where they are quantitatively known. Furthermore, because the non-monotonic dependence of E_0 on n_s cannot be explained by layer-thickness fluctuations and interface disorder, a more detailed modelling of the QCL's as a function of n_s is necessary, which should also include many-particle effects.

Acknowledgement

This work was supported in part by the Deutsche Forschungsgemeinschaft within the framework of the Forschergruppe 394.

References

[1] Beck M, Hofstetter D, Aellen T, Faist J, Oesterle U, Ilegems M, Gini E, and Melchior H 2002 Science 295 301–305
[2] Gmachl C, Sivco D L, Colombelli R, Capasso F, and Cho A Y 2002 Nature 415 883–887
[3] Köhler R, Tredicucci A, Beltram F, Beere H E, Linfield E H, Davies A G, Ritchie D A, Iotti R C, and Rossi F 2002 Nature 417 156–159
[4] Sirtori C, Kruck P, Barbieri S, Page H, Nagle J, Beck M, Faist J, and Oesterle U 1999 Appl. Phys. Lett. 75 3911–3913
[5] Becker C, Sirtori C, Page H, Glastre G, Ortiz V, Marcadet X, Stellmacher M, and Nagle J 2000 Appl. Phys. Lett. 77 463–465
[6] Gmachl C, Capasso F, Sivco D L, and Cho A Y 2001 Rep. Prog. Phys. 64 1533–1601
[7] Straub A, Mosely T S, Gmachl C, Colombelli R, Troccoli M, Capasso F, Sivco D L, and Cho A Y 2002 Appl. Phys. Lett. 80 2845–2847

Author Index

Abdulrhmann S   409
Abe S   21
Abrokwah J K   251
Aellen T   323, 455
Aguilar-Hernandez J   65
Ahlers F J   105
Ahmed M   409
Ahmed S   41
Ajili L   371, 439
Akazawa M   299
Akiyama H   379
Akiyama Y   169
Alfaro Lopez H M   69
Anders S   375, 385
Araki T   13, 17
Arokiaraj J   45
Asakawa K   133
Asaoka Y   25
Aufinger K   217

Baba M   379
Badilita V   405
Baeumler M   53
Baklenov O   1
Baldwin K   223
Bandara S V   393
Bauer A   109
Baur J   315
Beck M   323, 439, 455
Becker C   385
Beere H   371, 439
Belenky G   443
Beltram F   389
Berseth C-A   415
Bewley W W   331
Beyertt S-S   427
Bhattacharya P   117
Biasiol G   389
Blagnov P A   137
Blaser S   323, 371
Böck J   217
Bolognesi C   203
Borkovska L V   57

Bowers J E   351, 367
Boyd E   291
Brandt O   73, 141, 431
Brauch U   427
Breiter R   339
Brunner F   239
Brunner M   405
Bühlmann H J   419
Bulakh B M   57
Bungarzeanu C   415
Byun Y T   435

Cabanski W   339
Canedy C L   331
Carlin J F   405, 419
Casas Espinola J L   69
Chakrabarti S   117
Chao P C   211
Chen J Y   271
Chen T T   397
Chen Y   291
Chen Y F   397
Cheng C C   271
Chi T W   179
Chiu Y-J   367
Chiu Y S   397
Cho S   113
Choi K K   393
Chu S N G   223
Chuang H M   271
Chun Y-H   283
Craigo B   1
Curless J   1

Daicho S   25
David J P R   263, 267, 355
Davies G   371, 439
Dehaese O   97
Dekorsy T   109
DenBaars S P   367
Dhar S   141
Dini D   389
Diwo E   53

Djie H S   45
Döhler G H   427
Donchev V   191
Dorsaz J   363, 419
Droopad R   1, 251
Dupertuis M-A   183
Dvorak M W   203

Eastman L F   227, 287
Edwards J   1
Eichhorn F   109
Eisenbeiser K   1
Eisert D   315
Elgaid K   291
Eliseev P G   69
Elyukhina O V   93
Eschrich T   1
Even J   97

Faist J   323, 371, 439, 455
Fehrer M   315
Fejes P   1
Fellows J A   49
Ferdos F   149
Finder J   1
Fischer C   117
Fleissner J   339
Folliot H   97
Friedland K J   141
Fuchs F   447
Fujimoto T   307
Fukuzawa M   77, 85

Gambin V   101
Georgiev N   109
Germanova K   191
Geske J   351, 367
Ghosh S   431
Giehler M   459
Giesen A   427
Gini E   323, 455
Göbel E O   105
Gornik E   375, 385
Goto S   307
Gottwaldt L   105
Grahn H T   431, 459
Grant I   53

Grasser T   303
Green B   227
Groves C   263
Guézo M   97
Gunapala S D   393
Gwilliam R   41

Hadley P   125
Hahn B   315
Hale M   251
Han H-J   279
Härle V   315
Harris J S   101
Harrison C N   263
Harrison P   451
Hasegawa H   33, 145, 299
Hasegawa T   33
Hashimoto H   61
Hashizume T   299
Helm M   109
Hengehold R L   49
Hey R   459
Hill G   267
Hilsenbeck J   239
Hilt L   1
Hirayama Y   247
Hofstetter D   323, 371, 439, 455
Höntschel1 J   255
Hopkinson M   263
Horikoshi Y   29, 33
Houston P A   267
Hsiao C L   179
Hsieh K Y   179
Hsu J W P   223
Hu C F   179
Hu X   1

Iakovlev V   415
Ihara I   25
Ikeda M   307
Ikeda N   133
Ikeda S   307
Ikonić Z   451
Ilegems M   323, 359, 363, 405, 419, 455
Inagaki T   299
Indjin D   451
Ishimori W   409

Isshiki M 21
Ivanov S V 161

Jang J 187
Jang Y D 187
Jantz W 53
Jeong W G 187
Jhabvala M 393
Joray R 359, 363
Jordan D 1

Kaiser S 315
Kaneko T 25
Kaper V 227, 287
Kapon E 183, 415
Karlsson F 183
Karnutsch C 363, 427
Kawazu T 169
Kelsall R W 451
Khomenkova L 57
Kiefer R 235, 347, 447
Kim B W 113
Kim C S 331
Kim H 227, 287
Kim H-S 283
Kim J 113
Kim K W 401
Kim N J 187
Kim S-C 275, 283
Kim S-K 275, 283
Kim S-J 191
Kishimoto S 243
Kisin M V 443
Kissel H 81, 165
Kita T 173
Kitamura K 13
Klix W 255
Knapp H 217
Kobayashi N 231
Kobuse T 157
Kochelap V A 401
Kochowski S 37
Köhler K 235, 447
Köhler R 389
Koidl P 339
Kolobov A 65
Komirenko S M 401

Konagai M 157
Kondo N 169
Korsunska N O 57
Kostial H 459
Kovalenkov O V 137
Král K 153
Krishna S 117
Krispin P 101
Kulik J 1
Kumakura K 231
Kummel A C 251
Kunets V P 81

Labbé C 97
Lai M-J 125
Lalev G M 21
Lang D V 223
Larsson A 149
Le Corre A 97
Lee D 187
Lee H-S 275, 283
Lee M-K 283
Lee S-D 279
Leifer K 183
Lell A 315
Lenk F 239
Li C 137
Li H 1
Liang Y 1
Lientschnig G 125
Lim B-O 275, 283
Lin K W 271
Linder N 427
Lindle J R 331
Linfield E 371, 439
Litzenberger M 375
Liu J K 393
Liu K T 29
Liu W C 271
Lo I 179
Loualiche S 97
Lour W S 295
Luft J 427
Lugauer H-J 315
Luryi S 443

Macintyre D 291

Maezawa K 243
Majerfeld A 113
Makimoto T 231
Malloy K J 69
Mamiya H 13
Manfra M J 223
Mann C 447
Mariette H 173
Markevich I V 57
Marsal L 173
Maruyama T 13
Masselink W T 81, 109, 165
Masumoto K 21
Matsumoto O 307
Matsuzaki Y 157
Mazur Y I 165
McLelland H 291
Mei T 45
Meister T F 217
Melchior H 323, 455
Mereuta A 415
Mermelstein C 347
Merz J L 137
Meyer J R 331
Mintairov A M 137
Mircea A 415
Misra P 431
Miyashita S 247
Mizuno T 307
Mizuo K 17
Mizutani T 243
Mochizuki K 195
Molnar R J 223
Moore K 1
Moran D 291
Müller G 53
Müller M I 427
Müller S 235
Müller U 81

Nagahara S 173
Nakamura H 133
Nakamura Y 133
Nanishi Y 13, 17
Nekrutkina O V 161
Ng B K 355
Ng J S 267

Nishizawa J-i 9
Noda T 169, 379
Norris T 117

O'Steen M 1
Oesterle U 323, 359, 455
Ogura M 191
Ohashi R 157
Ohfuji Y 307
Ohizumi Y 61
Ohkouchi S 133
Ohno T 9
Ohno Y 243
Ohtsuka T 459
Oikawa K 307
Okada D 33
Okamoto T 409
Oktyabrsky S 137
Okuno Y L 351, 367
Ootomo S 299
Overgaard C 1, 251
Oyama Y 9

Palankovski V 303
Panajotov K 405
Park H-C 275, 283
Park H S 435
Park K 187
Passlack M 251
Paszkiewicz B 37
Paszkiewicz R 37
Patriarche G 113
Pena Sierra R 69
Pfeiffer L N 223
Pflügl C 375, 385
Pierz K 105
Pletschen W 347
Ploog K H 73, 101, 141
Pogany D 375
Polupan G 65
Prunty T 227, 287

Qin C 355
Quay R 235

Rafol S B 393
Ramdani J 1

Rattier M 363
Rattunde M 347
Rees G J 263, 267, 355
Rehm R 339
Reznitsky A A 161
Rhee J-K 275, 279, 283
Ritchie D 371, 439
Rochat M 371
Rollbühler N 235
Royo P 359
Rudra A 183, 415
Ryu M-Y 49

Sadeghi M 149
Sahr U 53
Saito Y 17
Saitoh T 33
Sakaki H 169, 379
Sano N 25
Santos P V 73
Sanz-Hervás A 113
Saraydarov M 191
Sasaki T 307
Sato T 145
Savkina R K 89
Scalari G 371, 439
Schmid W 427
Schmitz J 347
Schneider H 339, 447
Schönherr H-P 73
Schrenk W 375, 385
Schrottke L 459
Schultheis R 303
Schweitzer L 105
Sealy B J 41
Sedova I V 161
Selberherr S 303
Semtsiv M 109
Sexton J 251
Shealy J R 287
Shealy R 227
Sheng T T 179
Shigekawa N 259
Shin D-H 275, 283
Shiojima K 259
Sirtori C 385
Sitnikova A A 161

Smart J 227, 287
Smirnov A B 89
Sodesawa J 29
Sorba L 389
Sorokin S V 161
Stanley C 291
Stanley R P 359, 363, 419
Steiner T D 49
Stenzel R 255
Stiff-Roberts A D 117
Stintz A 69
Stolz W 105
Strasser G 375, 385
Strauss U 315
Streubel K 363, 427
Stroscio M A 443
Su Y K 29
Sul W-K 279
Sugimoto Y 133
Sugita S 29
Suruceanu G 415
Suto K 9
Syrbu A 415

Takagaki Y 73
Takeya M 307
Tamai I 145
Tan S W 295
Tarasov G G 81, 165
Tezuka K 9
Thayne I 291
Thompson R 227, 287
Thoms S 291
Tilak V 227, 287
Tokranov V 137
Tomm J W 165
Too P 41
Torchynska T V 65, 69
Tozer R C 355
Trampert A 141
Tredicucci A 389
Tsai M K 295
Tsuruoka T 61
Tu L W 179

Uchida S 307
Unuma T 379

Urayama J   117
Ushioda S   61

van der Tol J J G M   423
Verschaffelt G   405
Vinokurov D A   137
Vlasov A S   137
Vukusic J   149
Vurgaftman I   331

Wada O   173
Wagner J   347
Wagner S   303
Walther M   339, 347
Wang J   21
Wang S M   149
Wang X-L   191
Watari Y   29
Watkins S P   203
Wei Y Q   149
Weimann G   235, 339
Weimann N G   223
Weman H   183
West K W   223
Wei Y   1
Wiebicke E   73
Willenberg H   371, 439
Wu J F   179
Wu Y W   295
Würfl J   239
Wurzer M   217

Ya M H   397
Yabuki Y   307
Yamada A   157
Yamada M   77, 85, 409
Yamaguchi H   247
Yamaguchi T   17
Yamamizu H   29
Yan F   355
Yang Q   447
Yang Y J   295
Yee M C   267
Yeo Y K   49
Yi J C   435
Yi S I   251
Yokoyama K   25
Yokoyama Y   243
Yoshita M   379
Yoshizawa G   29
Yu K H   271
Yu Z   1, 251
Yun S-W   283

Zavada J M   401
Zdeněk P   153
Zellweger Ch   419
Zhao G   423
Zhao J H   355
Zhao Q X   149
Zhuchenko Z Ya   81, 165
Ziegler J   339